THEORY

OF

FLIGHT

by

RICHARD VON MISES

WITH THE COLLABORATION OF W. PRAGER
and GUSTAV KUERTI

WITH A NEW INTRODUCTION
by KURT H. HOHENEMSER

DOVER PUBLICATIONS, INC.
NEW YORK

Published in Canada by General Publishing Company, Ltd., 30 Lesmill Road, Don Mills, Toronto, Ontario.
Published in the United Kingdom by Constable and Company, Ltd., 10 Orange Street, London WC 2.

This Dover edition, first published in 1959, is an unabridged and unaltered republication of the English translation first published by McGraw-Hill Book Company in 1945. A new Introduction was specially written by Kurt H. Hohenemser for this edition.

International Standard Book Number: 0-486-60541-8
Library of Congress Catalog Card Number: 59-4203

Manufactured in the United States of America
Dover Publications, Inc.
180 Varick Street
New York, N. Y. 10014

To

HARVARD UNIVERSITY

INTRODUCTION TO DOVER EDITION

When in the rapidly progressing field of aeronautical sciences a textbook written fourteen years ago is reprinted without changes or additions some words of explanation seem in order. At the time the English adaptation of Richard von Mises' famous *Fluglehre* was prepared by the author and his collaborators, W. Prager and G. Kuerti, the turbo-jet engine and the rocket engine were being developed which, after their introduction, revolutionized aeronautics and soon pushed airplane flight speeds past the speed of sound. The *Theory of Flight*, written during the war when most of the new material was classified, is a prerevolution book. Airplane speeds are assumed to be sufficiently low so that air compressibility effects are unimportant. The airplane engine is of the reciprocating type. Airplane propulsion is produced by airscrews.

It is true that even in our "Jet Age" airplane prototypes are being developed for which the analytical methods presented in the *Theory of Flight* are entirely adequate. All of our private flying and almost all of our commercial and military air transportation is performed by propeller airplanes of sufficiently low subsonic speeds so that air compressibility requires only minor corrections to incompressible flow theory. It is also true that in aerodynamics as well as in other branches of science the understanding of the recent developments requires familiarity with the fundamentals. Compressible flow theory is based on the theory of incompressible flow which, therefore, remains an introductory subject of study. However, all these arguments in themselves offer hardly a sufficient explanation for the reprinting of a textbook which omits important areas of problems confronting the designer of modern high speed aircraft.

The real reasons which still make this book on outstanding tool in the education of our future aeronautical engineers must be found in its positive and unique features. There are many excellent books on fluid dynamics which treat the subject from a theoretical point of view. There also are many textbooks on airplane design with the emphasis on the practical problems of performance prediction, stability and control and the like. There is, however, no other introduction to aeronautical engineering problems which is equally well balanced with respect to both the theoretical and the practical side. The *Theory of Flight* avoids the formidable mathematical structure of fluid dynamics and yet conveys by often unorthodox methods a full understanding of the physical phenomena and of the mathematical concepts used in aeronautical engineering.

In addition to the excellent balance between the theoretical and practical aspects, the lasting value of the present work lies in the lucidity and depth of its aerodynamical chapters which are compact and directed toward the essentials. If it is the main objective of our engineering education to make the student thoroughly familiar with the fundamentals of engineering sciences rather than to give him a training in design practices in a special field, the *Theory of Flight* will continue to be an ideal introductory textbook in spite of its incompleteness with regard to some of the more recent developments.

Finally, it is appropriate to mention that the author of the book, Richard von Mises who died 1953 at the age of 70, devoted a good part of his extraordinarily diversified interests to the development of the aeronautical sciences and was one of the great pioneers in this field. The work of such an author has a flavor of authenticity not to be found in more conventional textbooks. Since the book summarizes much of the author's own earlier work it also has considerable value for the historically minded expert.

While the study of the *Theory of Flight* will give the student an excellent introduction to the theoretical concepts of airplane flight and to their practical applications, aeronautical engineering has progressed since the time of the writing of this book at a very rapid rate and entirely new fields of study must be added in order to understand the varied problems of modern high speed aircraft. There exists no single book which gives a comprehensive survey of these problems and which could be compared in scope to the *Theory of Flight** although there is a large and ever increasing volume of textbook literature on special aspects of aeronautics. In view of the recent and very extensive accumulation of knowledge not *one* book but a series of specialized works seems to be needed in order to supplement the basic introduction given in *Theory of Flight* and to guide the student into the various directions of modern aeronautics. While it is beyond the scope of a conventional preface, it may appear worth while to offer, for the benefit of the

---

* When preparing the sixth edition of von Mises' original German *Fluglehre* (published 1957) the writer of this introduction tried to modernize the book without changing its size or its character. The *Fluglehre* is written, however, on a more elementary level than the *Theory of Flight*.

reader, a few comments and suggestions on each of the five parts of this book with regard to recent progress and to mention a few more modern specialized textbooks written approximately on the same intermediate level as the *Theory of Flight*.

Part One begins with the equilibrium state of the atmosphere. The method of reduction of a climb to standard atmosphere described in Chapter I is not applicable to high-speed or turbine-powered aircraft. Chapter V, Air Resistance, may be supplemented by data on shock wave drag which becomes predominant at supersonic flight speeds. Data on parasite drag of airplane components including wave drag can be found in Hoerner, *Aerodynamic Drag* (1958). On the whole, Chapters II to V of Part One represent today, as they did fourteen years ago, a unique introduction to the dynamics of incompressible fluids. Because of the lucidity of the text and the originality of the methods these chapters are of lasting interest.

The same is true of Chapters VIII and IX of Part Two dealing with wing theory to which the author has substantially contributed. Part Two may be supplemented by concepts and data necessary for the treatment of high subsonic and supersonic flight. The theory of the lifting surface in these flight regimes makes use of methods developed in gas dynamics (compressible fluid flow). We mention the *Elements of Gas Dynamics* by Liepmann and Roshko (1957) and, as a more advanced book in the field, *Mathematical Theory of Compressible Fluid Flow* (1958), the basic work by Richard von Mises which was completed after the author's death by Geiringer and Ludford from his lecture notes.

Regarding aircraft propulsion treated in Part Three, very substantial progress has been made since the writing of this book. As far as it is still in use, the propeller has not developed much — except for increases in blade tip speed. The reciprocating engine, however, has been replaced in all large aircraft developments by either propeller shaft-turbines or by turbo-jets. The elements of the theory of jet propulsion are included in Kuechemann and Weber, *Aerodynamics of Propulsion* (1953). On the theory and design of aircraft gas turbines we mention C. W. Smith, *Aircraft Gas Turbines* (1956).

Part Four is a completely valid explanation of the basic performance problems of propeller airplanes. It may be supplemented by data applicable to jet airplanes given, for example, in Perkins and Hage, *Airplane Performance, Stability and Control* (1950).

Part Five contains the valuable Chapters XVIII and XIX, the first introducing the student to the theory of non-uniform flight, the second providing a lucid introduction to the general theory of dynamic stability. Part Five may be supplemented by concepts and data necessary for the treatment of automatic control problems. As an introduction to these problems we mention Murphy, *Basic Automatic Control Theory* (1957).

Finally we list as an introduction to the growingly important subject of rotary wing flight Gessow and Meyers, *Aerodynamics of the Helicopter* (1952).

Any student who has made himself familiar with the fundamentals and applications contained in the *Theory of Flight* will have acquired an excellent background for studying the additional and more specialized fields required in modern aeronautical engineering sciences.

St. Louis, Missouri
November 1958

KURT H. HOHENEMSER

PREFACE

Thirty years ago, in the summer of 1913, the author gave for the first time a university course on the mechanics of airplane flight. On the basis of these lectures and similar ones in the following years there developed a small elementary textbook "Fluglehre," which was published up to 1936 in five German editions and translated into several languages. When asked, about three years ago, to prepare an English edition the author found that a book of somewhat different type would be more useful under the present conditions. There already exist in this country several introductory texts on the mechanics of flight suitable for beginners. On the other hand, a mature student or an engineer well versed in the advanced stages of higher mathematics can readily work his way through the currently published research papers. An actual demand does, however, seem to exist for a book at an intermediate level, say, on the borderline between the last year of college and the first year of graduate work. It is assumed that the reader of the present text knows the principles of calculus and has some training in general mechanics insofar as the standard college education provides for these things. Only such parts of the theory of flight as are understandable at this level are presented in the book, the mathematical theory of fluid mechanics thus being excluded.[1] In order to keep the book within reasonable limits peripheral questions and many details in all topics have had to be omitted.

Although the book is meant as a text for students and engineers, it was not the author's purpose to supply the reader with ready-made formulas in which to substitute design data. Nor was it his intention to give a collection of results that the candidate for a degree should memorize. The book aims, as did the "Fluglehre," to develop interest in and understanding of the fundamental ideas that underlie the design and the operation of modern aircraft. The problems inserted at the ends of the sections, particularly those marked by an asterisk, are intended to give some direction to the efforts of any reader who would apply himself toward a better comprehension of the diversified questions involved in the theory of airplane flight.

Whoever ventures to write a textbook on a subject like the present one finds himself confronted with the dilemma of not knowing to what

[1] Mimeographed notes of Lectures on Advanced Fluid Mechanics by R. von Mises, K. O. Friedrichs, and S. Bergman are available through the Graduate School, Brown University, Providence, R.I.

extent he is expected to give an account of the commonly accepted ideas and how far he may be allowed to advance his own points of view. The expert reader will find that in many places, even where well-known arguments are discussed, the text deviates more or less from what is usually presented in conventional papers and books. This is mentioned here, not because the author wishes to claim the credit for novelty, but rather to state his responsibility. He is well aware of the sentence that the Marquis de Vauvenargues set at the head of his "Réflexions et maximes": "It is easier to say original things than to reconcile with one another things already said."

Sincere apologies must be offered with respect to evident imperfections in English style and diction. As this is the fourth language in which the author has had to teach, it was no easy task to write the book in English. In these busy times a competent expert who would go through the entire text remodeling and correcting it was not available. It goes without saying that kind advice offered by friends was gladly accepted, but the suggestions did not always agree with each other.

The work on this book was begun under a coauthorship agreement with Professor W. Prager of Brown University. According to the original agreement the two authors were to share equally in the work and in the credit for it. Unfortunately, during the work on Chaps. IX to XII it turned out that the burden of other duties made it impossible for Professor Prager to continue to collaborate on these lines. Since he had to drop out before the definitive text was established even for the first chapters, he has no responsibility for the text as it now stands. The author deeply regrets losing his valuable collaboration for the second part of the book. His place was taken by Dr. Gustav Kuerti, who also revised the whole manuscript and shared in the proofreading. The author is greatly indebted to Dr. Kuerti, whose devoted cooperation made it possible to publish the book without too much delay.

Thanks are also extended to many friends who in one way or another gave valuable help.

The author is glad, on this occasion, to express his fullest gratitude to the Graduate School of Engineering, Harvard University, which generously offered him the opportunity to continue his scientific work in this country and thus enabled him to publish the text presented here.

R. v. MISES.

HARVARD UNIVERSITY,
CAMBRIDGE, MASS.,
November, 1944.

CONTENTS

PAGE

PREFACE. . . . . *xi*

Part One

EQUILIBRIUM AND STEADY FLOW IN THE ATMOSPHERE

CHAPTER I. THE ATMOSPHERE AT REST

1. Density. Pressure. Equation of State 1
2. Equilibrium of a Perfect Gas under the Influence of Gravity. . . . 4
3. The Standard Atmosphere 8
4. Determination of True Altitude. Reduction of a Climb to Standard Atmosphere 13
5. Troposphere and Stratosphere. Influence of Humidity 18

CHAPTER II. BERNOULLI'S EQUATION. ROTATION AND CIRCULATION

1. Steady Motion 22
2. Bernoulli's Equation. 26
3. Dynamic Pressure. 31
4. Variation of Total Head across the Streamlines. Rotation 35
5. Circulation and Rotation. 39
6. The Bicirculating Motion 45

CHAPTER III. MOMENTUM AND ENERGY EQUATIONS

1. Flux of Momentum in Steady Flow 52
2. Momentum Equation for Steady Flow. 55
3. Moment of Momentum 58
4. Quasi-steady Flow. Relative Flow 63
5. Energy Equation 67

CHAPTER IV. PERFECT AND VISCOUS FLUIDS. TYPES OF FLOW

1. Viscosity. 74
2. Law of Similitude. Reynolds Number. 77
3. Laminar and Turbulent Motion. 81
4. Continuous and Discontinuous Motion. 85
5. Boundary Layer. 90

CHAPTER V. AIR RESISTANCE, OR PARASITE DRAG

1. Definitions. 95
2. Bluff Bodies 96
3. Round Bodies. 99
4. Streamlined Bodies 102
5. Skin Friction. 105
6. Parasite Drag of Major Airplane Components. 107

Part Two

THE AIRPLANE WING

CHAPTER VI. FUNDAMENTAL NOTIONS. GEOMETRY OF WINGS

PAGE
1. The Three Coefficients. 112
2. Geometry of Airfoil Profiles. Sets of Profiles. 115
3. Theoretically Developed Airfoil Sections. 121
4. Geometry of Airplane Wings 132

CHAPTER VII. EMPIRICAL AIRFOIL DATA

1. The Three Main Results. 139
2. Influence of Aspect Ratio 148
3. Historical Development of Wing Profiles. 157
4. Influence of the Shape of the Profile. 161
5. Influence of the Reynolds Number. Degree of Turbulence. 167

CHAPTER VIII. THE WING OF INFINITE SPAN

1. The Momentum Equation for Irrotational Flow. 170
2. The Lift on an Airfoil of Infinite Span. 174
3. The Pitching Moment of an Airfoil of Infinite Span 181
4. The Metacentric Parabola 186
5. Vortex Sheets, Another Approach. 188
6. Theory of Thin Airfoils 198

CHAPTER IX. THE WING OF FINITE SPAN

1. Curved Vortex Lines. 211
2. Vortex Sheet and Discontinuity Surface 219
3. The Flow Past a Wing of Finite Span 224
4. Prandtl's Wing Theory. 231
5. Elliptic Lift Distribution. 239
6. Biplane Theory. 244
7. General Lift Distribution. 250

CHAPTER X. ADDITIONAL FACTS ABOUT WINGS

1. Stalling 258
2. High-lift Devices 264
3. Pressure Distribution 271
4. Influence of Compressibility 275

Part Three

PROPELLER AND ENGINE

CHAPTER XI. THE PROPELLER

1. Basic Concepts. 285
2. Geometry of Propellers. 290
3. Propeller Characteristics. 296
4. Quantitative Analysis 302
5. Propeller Sets and Variable-pitch Propeller. Propeller Charts . . . 310

CHAPTER XII. OUTLINE OF PROPELLER THEORY

PAGE
1. Blade-element Theory. 317
2. Momentum Theory, Basic Relations. 326
3. Momentum Theory, Conclusions 334
4. Modified Momentum Theory. 339
5. The Two Theories Combined. 345
6. Additional Remarks. 350

CHAPTER XIII. THE AIRPLANE ENGINE

1. The Engine at Sea Level. 356
2. The Engine at Altitude 364
3. Engine Vibrations. 371

Part Four

AIRPLANE PERFORMANCE

CHAPTER XIV. THE GENERAL PERFORMANCE PROBLEM

1. Introduction 381
2. Power-required and Power-available Curves . . . . ` 385
3. Dimensionless Performance Analysis. 394
4. Discussion of Sea-level Flight. 398
5. Altitude Flight 409

CHAPTER XV. ANALYTICAL METHODS OF PERFORMANCE COMPUTATION

1. Analytic Expressions for the Power Curves. 419
2. Gliding. Level Flight with Given Power. 427
3. The Ideal Airplane: Power Available Independent of Speed. . . . 437
4. Numerical Data. Example 442
5. Small Variations. Choice of Propeller. 447
6. Power Available Varying with Speed. 450
7. Numerical Discussion. 455

CHAPTER XVI. SPECIAL PERFORMANCE PROBLEMS

1. Range and Endurance. 461
2. Take-off 469
3. Steep Gliding and Diving 475
4. Landing Operation. Landing Impact 483
5. Seaplane Problems 488

Part Five

AIRPLANE CONTROL AND STABILITY

CHAPTER XVII. MOMENT EQUILIBRIUM AND STATIC STABILITY

1. Pitching-moment Equilibrium. 497
2. The Contribution to the Pitching Moment from the Tail. 501
3. The Contribution from the Propeller and the Fuselage. 507
4. Static Stability and Metacenter. 514
5. Simplified Stability Discussion 522
6. Lateral Moments 527

CHAPTER XVIII. NONUNIFORM FLIGHT

PAGE
1. Introduction. Elementary Results 533
2. Lanchester's Phugoid Theory. 539
3. Longitudinal Flight along a Given Path 545
4. Effect of Elevator Operation 551
5. Asymmetric Motion. 555

CHAPTER XIX. GENERAL THEORY OF MOTION AND STABILITY

1. The General Equations of Motion of an Airplane 564
2. Steady Motion. Specification of Forces 570
3. Theory of Dynamic Stability. 574
4. Application to the Airplane. 580

CHAPTER XX. DYNAMIC STABILITY OF AN AIRPLANE

1. Longitudinal Stability of Level Flight 586
2. The Small Oscillations Following a Disturbance. 593
3. Lateral Stability 599
4. Numerical Discussion 603
5. Final Remarks. Autorotation. Spinning 608

BIBLIOGRAPHICAL AND HISTORICAL NOTES. 614

INDEX 621

THEORY OF FLIGHT

Part One

EQUILIBRIUM AND STEADY FLOW IN THE ATMOSPHERE

CHAPTER I

THE ATMOSPHERE AT REST

1. Density. Pressure. Equation of State. In kinetic theory a gas is considered as composed of discrete molecules in vigorous, irregular motion, continually colliding with one another. In fluid dynamics one substitutes for this picture the simpler one of continuously distributed matter moving without sudden changes of velocity.

At any point P of a region in space occupied by such a continuous medium, or continuum, the *density* can be defined in the following way: Consider the mass and the volume of the substance that is contained in a small region R surrounding the point P. The density at P then is the limiting value of the ratio of the mass to the volume as the linear dimensions of R tend simultaneously to zero.

In this book the density will be denoted by ρ. In the engineering system the unit of density is one slug per cubic foot. At 59°F. and standard atmospheric pressure (29.921 in. Hg) the density of dry air is 0.002378 slug/ft.³

The *specific weight* γ, defined as the weight per unit of volume, is the product of the density and the acceleration g of free fall: $\gamma = g\rho$. In the engineering system the unit of specific weight is the pound per cubic foot. With $g = 32.174$ ft./sec.², the specific weight of dry air at 59°F. and standard pressure is seen to be 0.07651 lb./ft.³

Consider two portions C_1 and C_2 of the continuum touching each other at the point P. Around P mark off an infinitesimal area dS (Fig. 1). Across dS the portion C_2 exerts an infinitesimal force on the portion C_1. Denote the magnitude of this force by dF. The *stress* that C_2 exerts at P across the surface S on the portion C_1 of the continuum then is defined as the vector whose magnitude is dF/dS and whose direction is that of the infinitesimal force dF. It follows from Newton's third law of motion that the stress which C_1 exerts at P across the surface S on the portion C_2 is given by a vector of the same magnitude and the opposite direction.

It is a fundamental assumption of the mechanics of continua that the stress transmitted at P across dS will not depend on the shape of the surfaces S_1 and S_2 as long as the tangential plane of both surfaces at P remains the same. We thus can speak of the stress transmitted across a surface element without defining the surfaces to which this element belongs and the portions C_1 and C_2 which they confine.

The stress transmitted across a surface element is in general oblique

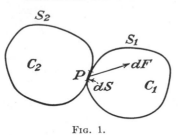

FIG. 1.

to it. The stress components perpendicular and parallel to the surface element are called *normal stress* and *shearing stress*, respectively. The normal stress can be a thrust or a tension.

In this book the continuum to be considered is the atmospheric air. Its mechanical properties are essentially the same as those of other gases and partly the same as those of liquids. The word "fluid" will be used to designate both gases and liquids. It is assumed as the characteristic property of fluids that in a state of rest *no shearing stresses are transmitted and that the normal stress on any surface element is a thrust*. Moreover, this assumption is maintained in most problems of fluid in motion. If we do this we call the fluid a *perfect fluid* (see Sec. II.2).

By considering a small portion of fluid enclosing a point P we can prove that the normal stress has the same value for any surface element through P if no shearing stress exists. Take a small tetrahedron $PQ_xQ_yQ_z$, three edges of which are parallel to the axes of a system of rectangular coordinates (Fig. 2). Let dS be the area of the face $Q_xQ_yQ_z$ oblique to these edges and α its angle with the y-z-plane. Then $dS_x = dS \cos \alpha$ is the area of the face parallel to the y-z-plane. Denote the normal stresses transmitted across the faces dS_x and dS by p_x and p, respec-

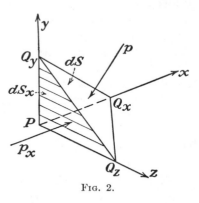

FIG. 2.

tively. The force exerted by the fluid outside the tetrahedron on the face dS_x has the direction of the x-axis and the intensity $p_x dS_x$. The force $p \, dS$ acting on the face dS is perpendicular to this face and thus makes the angle α with the x-axis. The forces acting on the two remaining faces are perpendicular to the x-axis. Thus the sum of the x-components of the forces acting on the portion of fluid under consideration is

$$p_x \, dS_x - p \, dS \cos \alpha = (p_x - p) \, dS_x$$

because $dS_x = dS \cos \alpha$. According to Newton's second law this sum must equal the product of mass times acceleration component a_x in the x-direction. The mass is the product of the density ρ and the volume, which is $\frac{1}{3} dS_x \overline{PQ_x}$. Therefore,

$$(p_x - p) \, dS_x = \tfrac{1}{3}\rho a_x \, dS_x \, \overline{PQ_x} \qquad \text{or} \qquad a_x = \frac{3}{\rho} \frac{p_x - p}{\overline{PQ_x}}$$

To secure a finite value of a_x we have to assume that $p_x - p$ tends toward zero if $\overline{PQ_x}$ becomes smaller and smaller, *i.e.*, if the tetrahedron reduces to the point P. Thus the two stresses p_x and p on the two surface elements in P must be equal. Since the direction of the x-axis can be chosen arbitrarily, there exists only one stress value in P, or *the state of stress at any point P of a perfect fluid, or of any fluid in equilibrium, is completely specified by the pressure p acting on any surface element through P*. This is still true if the fluid is subjected to so-called "body forces," which, like gravity, vary as the volumes of the fluid elements on which they are acting. Indeed, this would add on the right-hand side of the foregoing equation for a_x only a term equal to the quotient body force by mass, which is supposed to be a finite quantity, *e.g.*, equal to g in the case of gravity.

If the foot and pound are used as the units of length and force, the unit of pressure is one pound per square foot. The atmospheric pressure, however, is usually expressed in inches of mercury. The standard specific weight of mercury is 848.71 lb./ft.³; the standard pressure of 29.921 in. Hg corresponds therefore to 2116.2 lb./ft.² or 14.696 lb./in.²

It is known from physics that the density ρ (or the specific weight γ), the pressure p, and the absolute temperature T of a gas are connected by an equation called the *equation of state*. For a so-called "perfect gas" this equation has the form[1]

$$p = R\gamma T = Rg\rho T \tag{1}$$

Atmospheric air follows this law with sufficient approximation. In Eq. (1) R is a constant and T the absolute temperature, which is connected with the temperature Θ on the Fahrenheit scale by

$$T = \Theta + 459.4 \tag{2}$$

It follows from the dimensions of all other terms in Eq. (1) that the gas constant R has the dimension length/temperature. For dry air the value generally used is $R = 53.33$ ft./°F.

Problem 1. At 32°F. and standard pressure, 1 ft.³ of helium weighs 0.01113 lb. Determine the value of the gas constant R.

[1] In physics the specific volume v is generally used instead of the specific weight γ. These variables are connected by the relation $\gamma v = 1$.

Problem 2.   Using the standard values given in this section find the specific weight of dry air under a pressure of 41.0 in. Hg at 41°F.

Problem 3.   Determine the volume that 2.5 lb. of dry air will occupy under a pressure of 100.0 in. Hg at 80°F.

2. Equilibrium of a Perfect Gas under the Influence of Gravity.

Consider a horizontal cylinder of gas of small cross-sectional area, whose plane faces are perpendicular to its generatrices (Fig. 3).   Since gravity and the thrusts on the cylindrical surface are perpendicular to the gen-

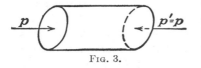

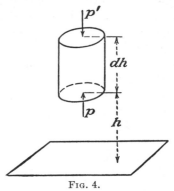

eratrices, the only forces in the direction of the generatrices are the thrusts on the bases.   The equilibrium of the cylinder then requires these thrusts to be equal.

Fig. 3.

Since the areas on which these thrusts act are equal, the pressure must have the same value at the two ends of the cylinder.   As this is true whatever the length of the cylinder or the direction of its horizontal generatrices may be, the following result is established: In a bulk of gas at rest, under the influence of gravity, the pressure has the same value at any two points of the same horizontal plane.

In order to determine how the pressure depends on the altitude h above some horizontal plane of reference (for example, the altitude above sea level), consider a vertical cylindrical column of gas whose ends are at the neighboring levels h and $h + dh$ (Fig. 4).   The forces acting on this column are gravity, the vertical thrusts on the horizontal bases, and the horizontal thrusts on the cylindrical surface.   Denoting the cross-sectional area of the column by dS and the pressure at the top and bottom of the cylinder by p' and p, respectively, we have the thrust $p'\, dS$ acting downward on the top and the thrust $p\, dS$ acting upward on the bottom.   The resultant of these thrusts is a vertical force

Fig. 4.

of the magnitude $(p - p')\, dS$ acting upward.   The weight of the cylinder is obtained as the product $\gamma\, dh\, dS$ of the specific weight γ of the gas and the volume $dh\, dS$ of the cylinder.   The equilibrium of the cylinder requires the sum of the vertical forces acting on the cylinder to be zero.   Thus,

$$p - p' = \gamma\, dh$$

The pressure p is supposed to be a continuously varying function of the altitude h.   As this altitude increases, the pressure p varies at the rate expressed by the derivative dp/dh of p with respect to h.   Accordingly, the difference $p' - p$ can be written as the product of the derivative dp/dh times the small distance dh between the two points where the pressures p and p' act.   Introducing this into the condition of equilibrium already obtained, we find

$$p - p' = -\frac{dp}{dh}\, dh = \gamma\, dh \qquad \text{or} \qquad \frac{dp}{dh} = -\gamma \qquad (3)$$

Since, by its nature, γ is positive, Eq. (3) shows that the pressure decreases at the rate γ with increasing altitude.   Furthermore, the pressure p having the same value at all points of the same horizontal plane, the derivative dp/dh also has a constant value in each horizontal plane. Equation (3) then shows the specific weight γ and, consequently, the density $\rho = \gamma/g$ to be constant in each horizontal plane.   We thus have the following result: *In a gas that is in equilibrium under the influence of gravity, the pressure p as well as the density ρ have constant values in each horizontal plane; with increasing altitude h the pressure p decreases at the rate γ or $g\rho$.*

The equation of state (1) and the condition of equilibrium (3) enable us to determine p and γ (or ρ) as functions of h, if we know how the temperature T varies with the altitude h.

If only a small range of altitudes has to be considered, we may assume the temperature to be constant within this range.   In this so-called "isothermal case" the product RT in (1) is constant.   We find, by substituting from (1) into (3),

$$\frac{dp}{dh} = -\frac{p}{RT} \qquad \text{or} \qquad \frac{dp}{p} = -\frac{dh}{RT}$$

where the factor of dh is a constant.   This equation can easily be integrated since the left-hand term is the differential of $\log p$* and the right-hand term the differential of h/RT.   Thus,

$$\log p = -\frac{h}{RT} + \text{const.}$$

Here h is the altitude above an arbitrarily chosen level of reference. Denote by p_0 the pressure at this level of reference, *i.e.*, the pressure corresponding to $h = 0$.   The constant of integration in the expression for $\log p$ is then seen to have the value $\log p_0$.   Accordingly,

$$\log \frac{p}{p_0} = -\frac{h}{RT} \qquad \text{or} \qquad h = -RT \log \frac{p}{p_0} \qquad (4)$$

This is called the (logarithmic) *barometric formula*.   It can be used for estimating comparatively small differences in altitude (up to a few

* Throughout this book the logarithms indicated by "log" are natural logarithms (with base e).

thousand feet), when the mean temperature and the values of the pressure at the top and at the bottom level are known. The equation of state (1) shows that, for constant temperature, p and ρ are proportional. It follows, therefore, from Eq. (4), that under isothermal conditions

$$\frac{p}{p_0} = \frac{\rho}{\rho_0} = e^{-h/RT} \tag{5}$$

In aeronautical problems, however, the assumption of constant temperature is not sufficiently accurate. In dealing with altitudes ranging up to 40,000 ft. or more, we must take into account the fact that within this range the temperature varies considerably. In general, the temperature will first decrease with increasing altitude, and then, beyond a certain altitude, it will rise again. This rise of temperature is known as the inversion of temperature; it is not taken into account in the usual aeronautical computations (see also page 19).

From the mathematical point of view the simplest assumption, next to that of constant temperature, is that the temperature T decreases at a constant rate with increasing altitude h,

$$\frac{dT}{dh} = -\lambda$$

where λ is a positive constant, called the *temperature gradient*. Integrating this relation and denoting by T_0 the temperature at the level of reference $h = 0$, we obtain

$$T = T_0 - \lambda h \tag{6}$$

Substituting (1) and (6) in (3), we find

$$\frac{dp}{dh} = -\frac{p}{R(T_0 - \lambda h)} \qquad \text{or} \qquad \frac{dp}{p} = -\frac{dh}{R(T_0 - \lambda h)}$$

In the last equation the left-hand term is the differential of log p and the right-hand term the differential of log $(T_0 - \lambda h)/\lambda R$. Integration of this relation therefore furnishes

$$\log p = \frac{1}{\lambda R} \log (T_0 - \lambda h) + \text{const.}$$

Denote by p_0 the pressure at the level of reference $h = 0$. The constant of integration is then seen to have the value log $p_0 - (1/\lambda R)$ log T_0. Thus,

$$\log \frac{p}{p_0} = \frac{1}{\lambda R} \log \frac{T_0 - \lambda h}{T_0} \qquad \text{or} \qquad p = p_0 \left(1 - \frac{\lambda h}{T_0}\right)^{\frac{1}{\lambda R}} \tag{7}$$

From this formula the pressure p at any altitude h can be found if the pressure p_0 and the temperature T_0 at some level of reference $h = 0$ are given and if the temperature gradient λ has a given constant value.

We may note that, as the value of the temperature gradient tends to zero, this formula tends toward Eq. (4). Indeed, as $\log (1 + x)$ for small x can be replaced by x, Eq. (7) supplies for small λ

$$\log \frac{p}{p_0} = \frac{1}{\lambda R} \log \left(1 - \frac{\lambda h}{T_0} \right) \sim - \frac{1}{\lambda R} \frac{\lambda h}{T_0} = - \frac{h}{RT_0}$$

Returning to the general case of a nonvanishing temperature gradient, consider now the variation of the density ρ with the altitude h. With $T = T_0 - \lambda h$, the equation of state (1) furnishes

$$p = Rg\rho(T_0 - \lambda h) \qquad \text{and} \qquad p_0 = Rg\rho_0 T_0 \tag{8}$$

where ρ_0 is the density at the level of reference $h = 0$. Substitute these expressions in (7), and divide both sides by $Rg(T_0 - \lambda h)$. Thus,

$$\rho = \rho_0 \left(1 - \frac{\lambda h}{T_0} \right)^{(1 - \lambda R)/\lambda R} \tag{9}$$

Equations (7) and (9) may be solved with respect to h, thus giving

$$h = \frac{T_0}{\lambda} \left[1 - \left(\frac{p}{p_0} \right)^{\lambda R} \right] = \frac{T_0}{\lambda} \left[1 - \left(\frac{\rho}{\rho_0} \right)^{\lambda R/(1 - \lambda R)} \right] \tag{10}$$

In the case of a constant, nonvanishing temperature gradient this relation takes the place of (4), which is valid under isothermal conditions only.

Comparison of the two expressions for h given by (10) shows that

$$\frac{p}{p_0} = \left(\frac{\rho}{\rho_0} \right)^{1/(1 - \lambda R)} \qquad \text{or} \qquad \frac{p}{\rho^\kappa} = \frac{p_0}{\rho_0^\kappa} \tag{11}$$

where

$$\kappa = \frac{1}{1 - \lambda R} \tag{12}$$

(In the isothermal case $\lambda = 0$, one has $\kappa = 1$ and therefore $p/\rho = p_0/\rho_0$, which also follows immediately from the equation of state with $T = \text{const.}$)

Any change of state of a gas during which Eq. (11) is satisfied is called *polytropic*. It is shown in thermodynamics that the so-called "adiabatic" change of state, i.e., a change of state without loss or gain of heat, is also represented by Eq. (11) and that the constant κ then equals the ratio of the specific heat at constant pressure to the specific heat at constant volume. For dry air this ratio has the value $\kappa = 1.405$. The corresponding value of λ is then found from Eq. (12) to equal $0.00535°F./\text{ft.}$

Problem 4. The atmospheric pressure at the peak of a mountain is found to be 9 per cent less than in the valley. The mean temperature is 45°F. Determine the height of the mountain.

Problem 5. Assuming isothermal conditions, determine the atmospheric pressure and the density at altitudes of 5000 and 10,000 ft., if at sea level the atmospheric pressure is 30.5 in. Hg and the temperature 64°F. Compute also the values of the pressure gradient $(-dp/dh)$ at these altitudes.

Problem 6. How will the assumption of a constant temperature gradient of 0.003°F./ft. affect the answers of Prob. 5?

Problem 7. Assuming polytropic conditions, investigate how the values of pressure and density at 5000 and 10,000 ft. are affected by a 5 per cent increase of (a) the absolute temperature at sea level T_0, (b) the temperature gradient λ, if originally

$$T_0 = 519°F., \qquad λ = 0.0036°F./ft.$$

*Problem 8. Equation (10) is used to determine the altitude of a plane from the observed values of $p_0 = 29.9$ in. Hg, $T_0 = 519°F.$, $λ = 0.00357°F./ft.$, and

$$p = 18.3 \text{ in. Hg.}$$

If the observed values of p_0 and p are accurate to within 1 per cent each, determine the accuracy with which h is obtained from Eq. (10).

*Problem 9. At two points of unknown altitudes the barometer shows $28\frac{1}{4}$ in. Hg and $31\frac{1}{2}$ in. Hg, respectively. What can be concluded from this concerning the altitudes of the two points, if it is known that the temperature varies between 43 and 52°F.?

*Problem 10. How are the equilibrium conditions changed if account is taken of the fact that the earth is a sphere and gravity a force directed toward the center of this sphere?

3. The Standard Atmosphere. The performance of an airplane depends on the density of the air in which it is flying. Since the density varies with the atmospheric conditions, observations concerning the performance of an aircraft can be compared with one another only after having been reduced to certain standard conditions. For this purpose the *standard atmosphere* has been adopted. It is based on the following assumptions, generally accepted in the United States:

1. The air is a perfect gas with the gas constant $R = 53.33089$ ft./°F.

2. The pressure at sea level is $p_0 = 29.921$ in. Hg.

3. The temperature at sea level is $\Theta_0 = 59°F.$ ($T_0 = 518.4°F.$).

4. Within the lower part of the atmosphere the temperature gradient has the constant value $λ = 0.003566°F./ft.$

5. Above the level at which the temperature $\Theta_1 = -67°F.$ is reached the temperature remains constant.

The numerical values in these assumptions correspond to average atmospheric conditions (see Fig. 10). Slightly different basic values are used in other countries. Assumptions (3) and (4) yield the tempera-

---

* Problems whose solution requires more than a simple application of the formulas **given** in the text are marked, here and later, by *.

ture distribution $\Theta = 59 - 0.003566h$. Solving with respect to h we see that the temperature $\Theta_1 = -67°F$. will be reached at the altitude $h_1 = 35,332$ ft. This level is sometimes referred to as the *isothermal level*. Above it the temperature is supposed to remain constant.

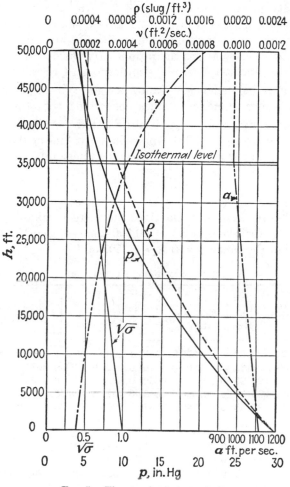

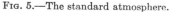

FIG. 5.—The standard atmosphere.

With the numerical values given in assumptions 1 to 4, Eqs. (7) and (9) furnish the following relations valid below the isothermal level:

$$\frac{p}{p_0} = (1 - 0.00000688h)^{5.256}$$

$$\frac{\rho}{\rho_0} = (1 - 0.00000688h)^{4.256}$$

(13)

Above the isothermal level we obtain from Eq. (5)

$$\frac{p}{p_1} = \frac{\rho}{\rho_1} = e^{-0.0000478(h-h_1)}$$

where $p_1 = 6.925$ in. Hg and $\rho_1 = 0.000727$ slug/ft.^{3} are the pressure and the density at the isothermal level h_1. In all these relations the altitude h must be expressed in feet. Figure 5 shows how pressure and density in the standard atmosphere vary with the altitude.

In performance calculations the ratio ρ/ρ_0 of the density ρ at the level h to the density ρ_0 at sea level plays an important role. This so-called "density ratio" is denoted by σ. Values of σ and $\sqrt{\sigma}$ are given in Table 1, together with other numerical values concerning the standard atmosphere. In Fig. 5 the graph of $\sqrt{\sigma}$ is also plotted. An inspection of this graph shows that up to 50,000 ft. approximate values of the square root of the density ratio can be obtained from the linear relation

$$\sqrt{\sigma} = 1 - \frac{h}{c} \tag{14}$$

with $c = 81,000$ ft. The formula gives the correct values $\sqrt{\sigma} = 1$ and 0.5, respectively, for $h = 0$ and $h = 40,500$ and less than 2 to 3 per cent deviation within the range $h = 0$ to $h = 50,000$ ft.

TABLE 1.—STANDARD ATMOSPHERE[1]

h ft.	θ °F.	p in. Hg	ρ slug/ ft.^{3}	γ lb./ft.^{3}	$\frac{p}{p_0}$	σ	$\sqrt{\sigma}$	ν ft.^{2}/sec.	a ft./ sec.
0	59.0	29.92	0.002378	0.07651	1.0000	1.0000	1.0000	0.000157	1116
2,500	50.1	27.31	0.002209	0.07107	0.9129	0.9288	0.9637		
5,000	41.2	24.89	0.002049	0.06592	0.8320	0.8616	0.9282	0.000177	1097
7,500	32.3	22.65	0.001898	0.06107	0.7571	0.7982	0.8934		
10,000	23.3	20.58	0.001756	0.05649	0.6876	0.7384	0.8593	0.000200	1077
12,500	14.4	18.65	0.001622	0.05219	0.6234	0.6821	0.8259		
15,000	5.5	16.88	0.001496	0.04814	0.5642	0.6291	0.7932	0.000228	1057
17,500	− 3.4	15.25	0.001378	0.04433	0.5097	0.5793	0.7611		
20,000	−12.3	13.75	0.001267	0.04075	0.4594	0.5327	0.7299	0.000261	1037
25,000	−30.2	11.10	0.001065	0.03427	0.3709	0.4480	0.6693		
30,000	−48.0	8.880	0.000889	0.02861	0.2968	0.3740	0.6116	0.000346	995
35,000	−65.8	7.036	0.000736	0.02369	0.2352	0.3098	0.5566		
40,000	−67.0	5.541	0.000582	0.01872	0.1852	0.2447	0.4947	0.000508	972
45,000	−67.0	4.364	0.000459	0.01474	0.1458	0.1926	0.4389		
50,000	−67.0	3.436	0.000361	0.01161	0.1149	0.1517	0.3895	0.000819	972

[1] For the meaning of ν and a see Secs. IV. 1 and X. 4.

The altimeters used on aircraft are aneroid barometers; they measure the actual pressure of the surrounding air. Such an instrument is calibrated by exposing it to various pressures and marking on the dial the altitudes that, in the standard atmosphere, correspond to these pressures. If the atmospheric conditions are different from those of the standard atmosphere, the altimeter reading furnishes, not the real altitude h, but a fictitious altitude h_p, called the "pressure altitude." Independently of the process of calibration, the pressure altitude can be defined as the altitude at which a given pressure p is found in the standard atmosphere. By solving Eq. (13) for h we find the pressure altitude

$$h_p = 145,300 \left[1 - \left(\frac{p}{p_0} \right)^{0.1903} \right] \tag{13'}$$

We shall see in the following section how the correct altitude of a climbing plane can be obtained from continued observations of pressure altitude and temperature.

For performance tests of airplanes a second fictitious altitude, the so-called "density altitude" h_ρ, is important. It is defined as the altitude at which a given density is found in the standard atmosphere. From the second equation (13) it follows that

$$h_\rho = 145,300 \left[1 - \left(\frac{\rho}{\rho_0} \right)^{0.2350} \right] \tag{13''}$$

The altimeter reading gives the pressure altitude. If, in addition, the actual temperature is known, the density altitude can be determined. Indeed, Eq. (13) or Table 1 or Fig. 5 will give the pressure p that, in the standard atmosphere, is found at the altitude h_p. The equation of state (1) then furnishes the density ρ that corresponds to this pressure and to the observed temperature T. The density altitude h_ρ finally is obtained from Eq. (13'') or Table 1 or Fig. 5 as the altitude at which the density ρ is found in the standard atmosphere.

Figure 6 shows the relation between h_ρ, h_p, and T, given by the equation of state in which h_p and h_ρ are introduced from (13). The relation then reads

$$\frac{(1 - 0.00000688h_p)^{5.256}}{(1 - 0.00000688h_\rho)^{4.256}} = gRT \frac{\rho_0}{p_0} = \frac{T}{T_0}$$

Abscissas and ordinates in this density and pressure altitude conversion chart represent the density altitude h_ρ and the temperature Θ on the Fahrenheit scale, respectively. Accordingly, any state of the air, defined by density and temperature, is represented by the point whose coordinates correspond to this density and temperature. Points representing states of equal pressure are joined by curves labeled according to the corresponding pressure altitude. By means of this chart, pressure altitude can be converted into density altitude, and vice versa, when the temperature is known. For example, take the density altitude corresponding to the pressure altitude $h_p = 10,000$ ft. and the temperature $\theta = 40°F$. On the line $h_p = 10,000$ ft. we locate the point whose ordinate represents $\theta = 40°F$. The abscissa of this point corresponds to $h_\rho = 11,000$ ft., which is the required density altitude.

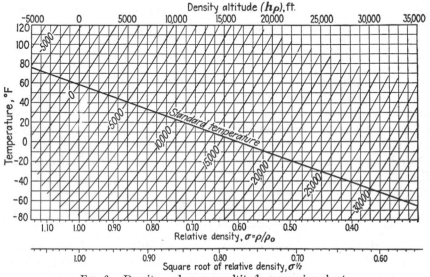

Fig. 6.—Density and pressure altitude conversion chart.

In Fig. 6 the points for which $h_p = h_\rho$ lie on a straight line that represents the temperature distribution in the standard atmosphere below the level of 35,300 ft.

Problem 11. Determine the altitudes at which the pressure in the standard atmosphere is 75, 50 and 25 per cent, respectively, of the pressure at sea level.

Problem 12. Determine the errors made in computing the square root of the density ratio σ by means of the approximate formula (14) for $h = 10,000$ ft. and $h = 20,000$ ft.

Problem 13. In the standard atmosphere the polytropic relation $p/\rho^\kappa = $ const. is valid below the isothermal level. Find the value of κ.

Problem 14. The German standard atmosphere is based on the values

$$p_0 = 10,363 \text{ kg./m.}^2$$

$\gamma_0 = g\rho_0 = 1.25$ kg./m.^{3}, $\theta = 10°C$., $\lambda = 0.005°C./m.$ Compare the values of p/p_0 and ρ/ρ_0 at the altitudes of 10,000, 20,000, and 30,000 ft. with those of the American standard atmosphere.

Problem 15. The following maximum velocities in level flight V and temperatures Θ have been observed at various altitudes h_p indicated by the altimeter:

$h_p =$	0	3,000	6,000	9,000	12,000	15,000	18,000 ft.
$\Theta =$	49	40	32	23	13	4	$-4°$F.
$V =$	305	314	324	330	312	297	260 ft./sec.

Draw a diagram representing V as a function of the altitude h_p in the standard atmosphere.

4. Determination of True Altitude. Reduction of a Climb to Standard Atmosphere. In a climbing test the pressure p (or the pressure altitude h_p) and the temperature are either observed at regular intervals or continually recorded by self-registering instruments. In connection with the evaluation of such records two problems occur. The first is to determine the *correct altitude above the ground that corresponds to a certain altimeter reading,* a question that arises, for example, in the evaluation of meteorological ascents. If the pressure p and the absolute temperature T have been recorded, the equation of state (1) and the condition of equilibrium (3) enable us to solve this problem. Substituting from (1) in (3), we find

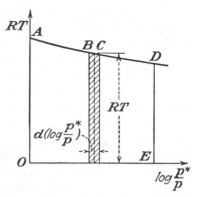

FIG. 7.—Determination of true altitude.

$$\frac{dp}{dh} = -\frac{p}{RT} \quad \text{or} \quad dh = -RT\frac{dp}{p}$$

Take the ground as the level of reference $h = 0$, and denote the pressure there by p^*. Since p^* is a constant, the expression $-dp/p$, which occurs on the right side of the last equation, can be written as $d[\log (p^*/p)]$.[1] Now plot the products RT, which have the dimension of a length and are computed from the observed values of T, against the values of $\log (p^*/p)$, as taken from the pressure observation. Thus, each point of the curve AD in Fig. 7 has as coordinates a pair of simultaneously observed values of RT and $\log (p^*/p)$, respectively. Since

$$dh = RT\, d\left(\log \frac{p^*}{p}\right)$$

the (infinitesimal) difference in level between two successive observations, represented by the neighboring points B and C of the graph, is given by

---

[1] The introduction of the constant p^* is helpful in order to render dimensionless the expression after the log sign.

the shadowed area. If, therefore, A is the point corresponding to the ground values of p and T and if D is the point representing the observations at altitude h, this h-value is given by the area $OADE$ of the graph. For example, let 10 in. be chosen as the unit for plotting log $(p*/p)$ and 1 in. as representing the value $RT = 10,000$ ft. Then, the area of 1 sq. in. will correspond to the altitude $h = (10,000/10) \times 1 = 1000$ ft.; and, in Fig. 7, the area $OAED$ will correspond to 4250 ft. If during the climb the pressure altitude has been observed instead of the pressure, the corresponding pressure values must be obtained first from Eq. (13'), Table 1, or Fig. 5.

The second problem is the *reduction of a climb diagram to standard atmospheric conditions*. When the climbing ability of a new type of airplane is tested, we are interested, not so much in the performance under the actual atmospheric conditions, but rather in the performance under the standard atmospheric conditions. This reduction of the actually observed performance to standard atmosphere is essential when tests made under different conditions are to be compared. Assume again that during the climb the pressure p and the absolute temperature T are recorded as functions of the time t. Moreover, the pilot is supposed to have used at any instant the maximum climbing rate of which the plane is capable under the conditions prevailing at that instant. We anticipate the fact (see Chap. XIV) that at any time the maximum climbing rate for a given plane depends on the actual density ρ only. From the records of p and T we wish to determine the time that the plane would need in the standard atmosphere to climb to a given altitude, if again the maximum climbing rate is used at any instant.

The condition of equilibrium (3) gives

$$dh = -\frac{1}{\gamma} dp$$

Dividing both sides by the interval of time dt, during which the plane climbs through dh, the actual rate of climb is found to be

$$w = \frac{dh}{dt} = -\frac{1}{\gamma}\frac{dp}{dt} \tag{15}$$

where the specific weight can be determined from the observed values of p and T according to the equation of state $\gamma = p/RT$. As the dependence of p and T upon the time t is known from the records of the climb, the right-hand side of (15) can be evaluated. For each instant t and the corresponding values of p and T, the actual rate of climb w is thus obtained. According to our assumptions this w is the maximum climbing rate compatible with the actual value of $\gamma = p/RT$. In the standard atmosphere the same value of w would be the climbing rate at the density altitude h_ρ, which corresponds to the ρ-value p/gRT with the observed values of p and T.   The diagram we want to find, representing the climb reduced to the standard atmospheric conditions, must therefore show the slope w at the point with ordinate h_ρ.   If we call t_ρ the abscissa of this point with the ordinate h_ρ in the reduced climb curve, this curve is determined by the condition that $dh_\rho/dt_\rho = w$, with w a known function of h_ρ.

The reduced diagram can now be obtained by the following graphical procedure:[1] In the right half of Fig. 8 the observed values of p and RT

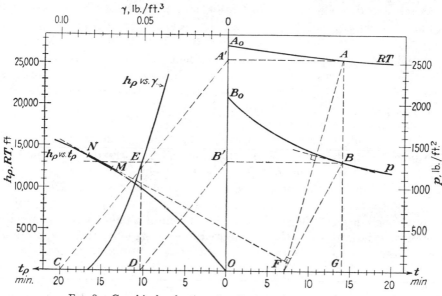

FIG. 8.—Graphical reduction of a climb to standard conditions.

are plotted against the time t.   In the left half the density altitude h_ρ is plotted vs. the specific weight $\gamma = g\rho$ that, in the standard atmosphere, is found at the altitude h_ρ.   This curve represents Eq. (13″) and thus is independent of the actual observations.   For RT, which has the dimension of a length, the same scale is used as for h_ρ.   It will be convenient to take the scale for p such as to have 1000 lb./ft.² represented by the same ordinate as the altitude $h_\rho = 10,000$ ft.   Then, on the axis of abscissas, the point C is marked that represents $\overline{OC} = 0.1$ lb./ft.³ in the γ-scale.

The values of temperature and pressure observed at a certain instant t are represented by the points A and B.   In order to find the correspond-

[1] MISES, R. v., *Zeitschrift für Flugtechnik und Motorluftschiffahrt*, **8**, 173–177 (1917).

ing specific weight $\gamma = p/RT$, project these points upon the ordinate axis, thus obtaining A' and B', and make $B'D$ parallel to $A'C$. Then $\overline{OD} = \overline{OC} \times (\overline{OB'}/\overline{OA'})$ represents the specific weight $\gamma = p/RT$ that prevailed in the actual atmosphere at the time t. Through D draw a parallel to the ordinate axis intersecting the h_p vs. γ-line in E. Then $\overline{DE}$ represents the density altitude h_p that corresponds to the observed values $p = \overline{GB}$ and $RT = \overline{GA}$. In order to determine the rate of climb at this altitude h_p, draw the tangent of the p-line at B. From A draw the perpendicular to this tangent intersecting the axis of abscissas at F, and draw FB. The slope of a straight line normal to FB is then RT/p times the slope of the tangent in B, thus representing the absolute value of the climbing velocity w according to (15). In this way one small element $\overline{MN}$ of the h_p vs. t_p curve (with the abscissas pointing to the left) can be found. If we start at the origin with the element corresponding to the points A_0, B_0 at $t = 0$ and proceed in this way, we obtain step by step the complete h_p vs. t_p curve. This solves the problem of reducing the climb to standard atmospheric conditions.

Instead of this graphical method the following numerical way of reducing a climb may be used:[1] Pressure and temperature are observed in short intervals of time Δt. The specific weight corresponding to these values is computed by means of the equation of state $\gamma = p/RT$. From the arithmetic mean γ_m of two consecutive values of γ and the difference Δp of the corresponding values of p the actual climb Δh between consecutive observations is then computed in accordance with the condition of equilibrium $\Delta h = -\Delta p/\gamma_m$. The density altitude corresponding to the values of p and T is obtained from Table 1 or Fig. 5. If Δh_p denotes the difference of two consecutive values of h_p, the assumption that the rate of climb depends only on the density ρ (or the specific weight γ) leads to $\Delta h/\Delta t = \Delta h_p/\Delta t_p$. From this relation the values Δt_p are computed. Since for $t = 0$ the density altitude generally has a non-vanishing value h_0, we compute the time Δt_0 necessary to climb from sea level to h_0 by assuming that the rate of climb has the same value as during the interval between the first two observations. The time t_p necessary in standard atmosphere to climb to an altitude h_p is finally obtained by summing the intervals Δt_p previous to the instant at which h_p is reached. Plotting h_p vs. t_p we obtain the diagram of the reduced climb.

The following table shows for the first three steps the arrangement of the computation outlined in the foregoing. The observed data are the same as those in Fig. 8.

---

[1] See *NACA Tech. Rept.* 216 (1925).

t min.	Δt min.	p in. Hg	Δp in. Hg	Θ °F.	T °F. abs.	γ lb./ft.³	γ_m lb./ft.³	Δh ft.	h_p ft.	Δh_p ft.	Δt_p min.	t_p min.
0		29.42		52.0	511	.0763			90			0.08
	2		−2.46				.0736	2360		2490	2.11	
2		26.96		44.5	504	.0709			2580			2.19
	2		−2.12				.0685	2190		2330	2.13	
4		24.84		39.0	498	.0661			4910			4.32
	2		−1.54				.0641	1700		1990	2.34	
6		23.30		37.5	497	.0622			6900			6.66

The results do not differ noticeably from those obtained by the graphical method. They are shown in Fig. 9, which includes both the

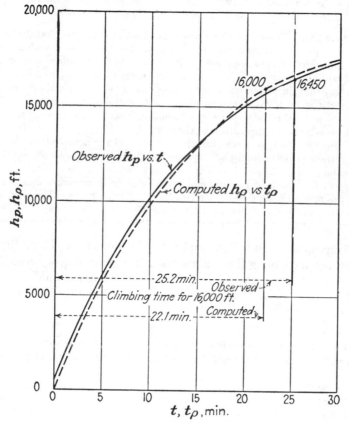

Fig. 9.—Reduction of a climb to standard conditions.

observed barograph h_p vs. t and the plot h_p vs. t_p as reduced to standard conditions.

Problem 16. During a meteorological ascent the following observations of pressure and temperature have been made:

p = 29.34   27.86   26.40   24.85   23.40   22.00   20.55 in. Hg
θ = 60.5   68.0   64.5   57.0   50.0   43.0   36.5°F.

The airport is 570 ft. above sea level. Determine the level at which the temperature is 40°F.

Problem 17. Reduce the following climb record to standard atmospheric conditions:

t = 0   1.5   3.0   4.5   6.0   7.5 min.
p = 30.82   28.17   25.96   24.03   22.75   21.53 in. Hg
θ = 84.0   69.5   57.5   50.5   45.0   41.5°F.

t = 9.0   10.5   12.0   13.5   15.0 min.
p = 20.58   19.92   19.28   18.54   18.03 in. Hg
θ = 37.0   30.5   25.0   23.5   21.0°F.

*Problem 18. The contract for a new type of plane stipulates that the plane shall be able to climb in standard atmosphere from sea level to 5000 ft. within 2 min. If the atmospheric conditions are different from those in the standard atmosphere, the range of 0 to 5000 ft. may be replaced by another range of 5000 ft. (from h to $h + 5000$) in such a way that the arithmetic mean of the densities at the ends of this range equals the arithmetic mean of the densities which in the standard atmosphere are found at sea level and at 5000 ft. Draw a chart furnishing this "equivalent range" for given values of pressure and temperature at the ground.

*Problem 19. To take account of the deviation of the actual atmospheric conditions from those in the standard atmosphere the altimeter dial is set so that the needle indicates the correct altitude $h = 0$ at the start from sea level. If the angular displacement of the needle is proportional to the change in pressure, find the relation between the true altitude h and the altitude h_i indicated by the altimeter. Assume that the temperature distribution is that of the standard atmosphere, while the value of the pressure at sea level differs slightly from that in the standard atmosphere.

5. Troposphere and Stratosphere. Influence of Humidity. The assumptions underlying the arguments in the preceding sections must be modified if very large altitudes are involved or if the humidity of the air is considerable. Some data about the physical conditions of the atmosphere follow:

The atmosphere is composed of nitrogen, oxygen, argon, carbon dioxide, and traces of other gases such as hydrogen, neon, and helium. For dry air at sea level the percentage by volume of the main constituents is

Nitrogen... 78.03
Oxygen.. 20.99
Argon... 0.94
Carbon dioxide.. 0.03

On the basis of weight or mass the per cent composition is

75.5 : 23.2 : 1.33 : 0.045

Within the lower part of the atmosphere, the *troposphere*, this composition is kept practically constant by winds and vertical currents. The troposphere extends to about 4 miles altitude at the poles and about 9 miles at the equator. Above the troposphere is the *stratosphere*, where the percentage of the heavier gases diminishes with increasing altitude.

The mean temperature at about 40°N. latitude in the United States is represented in Fig. 10.[1] It is seen from this figure that the assumption

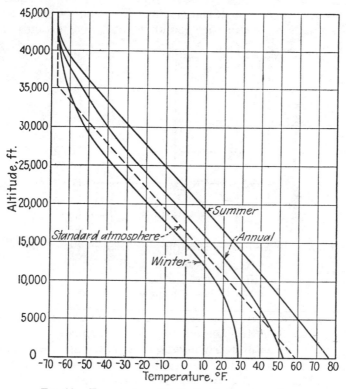

FIG. 10.—Temperature distribution in the atmosphere.

of a constant temperature gradient is justified within a large central portion of the troposphere. In the layers above and below this portion the temperature gradient has a smaller value than within the central portion. In summer the zone of practically constant temperature gradient is larger and the variation of the temperature gradient is smaller than in winter. In the lower regions of the stratosphere the temperature has a constant value of −67°F. This region of constant temperature extends to about 70,000 ft. Above this level the temperature increases

[1] *NACA Tech. Rept.* 147 (1922).

again so that it reaches the value of $-40°F$. at about 100,000 ft. This phenomenon, as mentioned above, is known as the *inversion of temperature*.

In addition to its constant constituents, atmospheric air contains water vapor in varying proportion. At $59°F$. and standard pressure, air saturated with water vapor is 0.65 per cent lighter than dry air. In the evaluation of wind-tunnel tests the humidity therefore must be taken into account whenever results of high accuracy are required. In order to determine humidity, the *wet-and-dry-bulb hygrometer* is used. This instrument consists of two thermometers. The bulb of one is kept wet by a strip of muslin which is tied round the bulb and the ends of which

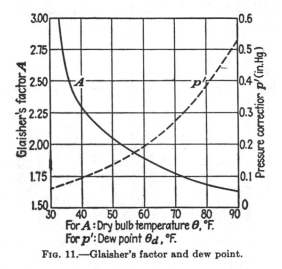

For A: Dry bulb temperature θ, °F.
For p': Dew point θ_d, °F.

FIG. 11.—Glaisher's factor and dew point.

hang into a vessel containing water. The wet bulb is cooled by the evaporation from its surface, so that the two thermometers show different readings. The less saturated with water vapor the air, the stronger the evaporation from the surface of the wet bulb, and consequently the bigger the difference between the two thermometer readings. From a great number of simultaneous observations of the dry- and wet-bulb temperatures and of a Daniel hygrometer, Glaisher derived an empirical formula for the *dew point*, i.e., the temperature to which the air must be cooled, at constant pressure, in order to become saturated. Denoting the temperature of the dew point by θ_d and the dry- and wet-bulb thermometer readings by θ and θ', respectively, one can write Glaisher's formula in the form

$$\theta_d = \theta - A(\theta - \theta')$$

where the empirical factor A (Glaisher's factor) depends on the **dry-bulb** temperature θ. Figure 11 shows A as a function of θ.

Moist air being lighter than dry air, the density of moist air at the temperature Θ and the pressure p equals the density of dry air at the temperature Θ and a certain *reduced pressure* p_r. We set $p_r = p - p'$, where the pressure correction p' depends on the dew point Θ_d of the moist air. Figure 11 includes a curve that gives empirical values for p' as a function of Θ_d. According to the definition of the reduced pressure the specific weight of moist air at the absolute temperature T and for the reduced pressure p_r is

$$\gamma = \frac{p_r}{RT}$$

where the gas constant R is again that of dry air. In the evaluation of a climb the influence of humidity can therefore be taken into account by using the reduced pressure p_r instead of the observed pressure p in the determination of ρ.

Problem 20.   What is the specific weight of air when the dry-bulb temperature is 70.5°F., the wet-bulb temperature 64.0°F., and the barometer reading 30.5 in. Hg?

Problem 21.   Evaluate the climb of Prob. 17 under the assumption that the air is everywhere saturated with water vapor.

*Problem 22.   The simultaneous observations of temperature and pressure made during a climb are in perfect agreement with the values to be expected in the standard atmosphere. For various altitudes h in the standard atmosphere, find by how much, at most, the true altitude of the plane may differ from h because of the humidity.

CHAPTER II

BERNOULLI'S EQUATION. ROTATION AND CIRCULATION

1. Steady Motion. Motions of fluids show an amazing variety of character. Throughout this book, which attempts to give only a first introduction to the problems of fluid mechanics connected with airplane design, one restriction will be maintained: The types of motion to be discussed will belong to the class of so-called "permanent" or "steady flow." This kind of flow can be defined by postulating that the state of motion remain unchanged at any place within the fluid. This implies that all quantities necessary for a complete description of the flow shall have constant values at any fixed point in the region occupied by the fluid. That is, in a steady flow, pressure, density, magnitude and direction of velocity, etc., are functions of the position in space but are independent of time. We thus have an unchanging pattern of motion carried by ever-changing particles.

At this point the reader should be cautioned against a possible misunderstanding regarding the definition of a steady flow. Although the velocity at any given spot is constant, the motion of a given fluid particle need not be uniform. Indeed, proceeding along its path, the particle continually adapts its velocity to that pertaining to its instantaneous position.

The theory of steady motion applies to most problems that are of importance in the theory of flight. At first glance this statement seems to be at variance with the fact that the flow produced by a rigid body (airplane) moving through a bulk of air at rest is by no means steady. For example, let us consider a point that is in the path of the moving body. As the body approaches, a certain velocity will develop at this point; but after the body has passed, the air at this spot will return more or less to its original state of rest. Thus we are far from having an unchanging state of motion at any fixed point. However, in the most important case of a body engaged in a uniform rectilinear translation through a fluid at rest, an observer moving with the body will have the impression of a steady flow of fluid around a body at rest. This apparent flow is called the *relative motion* of the fluid with respect to the moving body. Let us introduce a system of coordinates that is rigidly connected with the body and, consequently, partakes in its motion. At any point with constant coordinates with respect to this system, pressure, density, apparent (relative) velocity, etc., remain constant. *The relative flow of the fluid with respect to the body is a steady flow.* We may obtain this relative motion by superposing on the original motion a uniform rectilinear translation whose velocity is equal and opposite to that of the rigid body. This brings the body to rest while, at a great distance from it (mathematically speaking, at infinity), the fluid is given a velocity equal and opposite to the original velocity of the body. On the other hand, the superposition of such a motion without acceleration cannot influence the forces that the fluid exerts on the body. These forces are the same, whether the fluid is at rest and the body moves through it with a certain constant velocity or whether the body is at rest and the fluid streams

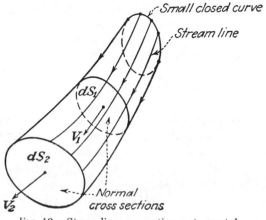

Fig. 12.—Streamlines generating a stream tube.

toward it with an equal and opposite velocity. The latter flow, *i.e.*, the relative motion of the fluid with respect to the body, is sometimes referred to as the *inverse flow*.

The conception of the *streamlines* is of the greatest value in the discussion of patterns of steady flow. In a steady motion the conditions at any given point remain constant. Hence the fluid particles that one after another reach a certain point move in the same manner along the same path. In this kind of flow the paths of the fluid particles are therefore permanent curves. At each point of such a curve the tangent has the direction of the velocity. These curves are called streamlines. In the case of an unsteady flow the streamlines, *i.e.*, the curves whose tangents have the velocity direction are, in general, not the pathways of the particles.

The streamlines of a steady flow furnish, moreover, some information concerning the magnitude of the velocity. In order to explain this let us consider a small closed curve and draw the streamlines passing through its points (Fig. 12). These lines form a so-called "stream tube." By the definition of the streamlines no fluid particle can enter or leave the tube through its walls. Accordingly, the mass of the fluid flowing through a cross section of the tube per unit of time must have a constant value along the tube. This constant is called the *flux*. If dS denotes the area of a normal section of the tube, ρ the density, and V the velocity of the fluid at this section, the flux equals $\rho V \, dS$. We therefore have the following theorem: *Along each stream tube the product of density, velocity, and cross-sectional area is constant.*

$$\rho_1 V_1 \, dS_1 = \rho_2 V_2 \, dS_2 \tag{1}$$

This relation is sometimes called the *condition of continuity*.

In many cases the density of the fluid changes but little within the field of flow under consideration. If we neglect completely such density changes, we say that the fluid is considered as *incompressible*. In this case the foregoing theorem takes the following form: *Along each stream tube the product of velocity and cross-sectional area is constant.* In other words, along each stream tube the velocity varies inversely to the cross-sectional area.

A still simpler form of the theorem is obtained in the case of the so-called "two-dimensional flow." In this case all streamlines are parallel to one plane, which may be chosen as the x-y-plane of a system of rectangular coordinates x, y, z. Moreover, the streamlines are supposed to be the same in each plane $z = $ const., the velocity distribution as well as the density, pressure, etc., depending on x and y only. We then can restrict our considerations to the flow in the x-y-plane, with the understanding that we are dealing with a fluid layer of unit thickness perpendicular to that plane. Accordingly, any stream tube that we shall consider will have rectangular cross-sections, one pair of sides being formed by the unit thickness, the other appearing in the x-y-plane as the distance of two neighboring streamlines (Fig. 13). The stream tube thus will be completely defined by the two neighboring streamlines in the x-y-plane. We call *stream filament* the strip of the x-y-plane lying between two such streamlines and *width of the filament* the normal distance dn between these streamlines (Fig. 14). The stream tube corresponding to the filament then has a rectangular cross-sectional area whose numerical value is dn. The foregoing theorems consequently can be modified as follows: *Along each stream filament in a two-dimensional steady flow the product $\rho V \, dn$ of density, velocity, and width is constant. In the case of an incompressible fluid the velocity varies inversely to the width of the filament.*

In a two-dimensional flow we can select streamlines in such a way that for all stream tubes determined by two consecutive streamlines the flux, *i.e.*, the mass dQ of the fluid flowing through a cross-section per unit of time, has the same value. The pattern formed by these selected streamlines gives, then, an approximate picture of the distribution of the values ρV. In the case of an incompressible fluid, these streamlines immediately picture the velocity distribution. At any point P the velocity has the magnitude $V = dQ/\rho\, dn$, where ρ is the constant density of the fluid and dn the width, at P, of the stream filament containing the point P.

Let f denote some quantity, such as density or pressure, connected with the streaming fluid. At any given point in a steady flow this quantity will have a constant value. We may, however, consider a fluid particle and ask how the value of f changes for this particle as it moves

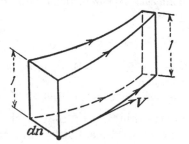

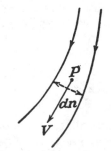

Fig. 13.—Stream tube in two-dimensional flow.   Fig. 14.—Stream filament.

along. Let s denote the distance that the particle has covered, traveling along its path from some conveniently chosen initial position to its position at the instant t. The motion of the particle along its path can then be described by stating how s depends on t. On the other hand, for each position of the particle given by the corresponding value of s there exists a certain value of f, that is, f can be considered as a function of the distance s, which itself is a function of the time t. Accordingly, we have

$$\frac{df}{dt} = \frac{df}{ds}\frac{ds}{dt}$$

But ds/dt equals V, the velocity of the particle; hence,

$$\frac{df}{dt} = V\frac{df}{ds} \tag{2}$$

In a steady flow the time rate of change of any quantity f connected with a fluid particle equals the product of the velocity V of this particle and the space rate of change of f in the direction of V. This theorem states one of the most important features of the steady motion. We shall see later (Sec. III.3) how it must be modified in the case of unsteady flow. Relation (2) and the statement expressing it will be referred to as *Euler's rule of differentiation.*

As an example assume the quantity f to be the magnitude V of the velocity of the particle. It is known from kinematics that the acceleration of a particle in curvilinear motion consists of a tangential component of magnitude dV/dt and of a normal component of magnitude V^2/R, where R is the radius of curvature of the path. From (2) the tangential component is found to be

$$\frac{dV}{dt} = V\frac{dV}{ds} = \frac{d}{ds}\left(\frac{V^2}{2}\right) \tag{2'}$$

Problem 1. Show that along a conical pipe the acceleration of an incompressible fluid in steady flow varies inversely to the fifth power of the diameter.

*Problem 2. Show that in the steady two-dimensional flow of an incompressible fluid the continuity condition (1) can be written as

$$\frac{dV}{ds} + V\frac{d\theta}{dn} = 0$$

where θ is the angle of the velocity vector with any fixed direction and dn is the differentiation in the direction normal to the streamline.

*Problem 3. Establish the analogous formula for a compressible fluid.

2. Bernoulli's Equation. It has been stated in Sec. I.1 that no shearing stresses are transmitted in a fluid at rest. In a fluid in motion, however, shearing stresses may occur. Consider, for example, two neighboring stream filaments in a two-dimensional flow (Fig. 15). If the filament B moves at a greater speed than A, we may expect it to exert a shearing stress on A that would accelerate A. The filament A then would exert on B an equal and opposite shearing stress, which would retard B. In sticky fluids the presence of such shearing stresses influences the flow pattern considerably. In air, however, the influence of the shearing stresses is often negligible, at least for the purpose of obtaining a first approximation to the solution of a problem of flow. The stress transmitted across any surface element may then still be assumed to be a pressure perpendicular to this element and of equal value for all elements through the same point. The hydrodynamic theory based on this assumption is called the theory of the ideal or *perfect fluid.* Within the framework of this theory numerous problems of the theory of flight can be adequately dealt with. However, care should be taken to check in each particular case how far the results furnished by the theory of the perfect fluid may be modified by the presence of shearing stresses. The discussion of this question will be left to Chap. IV. Let us now establish an important theorem valid for perfect fluids in steady flow.

Consider an element of the fluid having the shape of a small cylinder whose axis is formed by a short segment of a streamline (Fig. 16). When

ds denotes the length and dS the cross-sectional area of this cylinder, the mass of the cylindrical element of the fluid is $\rho\, dS\, ds$. According to Newton's second law of motion the product of the mass and the component of the acceleration of the element in any direction equals the sum of the forces acting on this element in this direction. The pressure on the cylindrical surface of the element, being perpendicular to the axis of the cylinder, has no component in the direction ds. According to Eq. (2') the acceleration component in question is $V\, dV/ds$. The projection, on the axis of the element, of the weight $g\rho\, dS\, ds$ is $-g\rho\, dS\, ds\, \cos\beta$, where β is the angle between the axis of the element and the upward vertical. Denoting by h the elevation of the element above some

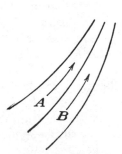

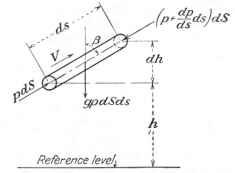

FIG. 15.—Shearing stress between neighboring stream filaments.

FIG. 16.   Forces upon a cylindrical fluid element in direction of flow.

horizontal plane of reference, we have $\cos\beta = dh/ds$ as is seen from Fig. 16. Writing the thrusts on the bottom and the top of the element as $p\, dS$ and $(p + dp/ds \cdot ds)\, dS$, respectively, we obtain from Newton's law

$$\rho\, dS\, ds\, V \frac{dV}{ds} = p\, dS - \left(p + \frac{dp}{ds}\, ds\right) dS - g\rho\, dS \frac{dh}{ds}\, ds$$

Here $p\, dS$ and $-p\, dS$ cancel out and the common factor $dS\, ds$ of all remaining terms can be omitted. Dividing by $g\rho$ and using

$$d(V^2) = 2V\, dV$$

we have, finally,

$$\frac{d}{ds}\left(\frac{V^2}{2g} + h\right) + \frac{1}{g\rho}\frac{dp}{ds} = 0 \qquad (3)$$

Let us first discuss this equation in the case of an incompressible fluid. Since, there, the density is constant, (3) can be written as

$$\frac{d}{ds}\left(\frac{V^2}{2g} + h + \frac{p}{\gamma}\right) = 0$$

where $\gamma = g\rho$ is the constant specific weight of the fluid. Along each streamline the expression in the parentheses has therefore a constant value, which we shall denote by H:

$$\frac{V^2}{2g} + h + \frac{p}{\gamma} = H = \text{constant along each streamline} \qquad (4)$$

This result is called *Bernoulli's equation* for an incompressible fluid. All terms of (4) have the dimension of a length; the first, $V^2/2g$, is called the *velocity head;* it can be described as the height from which a body must fall freely to acquire the speed V. The quantity p/γ is called the *pressure head;* it represents the height of a vertical column of fluid the pressure at the bottom of which exceeds by p the pressure at its top. Using these concepts we can express Bernoulli's equation as follows: *Along each streamline the sum of velocity head, pressure head, and elevation above some horizontal plane of reference is constant.*

This statement remains true for a compressible fluid, if the density can be assumed to be a single-valued function of the pressure, *i.e.*, if the density has the same value at any two points where the pressure is the same.[1] But the definition of the pressure head must then be modified in an appropriate way. Let us define the pressure head as

$$h_p = \int \frac{dp}{g\rho(p)} \qquad (5)$$

This means that the derivative of h_p with respect to p is $1/g\rho$. As an indefinite integral, h_p is determined except for an additive constant. This latter is also true for the geometrical height h where the level of reference can be chosen arbitrarily.

According to (5) we have

$$\frac{dh_p}{ds} = \frac{dh_p}{dp}\frac{dp}{ds} = \frac{1}{g\rho}\frac{dp}{ds} \qquad (6)$$

and therefore (3) can be written as

$$\frac{d}{ds}\left(\frac{V^2}{2g} + h + h_p\right) = 0 \qquad (7)$$

---

[1] Since the density generally depends on pressure and temperature, this assumption restricts the admissible types of change of state. In meteorological questions, where vast regions are considered, this assumption is rarely justified. Within restricted regions, however, the change of state can be assumed as either isothermal or polytropic. The density then becomes a single-valued function of the pressure: $\rho = \rho(p)$.

This shows that the expression in the parentheses has a constant value along each streamline. Thus the general form of Bernoulli's equation is

$$\frac{V^2}{2g} + h + h_p = H = \text{constant along each streamline} \qquad (8)$$

The statement expressing this relation in words is the same as that already given in connection with Eq. (4).

Equation (5) includes in particular the following important cases:

a. *The Fluid Is Incompressible.* The density ρ is then constant; consequently, Eq. (5) gives the pressure head

$$h_p = \int \frac{dp}{g\rho} = \frac{1}{g\rho} \int dp = \frac{p}{g\rho} + \text{const.} = \frac{p}{\gamma} + \text{const.} \qquad (9)$$

We are free to choose the integration limits such as to have the constant of integration equal to zero. With the expression thus obtained for h_p, Eq. (8) reduces to (4).

b. *The Change of State Is Isothermal.* If the pressures p_0 and p correspond to the densities ρ_0 and ρ, respectively, the equation of state, Eq. (1), Chap. I, for constant temperature, gives $p/p_0 = \rho/\rho_0$. Equation (5) accordingly furnishes the pressure head

$$h_p = \int \frac{dp}{g\rho} = \frac{p_0}{g\rho_0} \int \frac{dp}{p} = \frac{p_0}{\gamma_0} \log p + \text{const.}$$

Choose the constant of integration in such a way that $h_p = 0$ for a point where $p = p_0$. Then the pressure head in the isothermal case becomes

$$h_p = \frac{p_0}{\gamma_0} \log \frac{p}{p_0} \qquad (10)$$

c. *The Change of State Is Polytropic.* Here we have $p/p_0 = (\rho/\rho_0)^\kappa$ in accordance with Eq. (11), Sec. I.2. Equation (5), therefore, gives the pressure head as

$$h_p = \int \frac{dp}{g\rho} = \frac{p_0^{1/\kappa}}{g\rho_0} \int \frac{dp}{p^{1/\kappa}} = \frac{\kappa}{\kappa - 1} \frac{p_0^{1/\kappa}}{\gamma_0} p^{(\kappa-1)/\kappa} + \text{const.}$$

Again, choosing the constant of integration so that $h_p = 0$ for $p = p_0$, we obtain

$$h_p = \frac{\kappa}{\kappa - 1} \frac{p_0^{1/\kappa}}{\gamma_0} (p^{(\kappa-1)/\kappa} - p_0^{(\kappa-1)/\kappa}) = \frac{\kappa}{\kappa - 1} \frac{p_0}{\gamma_0} \left[\left(\frac{p}{p_0}\right)^{(\kappa-1)/\kappa} - 1 \right] \qquad (11)$$

as the pressure head in the polytropic case.

Applying Eq. (8) to a fluid in equilibrium ($V = 0$), it is seen that, with the choice of the constant of integration made under (b) and (c), the pressure head h_p is the difference in level between two points at which the pressure has the values p_0 and p, respectively. The pressure altitude h_p introduced in Sec. I.3 was defined in the same way, p_0 being the pressure at sea level.

In the majority of applications of fluid dynamics to the theory of flight it is permissible to neglect, not only the influence of the compressibility of the air, but also the effect of its weight. In this case the term h, which represents the effect of gravity in (4), must be omitted. Along each streamline we thus have

$$\frac{V^2}{2g} + \frac{p}{\gamma} = H = \text{const.} \tag{12}$$

Bernoulli's equation in this simple form will be used in most of our problems.

Problem 4. Water flows steadily downhill in a pipe of varying cross section. At a place where the pressure is 20.5 in. Hg, the speed is 5.6 ft./sec.; 4 ft. below this level the diameter of the pipe is 40 per cent greater. Neglecting the friction in the pipe, determine the pressure at this level.

Problem 5. An incompressible fluid is engaged in a steady two-dimensional flow parallel to a vertical plane. It is observed that along a certain stream filament the pressure is constant. If dn_0, dn, dn_1 denote the width of this stream filament at the levels 0, h, h_1, respectively, express $z = dn/dn_1$ as a function of $x = h/h_1$ and of $y = dn_0/dn_1$.

Problem 6. Show that for the steady flow of a gas under polytropic change of state the expression

$$V^2 + \frac{2\kappa}{\kappa - 1}\frac{p}{\rho}$$

is constant along each streamline, if the influence of gravity is neglected.

Problem 7. Following the adiabatic law ($\kappa = 1.4$) air flows out of a large container in which the pressure is 3000 lb./ft.² If the atmospheric pressure is 2120 lb./ft.², find the velocity of efflux. Container temperature is 85°F.

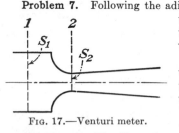

FIG. 17.—Venturi meter.

*Problem 8.** Study the flow of a gas in a converging pipe and, using the result of Prob. 6, show that the speed will increase in the direction in which the pipe converges whenever the velocity V is smaller than the "local velocity of sound" $\sqrt{\kappa p/\rho}$.

*Problem 9.** The Venturi meter, the arrangement of which is shown in Fig. 17, is used for measuring the amount of fluid discharged through a pipe. From the pressures p_1 and p_2 observed at the sections 1 and 2 of cross-sectional areas S_1 and S_2, respectively, determine the volume of fluid discharged per unit of time, assuming that (a) the fluid is incompressible and has the density ρ; (b) the fluid is a perfect gas in adiabatic flow ($\kappa = 1.4$), the density at the pressure p_1 being ρ_1.

*Problem 10.** In the two-dimensional case, derive Bernoulli's equation by considering a short segment of a stream filament instead of a cylindrical fluid element.

Point out how the proof given in the foregoing has to be modified on account of the variable width of the stream filament.

3. Dynamic Pressure. The constant H in Bernoulli's equation is called the *total head*. Though a constant for each streamline, it may vary from one streamline to the next. However, in most cases that are of interest in aeronautics, the streamlines originate from a region of constant pressure and velocity. Within this region the total head has the same value for all streamlines, if the effect of gravity can be neglected. Since the total head remains constant along each streamline, it is seen that under this assumption the total head will be constant throughout the field of flow.

As an example take the main problem of applied aerodynamics, the uniform rectilinear motion of a body (airplane) through a bulk of air at rest. As has already been pointed out, the flow of air produced by the

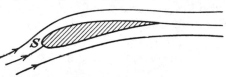

FIG. 18.—Stagnation point.

motion of the body is not steady. Bernoulli's equation therefore cannot be applied to this flow. However, it has been shown in the foregoing that the inverse flow, in which the air exerts the same forces on the body, is a steady flow to which, therefore, Bernoulli's equation can be applied. On the other hand, at a great distance from the body the velocity of the air in the inverse flow is constant and so is the pressure, if the effect of gravity can be neglected. The air flow may be considered as an incompressible fluid motion. If, then, the constant values of velocity and pressure at a great distance from the body are denoted by V_0 and p_0, the total head will have the constant value $V_0^2/2g + p_0/\gamma$ at any point in the inverse flow.

Let us now study a little more closely the inverse flow under the assumption that it is two-dimensional. Some streamlines of this inverse flow will then pass above the body, and some below it (Fig. 18). These two groups of streamlines are separated from one another by a streamline that meets the surface of the body at some point S and splits there into two lines, one following the upper part of the contour of the body, the other the lower part. At the point S the direction of the velocity is indeterminate, which indicates that the magnitude of the velocity is zero. The point S is therefore called a stop point or *stagnation point*. It is seen that in the two-dimensional case at least one such stagnation point must exist. It is easily understood that this result holds good in the three-dimensional case, also.

We now denote the pressure and the velocity of the undisturbed flow (at a great distance from the body) by p and V, respectively, and the pressure at the stagnation point by p_s. Since the velocity at the stagnation point vanishes, we get from the form (12) of Bernoulli's equation, assuming that the air can be considered as nearly incompressible,

$$p_s = p + \frac{\rho}{2} V^2 \quad \text{or} \quad \frac{\rho}{2} V^2 = p_s - p$$

In the preceding section the so-called "velocity head," which measures a velocity in terms of altitude, was introduced. In the present context it is useful to introduce another concept expressing a velocity in terms of pressure. The expression $q = \rho V^2/2$, which has the dimension of a pressure, is called the stagnation pressure or *dynamic pressure*. As is seen from the relation already obtained, the dynamic pressure q equals the difference between the pressures p_s and p at two points of a streamline where the velocity is equal to zero and V, respectively:

$$q = p_s - p = \frac{\rho}{2} V^2 \tag{13}$$

As will be seen in Chap. IV, the dynamic effect of a moving fluid is in most cases expressible in terms of the dynamic pressure. This means that, under otherwise unchanged circumstances, the expression $q = \rho V^2/2$ determines the magnitude of the force which a fluid exerts upon a body submerged in the flow.

Table 2 gives the stagnation pressures q corresponding to various velocities V and to the densities that, in the standard atmosphere, are found at sea level and at the altitudes of 20,000 and 40,000 ft. Since, with increasing altitude, the density of the air decreases, the stagnation

TABLE 2.—DYNAMIC PRESSURE q, ACCORDING TO (13)

V ft./sec.	$h = 0$ ft. $\rho = 0.002378$ slug/ft.^{3} $p = 2116.2$ lb./ft.^{2}		$h = 20,000$ ft. $\rho = 0.001267$ slug/ft.^{3} $p = 972.2$ lb./ft.^{2}		$h = 40,000$ ft. $\rho = 0.000582$ slug/ft.^{3} $p = 391.9$ lb./ft.^{2}	
	lb./ft.^{2}	% of p	lb./ft.^{2}	% of p	lb./ft.^{2}	% of p
100	11.9	0.563	6.33	0.651	2.91	0.743
200	47.6	2.25	25.3	2.60	11.6	2.96
300	107.0	5.06	57.0	5.86	26.2	6.69
400	190.2	9.00	101.4	10.43	46.6	11.89
500	297.2	14.05	158.4	16.30	72.8	18.57
600	428.0	20.22	228.1	23.47	104.8	26.75
700	582.6	27.52	310.6	31.95	142.7	30.45

pressure corresponding to a certain velocity will diminish with increasing altitude.

From Table 2 we learn that at sea level the change in pressure

$$p_s - p = q$$

occurring when a stream with a velocity of 200 ft./sec. (136.3 m.p.h.) strikes a body at rest, is slightly less than 2.5 per cent of the atmospheric pressure. Since the change in density accompanying such a small change in pressure is also small, we are justified in treating the air as an incompressible fluid as long as no higher velocities are involved. Even for speeds of about 400 ft./sec. the theory based on the assumption of incompressibility can be expected to give a reasonably good approximation, since for such speeds the dynamic pressure equals only about 10 per cent of the pressure in the undisturbed air. However, with the speed of modern planes approaching more and more the velocity of the propagation of sound (1000 ft./sec.), the effects of the compressibility of the air become increasingly important.

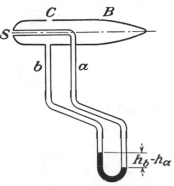

FIG. 19.—Pitot tube.

The so-called "Pitot tube," used for measuring the velocity of an air stream, is based on the relation expressed in Eq. (13). A slender, hollow body B of the form indicated in Fig. 19 is placed with its axis parallel to the stream and its open end upstream. The narrow tube a, which communicates with the outside flow at the stagnation point S of the body, is filled with air at rest. If the influence of the gravity of the air is neglected, the pressure within the tube a equals the pressure p_s at the stagnation point. A second tube b communicates with the interior of the hollow body B and is also filled with air at rest. Owing to a hole C in the side of B this still air is under the same pressure p_c as the air that, at C, streams along the surface of B. The tubes a and b are connected with the two ends of a U-tube which is filled with some manometric liquid of the specific weight γ_L.

Applying Bernoulli's equation to the flow around the body B, we obtain, by comparing the conditions at S and C,

$$p_s + 0 = p_c + \frac{\rho}{2} V_c^2$$

Now, for a properly designed slender body B the point C can be chosen so that the velocity V_c and the pressure p_c at C will not differ perceptibly from the velocity V and the pressure p in the undisturbed stream. We may therefore write

$$p_s = p + \frac{\rho}{2} V^2 \tag{14}$$

Thus the pressure difference $p_s - p$, which is indicated by the difference of the levels h_b and h_a of the differential manometer, equals the dynamic pressure corresponding to the velocity V of the undisturbed stream

$$\frac{\rho}{2} V^2 = p_s - p = \gamma_L(h_b - h_a) \tag{15}$$

This formula is correct under the assumption that the air can be considered as incompressible. If, however, the Pitot tube is used for measuring higher velocities, it may be more accurate to apply Bernoulli's equation in the form (8) with the value of h_p as given in (11) for adiabatic conditions. In this case $V^2/2g$ equals the difference of the h_p-values that occur at the stagnation point and in the undisturbed flow. As we have to introduce in (11) p_s for p and p for p_0, we find

$$\frac{V^2}{2g} = \frac{\kappa}{\kappa - 1} \frac{p}{\gamma} \left[\left(\frac{p_s}{p}\right)^{(\kappa-1)/\kappa} - 1 \right]$$

This gives, when solved for p_s,

$$p_s = p \left(1 + \frac{\rho V^2}{2p} \frac{\kappa - 1}{\kappa}\right)^{\kappa/(\kappa-1)} \sim p + \frac{\rho V^2}{2} + \frac{(\rho V^2)^2}{8\kappa p} \tag{15'}$$

The corrected stagnation pressures, *i.e.*, the values $p_s - p$ according to (15'), are given in Table 3. The deviations from the values of Table 2 become marked at higher velocities.

TABLE 3.—STAGNATION OVERPRESSURE $(p_s - p)$ CORRECTED FOR ADIABATIC FLOW

V ft./sec.	$h = 0$ ft. $\rho = 0.002378$ slug/ft.^{3} $p = 2116.2$ lb./ft.^{2}		$h = 20,000$ ft. $\rho = 0.001267$ slug/ft.^{3} $p = 972.18$ lb./ft.^{2}		$h = 40,000$ ft. $\rho = 0.000582$ slug/ft.^{3} $p = 391.92$ lb./ft.^{2}	
	lb./ft.^{2}	% of p	lb./ft.^{2}	% of p	lb./ft.^{2}	% of p
100	11.92	0.564	6.34	0.651	2.92	0.746
200	48.0	2.27	25.53	2.63	11.7	2.99
300	108.9	5.15	58.2	5.98	26.8	6.84
400	196.3	9.28	105.2	10.82	48.6	12.40
500	312.1	14.75	167.6	17.25	77.6	19.80
600	458.9	21.70	247.2	25.42	114.8	29.30
700	639.8	30.22	346.0	35.60	161.2	41.15

Problem 11. Assuming standard atmosphere, compare the dynamic pressures corresponding to the velocity $V = 250$ m.p.h. at sea level and at 10,000 ft. altitude.

Problem 12.   If the barometric pressure $p_0 = 2120$ lb./ft.^{2} and the pressure at the stagnation point $p = 2300$ lb./ft.^{2}, determine the speed V_0 of the undisturbed air stream under the assumption that the air is (a) an incompressible fluid of the density $\rho_0 = 0.0024$ slug/ft.^{3}; (b) a perfect gas in adiabatic flow ($\kappa = 1.4$), the density under the pressure p_0 being $\rho_0 = 0.0024$ slug/ft.^{3}

Problem 13.   By how much is the density of the air increased at the stagnation point in Prob. 12(b)?

4. Variation of Total Head across the Streamlines. Rotation.

It has been stated in the preceding section that for the most important group of fluid motions the total head H is constant not only along each streamline but throughout the field of flow. In order to discuss these motions more completely, it will be useful to study first the more general type of fluid motions for which the total head varies from one streamline to the next.

Bernoulli's equation was derived in Sec. II.2 by applying Newton's second law of motion to a cylindrical element of fluid the axis of which is formed by a short segment of a streamline. However, Bernoulli's equation is not a complete expression of Newton's law. Indeed, in deriving this equation the law of motion was applied to the force components in the direction of the axis of the cylindrical fluid element only, *i.e.*, to the direction of the velocity of the fluid. The next step will be to write down Newton's law for a direction that is perpendicular to the velocity of the fluid element. To simplify our discussion, we consider here only the case of a two-dimensional flow.

The infinitesimal fluid element has in this case the shape of a rectangular parallelepiped (see Fig. 13). It is assumed to be of unit thickness perpendicular to the plane of the flow and to have the small dimensions ds and dn in the direction of the tangent and normal of the streamline along which its center is moving (Fig. 20). As the positive direction of the normal of this streamline take that obtained by rotating the direction of the velocity counterclockwise through a right angle. According to whether the center of curvature of the streamline is on the side to which the positive normal points or on the opposite side, the radius of curvature R of the streamline will be regarded as positive or negative. With this convention the normal acceleration of the fluid element can be written as V^2/R, the sign of this expression indicating the direction of the acceleration. According to Newton's second law, the product of V^2/R and the mass of the fluid element equals the sum of the forces acting on the element in the direction of the normal of the streamline.

In Fig. 20 the plane of flow is assumed to be vertical. (Almost the same reasoning, however, could be applied to the flow in any inclined plane.) Denoting by α the angle between the positive normal of the streamline and the upward vertical, we obtain $-g\rho\, dn\, ds \cos\alpha$ as the projection of the weight of the fluid element on the normal. If h denotes as before the elevation of the element above some horizontal plane of reference, $\cos\alpha$ can be replaced[1] by dh/dn. The thrusts on the sides

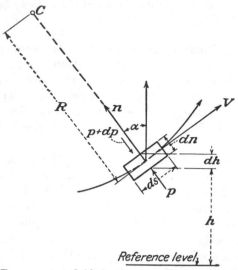

Fig. 20.—Forces upon a fluid element normal to the direction of flow.

of the element are $p\, ds$ and $(p + dp/dn \cdot dn)\, ds$. Since $\rho\, dn\, ds$ is the mass of the element, Newton's second law furnishes

$$\rho\, dn\, ds\, \frac{V^2}{R} = p\, ds - \left(p + \frac{dp}{dn}\, dn\right) ds - g\rho\, \frac{dh}{dn}\, dn\, ds$$

Here the terms $p\, ds - p\, ds$ cancel out; dividing the remaining terms by $g\rho\, dn\, ds$ we obtain

$$\frac{V^2}{gR} = -\frac{dh}{dn} - \frac{1}{g\rho}\frac{dp}{dn} \tag{16}$$

For the general case the total head has been defined in (4) as

$$H = \frac{V^2}{2g} + h + h_p$$

The difference of the values of H at two neighboring points accordingly will be

$$dH = \frac{V\, dV}{g} + dh + dh_p$$

---

[1] Where no confusion is to be feared, the ordinary d will be used in the sign of differentiation throughout this book, even if the function depends on more than one variable.

thus furnishing

$$\frac{d(h + h_p)}{dn} = \frac{dH}{dn} - \frac{V}{g}\frac{dV}{dn} \qquad (17)$$

On the other hand, Eq. (5) gives $dh_p/dp = 1/g\rho$. Consequently,

$$\frac{1}{g\rho}\frac{dp}{dn} = \frac{dh_p}{dp}\frac{dp}{dn} = \frac{dh_p}{dn} \qquad (18)$$

Hence (16) takes the form

$$\frac{V^2}{gR} = -\frac{dh}{dn} - \frac{dh_p}{dn} = -\frac{d(h + h_p)}{dn} = -\frac{dH}{dn} + \frac{V}{g}\frac{dV}{dn}$$

Multiplying by g/V, we find

$$\frac{V}{R} - \frac{dV}{dn} = -\frac{g}{V}\frac{dH}{dn} \qquad (19)$$

We conclude from this important relation that *fluid motions with the same total head value on all streamlines are characterized by the fact that the expression $V/R - dV/dn$ vanishes everywhere.*

In order to arrive at a first physical interpretation of this expression, we consider the motion of the infinitesimal fluid element to which Newton's second law was applied above. In Fig. 21, $\overline{AB}$ denotes the line joining the centers of the front and back surfaces of this element and $\overline{CD}$ the line joining the centers of its sides. As the element moves, the segment $\overline{AB}$ will glide with the velocity V along a streamline of curvature $1/R$ and will therefore rotate with the angular velocity V/R. According to the conven-

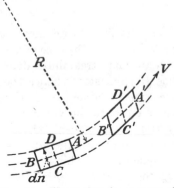

Fig. 21.—Illustrating the concept of rotation.

tion already adopted regarding the sign of R, the radius of curvature R must be considered as positive in the case of Fig. 21. A positive sign of the quotient V/R, consequently, indicates counterclockwise rotation. The segment $\overline{CD}$ will rotate, in general, with a different angular velocity, which can be obtained in the following way: If both C and D had the same velocity V parallel to AB, the angular velocity of CD would be zero. In general, the velocities of C and D differ by dV/dn times $\overline{CD}$. If this expression is positive, the point D, after an interval of time dt, will be ahead of C by dV/dn times $\overline{CD}\,dt$. Thus, the infinitesimal angle between $\overline{CD}$ and $\overline{C'D'}$ will be dV/dn times

dt, and the time rate of change of this angle is dV/dn. The figure shows that for a positive value of dV/dn the rotation of CD is clockwise. We have therefore $-dV/dn$ as the angular velocity of $\overline{CD}$. Consequently, the expression

$$\omega = \tfrac{1}{2}\left(\frac{V}{R} - \frac{dV}{dn}\right) \tag{20}$$

represents the arithmetic mean of the angular velocities of $\overline{AB}$ and $\overline{CD}$; it is called the mean rotation or, simply, the *rotation* of the fluid element under consideration. If this element were moving as a rigid body, $\overline{AB}$ and $\overline{CD}$ would have the same angular velocity and the rotation ω would equal this common value.

It has already been seen that for fluid motions with constant total head the expression $V/R - dV/dn$ vanishes. This result can now be expressed as follows: *When the total head has the same value for all stream-lines of a steady two-dimensional flow, the rotation of all fluid elements vanishes.* Such motions are therefore called *irrotational motions.* They are characterized by the equation

$$\frac{dV}{dn} - \frac{V}{R} = 0 \tag{20'}$$

If the fluid is an incompressible one, Eq. (19) can be easily interpreted, even in the case of a flow with varying total head. Consider two neighboring streamlines defining a stream filament. On each stream-line H is constant; hence the difference dH is constant anywhere along the stream filament; and, along the stream filament, $V\,dn$ has a constant value equal to the volume of fluid flowing per unit of time through any section of the stream tube that is defined by the stream filament. The right side of (19) therefore is constant along any stream filament. The left side of (19), which equals 2ω, thus being constant, we have the following theorem: *In a steady two-dimensional flow of an incompressible fluid the rotation of any fluid element maintains its value as this element moves along.*

This result covers a very particular case of a famous theorem, due to H. v. Helmholtz (1858). In fact, a similar, more general statement involving a quantity connected with the rotation, rather than the rotation itself, can be established if the flow is neither two-dimensional nor steady and even in the case of compressible fluids.

*Problem 14. A bulk of fluid rotates like a rigid body with constant angular velocity ω about a fixed axis. Neglecting gravity and treating the fluid as incompressible, show that the pressure at the distance r from the axis will have the value

$$p = p_0 + \frac{\rho}{2}\,\omega^2 r^2$$

where p_0 denotes the pressure at the axis.   If the axis of rotation is vertical and gravity is not neglected, what will be the shape of the free surface of the liquid?

*Problem 15.   Prove that in a steady two-dimensional flow of a compressible fluid the product of the rotation and the volume of a fluid element maintains its value as the element moves along.

5. Circulation and Rotation.   Preliminary to giving another interpretation of the left side of (19), *i.e.*, of the expression $V/R - dV/dn$, we consider an incompressible fluid flowing steadily in an annular channel bounded by two concentric circles (Fig. 22).   We may assume in this case that the streamlines are circles with the same center.   Since any stream filament then has a constant width, the velocity V has a constant magnitude along each streamline.   Neglecting gravity, we see from Bernoulli's equation (12) that the pressure, too, is constant along each streamline.   Information about the manner in which the velocity V and the pressure p vary across the streamlines can be obtained from Eq. (19).   If the flow is counterclockwise, the center of curvature of any circular streamline lies on the side to which the positive normal points.   If r then denotes the radius vector of a point P, with respect to the common center of the streamlines, r equals the radius of curvature R of the streamline through P.   On the other hand,

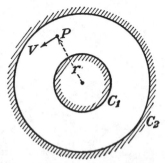

FIG. 22.—Circulating motion.

the positive normal of the streamline points toward the side of decreasing r, so that we have to write $dn = -dr$ and consequently obtain from (19)

$$\frac{V}{r} - \frac{dV}{dn} = \frac{V}{r} + \frac{dV}{dr} = \frac{g}{V}\frac{dH}{dr} \tag{21}$$

If the velocity V is a known function of r, this equation enables us to determine H and hence the pressure p as functions of r.

In the particular case where the total head H is constant throughout the field of flow, we have

$$\frac{V}{r} + \frac{dV}{dr} = 0 \qquad \text{or} \qquad V\,dr + r\,dV = 0 \tag{22}$$

Since the left side of the second equation is the differential of rV, this product is seen to have a constant value: $rV = \text{const.}$ For reasons that will appear later we denote this constant value by $\Gamma/2\pi$.   Since a flow with constant total head is an irrotational flow, we can state our result as follows.   *In a two-dimensional steady irrotational flow in concentric circles the velocity V is inversely proportional to the radius r of the circular streamlines:*

$$V = \frac{\Gamma}{2\pi r} \tag{23}$$

This pattern of motion is called a *circulating motion* and the constant Γ the intensity of the circulation or simply the *circulation*.

It should be emphasized once more that a circulating motion is irrotational although each particle describes a full circle. Irrotationality means that the fluid elements, though moving along circles, do not have any mean rotation but, in a certain way, perform a translation along circular pathways.

For a circulating motion the constant circulation Γ is seen from (23) to equal the product of the perimeter $2\pi r$ of a circular streamline and the

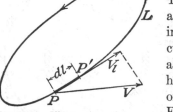

magnitude of V, the velocity of this line. The definition of circulation can be generalized for an arbitrary motion in the following way: Let L be a *circuit, i.e.,* a closed curve, which is situated entirely in the fluid and on which a certain sense of progression has been defined (see Fig. 23, where the sense of progression is marked by an arrow).

Fig. 23.—Illustrating the concept of circulation.

From an arbitrary point P on the circuit we proceed according to the sense of progression to a neighboring point P'. Denoting the infinitesimal length $\overline{PP'}$ by dl, multiply it by the projection V_l of the velocity at P on the line PP'. Forming the analogous product at P' and continuing in this way all round the circuit L, we define as *circulation around L* the sum (or the integral) of all the expressions $V_l\, dl$. The usual notation for this integral is

$$\Gamma = \oint_L V_l\, dl \tag{24}$$

Here the sense of progression along the circuit is indicated by the arrow on the circle which is drawn across the integral sign. In the case of the circulating motion already considered, the circulation along a circular streamline can be found as follows: The velocity V_l is at each point equal to V and constant along any one of the streamlines. The general definition applied to a circular streamline taken as the circuit L accordingly yields

$$\Gamma = V \oint dl = V \cdot 2\pi r$$

Suppose a circuit L is given. We may divide the area bounded by L into two parts by drawing a line AB across it (Fig. 24). The sense of progression on L defines a sense of progression along each of the bound-aries L_1 and L_2 of these two areas.   There exists a simple relation between the circulation Γ for the circuit L and the two circulations Γ_1 and Γ_2 for the circuits L_1 and L_2.   The sense of progression along the line AB as part of L_1 is opposite to the sense of progression along this same line as part of L_2.   Hence, on AB, the velocity component V_l computed for

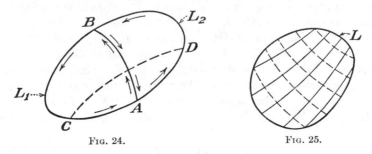

FIG. 24.   FIG. 25.

AB as a part of L_1 will be just the opposite of the value as computed for AB as a part of L_2.   Thus the contributions of the segment AB to the circulations Γ_1 and Γ_2 will be equal in magnitude and opposite in sign. If we add the circulations around L_1 and L_2, these contributions therefore cancel out and, as the sum of the remaining parts of L_1 and L_2 coincides with L, the sum of the integrals Γ_1 and Γ_2 equals the circulation Γ around L.   Applying the same prin-ciple to each of the areas bounded by L_1 and L_2, respectively, we may subdivide these areas by a line CD drawn across both of them.   Con-tinuing in this manner we can finally cover the area of L by a network formed by two families of curves (Fig. 25).   *The circulation around a circuit L can be expressed as the sum of the circulations around the individual meshes of a network that covers the area bounded by L.*

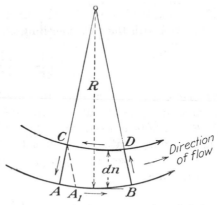

FIG. 26.—Circulation around an infini-tesimal mesh.

This is also true in the limit when the meshes are infinitesimal, each individual circulation becoming a differential and the sum an integral.

Thus, in order to compute the circulation around L we may, for example, use the network formed by the streamlines and the lines that intersect them at right angles.   Let $ABCD$, in Fig. 26, be a mesh of this network, AB and CD being segments of neighboring streamlines.   At any point on AC (or BD) the velocity is perpendicular to dl.   This segment accordingly does not furnish any contribution to the circulation around the mesh $ABCD$ since here the projection V_l of the velocity V on the element dl is zero.   Denoting the infinitesimal length of AB by ds and the velocity along AB by V, we see that this segment furnishes the contribution $V\,ds$ to the circulation around the mesh $ABCD$, since here V_l equals V and ds stands for the line element dl.   Along CD the velocity has the magnitude $V + dV/dn \cdot \overline{AC}$, where $\overline{AC} = dn$ is the infinitesimal distance between AB and CD.   If R is the radius of curvature of the streamline AB, the segment CD has the length $ds(1 - dn/R)$ as can easily be seen from Fig. 26, where CA_1 is parallel to DB so as to yield $\overline{CD} = \overline{A_1B} = \overline{AB} - \overline{AA_1}$ and $\overline{AA_1}:\overline{AB} = dn:R$.   Since along DC the sense of progression is opposite to the velocity, the contribution of DC to the circulation is the negative product of $(V + dV/dn \cdot dn)$ and $(1 - dn/R)\,ds$.   We thus find the circulation around the mesh $ABCD$ to be

$$V\,ds - \left(V + \frac{dV}{dn}\,dn\right)\left(1 - \frac{dn}{R}\right)ds$$

or

$$\left(\frac{V}{R} - \frac{dV}{dn} + \frac{1}{R}\frac{dV}{dn}\,dn\right)ds\,dn$$

Neglecting the last term in the parentheses which is small of higher order compared with the two preceding ones, we obtain

$$\left(\frac{V}{R} - \frac{dV}{dn}\right)ds\,dn$$

as the circulation around the mesh $ABCD$.   In this expression the product $ds\,dn$ represents the area dA of the mesh $ABCD$, and the factor $V/R - dV/dn$ is twice the rotation ω within this mesh.   According to the foregoing statement about the summation of the circulations for individual meshes, the circulation around the circuit L can therefore be written as

$$\Gamma = \int\int \left(\frac{V}{R} - \frac{dV}{dn}\right)ds\,dn = 2\int \omega\,dA \tag{25}$$

where the integration has to be extended over the area enclosed by L.

If, in particular, the motion is irrotational in the entire area bounded by L, it is seen from (25) that the circulation around L must vanish.   At first glance this statement seems to be at variance with a former result obtained in the case of the circulating motion of an incompressible fluid moving in an annular channel bounded by two concentric circles.   There we found that the circulation Γ along each circular streamline was different from zero although the motion was irrotational.   This apparent contradiction is eliminated by the following argument:

In deriving the formula (25) it was tacitly assumed that the entire area enclosed by the circuit L can be covered by a network consisting of streamlines and their orthogonal trajectories and that we have a continuous velocity distribution in this area.   If the first of these assumptions is not fulfilled, our reasoning evidently fails; if the second assumption is not satisfied, the velocity along CD cannot be written as

$$V + \frac{dV}{dn}\, dn$$

Formula (25) therefore will be valid only if

1. The area enclosed by the circuit L is entirely covered by the fluid;
2. The velocity distribution in this fluid mass is continuous.

In this case the area enclosed by the circuit L is covered by a regular network of streamlines and their orthogonal trajectories in such a manner that through any given point within this area there passes only one line of each of these two families of lines.

Now, if the circulating motion is performed in a channel bounded by two concentric circles C_1 and C_2, formula (25) is not applicable to any one of the circular streamlines (because of 1), since there is no fluid inside C_1.   If, on the other hand, the entire interior of a circular streamline were filled by the fluid, the velocity would be infinite at the center and the orthogonal trajectories would intersect each other.   Thus formula (25) is again unapplicable, because of 2.

The complete statement that can be inferred from formula (25) in the case of an irrotational motion is thus as follows: *If the entire area enclosed by a circuit L is covered by a regular irrotational stream pattern, the circulation around L is zero.*   There is no reason to assume that this is the case when L encloses a rigid body or any kind of discontinuity of the velocity distribution, *e.g.*, points through which pass more than one streamline or more than one of the orthogonal trajectories of the streamlines.

As an illustration consider the steady two-dimensional *flow around an airfoil* (Fig. 27).   Here all streamlines originate from a region far in front of the airfoil where pressure and velocity can be assumed constant. The total head H therefore is constant throughout the field of flow; *i.e.*, the motion is irrotational.   For any circuit like L_1, L_2, ..., which does not enclose the airfoil, the circulation must vanish.   But the circulation around the circuit L', which encloses the airfoil, may be different from zero, and there is no need of any special explanation of why this circulation does not vanish in an actual case.   However, our theorem furnishes some information concerning the values of the circulations around two circuits L' and L'', which both enclose the airfoil. Introducing a "bridge" or "barrier" AB we can form one circuit L consisting of L', L'', and the two sides of the barrier as indicated in Fig. 28. The shaded area enclosed by this circuit being entirely covered by an irrotational stream, the circulation around this circuit is zero. Now, if along the part L' we choose the counterclockwise sense of progression, the sense of progression along the part L'' will be clockwise. Furthermore, the contributions that the two sides of the barrier make to the circulation

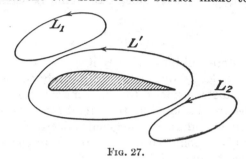

Fig. 27.

around L cancel out. The circulations around L' and L'' therefore have the sum zero, if these circuits are taken with opposite senses of progression as already explained. If the same sense of progression is used for both these circuits, the circulations around L' and L'' are seen to be equal. We thus have the following result: *The circulation has the same value for any two circuits bounding an annular area, if this entire area*

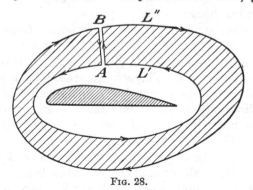

Fig. 28.

is covered by a regular irrotational stream pattern. In the particular case of the circulating motion we found, indeed, the same circulation Γ for any circular streamline.

In a more general way, one can easily show that in an irrotational flow the circulation is the same for any two circuits which can be continuously converted into each other in such a way that no fluid boundary is passed over in the transformation (see Prob. 18).

For the steady irrotational two-dimensional flow around an airfoil, the following results can be formulated: For each circuit that does not enclose the airfoil the circulation is zero if the velocity distribution outside the airfoil is assumed to be continuous. For any two circuits that both enclose the airfoil the circulation has the same value if there is no discontinuity of the velocity distribution between the two circuits.

Problem 16. The streamlines of a steady two-dimensional flow are parallel to the x-axis of a system of rectangular coordinates x, y. The magnitude of the velocity is proportional to y and has the value $V = 15$ ft./sec. at a distance of 3 ft. from the x-axis. Show that the rotation is constant, and determine its value.

Problem 17. In a steady two-dimensional flow the magnitude of the velocity is everywhere the same. Prove that the streamlines must be straight if the flow is to be irrotational and that they must be circles if rotational flow is admitted.

Problem 18. Two circuits L_1 and L_2, which both surround the airfoil section in Fig. 28, intersect each other at four points A, B, C, D. Show that the two circulations have the same value if the flow is irrotational and continuous.

*Problem 19. (*a*) Prove that in a steady two-dimensional flow of an incompressible fluid the circulation around any circuit moving with the fluid remains constant. (*b*) Using the result of Prob. 15 show that in a steady two-dimensional flow of a compressible fluid the circulation around any circuit moving with the fluid remains constant.

6. The Bicirculating Motion. For the circulating motion studied in the preceding section the velocity V at any point P is inversely proportional to the distance r of P from a fixed point O (the common center of the circular streamlines). There exists another two-dimensional irrotational stream pattern with the same property. The streamlines being straight lines through O, this flow is described as *radial* flow with center O. If this flow is to be irrotational, the rotation

$$\omega = \tfrac{1}{2}\left(\frac{V}{R} - \frac{dV}{dn}\right)$$

must vanish. Here R denotes the radius of curvature of the streamlines and dV/dn the derivative of V in the direction normal to the streamlines. Now, for the radial flow the streamlines are straight lines, and consequently $R = \infty$. The condition that the flow shall be irrotational therefore leads to $dV/dn = 0$; *i.e.*, along every line intersecting the streamlines at right angles V is constant. That is, the velocity has a constant magnitude along any circle with center at O, or the velocity V at a point P depends only on the distance r of P from O. In a layer of unit thickness the quantity Q of fluid that crosses per unit of time a cylindrical surface intersecting the plane of flow in a circle of center O and radius r is

$$Q = 2\pi r V$$

For an incompressible fluid the value Q of the flux must be independent of r. We therefore have $V = Q/2\pi r$. This velocity, inversely proportional to r, is everywhere directed either away from O or toward O. In the first case we must assume that, at O, fluid is continually infused into the flow and in the second that fluid is continually withdrawn from the flow at O. In the first case we speak of a *source* at O, in the second of a *sink*.

Since for a circulating motion around O as well as for a radial flow with the center O the velocity V at a point P is inversely proportional to the distance r of P from O, the same will be true for any flow obtained by superimposing a circulating flow around O on a radial flow with the center O. As can be seen from Fig. 29, the resultant velocity V forms

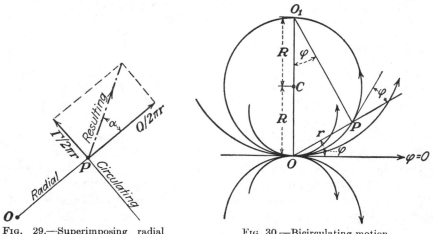

FIG. 29.—Superimposing radial and circulating flows.

FIG. 30.—Bicirculating motion.

with the radius vector OP an angle α, whose tangent equals Γ/Q and thus is independent of r. The streamlines of the flow under consideration, intersecting the radii through O at this constant angle α, are *logarithmic spirals* winding around O. In this case as well as in the case of a simple radial flow with the center at O the flux across a closed line C surrounding O is different from zero, thus indicating the presence of a source or sink within C. It can be proved that, if we exclude sources and sinks, *the circulating motion around O is the only two-dimensional irrotational flow for which the velocity V is inversely proportional to r.*

With a view to later application in airfoil theory we shall now study a flow for which the velocity V is *inversely proportional to the square of r.* Let us consider a flow pattern the streamlines of which are circles with a common tangent at O. We introduce a system of polar coordinates r, φ, the origin $r = 0$ coinciding with O and the ray $\varphi = 0$ falling on the common tangent of the circular streamlines (Fig. 30). Consider, in particular, the streamline of radius R and center C, and take, on this circle, a point P with the polar coordinate r, φ. Then the distance OP, as seen from the figure, has the value

$$r = 2R \sin \varphi \tag{26}$$

This is the equation of the circular streamlines in our system of polar coordinates. Since the angles that a chord of the circle forms with the two tangents at its end points are equal, the angle between the line OP and the tangent of the circle at P equals φ. Accordingly, all circular streamlines cut the ray OP under the same angle.

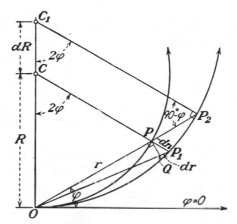

Fig. 31.—A stream filament of the bicirculating motion.

In Fig. 31, two neighboring circular streamlines of the radii R and $R + dR$ and the centers C and C_1, respectively, are shown. On the first circle take a point P with the polar coordinates r, φ, and denote by P_1 the point of intersection of CP and the second circle. P_1 thus lies on the normal of the circular streamline through P. According to our convention, the positive direction of the normal of this streamline is obtained from the direction of the velocity by a counterclockwise rotation through a right angle. For the direction of flow indicated in Fig. 31 the positive normal of the streamline through P is therefore directed toward C. The distance $\overline{PP_1}$ consequently must be denoted by $-dn$. If OP is prolonged up to the point P_2 on the circle of radius $R + dR$, the distance $\overline{PP_2}$ is the increase of r for constant φ and varying R. According to (26) we therefore have $\overline{PP_2} = 2\,dR \sin \varphi$. Neglecting terms of higher order one can state that the infinitesimal triangle PP_1P_2 has a right angle at P_1 and the angle φ at P_2 (Fig. 31); consequently,

$$-dn = \overline{PP_1} = \overline{PP_2} \sin \varphi = 2\,dR \sin^2 \varphi$$

Now, from (26) it follows that $\sin \varphi = r/2R$. Substituting this in the expression just obtained for dn, we find that

$$dn = -2\, dR\, \frac{r^2}{4R^2} = -\frac{dR}{2R^2}\, r^2$$

Thus, along each circular stream filament determined by fixed values of R and dR the width dn is proportional to r^2. Since for an incompressible fluid $V\, dn = $ const. along each stream filament, the velocity varies inversely proportional to r^2 along each stream filament. We may write

$$V = \frac{B}{r^2} \tag{27}$$

where B, so far, may vary from one circular streamline to the next.

By assuming the form of the streamlines, the direction of the velocity at any point and the variation of the magnitude of the velocity along any streamline have been fixed. We are still free to assume the velocity distribution across the streamlines and may choose it in such a way that B in (27) has the same value for all streamlines. In this case the velocity V will be inversely proportional to r^2 throughout the plane of flow.

Now, it can easily be shown that for constant B the flow is irrotational. According to (20), the condition for irrotational flow is

$$\frac{V}{R} - \frac{dV}{dn} = 0,$$

where R is the radius of curvature of the streamlines under consideration, i.e., in our case, the radius R of the circular streamline. When B l as the same value for all streamlines, the velocity $V = B/r^2$ depends on r only, and consequently

$$\frac{dV}{dn} = \frac{dV}{dr}\frac{dr}{dn} = B\frac{d}{dr}\left(\frac{1}{r^2}\right)\frac{dr}{dn} = -\frac{2B}{r^3}\frac{dr}{dn}$$

Here dr indicates the difference in length of the segments $\overline{OP_1}$ and $\overline{OP}$ in Fig. 31. Drawing PQ perpendicular to OP, we see that $dr = \overline{QP_1}$, if terms of higher order are neglected. PQ being parallel to the bisectrix of the angle OCP, the angle QPP_1 equals φ. We thus have

$$dr = \overline{QP_1} = \overline{PP_1}\sin\varphi$$

or $dr = -dn \sin \varphi$. Introducing $dr/dn = -\sin\varphi$ into the relation for dV/dn we obtain

$$\frac{dV}{dn} = \frac{2B}{r^3}\sin\varphi$$

On the other hand, from (26) and (27) we find

$$\frac{V}{R} = \frac{B}{r^2 R} = \frac{2B}{r^3} \sin \varphi$$

The difference $dV/dn - V/R$ therefore vanishes, and the flow is clearly irrotational.

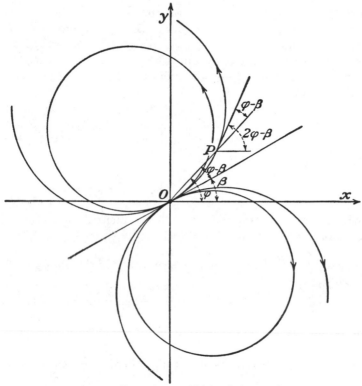

FIG. 32.—General case of bicirculating motion.

Summing up, one can state: *There exists a two-dimensional irrotational flow of an incompressible fluid such that the velocity V at any point P is inversely proportional to the square of the distance r of P from a fixed point O:* $V = B/r^2$. *The streamlines of this flow are circles with a common tangent at O. The direction of the velocity at P forms with OP the same angle as OP does with the common tangent of the streamlines at O.*

To simplify the computation it has been assumed so far that the common tangent of the streamlines at O coincides with the ray $\varphi = 0$. The general case is obtained by rotating the entire stream pattern through an arbitrary angle β (Fig. 32). The stream pattern thus obtained is completely determined by the parameters B and β. At the point P

with the polar coordinates r, φ the velocity V equals B/r^2 and forms with the radius vector OP the angle $\varphi - \beta$. Introducing rectangular coordinates with the x- and y-axis directed along the rays $\varphi = 0$ and $\varphi = 90°$, respectively, the x- and y-components of the velocity are

$$V_x = \frac{B}{r^2} \cos (2\varphi - \beta), \qquad V_y = \frac{B}{r^2} \sin (2\varphi - \beta) \tag{28}$$

For $\beta = 0$ we have the case studied above; for $\beta = 90°$ the y-axis is the common tangent of the streamlines at O. A flow represented by (28) with arbitrary values of B and β will be called a *bicirculating motion*. Each particular case being determined by the values of B and β, we call the vector of magnitude B and direction β the *bicirculation vector* of the bicirculating motion.

Let us break down the bicirculation vector into the components $B_x = B \cos \beta$ and $B_y = B \sin \beta$. According to (28), the bicirculating motion with the bicirculation vector $B \cos \beta$ directed along the x-axis has the velocity components

$$V'_x = \frac{B \cos \beta}{r^2} \cos 2\varphi, \qquad V'_y = \frac{B \cos \beta}{r^2} \sin 2\varphi \tag{29}$$

Similarly, the bicirculating motion with the bicirculation vector $B \sin \beta$ directed along the y-axis has the velocity components

$$\begin{aligned}
V''_x &= \frac{B \sin \beta}{r^2} \cos (2\varphi - 90°) = \frac{B \sin \beta}{r^2} \sin 2\varphi \\
V''_y &= \frac{B \sin \beta}{r^2} \sin (2\varphi - 90°) = -\frac{B \sin \beta}{r^2} \cos 2\varphi
\end{aligned} \tag{30}$$

Superimposing the two motions we obtain the velocity components

$$\begin{aligned}
V_x &= V'_x + V''_x = \frac{B}{r^2} (\cos 2\varphi \cos \beta + \sin 2\varphi \sin \beta) = \frac{B}{r^2} \cos (2\varphi - \beta) \\
V_y &= V'_y + V''_y = \frac{B}{r^2} (\sin 2\varphi \cos \beta - \cos 2\varphi \sin \beta) = \frac{B}{r^2} \sin (2\varphi - \beta)
\end{aligned} \tag{31}$$

as in (28). Thus it is proved that the motion obtained by superimposing two bicirculating motions with the same center O is a bicirculating motion whose bicirculation vector is the sum of the respective vectors of the original motions.

Consider again the case of a bicirculating motion whose bicirculation vector is directed along the x-axis ($\beta = 0$). Since the stream pattern is symmetrical with respect to the x-axis, the circulation Γ for a circle with the center in O is zero. Furthermore, since the stream pattern is antisymmetrical with respect to the y-axis, the flux through any such circle vanishes. As the fluid is incompressible and the flow is irrotational with a velocity distribution continuous everywhere except at the point O, it follows that the circulation along any closed line surrounding O and the flux across any such line are both zero.

So far, the bicirculating motions defined by (28) have been studied as examples of an irrotational flow for which the magnitude of the velocity is inversely proportional to r^2. It can be shown, by reversing the argument, that there exists no other two-dimensional irrotational stream pattern of this type. One can state that *the bicirculating motions discussed in this section are the only two-dimensional irrotational motions of an incompressible fluid for which the velocity at any point P is inversely proportional to the square of the distance of P from a fixed point O* (see Prob. 20).

All these statements about the circulating and bicirculating motions play an important part in the theory of airfoils to be developed in a later chapter.

*Problem 20. In order to show that the bicirculating motion is the only irrotational flow with $V = B/r^2$ one has to prove that under the latter condition (a) the velocity direction is constant along each radius vector; (b) the change $d\vartheta$ when one passes from one radius vector to the next is twice the change $d\varphi$ of the polar angle. Derive these two facts, using the continuity equation in the form given in Prob. 2 and the condition of irrotationality in the form (20′) with $1/R = d\vartheta/ds$.

*Problem 21. Prove the analogous statement for motions characterized by the condition $V = \text{const.}/r$.

CHAPTER III

MOMENTUM AND ENERGY EQUATIONS

1. Flux of Momentum in Steady Flow. The two basic equations of Chap. II, Bernoulli's equation and Eq. (19), have been obtained by applying Newton's second law of motion to an infinitesimal element of a fluid in a state of steady flow. In the present chapter Newton's law will be applied to a finite portion of such a fluid. The fact that the flow under consideration is steady will again play a decisive role.

In a steady field of flow consider a closed surface S. Assume that the region enclosed by this surface is entirely occupied by fluid and that the velocity distribution within S is continuous. The surface S will be called *control surface*. In order to apply Newton's second law to the fluid within the surface S, an expression is needed for the mass times acceleration product for this portion of the fluid. To find it, divide the fluid within S into (infinitesimal) elements, multiply the acceleration vector of each element by its mass, and finally form the sum (integral) of the vectors thus obtained.

If V_x denotes the x-component of the velocity of a fluid element, the x-component of its acceleration a_x is dV_x/dt. As the motion is steady, we may write according to Euler's rule of differentiation, Eq. (2), Chap. II,

$$a_x = \frac{dV_x}{dt} = V\frac{dV_x}{ds}$$

where V is the magnitude of the velocity and ds an element of the streamline described by the fluid element under consideration.

We now consider a stream tube of variable infinitesimal cross section dA, piercing the control surface S at the points marked by 1 and 2 in Fig. 33. The mass of the fluid contained in an element of length ds of this stream tube is $\rho\,ds\,dA$. The product of this mass and the x-component of its acceleration has therefore the value

$$\rho\,ds\,dA \cdot V\frac{dV_x}{ds} = (\rho V\,dA)\frac{dV_x}{ds}\,ds$$

The sum (integral) of such expressions extended over the entire region enclosed by the surface S has to be computed. To do this consider first the contributions furnished by the elements of the stream tube 1-2. As has been seen in Eq. (1), Chap. II, the expression $dQ = \rho V\,dA$ has a constant value along each stream tube, representing the mass of the fluid that flows per unit of time through any cross section of the stream tube. The sum of the products mass times x-component of acceleration, extended over all fluid elements within the stream tube under consideration, has therefore the value

$$\int_{(1)}^{(2)} dQ \frac{dV_x}{ds} ds = dQ \int_{(1)}^{(2)} \frac{dV_x}{ds} ds = dQ[V_x^{(2)} - V_x^{(1)}] \tag{1}$$

where $V_x^{(1)}$ and $V_x^{(2)}$ denote the x-components of the velocities at 1 and 2, respectively.

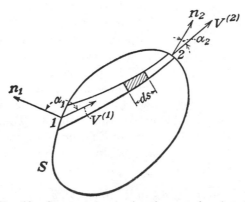

FIG. 33.—Stream tube piercing the control surface S.

Denote by $V^{(1)}$ the velocity with which the fluid enters at 1 the region enclosed by S, by $V^{(2)}$ the velocity with which the fluid leaves this region at 2, and by α_1 and α_2 the angles between the outward normals of S at the points 1 and 2 and the velocities $V^{(1)}$ and $V^{(2)}$, respectively. If dS_1 and dS_2 are the areas of the elements that the stream tube cuts out of S, the flux dQ can be expressed as

$$dQ = \rho V \, dA = \rho_2 V^{(2)} \, dS_2 \cos \alpha_2 = -\rho_1 V^{(1)} \, dS_1 \cos \alpha_1$$

The negative sign must be given to the last term because the angle α_1 between the outward normal of S at 1 and the velocity of inflow $V^{(1)}$ is obtuse. Thus $\cos \alpha_1$ is negative while the flux dQ is positive by definition. Substituting these values of dQ in (1), we obtain

$$\rho_2 V^{(2)} \, dS_2 \cos \alpha_2 V_x^{(2)} + \rho_1 V^{(1)} \, dS_1 \cos \alpha_1 V_x^{(1)}$$

as the contribution of the flow elements within one stream tube to the sum of the products mass times x-component of acceleration. This sum has to be extended over all fluid elements within S; i.e., the contributions of all stream tubes piercing S have to be added. If this is carried out, each surface element of S will appear either as a dS_1, through which the fluid enters the volume enclosed by S, or as a dS_2, through which the fluid leaves this volume. In the foregoing expression, however, both kinds of elements dS_1 and dS_2 appear in the same form, each multiplied by the values of ρ, V, V_x, and cos α taken at the position of the respective element. Accordingly, there is no need to distinguish between surface elements through which fluid enters and surface elements through which fluid leaves the volume within the control surface. The required sum of the products mass times x-component of acceleration, extended over the total mass within S, can be written as

$$\int\int \rho a_x \, dA \, ds = \int_{(S)} \rho V V_x \cos \alpha \, dS$$

Here α is the angle between the direction of the velocity V at a point of S and the outward normal of S at the same point.

The product $V \cos \alpha$ represents the projection V_n of the velocity V on the outward normal of S and is called the *normal velocity*. It is positive at points where the fluid leaves the volume enclosed by S and negative at points where the fluids enters this volume. Thus the following statement has been proved: *The sum of the products mass times x-component of acceleration, extended over all fluid elements within S, equals the surface integral, extended over S, of the product $\rho V_n V_x$:*

$$\int\int \rho a_x \, dA \, ds = \int_{(S)} \rho V_n V_x \, dS \tag{2}$$

It must be kept in mind that this result was derived under the assumptions that (1) the flow is steady and (2) the entire region within S is occupied by fluid possessing a continuous velocity distribution. On the other hand, it should be noted that no particular assumption was made about the nature of the fluid (whether or not it is a perfect fluid, etc.).

If one wishes to obtain the sum of the products mass times y-component (or z-component) of acceleration, one has to replace V_x in (2) by V_y (or V_z). Actually, as the direction of the x-axis can be chosen arbitrarily, the result obtained in the foregoing is valid for any direction.

The right side of (2) is called the *flux of x-momentum* across the surface S. In fact, $\rho V_n \, dS$ is the mass that per unit of time leaves or, in the case of negative V_n, enters the volume inside S through the surface element dS. The product of this mass and the velocity component V_x is the x-momentum of this mass. The integral (2) accordingly represents the excess of leaving momentum over entering momentum per unit of time.

It is often convenient to let the control surface S consist of a tubular part, formed by the streamlines through the points of a closed curve, and two cross sections of this tube (Fig. 34). Denoting by $V_x^{(1)}$ and $V_x^{(2)}$ the x-components of the velocities at the entrance section 1 and the exit section 2, we obtain from Eq. (1), since $V_n = 0$ along the tubular part of S, the following alternative expression of the flux of x-momentum across S:

$$\int V_x^{(2)} \, dQ - \int V_x^{(1)} \, dQ \tag{2'}$$

The first integral is extended over the exit section and the second over the entrance section.

The result expressed by (2) is valid also in the case where the control surface, *i.e.*, the boundary of the region under consideration, consists of more than one closed surface. In the case of Fig. 35, the flow of x-momentum across the boundaries of the region between S' and S'' is

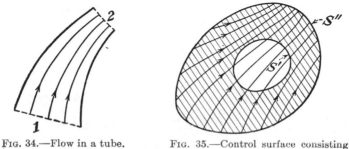

FIG. 34.—Flow in a tube.   FIG. 35.—Control surface consisting
of two portions S' and S''.

given by the integral (2) extended over both surfaces S' and S''. In evaluating this integral one must keep in mind that positive values of V_n indicate that the fluid is leaving the region under consideration. Under all circumstances the region bounded by the control surface must be completely filled by fluid with a continuous velocity distribution.

2. Momentum Equation for Steady Flow. Let us now apply Newton's second law of motion to the fluid occupying the region enclosed by a control surface S. The thrusts that the fluid outside S exerts across the elements of S on the fluid within S is statically equivalent to a resultant force plus a resultant couple. Let P_x be the x-component of this resultant force. The x-component of the weight of the fluid within S will be called W_x. The sum of all forces must equal the sum of all products mass times acceleration. Thus, taking into account Eq. (2), we obtain

$$P_x + W_x = \int_{(S)} \rho V_n V_x \, dS \tag{3}$$

As is seen from this equation, *the sum of the x-components of the external forces (thrusts, weight) acting on the fluid enclosed in a control surface S equals the flux of x-momentum across this surface.* The same relation exists, of course, between the y- (or z-) components of the forces and the flux of y- (or z-) momentum. The theorem expressed by (3) is called the *momentum theorem of fluid dynamics.*

It should be emphasized again that in deriving (3) the flow was assumed to be steady and the entire region within S to be occupied by a fluid with a continuous velocity distribution. On the other hand, the momentum theorem is valid for nonperfect fluids as well as for perfect fluids. In the first case P_x must include the shearing stresses acting on S as well as the normal stresses.

Let us discuss two typical applications of the momentum theorem. Consider a fluid flowing steadily through a tube (Fig. 36). In order to find an expression for the force that the moving fluid exerts on the

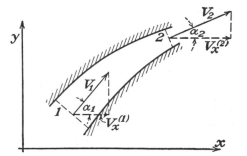

Fig. 36.—The control surface consists of the wetted tube surface and the cross sections 1 and 2.

walls of the tube between the points 1 and 2, let the control surface S consist of the interior surface of the respective part of the tube and the two cross sections 1 and 2. Denote by X the x-component of the force exerted by the fluid on the walls of the tube. The x-components of the force that the tube exerts on the fluid is then $-X$. Further denote by $P_x^{(1)}$ and $P_x^{(2)}$ the x-components of the thrusts that the fluid outside the portion 1-2 of the tube exerts on the fluid mass under consideration across the sections 1 and 2, respectively. Note that the exterior surface of the tube wall is subjected to the atmospheric pressure p_0. If X is to represent the x-component of the resultant of the pressure exerted on both the exterior and the interior surfaces of the tube walls, one must define $P_x^{(1)}$ and $P_x^{(2)}$ as the thrusts due to the overpressure $p - p_0$. If $P_x^{(1)}$ and $P_x^{(2)}$ were taken as the total pressure forces, X would mean the force exerted by the fluid inside the walls only. In both cases the thrust P_x appearing in Eq. (2′) consists of $P_x^{(1)}$, $P_x^{(2)}$, and $-X$, the latter being the reaction of the wall on the fluid.

Upon using the expression (2′) for the flux of x-momentum, the momentum equation takes the form

$$-X + P_x^{(1)} + P_x^{(2)} + W_x = \int\limits_{(2)} V_x^{(2)} \, dQ - \int\limits_{(1)} V_x^{(1)} \, dQ$$

Solving with respect to X we find

$$X = P_x^{(1)} + P_x^{(2)} + W_x + \int\limits_{(1)} V_x^{(1)} \, dQ - \int\limits_{(2)} V_x^{(2)} \, dQ \tag{4}$$

If the distribution of pressure and velocity over the cross sections 1 and 2 is known, the force that the fluid exerts on the tube can be determined from this formula and the two corresponding formulas for the y- and z-directions. Consider, for example, the case where the cross-sectional area is so small that a uniform distribution of pressure and velocity over the cross section can be assumed. Let p_1 and p_2 denote the values of the pressure and V_1 and V_2 the magnitudes of the velocity at 1 and 2, respectively. Denote further by α_1 and α_2 the angles between the x-axis and the directions of the velocity at 1 and 2, respectively, and by Q the mass of the fluid that per unit of time flows through a cross section of the tube. If S_1 and S_2

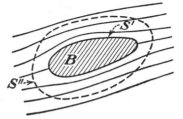

Fɪɢ. 37.—The control surface consists of S'' and the surface S' of B.

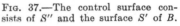

are the cross-sectional areas at 1 and 2, formula (4) can be written as

$$X = (p_1 S_1 + Q V_1) \cos \alpha_1 - (p_2 S_2 + Q V_2) \cos \alpha_2 + W_x$$

Denoting the density at 1 and 2 by ρ_1 and ρ_2, respectively, one has $Q = \rho_1 V_1 S_1 = \rho_2 V_2 S_2$, and therefore

$$X = (p_1 + \rho_1 V_1^2) S_1 \cos \alpha_1 - (p_2 + \rho_2 V_2^2) S_2 \cos \alpha_2 + W_x \tag{5}$$

If p_1 and p_2 denote the overpressures at the sections 1 and 2, this expression for X takes care of the atmospheric pressure acting on the exterior surface of the tube as well. In some cases the overpressures at the sections 1 and 2 are negligibly small. Setting $p_1 = p_2 = 0$, we then obtain, if the weight influence is also neglected,

$$X = Q[V_x^{(1)} - V_x^{(2)}] \tag{6}$$

Another important application of the momentum theorem leads to an expression for the force exerted by a moving fluid on a body B that is immersed in the fluid and kept at rest by suitable forces (Fig. 37). Let us apply the momentum theorem to the fluid between the surfaces S' of B and an outer surface S'' that surrounds B. The control surface S, that is, the boundary of the portion of the fluid under consideration, thus consists of the surfaces S' and S''. As there is no flow across the rigid surface S' of B, the flux of momentum across S is given by the integral (2) extended over the surface S''. Denote by X the x-component of the force that the fluid exerts on B, by P_x the x-component of the force that the fluid outside S'' exerts on the fluid within S'', and by W_x the x-component of the weight of the fluid included between S' and S''. The momentum theorem then furnishes

$$X = P_x + W_x - \int_{(S'')} \rho V_n V_x \, dS \qquad (7)$$

If for S'' another surface surrounding the body B is substituted, all right-hand terms will change in such a manner as to give the same sum X.

Problem 1. An incompressible perfect fluid of density ρ flows through a semi-circular pipe of constant small cross section (Fig. 38). If the pressure is p and the

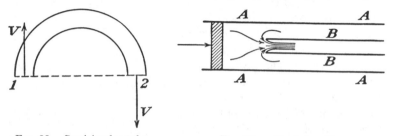

FIG. 38.—Semicircular tube.   FIG. 38a.—Borda's orifice.

velocity V, find the force that the fluid exerts on the walls of the tube. (Neglect the influence of gravity.)

Problem 2. How is the answer to Prob. 1 modified if the cross-sectional area at the exit is twice that at the entrance? (Denote by p, V, S the values corresponding to the entrance section.)

Problem 3. What will be the answer to Prob. 2 in the case of a compressible fluid flowing under adiabatic conditions?

*Problem 4.** The incompressible fluid within the large tube AA in Fig. 38a, is forced by a piston to the left to flow out into the open air through the smaller tube BB, introduced in the interior of AA. The cross-sectional areas are A and B, respectively. The free jet inside BB has the cross-section φB. Use the momentum equation (in connection with the Bernoulli equation and continuity equation) to find the ratio φ as a function of the ratio A/B. Show that φ becomes $\frac{1}{2}$ when A/B tends to infinity (Borda's mouthpiece).

3. Moment of Momentum. Consider, as in the preceding section, a body that is kept at rest in a fluid in steady flow. The thrusts that the fluid exerts on the surface elements of this body form a system of forces in space. It is shown in statics of rigid bodies that such a system is equivalent to a resultant force and a resultant couple. In the preceding section it was seen that the momentum theorem furnishes certain information concerning this resultant force. In the present section an analogous theorem concerning the resulting couple will be established.

Let S be a control surface enclosing a region that is entirely filled by a fluid possessing a continuous velocity distribution. Consider again a stream tube of variable infinitesimal cross section dA, piercing the control surface S at the points 1 and 2 (Fig. 33). The fluid in an element of length ds of this stream tube has the mass $\rho\, ds\, dA$. The products of this mass and the components of its acceleration with respect to the axes of a system of rectangular coordinates x, y, z therefore have the values

$$\rho\, ds\, dA\, \frac{dV_x}{dt} = (\rho V\, dA)\, \frac{dV_x}{ds}\, ds; \quad (\rho V\, dA)\, \frac{dV_y}{ds}\, ds; \quad (\rho V\, dA)\, \frac{dV_z}{ds}\, ds \quad (8)$$

These three expressions are the components dJ_x, dJ_y, dJ_z of the vector $\overrightarrow{dJ}$, which is obtained by multiplying the acceleration vector of the fluid element with the mass of this element. Consider the vector $\overrightarrow{dJ}$ as attached to the instantaneous position x, y, z of the element, and determine the moment of this vector with respect to the z-axis. This moment, which depends only on the x- and y-components of the vector $\overrightarrow{dJ}$, equals $x\, dJ_y - y\, dJ_x$, as is easily seen from Fig. 39. Introducing from (8) the

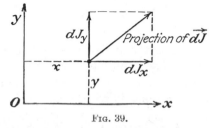

Fig. 39.

expressions for the components of $\overrightarrow{dJ}$, we obtain

$$x\, dJ_y - y\, dJ_x = \rho V\, dA \left(x\, \frac{dV_y}{ds} - y\, \frac{dV_x}{ds} \right) ds$$

But

$$\frac{d}{ds}\, (xV_y - yV_x) = \left(x\, \frac{dV_y}{ds} - y\, \frac{dV_x}{ds} \right) + \left(V_y\, \frac{dx}{ds} - V_x\, \frac{dy}{ds} \right)$$

Here the last parenthesis vanishes as is seen by multiplying each of its two terms by V and replacing $V\, dx/ds = dx/dt$ by V_x and $V\, dy/ds$ by V_y. Consequently,

$$x\, dJ_y - y\, dJ_x = \rho V\, dA\, \frac{d}{ds}\, (xV_y - yV_x)\, ds \quad (9)$$

In order to obtain a formula for the sum (integral) of these expressions extended over all fluid elements within S, the same procedure can be followed as in the preceding section. First sum up the contributions furnished by the elements of the stream tube 1-2. Denoting the constant value of $\rho V\, dA$ by dQ and distinguishing the values of $xV_y - yV_x$ at 1 and 2 by the superscripts (1) and (2), respectively, we obtain

$$\int\limits_{(1)}^{(2)} dQ \frac{d}{ds} (xV_y - yV_x)\, ds = dQ[(xV_y - yV_x)^{(2)} - (xV_y - yV_x)^{(1)}]$$

As has been pointed out in the preceding section,

$$dQ = \rho V\, dA = \rho_2 V^{(2)}\, dS_2 \cos \alpha_2 = -\rho_1 V^{(1)}\, dS_1 \cos \alpha_1$$

where $V^{(1)}$ is the velocity with which the fluid enters at 1 the region within S and $V^{(2)}$ the velocity with which the fluid leaves this region at 2; dS_1, dS_2 denote again the areas of the surface elements that the stream tube 1-2 cuts out of S, and α_1, α_2 the angles between the velocities $V^{(1)}$, $V^{(2)}$ and the outward normals of S at the points 1 and 2, respectively. Upon using these expressions for dQ, our previous equation takes the form

$$\int\limits_{(1)}^{(2)} dQ \frac{d}{ds} (xV_y - yV_x)\, ds = \rho_2 V^{(2)}\, dS_2 \cos \alpha_2(xV_y - yV_x)^{(2)}$$
$$+ \rho_1 V^{(1)}\, dS_1 \cos \alpha_1(xV_y - yV_x)^{(1)}$$

Summing the analogous contributions from all the stream tubes which pierce S and observing again that there is no need to distinguish between surface elements dS_1 through which fluid enters the region within S and surface elements dS_2 through which fluid leaves this region, one obtains

$$\int\limits_{(S)} \rho V \cos \alpha(xV_y - yV_x)\, dS \qquad \text{or} \qquad \int\limits_{(S)} \rho V_n(xV_y - yV_x)\, dS \qquad (10)$$

as the sum of the expressions $x\, dJ_y - y\, dJ_x$, extended over all fluid elements within S. Here $V_n = V \cos \alpha$ is again the normal velocity with which fluid leaves (or enters) the region bounded by S.

The expression (10) is the *flux* of *z-moment of momentum across* S. In fact, $\rho V_n\, dS$ is the mass of fluid that, per unit of time, leaves the region enclosed by S through the surface element dS. The product of this mass and V_x (or V_y) is the x-component (or y-component) of its momentum. Multiplying the y-component of the momentum with the abscissa x of the surface element dS and subtracting from this product the product of the x-component of the momentum and y, we obtain the moment of momentum with respect to the z-axis. This quantity of moment of momentum leaves per unit of time the region within S through the surface element dS. The expression (10) therefore gives the excess of leaving over entering z-moment of momentum for the region enclosed by S.

The momentum theorem of the preceding section consists in equating the vector sum of the $\overrightarrow{dJ}$ to the sum of forces acting on the fluid mass within the control surface. Analogously, the sum of the moments of the $\overrightarrow{dJ}$ has to be equated to the sum of the moments of these forces. The forces present are the weight of the fluid enclosed by S and the thrusts that the fluid outside S exerts on the fluid within S. If x_0, y_0, z_0 denote the coordinates of the center of gravity of the fluid within S and W_x, W_y, W_z the components of the total weight of this portion of fluid, the sum of the z-moments of the forces of gravity equals $x_0 W_y - y_0 W_x$. Denoting by M_x, M_y, M_z the moments with respect to the axes of coordinates of the thrusts acting on the surface S, we therefore obtain

$$M_z + x_0 W_y - y_0 W_x = \int\limits_{(S)} \rho V_n (x V_y - y V_x)\, dS \qquad (11)$$

Thus, *the sum of the z-moments of the forces (thrust, weight) acting on the fluid within a closed control surface S equals the flux of z-moment of momentum across this surface.* Corresponding relations hold for x- (or y-) moments and the flux of x- (or y-) momentum. The theorem expressed by (11) is called the *moment of momentum theorem of fluid dynamics.*

In the case where the system of forces acting on the fluid within the control surface is equivalent to a single force, its line of action is determined in a well known way by the resultant force as found in (3) and the moment given by (11). This will be used later for the flow around a wing that moves parallel to its symmetry plane (Chap. VIII).

A simple example presents itself in the case shown in Fig. 36. From Eq. (5) and the analogous relations for the other axes one concludes that, except for the weight influence, the force exerted on the wall of the narrow tube equals in magnitude and direction the resultant of the two forces of magnitude $S_1(p_1 + \rho V_1^2)$ and $-S_2(p_2 + \rho V_2^2)$ respectively, acting in the direction of V_1 and V_2, respectively. If the moment of momentum equation (11) is applied, this result is supplemented in the following way: Let M_z' be the z-moment of the force on the wall and M_z'' the moment of the forces acting on the fluid on the remainder of the control surface. Then $M_z = -M_z' + M_z''$ and (11) can be written as

$$M_z' = M_z'' + S_1 \rho V_1 (x V_y - y V_x)^{(1)} - S_2 \rho V_2 (x V_y - y V_x)^{(2)}$$
$$+ x_0 W_y - y_0 W_x \qquad (12)$$

This, in connection with (5), shows that the force on the wall is statically equivalent to the resultant of the forces $S_1(p_1 + \rho V_1^2)$ and $-S_2(p_2 + \rho V_2^2)$ if each is considered acting along the tangent of the tube at the points 1 or 2, respectively. That is, in a two-dimensional case the line of action of the force on the wall passes through the intersection of the two tangents. This is true, of course, only if the weight influence is disregarded. If p_1 and p_2, and thus M_z'', are also negligible, so that Eq. (6) can be used, the reaction of the moving fluid on the wall is statically equivalent to the sum of the forces QV_1 and $-QV_2$ acting along the two tangents of the stream filament at the beginning and at the end of the tube segment under consideration. This often leads to the popular, but incorrect, idea that the moving fluid exerts an "action" (forward push) at the entrance and a "reaction" (backward push) at its exit. In reality, the sum of the forces QV_1 and $-QV_2$ is only a mathematical expression for the resultant of all pressure forces acting along the convex surface of the tube.

As another example consider the body B submerged in an axially symmetrical tube with horizontal z-axis (Fig. 40). It is required to find

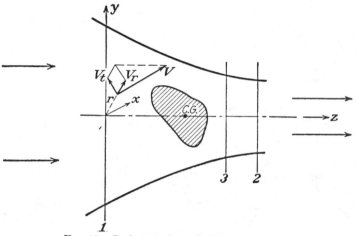

Fig. 40.—Body in a channel of axial symmetry.

the moment M_z with respect to this axis of the force acting on B. Assume the control surface S to consist of the inner wall of the tube between 1 and 2, of two planes perpendicular to the axis at 1 and 2, and of the surface of B. The moment of the weight will be zero if the body B has its center of gravity on the axis. The lines of action of all pressure forces on the wall intersect the axis, those of the pressure forces on the two plane surfaces are parallel to the axis, and hence the moment of both is zero. Consequently,

$$M_z = -\int \rho V_n (xV_y - yV_x)\, dS \qquad (13)$$

The contributions to this integral due to the walls of the tube and to the surface of the body are zero since V_n vanishes there. In the planes 1, 2 we can use polar coordinates and resolve the velocity into an axial, a radial, and a tangential component: V_a, V_r, V_t. Then $V_n = -V_a$ in 1 and $V_n = V_a$ in 2, while $xV_y - yV_x = rV_t$. Thus,

$$M_z = \int_{(1)} \rho r V_a V_t \, dS - \int_{(2)} \rho r V_a V_t \, dS \qquad (14)$$

This is the form of the moment of momentum equation as used in the theory of turbines and of propellers. It holds true for perfect as well as for imperfect fluids.[1] It also follows from (14) that for two normal planes, both on the same side of the body B, such as 2 and 3 in the figure, the integral $\int \rho r V_a V_t \, dS$ has the same value.

Problem 5. An incompressible perfect fluid of density ρ flows through a pipe of constant cross section S. The longitudinal section through the pipe forms a quadrant of a circle (rectangular pipe bend). Neglecting the influence of gravity, find the magnitude and the line of action of the force that the fluid exerts on the pipe, if the pressure is p and the velocity V.

Problem 6. How is the answer to Prob. 5 modified if the fluid is compressible and the pressure at the exit section equals 80 per cent of that at the entrance section? Assume adiabatic conditions. Determine V_1 and V_2 for given T_1.

4. Quasi-steady Flow. Relative Flow. The theorems of momentum and moment of momentum derived in the preceding sections presuppose a steady flow. The assumption of steady flow has been used twice. First, the x-component of the acceleration dV_x/dt has been written as $V \, dV_x/ds$. This is possible only when Euler's rule of differentiation holds in the form given in Eq. (2), Chap. II, *i.e.*, when the flow is steady. Second, summing the contributions from the elements of a stream tube we have assumed the flux $dQ = \rho V \, dA$ to be constant along the stream tube. This also is true only in the case of a steady flow, where the paths described by the fluid particles coincide with the streamlines.

In a steady flow the state of motion remains unchanged at any fixed point within the fluid. This means that pressure, density, magnitude and direction of velocity, etc., are functions of the position in space independent of time. In the case of an unsteady flow these quantities depend explicitly on the time t as well as on the position in space, which may be given by the rectangular coordinates x, y, z of the point under consideration. Let f be a quantity connected with a particle of the moving fluid. In the general case, f is a function of x, y, z, and t, to be written as $f(x,y,z,t)$. For the derivatives we had now better use the notations $\partial f/\partial x$, etc. The partial derivative $\partial f/\partial t$ then denotes the time rate at which the quantity f changes at the fixed point x, y, z; it is called the *local time rate of change of f*. In most mechanical problems, however, another time derivative of f is of interest, *viz.*, the time rate at which the quantity f, connected with a definite fluid particle, changes as this particle moves along. This rate is called the *material time rate of*

---

[1] Except for the (practically unimportant) moment of the shearing stresses along the wall and the two plane cross sections.

change of f, since it refers to a distinct material point, and will be denoted by df/dt. This derivative has been used in the foregoing in the definition of the acceleration.

Euler's rule of differentiation as established in Chap. II states that, in the case of a steady flow, one has $df/dt = V \, \partial f/\partial s$, where V is the magnitude of the velocity of the fluid particle under consideration and $\partial f/\partial s$ the space rate of change of f in the direction of the velocity V. This expression for the material time rate of change in a steady flow corresponds to the fact that the moving particle continually adapts itself to the value prevailing at its instantaneous position. In a steady flow the material time rate of change of f will vanish whenever the particle under consideration has the velocity zero. This is no longer true in an unsteady flow. There the material time rate of change of f for a particle at rest is given by $\partial f/\partial t$, and for a particle with a velocity V it is the sum of the *local* part $\partial f/\partial t$ and the *convective* part $V \, \partial f/\partial s$.

$$\frac{df}{dt} = \frac{\partial f}{\partial t} + V \frac{\partial f}{\partial s} \tag{15}$$

This is the form of Euler's rule of differentiation in the general case. In (15), f may be any quantity connected with a particle of the moving fluid, for example, the velocity V, or the x-component V_x of the velocity, or the density ρ.

According to (15) the x-component of the acceleration of a fluid particle in an unsteady flow is given by

$$\frac{dV_x}{dt} = \frac{\partial V_x}{\partial t} + V \frac{\partial V_x}{\partial s} \tag{16}$$

Applying Newton's second law of motion to the fluid enclosed by the control surface S, one has to multiply the volume dv of each fluid element within S by its density ρ and the x-component dV_x/dt of its acceleration and form the sum (integral) of all these products $\rho \, dv \, dV_x/dt$. In the case of a steady flow, $dV_x/dt = V \, dV_x/ds$, and the integral in question has the value $\int_{(S)} \rho V_n V_x \, dS$, as has been shown in Sec. 2. In the case of an unsteady flow the additional term $\partial V_x/\partial t$ in the expression (16) for the x-component of the acceleration leads to a corresponding supplementary term in the momentum theorem. It can easily be shown that in the case of an incompressible fluid this term is

$$\rho \int_{(R)} \frac{\partial V_x}{\partial t} \, dv$$

where the integration has to be extended over the entire region R enclosed by S. Thus the momentum theorem (3) actually takes the form

$$P_x + W_x = \rho \int\limits_{(S)} V_n V_x \, dS + \rho \int\limits_{(R)} \frac{\partial V_x}{\partial t} \, dv \qquad (17)$$

where W_x is the x-component of the weight of the fluid within S and P_x the x-component of the resultant of the thrusts that the surrounding fluid exerts on S.

Let us now consider a particular type of an unsteady flow of an incompressible fluid. Assume that at any fixed point all characteristic quantities (pressure, magnitude and direction of velocity) reassume the same values at certain instants $t_1, t_2, t_3, \ldots$. Such a flow is called *quasi-steady*. A particular case of a quasi-steady flow is the periodic flow where all quantities are *periodic* functions of the time. The study of quasi-steady flow is important, since in many cases careful observation of an apparently steady flow will reveal small and rapid fluctuations of pressure, magnitude and direction of velocity, etc. (see Sec. IV.4). Such a *turbulent flow* may be considered as a quasi-steady flow for which the above-mentioned intervals of time $t_2 - t_1$, $t_3 - t_2$, $\ldots$ are very short. In fact, they are so short, *i.e.*, the fluctuations are so rapid, that crude observation will not show that the flow is not strictly steady. The reading of routine instruments then gives time mean values rather than instantaneous values of the quantities under consideration.

Let us apply the momentum theorem to a quasi-steady flow of an incompressible fluid. Equation (17) is valid at any instant. However, with a view to the situation already explained one may be interested to obtain a relation valid for the mean values rather than for the instantaneous values of the quantities appearing in (17). The time mean value (or time average) of a quantity f during the time interval $t_2 - t_1$ is obtained by dividing the integral $\int\limits_{t_1}^{t_2} f \, dt$ by $(t_2 - t_1)$. Consequently the mean value of $\partial f/\partial t$ is zero, since the integral of $\partial f/\partial t$ equals $f(t_2) - f(t_1)$ and thus vanishes because $f(t_2) = f(t_1)$. Taking mean values on both sides of (17), we therefore see that the term containing $\partial V_x/\partial t$ will disappear. The resulting equation has again the form of Eq. (3), but all quantities now denote mean values instead of instantaneous values. Similar reasoning can be applied to Eq. (11), which expresses the theorem of moment of momentum. We obtain the following result: *Equations (3) and (11) are valid in the case of a quasi-steady flow of an incompressible fluid, if all quantities appearing there are understood to be time mean values.*[*]

[*] It should be clearly understood that on the right side of (3) the mean value of the product $V_n V_x$ has to be taken and not the product of the mean values of V_n and V_x, which may have a different value. An analogous remark applies to the right side of Eq. (11).

The following is an important application of this result: The flow produced by an airplane moving with constant speed through a bulk of air at rest is not steady. In Sec. II.1, however, it was noted that the relative flow of the air with respect to the airplane is steady and that consequently the theorems concerning a steady flow can be applied to this relative flow. Now, this is true only if one disregards the fact that the "air at rest" through which the airplane is moving is actually in a state of turbulent motion with fluctuating velocities but vanishing mean values of the velocities (see Chap. IV). In the inverse flow, the air streaming toward the airplane is therefore in a fluctuating rather than in a steady state of motion. Moreover, the entire airplane cannot be considered as a rigid body in a uniform rectilinear translation. If one thinks of the turning propeller, one realizes that the relative flow of the air with respect to the airplane is not strictly steady—even if the turbulence is disregarded—but rather quasi-steady, all characteristic quantities reassuming the same values after each complete turn of the propeller. If applied to this relative flow, the theorems of momentum and moment of momentum therefore must be used in the form given in the present section.

Let us now, instead of the aggregate consisting of the flying airplane and its propeller, consider an airscrew rotating at constant velocity on the ground without any forward speed. Then the relative flow with respect to the airscrew can be expected to be steady. However, our theorems concerning momentum and moment of momentum do not apply to the relative flow with respect to a rotating system of reference, even if this flow happens to be steady, since these theorems have been derived from Newton's second law of motion, which is not valid for the motion relative to a rotating system of reference. There is, however, an easy way to adapt our theorems to the type of problem under consideration. Let us assume that the control surface consists, for example, of a circular cylinder whose axis coincides with the propeller axis and of two parallel planes normal to it. If ds' denotes an element of the path that a fluid particle describes with respect to the rotating system of reference and V' its relative velocity, Euler's rule of differentiation for any f that stays constant at each point of the moving system can be written as

$$\frac{df}{dt} = V' \frac{\partial f}{\partial s'} \tag{18}$$

If V_x is the component of the absolute velocity of a particle with respect to a fixed x-axis, then, under the conditions of our problem, V_x is at each point of the moving system either constant or periodical with respect to time. For example, if the x-axis is parallel to the propeller axis, V_x will be constant with respect to time and (18) can be applied immediately to $f = V_x$. Since the x-component of acceleration is dV_x/dt, this can now be replaced by $V'\, dV_x/ds'$ instead of by $V\, dV_x/ds$ as was done in the beginning of this chapter. In the same way as Eq. (3) was found, one can therefore now derive

$$P_x + W_x = \rho \int_{(S)} V'_n V_x \, dS \tag{19}$$

The analogous modification of (11) leads to

$$M_z + x_0 W_y - y_0 W_x = \rho \int_{(S)} V'_n (x V_y - y V_x) \, dS \tag{20}$$

In the case of a force or moment component that is not constant but periodic with respect to time, Eqs. (19) and (20) still hold if the quantities occurring therein are understood to denote time mean values. Note that in these equations V_x and V_y are the components of the absolute velocity with respect to fixed x- and y-directions, whereas V'_n denotes the component of the relative velocity normal to the control surface S.

Problem 7. A horizontal jet of water hits a vertical plate that moves in the direction of the jet at the velocity $u = 10$ ft./sec. The diameter of the jet is 2.4 in., and its velocity is $V = 30$ ft./sec. Compute the force exerted on the plate and the power P of the force. If V is considered as constant and u as variable, for which u is the power a maximum and how great is P_{max}?

*Problem 8. A rocket moves in a horizontal direction at a constant velocity u under the influence of gas masses discharging into the atmosphere through an opening of area A in the rear. The overpressure inside the rocket is p, and the flow may be considered as adiabatic. Compute the force and the power, and discuss the possible values of u.

5. Energy Equation. Consider a perfect incompressible fluid in a flow that may be steady or unsteady. If ρ denotes the density and V the magnitude of the velocity of a fluid element of the volume dv, the expression $\rho V^2 \, dv/2$, that is, the product of half the mass and the square of the velocity, is the *kinetic energy* of the fluid element. Furthermore, if besides the thrusts exerted by neighboring particles gravity is the only force acting on the fluid element under consideration, the product $g\rho h\, dv$ of the weight $g\rho\, dv$ times the elevation h above some level of reference is known as the *potential energy* of the element.

Consider now a finite portion of the fluid. Let S denote the surface of this portion at a certain instant t (Fig. 41). The *total energy* E of the fluid within S at the instant t is defined as the sum (integral) of the kinetic and potential energies of all fluid elements within S.

$$E = \rho \int \left(\frac{V^2}{2} + gh \right) dv \tag{21}$$

In the same way as in the dynamics of a particle the time rate of change of this energy must equal the rate at which work is done by the forces, other than gravity, which act on the portion of fluid under consideration, *i.e.*, the *power P* imparted to the fluid by those forces. In the case of a perfect incompressible fluid, P is nothing else than the rate at which work is done by the normal thrusts which the fluid outside S exerts on the fluid within S. If p denotes the pressure, the thrust exerted on the surface element dS is p dS. During the interval of time dt, the point of application of this thrust is displaced by V dt in the direction of the velocity V. In order to obtain the work done by the thrust during this displacement, the thrust is multiplied by the projec-

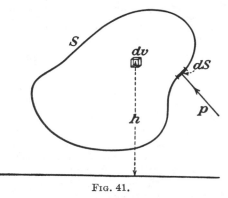

Fig. 41.

tion of the displacement on the line of action of the thrust. With respect to the sign of this product, it should be remembered that the thrust has the direction of the inward normal of the surface element dS. Since V_n is by definition the velocity component in the direction of the outward normal, the work is $-pV_n\, dS\, dt$. The total work done by the thrusts acting on the surface S is accordingly $-dt\int pV_n\, dS$. The rate at which these thrusts work is obtained by dividing this expression by dt. We therefore have the following equation:

$$\frac{dE}{dt} = \rho\frac{d}{dt}\int\left(\frac{V^2}{2} + gh\right)dv = -\int pV_n\, dS = P \qquad (22)$$

This so-called "energy equation" is valid for an incompressible perfect fluid, no matter whether the flow is steady or not. It does not include any independent statement but can be derived by formal transformations from Newton's second law under the two assumptions that the inner stress at each point is a normal pressure p, equal for all directions, and that the volume of each fluid particle remains constant.

Let us now consider the particular case of a steady flow. Applying Euler's rule of differentiation, Eq. (2), Chap. II, to $f = V^2/2 + gh$, we find

$$\frac{d}{dt}\left(\frac{V^2}{2} + gh\right) = V\frac{\partial}{\partial s}\left(\frac{V^2}{2} + gh\right)$$

Hence,

$$\rho\frac{d}{dt}\int\left(\frac{V^2}{2} + gh\right)dv = \rho\int\frac{d}{dt}\left(\frac{V^2}{2} + gh\right)dv = \rho\int V\frac{\partial}{\partial s}\left(\frac{V^2}{2} + gh\right)dv$$

In Sec. 1 of this chapter we considered the volume integral of $\rho V \, dV_x/ds$ extended over the region enclosed by a surface S. Dividing this region into stream tubes and summing the contributions of these stream tubes it was proved that this sum has the value $\int\rho V_n V_x \, dS$. At present we have to deal with the integral of a similar expression in which $V^2/2 + gh$ takes the place of V_x. By the same reasoning, it can be shown that this integral has the value $\rho\int V_n(V^2/2 + gh) \, dS$. The constant density ρ has been written in front of the integral since we are at present dealing with an incompressible fluid. Applying the energy equation (22) to a steady flow, we obtain, therefore,

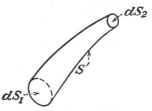

Fig. 42.—Stream tube.

$$\rho\int V_n\left(\frac{V^2}{2} + gh\right)dS = -\int pV_n \, dS = P \tag{23}$$

or, collecting all terms under the same integral sign and dividing by ρ,

$$\int V_n\left(\frac{V^2}{2} + gh + \frac{p}{\rho}\right)dS = 0 \tag{23'}$$

The expression in the parentheses in (23') is the same as that occurring in Bernoulli's equation. In fact, let us apply Eq. (23') to a stream tube, allowing the surface S to consist of two cross sections dS_1, dS_2 and of the tubular surface S' (Fig. 42). As on S' the normal velocity V_n is zero, the integral of (23') reduces to the two terms corresponding to dS_1 and dS_2. We thus obtain from (23')

$$V_n^{(1)} \, dS_1\left(\frac{V^2}{2} + gh + \frac{p}{\rho}\right)^{(1)} + V_n^{(2)} \, dS_2\left(\frac{V^2}{2} + gh + \frac{p}{\rho}\right)^{(2)} = 0$$

where the signs (1) and (2) refer to the elements dS_1 and dS_2, respectively. The normal velocity $V_n^{(1)}$ at the entrance section is negative and $V_n^{(2)}$ at the exit positive. As the flux $\rho V \, dS$ is constant along the stream tube, we have $-\rho V_n^{(1)} \, dS_1 = \rho V_n^{(2)} \, dS_2$. The foregoing equation therefore yields

$$\left(\frac{V^2}{2} + gh + \frac{p}{\rho}\right)^{(1)} = \left(\frac{V^2}{2} + gh + \frac{p}{\rho}\right)^{(2)} \tag{24}$$

which is identical with Bernoulli's equation.

In the case of a quasi-steady flow, Eq. (23) is still valid if all quantities $V_n V^2$, $V_n gh$, and pV_n occurring there are considered to be mean values instead of instantaneous values. This can be proved in the same way as the corresponding extension was proved in the preceding section for the forces and moments. Furthermore, if the flow under consideration is steady or quasi-steady with respect to a moving system of

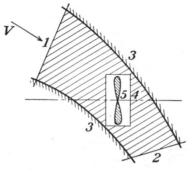

FIG. 43.—Arrangement of control surfaces in computation of propeller power.

reference, the use of Euler's equation in the form (18) leads to the following form of the energy equation equivalent to (23):

$$\rho \int V_n' \left(\frac{V^2}{2} + gh\right) dS = - \int V_n p \, dS = P \tag{25}$$

where V_n' denotes the relative normal velocity and where in the case of a quasi-steady flow mean values of all terms must be used.

The following is an important application of these results. Consider a propeller that turns with constant speed in a tunnel of arbitrary shape. In order to obtain an expression for the *propeller power P*, that is, the rate at which the propeller imparts work to the fluid, choose two cross sections 1 and 2 of the tunnel, thus delimiting a section 1-2 of the tunnel that contains the propeller (Fig. 43). The curved surface of this part of the tunnel will be denoted by 3. Consider further a small circular cylinder 4, which surrounds the propeller, and finally call the propeller surface 5. Supposing that both the flow with respect to the tunnel and the relative flow with respect to the turning propeller are quasi-steady, we apply, first, the energy equation (23) to the (shadowed) region bounded by the surfaces 1, 2, 3, 4 and, second, Eq. (25) to the region between the cylinder 4 and the surface 5 of the propeller. Summing these two equations and dividing by ρ we find

$$\int\limits_{(1,2,3,4)} V_n \left(\frac{V^2}{2} + gh + \frac{p}{\rho} \right) dS + \int\limits_{(4,5)} V'_n \left(\frac{V^2}{2} + gh \right) dS = -\frac{1}{\rho} \int\limits_{(4,5)} V_n p \, dS$$

The surface 4 appears in all these integrals. But the outward normal of 4 as a part of the boundary of (1,2,3,4) is the inward normal of 4 if considered as part of (4,5). Therefore, the term $V_n p/\rho$ in the first and $-V_n p/\rho$ in the integral to the right cancel each other as far as surface 4 is concerned. What is left reads in rearranged form

$$\int\limits_{(1,2,3)} V_n \left(\frac{V^2}{2} + gh + \frac{p}{\rho} \right) dS + \int\limits_{(4)} (V_n + V'_n) \left(\frac{V^2}{2} + gh \right) dS$$

$$+ \int\limits_{(5)} V'_n \left(\frac{V^2}{2} + gh \right) dS = -\frac{1}{\rho} \int\limits_{(5)} V_n p \, dS \quad (26)$$

The expression on the right-hand side of (26) is, except for the factor $1/\rho$, the propeller power P.

Since along the surface 3 the normal velocity V_n is zero, this surface does not furnish any contribution to the first integral in (26). Similarly, the relative normal velocity V'_n being zero along the propeller surface 5, the last left-hand integral is zero. Furthermore, for any point A on the surface of the cylinder 4 the absolute velocity can be obtained by superposition of the relative velocity at A and the velocity that A would have if it were rigidly connected with the turning propeller. Clearly, the latter velocity is at each point of 4 tangential to the surface of the cylinder 4. Along this surface the relative and absolute normal velocities are therefore equal in magnitude. Since for the two regions of integration (1,2,3,4) and (4,5) the outward normals of the cylinder 4 are defined with opposite directions, it follows that $V_n + V'_n = 0$, and therefore the second integral to the left in (26) vanishes, also. What remains on the left-hand side is the first integral extended over 1 and 2 only. The integral to the right is, except for the sign, what we call P, namely, the power imparted to the fluid by the propeller. Denoting by (1) and (2) the values in the points of the cross-sections 1 and 2, we thus find

$$\int \left[V_n \left(\frac{V^2}{2} + gh + \frac{p}{\rho} \right) \right]^{(1)} dS_1 + \int \left[V_n \left(\frac{V^2}{2} + gh + \frac{p}{\rho} \right) \right]^{(2)} dS_2 = \frac{P}{\rho}$$

Multiplying by ρ and introducing dQ for the flux through a surface element, *i.e.*, for the expression $-\rho V_n^{(1)} \, dS_1$ along the section 1 and $\rho V_n^{(2)} \, dS_2$ for the section 2, we finally obtain

$$P = \int \left(\frac{V^2}{2} + gh + \frac{p}{\rho}\right)^{(2)} dQ - \int \left(\frac{V^2}{2} + gh + \frac{p}{\rho}\right)^{(1)} dQ \qquad (27)$$

or

$$P = \int (H_2 - H_1)g \, dQ \qquad (28)$$

where H_1 and H_2 are the values of the total head (sum of velocity head, elevation, and pressure head) at the entrance and the exit, respectively. Owing to the quasi-steady character of the motion, the time mean values of the integrals and of P are meant in Eqs. (27) and (28).

These equations supply a remarkable addition to Bernoulli's equation, which states that in a steady motion H does not change. In fact, if the flow around the propeller were steady, no power exchange between the fluid and the propeller would be possible. The quasi-steady character of the motion is essential in producing an exchange of power. In a general form the result (27) can be stated as follows: *If a perfect fluid in a state of quasi-steady flow interferes with rigid bodies each of which has a steady or quasi-steady motion, then the time average of the power exchanged between the fluid and the rigid bodies equals the mean value of the difference of the total heads, upstream and downstream, integrated over the fluid weight passing per unit of time through a cross section.*

This result is often derived in an incorrect way. The correct proof requires the introduction of the notion of quasi-steady flow and the use of the auxiliary surface 4 in Fig. 43.

It should be added that the theorem has been established here for incompressible fluids only. In fact, in the form given the energy equation is valid only for constant ρ. For compressible fluids some modifications are necessary, an indication of which may be given here, without attempting a complete proof. In the expression for the total energy E a supplementary term, the so-called "inner energy" (or expansion energy) $\rho e \, dv$ must be introduced besides the kinetic energy $\rho V^2 \, dv/2$ and the gravitation energy $g\rho h \, dv$ of the fluid element. Here the inner energy e per unit of mass is defined by $de = -p \, d(1/\rho)$. The rate at which work is done by the thrusts acting on a portion of the fluid then equals the time rate of change of the expression $\int \rho(V^2/2 + gh + e) \, dv$. We thus have

$$\frac{d}{dt} \int \rho \left(\frac{V^2}{2} + gh + e\right) dv = -\int pV_n \, dS \qquad (29)$$

instead of (22). Proceeding as in the foregoing, we find

$$\int \rho V_n \left(\frac{V^2}{2} + gh + e\right) dS = -\int pV_n \, dS$$

or

$$\int \rho V_n \left(\frac{V^2}{2} + gh + e + \frac{p}{\rho} \right) dS = 0 \tag{30}$$

In Sec. II.2 the pressure head h_p was defined by $g\, dh_p = dp/\rho$. Accordingly,

$$de + d\left(\frac{p}{\rho}\right) = -p\, d\left(\frac{1}{\rho}\right) + d\left(\frac{p}{\rho}\right) = \frac{dp}{\rho} = g\, dh_p$$

or

$$e + \frac{p}{\rho} = gh_p + \text{const.}$$

By substituting this in (30) it is seen that the value of the constant is unessential since $\int \rho V_n\, dS = 0$. We thus find

$$\int \rho V_n \left[\frac{V^2}{2} + g(h + h_p) \right] dS = 0 \tag{31}$$

as the equation replacing (23) in the case of a compressible fluid. Since in this case gH is defined just like the expression in the brackets in (31), a way similar to the one followed above would lead to statement (28) in the case of compressibility, also. It may be added that the energy equation (29) is again a mathematical consequence of the equations of motion (Newton's second law) under the assumption of pure normal stresses in the fluid.

If shearing stresses are admitted, an additional term representing a *loss of total head* appears in the energy equation in the cases of compressible as well as of incompressible fluids.

Problem 9. In the buckets of a free jet wheel a jet of water, coming from an altitude H, is deflected through an angle β. The wheel has a circumferential velocity u smaller than the jet velocity. Using the momentum equation, compute the force exerted on the wheel and its power. Compute also the velocity of the water that leaves the buckets, and prove in this way that the energy equation for quasi-steady flow is fulfilled in this case, whatever the values of β and u.

*Problem 10. Using the hints given at the end of Sec. III.5, try to complete the derivation of the quasi-steady-flow energy equation for the case of a compressible fluid.

CHAPTER IV

PERFECT AND VISCOUS FLUIDS. TYPES OF FLOW

1. Viscosity. So far we have dealt almost exclusively with perfect fluids in which the stress transmitted across any surface element is normal to this element. The next step in the approach to reality will be the admission of shearing stresses of a certain nature. As has been pointed out in Sec. II.2, such shearing stresses can be expected whenever different velocities occur in two neighboring stream filaments, the stresses tending to diminish the difference.

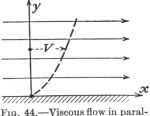

Fig. 44.—Viscous flow in parallel straight lines.

As the simplest example, consider a two-dimensional incompressible flow with parallel straight streamlines (Fig. 44). In the plane of flow assume a system of rectangular coordinates x, y with the x-axis parallel to the streamlines. Since the stream filaments have constant width, the velocity V is constant along each streamline and thus depends on y only. The derivative dV/dy constitutes a measure for the rate of change of V, in passing from one filament to the neighboring filament. With respect to the shearing stress τ transmitted between neighboring stream filaments, the simplest assumption, consistent with the idea just outlined, is that of proportionality between τ and dV/dy, that is,

$$\tau = \mu \frac{dV}{dy} \tag{1}$$

Here the so-called "coefficient of viscosity" μ is a physical constant of the fluid under consideration. As the quotient of this coefficient and the density ρ occurs frequently in the formulas, it will be denoted by a separate letter:

$$\nu = \frac{\mu}{\rho} \tag{2}$$

This quotient ν is known as the coefficient of kinematic viscosity or simply the *kinematic viscosity* of the fluid. Since the shearing stress has the dimension of force/area or (mass $\times$ acceleration)/area and dV/dy that of velocity/length or 1/time, it is seen that the coefficient of viscosity μ must have the dimension of mass/(length $\times$ time): $[\mu] = ML^{-1}T^{-1}$. The dimension of the kinematic viscosity ν is obtained by dividing that

74

of μ by the dimension of the density ρ, which is mass/volume. Hence, $[\nu] = L^2T^{-1}$; that is, ν has the dimension of the *product velocity times length*.

For dry air at standard pressure and 59°F. the value of μ is 3.72×10^{-7} slug/ft.-sec. With the exception of extremely high and extremely low pressures, such as are never found in the free atmosphere, the coefficient of viscosity of air μ can be considered as independent of the pressure. It varies, however, with the temperature. According to an earlier formula by Lord Rayleigh the ratio of the values μ' and μ'' at the absolute temperatures T' and T'', respectively, follows the law

$$\frac{\mu'}{\mu''} = \left(\frac{T'}{T''}\right)^{0.75} \tag{3}$$

Consequently, the kinematic viscosity $\nu = \mu/\rho$ can be found, since the equation of state supplies $\rho'/\rho'' = p'T''/p''T'$:

$$\frac{\nu'}{\nu''} = \frac{\mu'}{\mu''}\frac{\rho''}{\rho'} = \left(\frac{T'}{T''}\right)^{1.75}\left(\frac{p''}{p'}\right) \tag{3'}$$

The standard value of the kinematic viscosity of dry air is

$$\nu = 1.57 \times 10^{-4} \text{ ft.}^2/\text{sec.}$$

A more recent investigation gives for μ in slugs per foot-second the empirical expression[1]

$$\mu = 234.1 \frac{(\Theta + 459.4)^{3/2}}{\Theta + 682.6} 10^{-10} \tag{4}$$

where Θ is the temperature in degrees Fahrenheit. From this formula the values in the last column of Table 1 have been computed. Table 4 and Fig. 44a show how ν decreases at higher pressures.

TABLE 4.—KINEMATIC VISCOSITY ν IN FT.^{2}/SEC.
$p_0 = 2116.2$ lb./ft.$^2 = 29.92$ in. Hg
Values of $\nu \times 10^4$

Θ, deg. F. p/p_0	14	32	50	68	86	104	122
1	1.328	1.422	1.517	1.615	1.715	1.818	1.923
10	0.133	0.142	0.152	0.162	0.172	0.182	0.192
20	0.066	0.071	0.076	0.081	0.086	0.091	0.096

It is worth noting that the kinematic viscosity of water is lower than that of air. At standard pressure and temperature, ν for water is about 10^{-5} ft.^{2}/sec., that is, only about one-sixteenth the value for dry air. With increasing temperature the viscosity of water decreases

[1] MILLIKAN, R. A., *Ann. der Phys.*, **41**, 759 (1913).

rapidly.   Near the boiling point it is not more than one-sixtieth the ν for air at the same temperature and pressure.

Clearly, Eq. (1) must be modified in case of more general types of flow.   A discussion of the general theory of viscous fluids is beyond the scope of this book; suffice it to say that, in the same way as in the simplest case already studied, the state of stress at any given point can be considered as produced by the

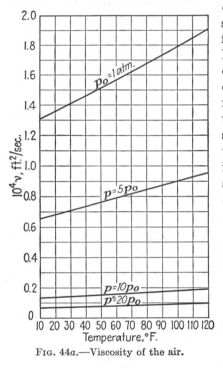

FIG. 44a.—Viscosity of the air.

superposition of pressures p, equal for all directions, and viscous stresses. It can be shown that for general types of flow the viscous stresses cannot consist of shearing stresses only but must include normal stresses as well, this, however, in such a way that the sum of these normal stresses for any three orthogonal directions will vanish.   In analogy to (1) all viscosity stresses are supposed to depend linearly on the space derivatives of the velocity components, and no physical constant other than the coefficient of viscosity μ appears in these relations.   This implies, first, that the shearing stresses are given as the products of μ and certain linear expressions in the space derivatives of the velocity components and, second, that the normal stresses are the sum of the pressure p and the products of μ and other linear expressions in those derivatives.   In the case of the flow already considered, the only nonvanishing derivative is dV/dy.   We then have the shearing stress $\tau = \mu\,dV/dy$ and no viscous normal stress in the x- and y-direction.

The main fact to be used in the following section can be summarized in the statement: *The forces due to viscosity appear as products of μ and expressions that have the dimension area $\times$ (velocity/length)* (derivatives of the velocity components with respect to the coordinates).   The complete mathematical analysis of these principles leads to a system of partial differential equations, the so-called "Navier-Stokes equations," which determine the distribution of velocity and pressure in a viscous fluid when suitable boundary conditions are given.

It should be kept in mind that the "viscous fluid" as well as the "perfect fluid" are idealizations.   In introducing the viscous fluid the presence of shearing stresses is admitted, and thus a broader hypothesis is used, which can be expected to give a better approximation of reality. However, we are not entitled to call "real fluid" what is still only an idealization. In this book the term will be used only when reference is made to observed facts.

Problem 1.   Compute the values of the kinematic viscosity ν in standard atmosphere at 5,000, 10,000, and 15,000 ft.
Problem 2.   Between two parallel plane plates there is a layer of lubricant $\frac{1}{100}$ in. thick.   The viscosity of the lubricant is $\mu = 0.004$ slug/ft.-sec.   If one plate slides over the other with a velocity of 120 ft./sec. and if the lubricant adheres to both plates, find the shearing stress that the lubricant exerts on the plates. (*See also* Prob. 7.)
*Problem 3.   Find the law according to which the temperature in the atmosphere should vary with the altitude in order to render the kinematic viscosity of the air constant.   Use Eq. (3') for the change of ν.

2. Law of Similitude. Reynolds Number. The theory of the Navier-Stokes equations which control the motion of a viscous fluid is beyond the scope of this book. However, the concepts introduced in the preceding section make it possible to arrive at some general conclusions about viscous fluids without explicitly using these equations.

Consider a rigid body in a state of uniform rectilinear motion through an incompressible viscous fluid at rest. In order to obtain some information about the forces that the fluid exerts on the body, the inverse flow may be studied in which the body is kept at rest and the fluid streams toward it with a velocity V equal and opposite to the absolute velocity of the body. Suppose that the body has a *given shape* and a *given orientation with respect to the undisturbed inverse flow*, i.e., with respect to the velocity vector as observed at a great distance in front of the body. If, for example, the body has the form of a circular cylinder, its shape can be described by stating the ratio of radius and length of the cylinder and the orientation by giving the angle that the axis of the cylinder forms with the direction of the velocity V of the undisturbed flow. Let us neglect the influence of gravity and assume that any force which the fluid exerts on the body (of given shape and orientation) depends (1) on the nature of the fluid as described by its coefficient of viscosity μ and its density ρ, (2) on the velocity V of the undisturbed flow, and (3) on the size of the body. In speaking of a force exerted by the fluid, any well-defined component F of the resultant of the stresses that the fluid exerts on the surface of the body will be meant; for example, the projection of this resultant on the axis of the cylinder.

Since the shape of the body is given, its size can be stated by giving some conveniently chosen linear dimension l of the body. In the case of a circular cylinder, possible choices for l are the length, the diameter, the radius, the perimeter of the circular cross section, etc. As the fluid is supposed to be incompressible, the density ρ is constant. The coefficient of viscosity μ is always considered constant throughout the fluid.

Using as the fundamental dimensions those of mass (M), length (L), and time (T), the dimension of a force F is $[F] = MLT^{-2}$. We now ask for combinations of μ, ρ, V, l having the same dimension as F. The product $\mu^a\rho^b V^c l^d$ has the dimension

$$[\mu^a\rho^b V^c l^d] = \left(\frac{M}{LT}\right)^a \left(\frac{M}{L^3}\right)^b \left(\frac{L}{T}\right)^c L^d = M^{a+b}L^{-a-3b+c+d}T^{-a-c}$$

To make this the dimension of a force, one must have

$$a + b = 1, \qquad -a - 3b + c + d = 1, \qquad -a - c = -2$$

From these three equations the exponents b, c, d can be expressed as functions of a. We find $b = 1 - a$, $c = 2 - a$, $d = 2 - a$. Thus the expression

$$\mu^a\rho^{1-a}V^{2-a}l^{2-a} = \rho V^2 l^2 \left(\frac{\rho V l}{\mu}\right)^{-a}$$

has the dimension of a force, no matter what value is given to the exponent a. The implication is that any expression for a force F must consist of the factor $\rho V^2 l^2$ and another factor which is a combination of terms, each being the product of a constant and a power of $\rho V l/\mu$. Such a combination is practically an arbitrary function of $\rho V l/\mu$ or of Vl/ν. Instead of the factor l^2 in the expression $\rho V^2 l^2$ any area can be used to characterize the size of the body. Let us call this area A, and let us call $2C_F$ the function of Vl/ν. Then

$$F = \frac{\rho}{2} V^2 A C_F\left(\frac{Vl}{\nu}\right) = qAC_F \tag{5}$$

where q is the dynamic pressure corresponding to the velocity V. The magnitude

$$\frac{Vl}{\nu} = \frac{\rho Vl}{\mu} = Re \tag{6}$$

is known as *Reynolds number*. The function C_F in (5) is called the *force coefficient;* it depends on the Reynolds number. Both Re and C_F are dimensionless quantities. Note that Re is infinite for a perfect fluid motion since ν has here the value zero.

Equation (5) can be given the following interpretation: Assume that we are experimenting with a set of circular cylinders of a given shape (cylinders with the same value of the ratio l/r). The cylinders are supposed to have the same orientation, *i.e.*, their axes are supposed to form the same angle α with the direction of the undisturbed flow. As the characteristic area of a cylinder we may choose its cross-sectional area A; as the characteristic length, its length l. Experimenting with various velocities V and with fluids of various viscosities μ and densities ρ (*e.g.*, air and water, or air at different temperatures or under different pressures), each time the value of the force F, which the fluid exerts on the cylinder, may be observed. Equation (5) then shows that the quotient

$$\frac{F}{qA} = \frac{2F}{\rho V^2 A} = C_F \tag{5'}$$

will remain constant as long as $Re = Vl/\nu$ keeps its value. In other words: *In order to obtain the value of the force F for any combination of μ, ρ, V, l it is sufficient to perform a single series of experiments, keeping, for example, the values of μ, ρ, and l constant and varying only V.* The task of collecting experimental information concerning F thus is reduced to a single sequence of experiments. This reduction is considerable since, without the use of (5), one would have to carry out a four-dimensional manifold of experiments, varying μ, ρ, l, and V independently.

The content of Eq. (5) can be stated in the following way: *If a body of given shape and given orientation is exposed to a uniform flow of an incompressible fluid (or moves with constant velocity through an incompressible fluid at rest), each component of the force which the fluid exerts on the body can be expressed as the product of the dynamic pressure, a certain area of reference, and a coefficient the value of which depends exclusively on the Reynolds number Re.* This is equivalent to the statement that the force coefficient, *i.e.*, the quotient F/qA, in any case depends on the Reynolds number and on the shape and orientation of the body. Note that numerical values of the Reynolds number and of the force coefficient have no precise meaning unless it is stated which length and area have been chosen as reference length and reference area.

The argument leading to Eq. (5) can be extended in the following way, to give important information about the *stream pattern around the body:* Let P be a point whose coordinates x, y, z with respect to the rigid body are given in terms of the reference length l: $x = \xi l$, $y = \eta l$, $z = \zeta l$ where ξ, η, ζ have given values. Consider a certain component v of the velocity at P, for example, its x-component. Again neglecting the influence of gravity, one has to assume that this velocity component will depend on μ, ρ, V, l only. The product $\mu^a \rho^b V^c l^d$ will have the dimension of a velocity if

$$[\mu^a \rho^b V^c l^d] = M^{a+b} L^{-a-3b+c+d} T^{-a-c} = LT^{-1}$$

Accordingly,

$$a + b = 0, \qquad -a - 3b + c + d = 1, \qquad -a - c = -1$$

or

$$b = -a, \qquad c = 1 - a, \qquad d = -a$$

The expression

$$\mu^a \rho^a V^{1-a} l^{-a} = V \left(\rho \frac{Vl}{\mu} \right)^{-a} = V(Re)^{-a}$$

therefore has the dimension of a velocity independently of the value of a. The velocity component v under consideration accordingly equals the product of V and a function of the Reynolds number. Since the stream pattern (shape and arrangement of the streamlines) remains the same as long as for any given point and direction the ratio v/V is constant, we see that *the stream pattern depends exclusively on the value of the Reynolds number.*

The same method can be used with respect to quantities other than force or velocity components. Applying it, for example, to the moment of the force with respect to a given point of the body, we find that this moment must have the form of a product of $\rho V^2 l^3$ and a function of Re. This function usually is written as $2C_M(Re)$. The moment then appears as the product of the *moment coefficient* C_M and $qA^{3/2}$ or qAl_1, where l_1 is a reference length that may be different from the reference length used in forming Re.

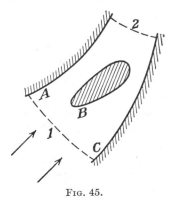

FIG. 45.

Pressures appear in our problem only as differences (for example, the force F considered above depends, not on the actual values of the pressure on both sides of the body, but rather on the difference of these values). Denoting the pressure in the undisturbed stream by p_0, we therefore apply our method to the difference $p - p_0$ of the pressure p at P minus the pressure p_0. Proceeding as before we find that this difference can be written as the product of the dynamic pressure q and a function of Re.

Introducing the concept of *boundary conditions* we arrive at a suitably general statement of the preceding results. According to this concept a particular steady flow of a viscous fluid, *i.e.*, a particular solution of the Navier-Stokes equations, is determined by certain boundary conditions, *viz.*, the form of the walls A, B, C, . . . , which the flow cannot traverse, and by the velocity distributions at an entrance surface 1 and at an exit surface 2. (Fig. 45.) The velocity value at a suitably chosen point of the entrance and a particular linear dimension of the wall are used as reference values for determining the Reynolds number of the problem. From the boundary conditions of the flow under consideration, boundary conditions of a second flow may be derived by multiplying all coordinates by a certain factor and all velocities by another factor. Boundary conditions that can be derived from one another in this way are said to be *similar conditions*. They may however, have different Reynolds numbers.

If, for two stream patterns, the similarity relation holds not only for the velocities at corresponding points of the boundaries but also for the velocities at corresponding points in the interior, the two flows are said to be *similar flows*. The result of the foregoing argument can then be expressed in the following way: *If in two cases of steady motion of a viscous fluid the boundary conditions are similar and if, moreover, the Reynolds number has the same value in the two cases, the two flows will be similar; each force component will be proportional to qA, each moment proportional to qAl, etc.* If, on the other hand, the Reynolds number is different in the two cases, we may have different stream patterns (and different values of force and moment coefficients) in spite of the fact that the boundary conditions are similar.

In view of the importance of this theorem, the assumptions under which it has been established may be stated once more: The fluid is incompressible; the coefficient of viscosity μ is constant throughout the fluid; the effects of gravity are negligible.

An important restriction of the range covered by the theorem is indicated by the following: As mentioned before, the pressure differences $p - p_0$ in two similar flows are proportional to q. Hence, if at one pair of corresponding points of the two flows (*e.g.*, at infinity) the pressure has equal values, it may happen that though the pressure is everywhere positive in one flow the theory leads to negative values at some points of the similar flow. Since negative pressure values are physically impossible, the similar flow cannot be realized and one may have to look for another solution of the Navier-Stokes equations.

Problem 4. A sphere of 4 in. diameter is exposed to a uniform air stream of the velocity $V = 150$ ft./sec. Using the diameter as the characteristic length, determine the Reynolds numbers at sea level and at an altitude of 20,000 ft. in the standard atmosphere.

Problem 5. The model of an airplane used in a wind-tunnel test is made one-fifteenth the size of the real airplane. Why is it impossible to fulfill rigorously the similitude conditions in the test if the real airplane is supposed to fly at a speed of 200 m.p.h. and if the testing is done with air under normal pressure? What pressure in the wind tunnel should be used if the highest available speed is 180 m.p.h.?

3. Laminar and Turbulent Motion. From the mathematical point of view any particular steady flow of a viscous fluid appears as a particular solution of the Navier-Stokes equations for steady flow, characterized by a certain set of boundary conditions. Only in a very restricted number of cases is it possible to overcome the mathematical difficulties of determining solutions corresponding to given boundary conditions. One of these cases is the uniform flow in a cylindrical tube of circular cross section. Denoting the average velocity by V and the diameter of the tube by d, one may use as Reynolds number $Re = Vd/\nu$. Integration of the Navier-Stokes equations then furnishes the following results: (1) The velocity is everywhere parallel to the axis, and its distribution over the diameter of the tube follows a parabolic law such that the maximum velocity occurring in the axis of the tube equals twice the average velocity V (curve a, Fig. 46). (2) The pressure head is constant along each cross section and decreases at the rate

$$\frac{V^2}{2gd}\frac{64}{Re} \quad \text{or} \quad \frac{q}{\gamma d}\frac{64}{Re}$$

per unit of length of the tube. In these expressions the factor $64/Re$ is called the *friction factor*.

Experiments show that these laws for the motion in a tube, known as Poiseuille's laws, are in accordance with the observed facts as long as Re does not exceed a certain critical value of about 2000. For air

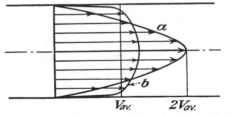

Fig. 46.—Velocity distribution over the cross section of a pipe in viscous flow; a, laminar; b, turbulent.

($\nu = 1.57 \times 10^{-4}$ ft.^{2}/sec.) and a tube of 1 in. diameter, this means that the average velocity should not exceed 3.77 ft./sec. Beyond the limit set by this critical value of the Reynolds number the behavior of a real fluid changes radically. (1) The velocities are much more uniformly distributed over the cross section, the maximum velocity being only a few per cent greater than the average velocity. Near the wall of the tube the velocity decreases very rapidly, assuming the value 0 at the wall (curve b, Fig. 46). (2) The friction factor computed from the actual pressure drop (*i.e.*, the observed decrease in pressure head per unit length of the tube multiplied by $2gd/V^2$) is no longer inversely proportional to Re but almost constant, or rather very slowly decreasing, so as to become about 200 times as great as Poiseuille's value for $Re = 500,000$. (3) The motion is no longer strictly steady but rather quasi-steady in the sense defined in Sec. III.3. The velocity distribution shown by curve b of Fig. 46 corresponds to time averages of continually fluctuating velocity values. Furthermore, there exist small fluctuating velocity components perpendicular to the axis of the tube. The flow has a *turbulent* character, thus differing entirely from the steady flow along parallel straight streamlines for which Poiseuille's laws can be derived as a consequence of the Navier-Stokes equations.

The latter fact indicates that the theory of viscous fluids leading to the Navier-Stokes equations need not be considered incompatible with observations (1) and (2) as described for the turbulent flow. It is possible that the mathematical problem defined by the Navier-Stokes equations together with the given set of boundary conditions admits of a second solution, distinct from the strictly steady flow studied by Poiseuille and corresponding to a quasi-steady flow which is in agreement with the experimentally observed distribution of the time averages of velocity and pressure. The question whether this is the case or not is as yet undecided. All solutions of the Navier-Stokes equations known to date correspond to so-called "laminar" flows. This term designates regular and smooth steady motions which do not show the irregular fluctuations characteristic of turbulent flows and of which the Poiseuille flow is the simplest example.

Let us now consider the flow of a perfect fluid in a cylindrical tube of circular cross section. Assuming that the fluid entering the tube comes from some region (reservoir) where the Bernoulli constant H has the same value throughout, a flow with the same H at all points, *i.e.*, an irrotational flow, must be expected. In the case of parallel straight streamlines this means that the velocity will have the same value at all points of the cross section. This uniform velocity distribution of the perfect fluid flow agrees much better with observations under turbulent conditions than the velocity distribution of a laminar viscous flow. To a certain extent this may be explained by the fact that the perfect fluid can be considered as a viscous fluid with the kinematic viscosity $\nu = 0$, thus rendering $Re = \infty$. It is not surprising that for large values of the Reynolds number the actual flow pattern approaches in a certain sense the pattern corresponding to a perfect fluid; on the contrary, it would be rather surprising if the parabolic velocity distribution of the laminar flow with $V_{max} = 2V$ were valid for no matter how large Reynolds numbers, but suddenly changed to the uniform velocity distribution of the perfect fluid for $Re = \infty$.

If these results are extended to more general types of flow (curved channels, varying cross section, etc.), the following experimentally confirmed facts can be stated: For small values of Re (small diameter, small velocity, or high viscosity) the flow actually is steady; the streamlines are smooth, nearly parallel to the walls, and the velocity increases slowly from its zero value at the walls to much higher values in the interior; the velocity distribution corresponds more or less to Poiseuille's parabolic law. Most practical cases of flow are beyond this domain of laminar motion. For higher values of Re the apparently steady flow is, in fact, continually changing in detail; at any given point the velocity components are fluctuating about certain mean values and the fluid particles move along wavy paths. If the small fluctuations are disregarded and attention is given only to the average values at each point, there appears a marked resemblance to the irrotational flow pattern of a perfect fluid. The mean values of the velocity are distributed very much like the instantaneous velocities in a perfect fluid; they are slowly varying up to the closest neighborhood of the walls and then drop to zero across a very thin layer (boundary layer; see next section). This turbulent kind of motion is encountered in practically all aeronautical problems.

There exists so far no mathematical theory that would account satisfactorily for the actually observed types of fluid motion. Serious discrepancies between the facts and the theory of both perfect and viscous fluids appear in almost all cases. However, in accordance with what has just been said, there is a way out of these difficulties that enables us to handle with a certain degree of approximation most problems that are of interest in aeronautics. A hypothesis that has proved to be sufficiently applicable in a great number of cases is that *the fluid motion at large, i.e., everywhere except in the immediate neighborhood of rigid bodies, may be considered approximately as a perfect fluid motion, at least as far as the velocity distribution is concerned.* That is, the virtually steady mean values of the velocity may be assumed to fulfill the mathematical relations of a steady perfect fluid motion. Although not always formulated explicitly, this *hydraulic hypothesis* is adopted more or less tacitly in almost every analysis of a real fluid motion. For example, the flow around an airfoil in a wind tunnel is doubtless a turbulent flow of a viscous fluid. But if the small oscillations are disregarded, the remaining steady velocity values agree very well with those computed from the theory of perfect fluids (Chaps. VIII and IX).

The hydraulic hypothesis does not contend that the viscosity effects are negligible. On the contrary, the influence of viscosity is considered as twofold. In the neighborhood of the walls the viscosity is supposed to be essential in producing a thin boundary layer across which the velocity increases rapidly from the value zero at the wall to the value prevailing in the free stream. There, in turn, the viscosity is responsible for the continual fluctuations or for the turbulent character of the motion. It is left undecided whether the instantaneous (fluctuating) velocities of the real fluid follow the Navier-Stokes equations or not. The hydraulic hypothesis states only that the mean velocity values satisfy, to a certain extent, the perfect-fluid equations.

In most cases the perfect fluid motion that is considered as representing the average values of the actual velocities is irrotational. It may be noted in this connection that, in the case of an incompressible fluid, any irrotational velocity distribution is in agreement with the Navier-Stokes equations. Only the boundary condition of vanishing velocity at the walls is not fulfilled by this solution. But the combination of a boundary layer, in which the magnitude of the velocity drops rapidly to the value zero at the wall, with an irrotational steady flow in the stream could supply a solution of the Navier-Stokes equations. However, such a solution would still not account for the turbulent fluctuations of the velocity in the free stream, nor would it cover the case of a compressible fluid. This remark shows again how far we are from a complete understanding of the phenomena involved.

The hydraulic hypothesis is primarily concerned with the velocity distribution, but the pressure obtained from the theory of perfect fluids according to Bernoulli's equation agrees in general also with the observations. This will be seen in more detail when particular problems are studied.

Problem 6. In the steady motion of a fluid, either laminar or turbulent, through a long cylindrical pipe line there will be a drop of total head H proportional to the axial length. If J is the drop per unit of length, prove, using the result of Sec. 2, that the factor

$$f = \frac{2gJd}{V^2}$$

can depend on the Reynolds number Vd/ν only.

Problem 7. Two parallel plates at a distance h move in the same direction, parallel to their plane, with the velocities V_1 and V_2. Show that, under the assumption of Eq. (1), the velocity distribution for a viscous fluid between the two plates is linear and that the shearing stress acting on each plate equals $\pm \mu (V_2 - V_1)/h$.

*Problem 8. In the case of the two-dimensional motion of a viscous fluid the shearing stress on the faces of an element $ds\,dn$ is given by the equation

$$\tau = \mu \left(\frac{\partial V}{\partial n} + \frac{V}{R} \right)$$

which replaces Eq. (1). In addition, there are tensile stresses due to the viscosity

$$2\mu \frac{\partial V}{\partial s} \quad \text{and} \quad 2\mu V \frac{\partial \vartheta}{\partial n}$$

in directions ds and dn, respectively (for the notation see Sec. II.5). Explain these formulas, and apply them to discuss the motion in concentric circles between two cylinders that rotate at velocities V_1 and V_2.

4. Continuous and Discontinuous Motion.

The hydraulic hypothesis becomes particularly significant in cases where continuous and discontinuous flow patterns are in competition. Of course, in a real fluid no actual discontinuity, in the strict mathematical sense of the word, will ever exist. Observing, however, the flow pattern behind a cylinder immersed in water one often sees a region of progressing flow distinctly separated from a region of "dead water" where the fluid essentially moves "on the spot." To a certain extent, this phenomenon is predicted by the theory of perfect fluids, as may be seen from the following example.

The problem of the two-dimensional steady irrotational flow of a perfect fluid around a cylinder of elliptical cross section can be solved under the assumption that the ellipse representing the cross section is one of the streamlines. The solution yields a stream pattern as indicated approximately in Fig. 47a and b. There are two stagnation points C, D at the ends of the axis that has the direction of the undisturbed stream and two points A, B of maximum velocity at the ends of the other axis. Since the flow is irrotational, the Bernoulli constant H has the same value for all streamlines. We therefore have minimum values of the pressure at the points of maximum velocity (the effect of gravity being neglected). If pressure and velocity in the undisturbed stream are denoted by p_0 and V_0, respectively, the pressure corresponding to the maximum velocity $V_{\max}$ is given by

$$p_{\min} = p_0 - \frac{\rho}{2}\left(V_{\max}^2 - V_0^2\right)$$

Now, in an irrotational flow $V/R - dV/dn = 0$ as shown in Sec. II.4. Hence, for a velocity of a given magnitude V, the smaller the radius of curvature R of the streamlines, the greater the space rate of change of velocity in a direction perpendicular to the streamlines. Accordingly, in the case of Fig. 47b a higher value of $V_{\max}$ is to be expected than in the case of Fig. 47a. If the ratio $\overline{AB}/\overline{CD}$ increases more and more, it may happen that $p_{\min}$ as computed from the foregoing equation drops to zero or even below zero; but negative pressures cannot occur in a real fluid. H. v. Helmholtz showed in 1868 that in such cases another solution of the perfect fluid equations exists, viz., the discontinuous flow shown in Fig. 47c. In this solution only a part of the elliptic boundary coincides with a streamline. Then a region of still water called *wake* is formed behind the obstacle. As the boundary of the wake is crossed, the magnitude of the velocity jumps from the value zero prevailing within the wake to a finite value outside the wake. Thus the velocity distribution is discontinuous. A fluid motion of this type, very often observed in actual flow, is called a *discontinuous flow* or *Helmholtz flow*. The points A and B where the boundaries of the wake detach themselves from the boundary of the body are known as *separation points*.

If the elliptical cross section of the cylindrical obstacle degenerates into a portion of a straight line, that is the cylinder into a plate set at right angles to the stream (Fig. 47d), the continuous flow would have infinite velocities at the edges of the plate independently of the magnitude of the velocity V_0 in the undisturbed stream.   In this case a discontinuous flow is to be expected and actually observed under all circumstances.   The exact nature of the conditions, however, under which the discontinuous flow pattern takes place in the case of elliptical

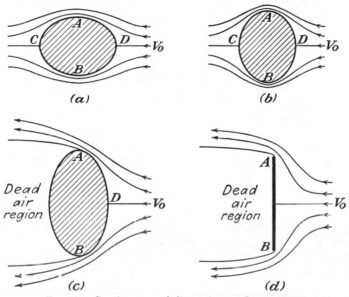

FIG. 47.—Continuous and discontinuous flow patterns.

cylinders or of obstacles of a more general form has not yet been sufficiently explored.   Some aspects of this question will be discussed in the following section.   A general result of the observations is that there exists a certain *tendency toward separation:* a discontinuous flow is already established before the conditions would lead to negative pressure values in the continuous flow.   For example, a wake develops in general behind a circular cylinder although a continuous symmetrical flow with positive pressure everywhere would be possible.

It has been observed in experiments that the wake behind a circular cylinder extends over a smaller or greater portion of the rear of the cylinder, depending on the value of the Reynolds number.   This is not in contradiction to the hydraulic hypothesis since the theory of perfect fluids furnishes a series of discontinuous solutions with the wake extending over an arc of the circular cross section, from about 120° up to 250°

(see Sec. V.3 and Fig. 58). From the point of view of the theory of perfect fluids, the problem then appears as a *stability question*. This means that of all the different flows compatible with the conditions of the problem the most stable one or the only stable one will materialize.

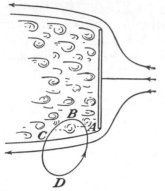

FIG. 48.—Discontinuous flow.

The wake, as it is described by the theory of perfect fluids, consists of a fluid at rest under a pressure equal to the undisturbed pressure p_0 that prevails in the fluid at an infinite distance from the obstacle. The real fluid, instead, shows a region where the fluid moves vigorously "on the spot" without progressing (Fig. 48) and where the pressure is considerably smaller than p_0. This pressure drop increases the force that the fluid exerts on the body (see Sec. V.3). If we consider the motion in the wake as a kind of oscillation superimposed on a state of rest, the entire stream pattern observed around the obstacle corresponds, in the sense of the hydraulic hypothesis, to the Helmholtz motion of a perfect fluid.

This discontinuous perfect fluid motion is irrotational. Neither the progressing fluid particles nor, of course, those at rest have any rotation. For any circuit lying entirely in the wake or entirely in the region occupied by progressing fluid, the circulation is zero. However, for a curve cutting across the boundary of the wake the circulation will not be zero. Consider, for example, the circuit indicated in Fig. 48. The part ABC within the wake does not give any contribution to the circulation since, for a perfect fluid, the velocity is zero here, and the contribution furnished by the part CDA is mainly negative. The circulation along this circuit, therefore, is different from zero. If the velocity distribution within the circuit were continuous, the conclusion would be that the fluid within the circuit must possess rotation. But, in Sec. II.5, the relation between rotation and circulation was derived in a way that does not hold when the velocity distribution within the circuit is discontinuous. One can, however, formally maintain this relation by attributing a finite vorticity to the (mathematical) points at the boundary between wake and stream. The boundary is then considered as a *vortex sheet*, *i.e.*, an infinitesimally thin layer of fluid consisting of particles with infinite values of rotation. It will be seen in Chap. IX how this concept can be usefully applied to certain problems.

Besides the continuous and discontinuous flows past a cylinder considered so far, the theory of perfect fluids furnishes still another flow pattern which is not strictly steady but can be described as a periodic motion closely connected with the discontinuous steady flow. In certain cases a motion can be observed in real fluids that, though not conforming exactly to this third type of perfect fluid motion, yet corresponds to it in the sense of the hydraulic hypothesis. As in the cases of the continuous and discontinuous steady motions this third type also shows smooth streamlines in front of the obstacle. In the rear, however, one finds so-called "single vortices" (Fig. 49) which alternately leave the obstacle at two points of separation, in such a manner that a definite period of time elapses between the formation of successive vortices. Here the term vortex is misleading insofar as this type of motion again is irrotational. There exist, however, singular points of the velocity field such that the circulation along a circuit surrounding one of these singular points is distinct from zero. In the neighborhood of such a

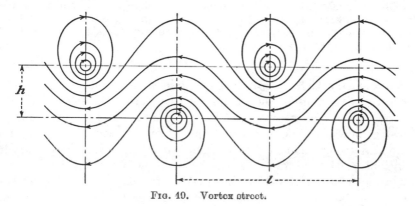

Fig. 10.   Vortex street.

singular point the fluid motion is of the circulating type discussed in Sec. II.5 and therefore irrotational. However, with a certain modification of the meaning of the term vortex such a point may be called a vortex point and a line that passes through such a point perpendicular to the plane of flow, a vortex line. This concept will be discussed in Sec. VIII.5. As the vortices alternately leave the obstacle, they move downstream, forming two parallel rows of equally spaced vortices of equal and opposite values of circulation. This *vortex street* was first observed in 1908 by H. Bénard, and its theory was given in 1910 by Th. v. Kármán. Kármán not only showed that a vortex street is compatible with the theory of perfect fluids; he also succeeded in carrying out a stability investigation with the result that the arrangement described in the foregoing is stable only if the ratio between the distance h of the two rows and the distance l of two consecutive vortices of the same row has the value 0.281 (at least, at a certain distance from the obstacle). Experiments are in good agreement with this theoretical result.

To sum up, it may be stated that the theory of perfect fluids supplies a variety of solutions for the flow around an obstacle: continuous flow, discontinuous flow patterns with various breadths of the wake, vortex street. *In accordance with the hydraulic hypothesis, each of these solutions gives an approximate picture of a flow pattern that can be observed in a real fluid.* Which of these patterns takes place in a particular case is a question that has not yet been sufficiently cleared up. From the standpoint of the theory of perfect fluids this question appears to be one of stability. However, the viscosity seems to play a certain role, and it may be that in each case, *i.e.*, for each value of Reynolds number, only one definite flow pattern conforms to the theory of viscous fluids. An essential concept of this theory is the previously mentioned boundary layer.

Problem 9. Show that if in the case of a Helmholtz flow of a perfect fluid (Fig. 47c or d) the pressure in the wake equals the undisturbed pressure p_0, the coefficient of the force acting on the body, referred to the frontal area, cannot be greater than 1. What limit is set to this coefficient if the pressure in the wake is known to be not smaller than $p_0 - \eta q$, with $\eta > 0$?

5. Boundary Layer. In a perfect fluid motion the velocity would keep a finite value up to the solid walls that border the stream. In the laminar motion of a viscous fluid, both the theory and the experimental evidence show that the velocity drops to zero at the walls at a finite rate which, in a certain way, is independent of the value of the viscosity coefficient ν. For example, in the case of the Poiseuille motion in a circular tube of diameter d with an average velocity V, the rate is found to equal $8V/d$. If the Reynolds number of the flow increases more and more beyond the limit that is set to a laminar flow, the velocity distribution across the stream must approach the flow pattern of a perfect fluid which is characterized by the value ∞ for Re. This means that at high values of Re the velocity will rise rapidly, *i.e.*, within a very narrow space, from the value zero at the wall up to the value V prevailing in the free stream. The thin fluid layer within which this rapid increase takes place is called the *boundary layer*. Its thickness must be expected to diminish with increasing Re, other things being equal. The presence of a boundary layer is thus connected with the fact that the motion in the free stream is turbulent. This does not exclude the assumption that the motion within the layer follows the laws of laminar flow.

L. Prandtl was the first to recognize the great importance of the boundary layer in hydrodynamic theory. In 1905 he started the investigation of the following problem: The free stream velocity V along the border line of the stream, not necessarily constant, is supposed to be given. Furthermore, this velocity V is assumed to be connected with the pressure p by Bernoulli's equation in the form $p + \rho V^2/2 = $ const.

so that the pressure values along the wall can be considered as being known, too. Finally, certain initial conditions at some upstream point of the wall are prescribed. Consider, for example, *a plane plate immersed parallel to the streamlines of a uniform flow* (Fig. 50). Here the stream velocity V and therefore the pressure p are constant along the boundary, and at the leading edge A the initial condition is that the velocity has the same value right up to the plate, *i.e.*, that the thickness of the boundary layer here is zero. From these data the theory of viscous fluids, applied to the fluid in the boundary layer, furnishes the *velocity profiles* (Fig. 50), *i.e.*, the distribution of the velocity values v across the boundary layer, rising in each cross section from zero at the wall to V in the free stream. In this example we see that the velocity profile is flattened out as we proceed along the plate in the direction of the stream; *i.e.*, the boundary layer increases in thickness from the initial point toward the trailing end.

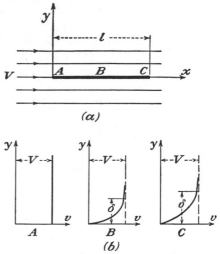

Fig. 50.—Velocity profiles of the boundary layer along a flat plate.

In order to give a definite meaning to the term "thickness of the boundary layer," one may choose to define as thickness the distance δ from the wall to a point where the velocity $v = 0.94V$ (the results would not be changed considerably if a slightly different factor were chosen instead of 0.94). It is then found that, in the case of the plate immersed in a uniform flow,

$$\delta = 3.65 \sqrt{\frac{\nu x}{V}} \tag{7}$$

where x denotes the distance from the leading edge. Thus at the trailing edge of the plate, $x = l$, the maximum thickness of the boundary layer has the value

$$\delta_l = 3.65 \sqrt{\frac{\nu l}{V}} = \frac{3.65 l}{\sqrt{Re}} \tag{8}$$

where $Re = Vl/\nu$ is the Reynolds number referring to the stream velocity V and the plate length l. As an example, consider a plate 1 ft. long in an air stream of $V = 30$ ft./sec. With $\nu = 0.00015$ ft.²/sec., we have $Re = 200,000$ and consequently $\delta_l = 0.0082$ ft. $= 0.098$ in.

The velocity distribution having been found, the *drag* can be computed since, according to (1), the shearing stress is determined by the value of dV/dy taken for $y = 0$. The formula for the drag will be given in Sec. V.5. The results are in good agreement with the observations up to Reynolds numbers of 250,000 to 300,000.

For greater values of the Reynolds number Prandtl's theory, which leads to formula (8) for δ_l, fails to agree with the observations. In this theory it is assumed that the free stream is turbulent so as to follow, according to the hydraulic hypothesis, approximately the laws of perfect

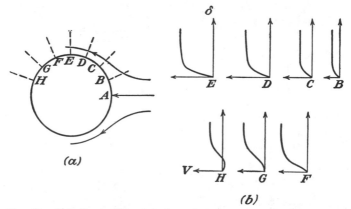

Fig. 51.—Velocity profiles of the boundary layer on a circular cylinder.

fluid motion, but the boundary layer is supposed to be laminar, *i.e.*, to correspond to a regular steady solution of the Navier-Stokes equations. These assumptions seem to be correct only up to a critical value of $Re \sim 250,000$. Beyond this limit the flow in the boundary layer, too, becomes turbulent, as can be verified by careful observation. In 1921, Kármán gave a half-theoretical, half-empirical solution of the problem of the plane plate in a uniform stream under the assumption that the boundary layer is turbulent. He showed that in this case the thickness of the boundary layer increases with the $\frac{4}{5}$ power of x instead of the $\frac{1}{2}$ power in the laminar case; the drag has accordingly to be computed by a different formula (which will also be given in Sec. V.5). This theory is in accordance with observations for values of the Reynolds number greater than 300,000 to 400,000. Between the two cases there is a certain region of transition under intermediate conditions.

The flat plate furnishes the simplest example in the theory of the boundary layer. Greater difficulties arise in less simple cases, for example, in the *flow around a circular cylinder* (Fig. 51). Here, whichever solution may be considered among those offered by the theory of perfect fluids, the stream velocity V in the neighborhood of the body is not constant. In the case of the symmetric continuous flow around the circle one has $V = 0$ at the stagnation point A; then follow increasing values of V up to a maximum, reached halfway, and finally decreasing values of V toward the rear. According to Bernoulli's equation the pressure decreases while the velocity is increasing and, conversely, increases again toward the rear where the velocity is decreasing. The velocity profiles corresponding to the points B to H of Fig. 51, computed on the basis of this pressure distribution according to Prandtl's laminar boundary-layer theory, show that the fluid in the boundary layer gradually slows down between E and H until, at H, negative velocities are obtained in the proximity of the wall. This means that somewhere in front of H the stream will detach itself from the surface of the cylinder. The point at which this happens is the *separation point G*. Beyond this separation

FIG. 52.—Stream pattern in the neighborhood of a separation point.

point the velocity in the immediate neighborhood of the wall is negative, as in the flow shown in Fig. 52.

Thus it might seem as if the boundary-layer theory might furnish a solution of the separation problem, *i.e.*, answer the question under what conditions and, in a given case, at what point separation sets in. But, so far, the expectations in this respect have not been fulfilled. Whether a laminar or a turbulent boundary layer is assumed, in both cases the location of a separation point can be found only when the pressure distribution along the body, at least up to the critical point, is known. But how can one determine this pressure distribution? It was seen in the preceding section that the perfect-fluid theory supplies various stream patterns for the flow around a circular cylinder, a continuous one and discontinuous patterns with various separation points, etc. In the original boundary-layer theory of Prandtl the pressure values corresponding to the irrotational continuous flow were used. But the result obtained in this way for the location of the separation point did not agree with the experimental evidence. A satisfactory agreement can be reached only if the pressure values that enter into the boundary layer computation as given quantities are taken from actual measurements. Although this is certainly a confirmation of the underlying differential equations, it cannot be considered as an answer to the real question.

If the hydraulic hypothesis is applied to this problem, one may perhaps expect a solution on the following lines: The perfect fluid flow with separation at a certain point S has a pressure distribution that, when introduced in the boundary-layer theory, will in general lead to a boundary layer that separates from the body at a different point S'. It may happen that for one definite choice of S the point S' computed in this way will coincide with S. This would then be the theoretically determined location of the separation point. It may even be that two different solutions of this kind present themselves according to whether the laminar or the turbulent boundary-layer theory is used. The actual observations on circular cylinders suggest such a duplicity (see Sec. V.3).

The present situation is such that the theoretical determination of a separation point on the basis of the boundary-layer theory has not yet been carried out. The only cases in which the point of separation can be exactly predicted are those in which the perfect-fluid theory leads to a unique solution (Helmholtz flow) as in the case of a sharp or nearly sharp corner. For further remarks on this subject, see Secs. V.3 and X.1.

*Problem 10.   A plate extending from $x = 0$ to $x = l$ is submerged in a flow that comes from the left with the velocity V at $x = 0$ for all y.   Call v the x-component of the velocity at any point $0 \leq x \leq l$, $y \geq 0$.   Then, using the momentum equation and relation (1), show that the shearing stress at any point of the plate equals

$$\tau = \mu \left(\frac{\partial v}{\partial y} \right)_{y=0} = \rho \frac{\partial}{\partial x} \int_0^\infty v(V - v) \, dy$$

If v is assumed in the form $v = V \times f(\eta)$ where $\eta = y/h$ and h depends on x only (i.e., all velocity profiles equal except for the scale of y), prove that the value of h is determined by

$$h^2 = \frac{2f'(0)}{\kappa} \frac{\nu x}{V} \text{ with } \kappa = \int_0^\infty f(1 - f) \, d\eta$$

For example, with $f = \eta$ for $\eta \leq 1$ and $f = 1$ for $\eta \geq 1$, the boundary-layer thickness h would become

$$h = \sqrt{12 \frac{\nu x}{V}} = 3.46 \sqrt{\frac{\nu x}{V}}$$

almost coinciding with Eq. (7).

CHAPTER V

AIR RESISTANCE OR PARASITE DRAG

1. Definitions. Consider a body with two planes·of symmetry orthogonal to each other. The line of intersection of these planes is called an axis of symmetry. If the body moves in the direction of this axis through a fluid at rest, the normal and shearing stresses which the fluid exerts on its surface will combine to a resultant force of which the line of action coincides with the axis of symmetry. This force, the direction of which will be opposite to that of the velocity of the body, is called *air resistance* or parasite drag; we shall denote it by D_p. In Sec. 6 of this chapter the notion of the parasite drag will be extended in a certain way.

According to Eq. (5), Chap. IV, the parasite drag can be written as

$$D_p = C_p \frac{\rho}{2} V^2 A \tag{1}$$

where the *coefficient of parasite drag* C_p depends on the Reynolds number and on the shape of the body but is independent of its size. Unless explicitly stated otherwise, we use as the area of reference A the area of the projection of the body on a plane perpendicular to the axis of symmetry. This area is called the *frontal area* of the body; in most cases it coincides with its greatest cross-sectional area. The reference length used in forming Re will be indicated in each particular case.

In presenting the results of drag measurements it is often useful to compare the values of drag found for bodies of various shapes with the drag of a body of some standard shape, *e.g.*, a circular disk. Let C_p' be the drag coefficient corresponding to this standard shape and to a specified value of Re. The parasite drag for some other body then can be written as

$$D_p = C_p \frac{\rho}{2} V^2 A = C_p' \frac{\rho}{2} V^2 \frac{C_p}{C_p'} A = C_p' \frac{\rho}{2} V^2 A' \tag{2}$$

where $A' = A C_p / C_p'$ is called the *equivalent frontal area*. It equals the frontal area of a body of the chosen standard shape that has the same resistance as the body under consideration.

In order to simplify the formula for D_p, we may think of an unspecified standard shape for which C_p' would have the value 1. Formula (2) then becomes

$$D_p = \frac{\rho}{2} V^2 A_p \tag{3}$$

where $A_p = C_pA$ is called the *parasite area* of the body whose frontal area is A. For a circular disk moving perpendicular to its plane the parasite-drag coefficient does not differ appreciably from 1; the parasite area therefore is approximately the area of an equivalent circular disk.

The coefficient of parasite drag C_p can be given the following physical interpretation: Introducing the dynamic pressure $q = \frac{1}{2}\rho V^2$, we write (1) as $D_p = qC_pA$. Hence

$$C_p = \frac{D_p}{qA} \tag{4}$$

Now, for a disk moving in a direction perpendicular to its plane, the parasite drag is due to the fact that the pressure on the front side is higher than the pressure on the rear side. The ratio $p' = D_p/A$ gives the average overpressure on the front side. Accordingly, the coefficient $C_p = D_p/qA = p'/q$ is the ratio of the average overpressure p' to the dynamic pressure q. The same interpretation applies to all bluff bodies, where the drag is largely due to the excess of pressure on the front side. For slender streamlined bodies, however, the shearing stresses transmitted by the fluid are largely responsible for the drag; here the interpretation loses its physical significance.

In the following sections experimental results concerning the parasite drag of various types of body will be discussed. The information regarding the values of C_p in dependence of the value of Re and the shape of the body will be useful in the computation of airplane performance (Chaps. XIV and XVI). Moreover, a comprehensive knowledge of the aerodynamic phenomena connected with air resistance is fundamental in all branches of flight theory.

2. Bluff Bodies. The term bluff body will be used to denote sharp-edged bodies opposing a plane or flat surface to the stream. A typical representation of this group is the *circular disk* put at a right angle to the stream. Here the discontinuous flow pattern of the type discussed in Sec. IV.4 will take place. Behind the disk a broad wake is observed (Fig. 48). The coefficient of parasite drag *is found to be almost independent of the Reynolds number*, at least within the range covered by experiments, *i.e.*, up to $Re = 5,000,000$, referring to the diameter. The first significant measurements were made by G. Eiffel at the beginning of this century. Experimenting with disks of various diameters he found an average value of $C_p = 1.07$. More recent experiments give the average value $C_p = 1.11$, which is widely adopted today.

In the case of a plane plate of any form the drag is entirely due to the excess of pressure on the front side. The physical interpretation of C_p as the ratio of the average overpressure on the front side to the dynamic pressure is therefore applicable. Now, at the stagnation point on the front side, the pressure is $p_s = p_0 + q$, where p_0 denotes the pressure in the undisturbed stream at a great distance in front of the plate, influence of gravity being neglected. According to Bernoulli's equation, the pressure will be smaller than p_s at any other point on the front side. If the fluid in the wake were entirely at rest, a condition that would correspond to the behavior of a perfect fluid, there would be a constant pressure throughout the wake. Behind the body the fluid outside the wake reassumes the velocity of the undisturbed stream and therefore the pressure the value p_0. Consequently, if the pressure in the wake is supposed to be constant, it can have only the value p_0. The excess of pressure on the front side of the plate would have the maximum value $p_s - p_0 = q$ at the stagnation point and smaller values at all other points. It follows that for the dis-

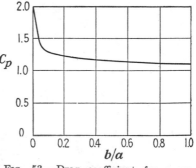

continuous flow of a perfect fluid around a plane plate the average over-pressure must be smaller than q and the coefficient of parasite drag accordingly smaller than 1. The exact theoretical value of C_p for a circular disk has not yet been determined; it would probably lie between 0.8 and 0.9. The difference between this and the observed value is due to the fact that in a real fluid the pressure in the wake is smaller than p_0, equaling about

Fig. 53.—Drag coefficient for a rectangular plate.

$p_0 - 0.3q$. This difference in pressure is due to the fact that the fluid in the wake is not at rest but in a vigorous eddying motion. On the front side of the disk the observed pressure distribution is in good agreement with that supplied by the theory of a perfect fluid.

For a *rectangular plate* the value of C_p will depend on the ratio of the longer side a to the shorter side b of the rectangle. For the square ($a/b = 1$) the drag coefficient C_p is about the same as for a circle, viz., $C_p = 1.05$ for $Re = 100,000$ to $C_p = 1.27$ for $Re = 700,000$ (according to the results of G. Eiffel). For a very slender rectangle ($a/b \to \infty$) the drag coefficient tends toward 2. Results obtained by K. Wieselsberger[1]) are represented in Fig. 53, where C_p is plotted vs. b/a. In those experiments the Reynolds number referred to a was about 50,000 to 150,000.

Eiffel also performed experiments with two parallel circular disks of equal diameter, put perpendicular to the line joining their centers (Fig. 54). If the distance l of the disks is small, the total drag is found

[1] Göttinger Ergebnisse, **2**, 33 (1923).

to be smaller than that of a single disk of the same diameter d. It seems that the presence of the second disk reduces the amount of eddies formed in the wake and thus decreases the resistance. For $l/d = 0$ the pair acts as a single circular disk, giving $C_p = 1.11$. With increasing l/d the coefficient C_p first decreases. For $l/d = 1$, C_p is about 0.93. The minimum value $C_p = 0.78$ is reached for $l/d = 1.5$. Upon increasing l/d further, higher values of C_p, for example, $C_p = 1.04$ for $l/d = 2$ and $C_p = 1.52$ for $l/d = 3$, are found. For $l/d \to \infty$ the drag of the pair of disks must tend toward twice the drag of a single disk, thus the drag coefficient toward $C_p = 2.22$.

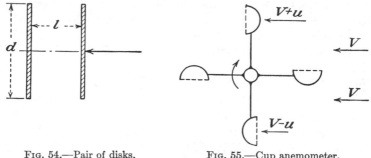

FIG. 54.—Pair of disks.   FIG. 55.—Cup anemometer.

A closed *hemisphere* with the plane side opposed to the stream has a drag coefficient of about 1.2, not much different from that of a disk. An open hemispherical cup turned toward the wind with the concave side has a drag coefficient of about 1.35 to 1.40. The slightly smaller value of $C_p = 1.33$ is used for *parachutes*. A hemispherical cup with the concave side at the rear behaves similar to a sphere (see the following section), the drag coefficient being about $C_p = 0.30$ to 0.40. NACA tests with anemometer cups[1] (Fig. 55) gave $C_p = 1.39$ in the first case and $C_p = 0.28$ in the second.

For a *circular cylinder* moving in the direction of its axis the drag coefficient depends on the ratio of length l and diameter d. For $l/d = 0$ it has the same value as the circular disk, $C_p = 1.11$. With increasing l/d the values of C_p first decrease, for example, $C_p = 0.91$ for $l/d = 1$. The minimum value $C_p = 0.82$ is reached for about $l/d = 2.5$. A further increase of l/d brings an increase in C_p, probably due to the shearing stresses on the surface of the cylinder. For $l/d = 7$, Eiffel found $C_p = 0.97$.

Problem 1. Assume that a parachute reaches its terminal velocity, where the air resistance is balanced by the weight, near sea level. If the gross weight (parachute + load) equals 250 lb., how large must be the frontal area if the terminal velocity is supposed to correspond to the velocity of free fall from 10 ft. altitude?

[1] *NACA Tech. Rept.* 513 (1935).

Problem 2. What influence has a temperature change of 10°F. on the limiting velocity of a parachute at sea level?

*Problem 3. How can the rotational velocity u of an anemometer be computed for a given stream velocity V, if it is assumed that the two forces acting on two opposite cups, one at the convex and the other at the concave side, are in equilibrium (see Fig. 55)?

3. Round Bodies. The term round body is used to denote cylinders, spheres, ellipsoids, and similar bodies. For bluff bodies the drag coefficient C_p was seen to be almost independent of Re. The early experiments of G. Eiffel and others seemed to indicate that this was true for round

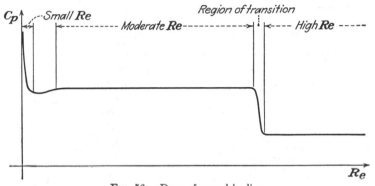

Fig. 56.—Drag of round bodies.

bodies, also. Later experiments apparently confirmed this independence of C_p but furnished much higher numerical values. The reason for this discrepancy was understood when Prandtl extended his experiments over a larger range of Reynolds numbers. It was then seen that there exists a comparatively narrow range of Reynolds numbers, within which C_p drops from a higher value valid for *moderate Reynolds numbers* to a smaller value valid for *high Reynolds numbers*. We thus have, for all round bodies, two regions of almost constant C_p, separated by a narrow region of transition as indicated schematically in Fig. 56. (For rather small values of Re that are not encountered in aeronautical problems but are important in certain problems of physics, C_p increases rapidly with decreasing Re. In the following this region of small Reynolds numbers will be disregarded.) The behavior of the drag coefficient suggests that two different stream patterns exist which occur below and above the region of transition, and probably some intermediate patterns in the region of transition. As the location of this region is determined by the value of Re, the viscosity of the fluid must play a decisive role in the phenomenon. According to recent investigations the flow in the boundary layer is laminar for moderate values of Re

and turbulent for high values of Re. The wake is broader in the first case than in the second, thus giving rise to a greater drag.

In the case of a *sphere*, if the diameter is used as the reference length, the range of moderate Reynolds numbers extends from $Re \sim 20,000$ to $Re \sim 200,000$, the corresponding value of the coefficient of parasite drag being $C_p \sim 0.5$. The range of high Reynolds numbers begins with $Re \sim 300,000$. The corresponding smaller value of C_p was determined as 0.2 by earlier investigators. Today a value of about $C_p = 0.1$ is believed to be more correct. The reason for this great difference was found in the fact that very small, apparently insignificant modifications of the suspension of the sphere have a decisive influence. It may still be

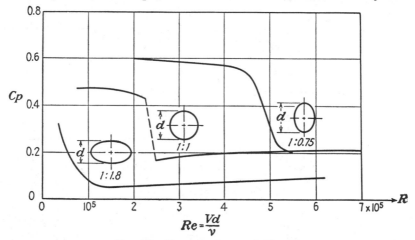

Fig. 57.—Drag of sphere and ellipsoid.

advisable not to rely too much on the smaller value, since practical conditions may correspond rather to the conditions under which $C_p = 0.2$ has been found. Eiffel's figures as well as those of the NACA are close to this value.

Experimental results for the drag coefficient of a sphere and two types of *ellipsoid* with axial symmetry are shown in Fig. 57. The Reynolds number used as abscissa refers to the diameter and the drag coefficient to the area of the maximum circular cross section. The almost sudden change of the regime is seen in all three cases.

Similar conditions prevail in the case of a *circular cylinder moving perpendicular to its axis*. Here the C_p vs. Re-curve depends on the ratio of the length l to the diameter d. For values of l/d between 5 and 30 the coefficient C_p, referred to the diameter, was found to be about 0.8 to 1.2 for moderate Reynolds numbers below about 500,000 and between 0.36 and 0.42 for high Reynolds numbers. With increasing l/d the region of transition moves in the direction of increasing Reynolds numbers. According to an investigation of F. Eisner,[1] for $l/d = 2.25$ this region extends approximately from $Re = 100,000$ to $Re = 200,000$; for $l/d = 4.5$, from $Re = 150,000$ to $Re = 300,000$.

It has been mentioned before that the wake is broader for moderate Reynolds numbers than for high Reynolds numbers. In the case of a circular cylinder it has been observed that the separation points of these two stream patterns correspond approximately to the two limiting positions found in the theory of perfect fluids, *i.e.*, to a wake extending over an arc of 250° in the first and of 120° in the second case (see Fig. 58). In the region of transition between these two flow patterns intermediate positions of the separation points may occur. In front of the cylinder the pressure distribution corresponds largely to that given by the theory of perfect fluids. In the rear of the cylinder the theoretical pressure in the wake equals the pressure p_0 in the undisturbed stream. Actually the pressure in the rear is smaller than p_0, because of the violent motion in the wake, and a comparatively large part of the drag is due to this

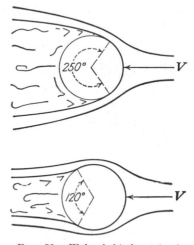

FIG. 58.—Wake behind a circular cylinder for moderate and large Reynolds numbers.

underpressure. As in the flow pattern corresponding to moderate Reynolds numbers the wake is broader, it is understandable that here larger C_p values occur than for high Reynolds numbers, where a flow pattern with a narrower wake takes place. As already mentioned, the transition from the first flow pattern to the second has been brought in relation to the fact that the boundary layer becomes turbulent at higher Re. Direct observations of the velocity distribution near the cylinder wall seem to confirm this assumption.

Problem 4. Two spheres of diameters 3.8 and 7.8 in., respectively, are moving with a velocity of 100 ft./sec. under standard conditions at sea level. Assuming $C_p = 0.5$ for $Re = 200,000$ and $C_p = 0.1$ for $Re = 400,000$, determine which of the spheres encounters the greater air resistance.

Problem 5. Determine the air resistance per foot of length of a taut wire of $\frac{1}{8}$ in. diameter, moving perpendicularly to its axis with a velocity of 250 m.p.h. under standard conditions, at sea level and at 30,000 ft. altitude.

[1] EISNER, F., "Widerstandsmessungen an umströmten Zylindern," p. 13, Berlin, 1929.

4. Streamlined Bodies. In the cases of bluff and round bodies the air resistance is largely due to the underpressure in the wake. F. W. Lanchester was the first to draw attention to this fact (1907). He introduced the concept of the *streamlined body*. A body is called streamlined if the flow past it is free (or practically free) of discontinuity surfaces separating a dead-air region from the stream. Of practical importance are mainly two types of streamlined body, cylinders moving perpendicular

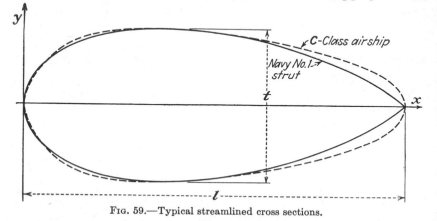

Fig. 59.—Typical streamlined cross sections.

to their axis and bodies of revolution proceeding in the direction of the axis. The cylindrical form is used for struts, the other for airships, airplane fuselages, and sometimes for shells. The usual sections of both forms (normal to the axis in the case of the cylinder and through the axis in the case of the body of revolution) are of a very similar shape (see Fig. 59). The sections possess one axis of symmetry, which has the direction of the undisturbed stream; the front part is thicker than the rear part, which is more or less pointed. The ratio of the length l to the greatest thickness t is called the *fineness ratio*. Usual values of this ratio vary from 2 to 6. For well-designed forms the drag coefficient may be as low as 0.06 for the cylinder and 0.04 for the body of revolution.

Table 5 gives the so-called "offsets" of two typical forms of streamlined bodies, the Navy Strut No. 1 and the C-Class Airship (Fig. 59).

For strut sections, fineness ratios ranging from about 3.0 to 3.5 give the smallest values of the coefficient of parasite drag. For bodies of revolution, the corresponding value of the fineness is about 2.1. This means that, for a set of bodies of revolution possessing the same maximum thickness t and the same offsets $2y/t$, the body with $l = 2.1\,t$ will have the least resistance. The problem of determining the fineness ratio of a streamlined body of least resistance for a given thickness t occurs in the design of nacelles, where t is prescribed by the dimensions of the engine or other parts to be accommodated in the nacelle. In airship design, however, the problem is rather that of determining the fineness of a streamlined body of least resistance for a given volume v. Now, for given offsets, the volume v is proportional to lt^2 or to t^3f, where $f = l/t$ denotes the fineness. The frontal area A is proportional to t^2 or to

TABLE 5.—OFFSETS FOR NAVY STRUT NO. 1 AND C-CLASS AIRSHIP
(For notations see Fig. 59)

x/l	$2y/t$		x/l	$2y/t$	
	Strut	Airship		Strut	Airship
0	0	0			
0.0125	0.260	0.200	0.35	1.000	0.999
0.0250	0.371	0.335	0.40	0.995	0.990
0.0500	0.525	0.526	0.50	0.950	0.950
0.0750	0.636	0.658	0.60	0.861	0.885
0.100	0.720	0.758	0.70	0.732	0.790
0.125	0.785	0.835	0.80	0.562	0.665
0.150	0.836	0.887	0.90	0.338	0.493
0.200	0.911	0.947	0.95	0.190	0.362
0.250	0.959	0.982	0.98	0.078	0.225
0.300	0.988	0.998	1.00	0	0

$v/f^{\frac{2}{3}}$. For given dynamic pressure q and given volume v the drag $D = C_p \, qA$ therefore is proportional to $C_p/f^{\frac{2}{3}}$. In Fig. 60, curve A shows the dependence of average C_p-values on the fineness f, and curve B represents the expression $C_p/f^{\frac{2}{3}}$. It is seen that for the curve A the

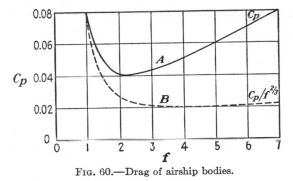

FIG. 60.—Drag of airship bodies.

minimum lies at about $f = 2.1$, while that of B occurs at about $f = 4.5$ and is less pronounced than that of A. This means that the streamlined body of a given volume and least resistance should have a fineness ratio of about 4.5 and that even considerably more fineness, for example, 6 or 7, will not appreciably increase the resistance per unit of volume.

A similar problem arises in the determination of the best value of the fineness ratio of streamlined wires. Since a wire of circular cross section (circular cylinder of great length) in most practical cases would still be within the range of moderate Reynolds numbers and thus would encounter the high resistance corresponding to $C_p = 1.2$, the use of special wire with streamlined section is advisable. Here the problem is to determine the fineness ratio of a wire of smallest resistance for a prescribed tensile strength, *i.e.*, for a given cross-sectional area. For given offsets, the cross-sectional area A' is proportional to lt or to $t^2 f$, where f again denotes the fineness. The frontal area A

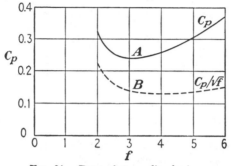

Fig. 61.—Drag of streamlined wire.

of a definite length of wire is proportional to t or to $A'/\sqrt{f}$. For a given dynamic pressure q and given cross-sectional area the drag $D = C_p q A$ is therefore proportional to $C_p/\sqrt{f}$. In Fig. 61 curve A represents values of C_p and curve B the corresponding values of $C_p/\sqrt{f}$. It is seen that the minimum of the curve B is situated at about $f = 4$. This fineness ratio actually is adopted for streamlined wires. However, the experimental basis for this computation is far from satisfactory. The observed values scatter considerably; they depend on surface properties and are very sensitive to the exact attitude of the axis of the cross section.

For well-streamlined strut sections of a fineness 3 to 4 the coefficient of parasite drag C_p can be taken as about 0.08 at a Reynolds number, based on thickness, in the range of about 100,000 to 200,000. Streamlined wires, of much less perfect form and working under smaller Reynolds numbers, practically have C_p values up to 0.2 and 0.3 and more.

The smallest values of the coefficient of parasite drag so far observed are those of well-designed bodies of revolution. For a fineness ratio of 2.0, values as low as $C_p = 0.04$ have been found. This means that the parasite area amounts to only 4 per cent of the greatest cross-sectional area. The value 0.04 applies to big rigid dirigibles the form of which can be made to correspond perfectly to the offset values of the design.

The case of airplane fuselages will be discussed in Sec. 6 of this chapter. While the drag of bluff and round bodies is almost entirely due to the excess of pressure on the front side, an essential part of the drag of stream-lined bodies originates in the shearing stresses that the fluid exerts on the surface of the body. The consequence is that the drag increases more slowly than with the second power of the velocity. In other words, the drag coefficient is not independent of the Reynolds number, but slightly decreasing. A usual assumption, based on various experi-ments, is that C_p decreases proportional to $(Re)^{-n}$ with $n = 0.1$ to 0.15. This means a drop of C_p of 20 to 30 per cent when Re is increased ten times.

Problem 6. A straight strut of 12 ft. length moves perpendicular to its axis at 300 ft./sec. near sea level. If a circular cross section of 4 in. diameter is used in one case and a streamlined section of 3.2 in. maximum thickness in another case, what is the difference in drag and in the amount of power required?

Problem 7. Determine the best dimensions of an airship body of 150,000 ft.^{3}, and compute the actual value of the drag and the power required for a cruising velocity of 110 ft./sec. at an altitude of 8000 ft.

*Problem 8. A small bomb of 6 lb. weight and 6.8 in.^{2} cross section has a parasite-drag coefficient $C_p = 0.38$. At each moment the difference between air resistance and drag determines the vertical acceleration. Compute the velocity of the bomb when it strikes sea level after being released at the altitude $h_1 = 10,000$ ft. Assume that the relation between density ratio σ and altitude h can be approximated by the hyperbolic law

$$\sigma = \frac{\sigma_1 h_1}{h(1 - \sigma_1) + \sigma_1 h_1}$$

where σ_1 is the σ-value at $h = h_1$.

5. Skin Friction. In the case of a flat plate moving in its own plane the drag is entirely due to the shearing stresses that the fluid exerts on the surface of the plate. A drag of this nature is called *skin friction*. The amount of skin friction developed on both sides of the plate can be determined by means of the boundary-layer theory discussed in Sec. IV.5.

Let us consider a rectangular plate of length l (in the direction of the flow) and breadth b and suppose that the relative motion of the fluid with respect to the plate is two-dimensional, the plane of flow being perpendicular to the plate. According to Prandtl's theory of the laminar boundary layer, as discussed in Sec. IV.5, the velocity distribution in the neighborhood of the plate can be computed on the basis of the Navier-Stokes equations. From this velocity distribution the derivative dV/dy of the velocity V in the direction normal to the plate is determined along the surface of the plate, and the shearing stresses τ that the fluid exerts on the plate are then found from $\tau = \mu(dV/dy)_{y=0}$. Finally, the drag of the plate is obtained by integrating these shearing stresses. The result of the computation, which cannot be reproduced here, is

$$D_p = 2\mu \int_0^l \left(\frac{dV}{dy}\right)_{y=0} dx = 1.33b \sqrt{\mu\rho l V^3} \tag{5}$$

In this formula the factor 2 in front of the integral sign is necessary because the fluid exerts the shearing stresses $\tau = \mu(dV/dy)_{y=0}$ on both sides of the plate. For a suitable definition of the coefficient of parasite drag C_p, the area of the "wetted surface" of the plate, $A = 2bl$, can be used as reference area. The drag coefficient then is obtained as

$$C_p = \frac{2D_p}{\rho V^2 A} = 1.33 \sqrt{\frac{\mu}{\rho V l}} = \frac{1.33}{\sqrt{Re}} \tag{6}$$

This result is found to agree well with the experimental evidence up to values of the Reynolds number of about 250,000 to 300,000.

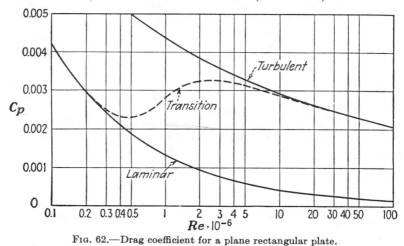

Fig. 62.—Drag coefficient for a plane rectangular plate.

At greater values of the Reynolds number, formula (6) fails because the boundary layer becomes turbulent. Kármán's theory mentioned in Sec. IV.5 furnishes the drag coefficient

$$C_p = \frac{2D_p}{\rho V^2 A} = \frac{0.072}{\sqrt[5]{Re}} \tag{7}$$

In a more recent discussion based on rather general laws of turbulence Kármán proposed a more exact relation between C_p and Re,

$$\frac{0.242}{\sqrt{C_p}} = \log_{10} (C_p Re) \tag{8}$$

For $Re = 4,000,000$, formula' (7) gives $C_p = 0.00344$; formula (8), $C_p = 0.00342$. These values are in close agreement with experimental results (Fig. 62). The phenomenon is further complicated by the fact that for a sufficiently long plate the boundary layer is laminar near the leading edge and only at some distance from the leading edge becomes turbulent. With increasing velocity of the free stream the point at which this transition from the laminar to the turbulent boundary layer takes place shifts toward the leading edge of the plate.

The formulas just given enable us to estimate the order of magnitude of the skin-friction coefficient for a wetted surface that behaves approximately as a flat plate, for example, a slender symmetrical profile or the outer surface of a fuselage (see the following section).

The skin friction obviously cannot be independent of the roughness of the wetted surface. At first glance, this statement seems to contradict the fact that for a given shape the drag coefficient depends only on the Reynolds number. But the roughness of the surface of a body has to be considered as a part of its shape. The law of similitude as developed in Sec. IV.2 states that, independent of the size of a body, the drag coefficient does not change as long as the value of the Reynolds number and the shape of the body remain exactly the same. This means that the *relative roughness* counts, *i.e.*, the ratio of the dimensions of the irregularities of the surface to the characteristic length of the body. The C_F-values already given correspond to practically smooth surfaces, where the relative roughness is negligible. Experience shows that in the case of a laminar boundary layer the drag coefficient is almost independent of the relative roughness. With a turbulent boundary layer the drag coefficient increases with the relative roughness.

Problem 9. A plane plate of length l and of 5 ft. width across the stream is subject to a uniform flow at a speed of 120 ft./sec. in the direction of l. Decide to what length l the boundary layer can be considered to be laminar. Compute the drag in this case. Assume air under standard conditions.

Problem 10. The side walls of an airplane fuselage have an average height of 8 ft. and a length of 40 ft. Compute the skin friction and the amount of power required at velocities of 200 and 300 ft./sec., respectively, at altitudes of 10,000 and 30,000 ft.

*Problem 11. Prove that under the assumptions of Prob. 10, Chap. IV, the drag coefficient of a plate immersed in uniform stream equals

$$C_F = 2 \sqrt{\frac{2\kappa f'(0)}{Re}}$$

Show that on the assumption $f = \eta$ the result agrees very well with the value observed in laminar flow but that an indefinitely small change in f, for example,

$$f = \eta \text{ for } \eta < 0.99; \qquad f = 1 - 0.01 \left(\frac{0.99}{\eta}\right)^m \text{ for } \eta > 0.99$$

may increase C_p up to ∞.

6. Parasite Drag of Major Airplane Components. With a view to practical airplane design the notion of parasite drag can be extended in

the following way.   A body may not fulfill exactly the symmetry conditions stated at the beginning of Sec. 1 of this chapter but still may
experience an air reaction similar to an air resistance, or drag, when
moving with uniform velocity in a rectilinear path through a bulk of
air at rest.   That is, the normal and shearing stresses that the air exerts
on the surface of such a body may be equivalent to a resultant force
practically parallel to the velocity vector.   This force is then called
the parasite drag of the body under consideration, and the notions of
parasite-drag coefficient and parasite area are applied in the way explained
in Sec. 1.   The designer is interested in getting information that enables
him to estimate the parasite drag of the complete airplane.

The most conspicuous part of an airplane—except for wings and
propellers, which will be considered separately—is the *fuselage*.   The

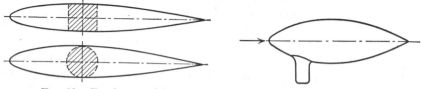

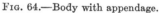

FIG. 63.—Fuselage models.   FIG. 64.—Body with appendage.

parasite-drag coefficient of an average fuselage moving parallel to its
longitudinal axis is generally greater than that of a streamlined body of
revolution of similar shape because of the inevitable deviations from the
ideal streamlined form.   For large planes this discrepancy is less significant than for small planes.

Figure 63 shows the longitudinal sections of two ideal airplane-body
models, one with circular and one with square cross section.   Both had
the same frontal area A of about 21 sq. in. and were tested[1] at a Reynolds
number, based on $\sqrt{A}$, of about 150,000.   The parasite-drag coefficient
was found to lie between 0.045 and 0.055, the larger values for the square
section.   Since the drag is caused to a high extent by skin friction, a
favorable scale effect can be expected, i.e., a decrease of C_p with increasing Re.   Some experimental investigations lead to the conclusion[2] that
C_p reduces to 90, 80, 70 per cent of the model value when the Reynolds
number increases to 2.5, 10, 40 times the figure just given.   This is
about the same as to say that C_p is proportional to the $-\frac{1}{10}$ power of Re.

The actual fuselage of an airplane will have a much higher drag
coefficient because of its less favorable form and various added elements,
appendages, and protuberances.   If to a well-streamlined body some
appendage is added, say, a small cylinder as shown in Fig. 64, *the resultant*

[1] *Göttinger Ergebnisse* **2**, 68 (1923).
[2] *NACA Tech. Rept.* 236 (1926).

drag is considerably larger than the sum of the two drag forces that are found when each part is tested independently. The surplus depends on the location of the disturbing part and reaches its maximum when the disturbance is close aft the maximum cross section of the main body. The drag contribution due to the interference is on the average about 30 per cent and in the most unfavorable position more than 50 per cent of the drag that the small body experiences when tested separately. Thus, with $C_p = 0.04$ for the streamlined body and $C_p = 0.4$ for the cylinder, the drag of the fuselage would be doubled by adding a cylinder with a frontal area of one-fifteenth the cross-sectional area of the fuselage.

Instead of summing up the contributions of all protuberances, etc., one can estimate the total drag acting on an average airplane fuselage on the basis of general experience. For a carefully streamlined body, without windshield or uncovered cockpit, with no major protruding parts, a C_p-value between 0.09 and 0.12 seems adequate. The coefficient may increase to 0.15 to 0.18 for smaller planes with a frontal area down to 15 sq. ft. Still larger values, up to 0.30, result if one departs considerably from the streamlined form for the purpose of accommodating the engine or of improving the sight for the pilot. From a number of statistical data compiled by the California Aeronautical Laboratory the following indications can be drawn:[1] $C_p = 0.070$ to 0.105 for large transports and bombers, no nose engine or turrets; $C_p = 0.090$ to 0.130 for small planes, including nose engine and closed cockpit. These figures correspond to Reynolds numbers of one to two millions, based on the chord length of the wing. In all this argument it is supposed that the airplane moves parallel to its longitudinal axis. The forces present when the direction of motion deviates from this axis will be discussed in Sec. XVIII.3 and later, in connection with stability problems.

The body or hull of *flying boats* must be adapted to certain hydrodynamic conditions which restrict the possibility of streamlining. The best values found in wind-tunnel tests with boat models at Teddington Laboratory were $C_p = 0.11$ to 0.15.

Well-shaped *engine nacelles* when tested in free air show a parasite-drag coefficient of about 0.10 to 0.15. The actual drag acting on a nacelle mounted on a wing depends to a great extent on its location. It goes up to twice the free-air value in the case of a nacelle situated on the upper side of the wing near the leading edge and down to less than half that value in the most favorable position on the underside of the wing.

The skin friction of the *tail surfaces* (stabilizer, rudder, and elevator) can be fairly well estimated from Eq. (7). Under average conditions the Reynolds number will be found within the limits 10^5 to 10^6; this

[1] MILLIKAN, C. B., "*Aerodynamics of the Airplane*," p. 98, New York, 1941.

gives, according to (7), a parasite-drag coefficient (with respect to the surface area) of 0.0046 to 0.0072. However, to the skin friction must be added the air resistance due to the thickness of the actual profile and the drag exerted on the levers, struts, wires, etc., that are necessary for the operation. Since the size of these elements increases with the surface area, one may refer the total drag of the tail to this area. Then C_p-values of 0.011 to 0.022 may correspond to normal conditions.

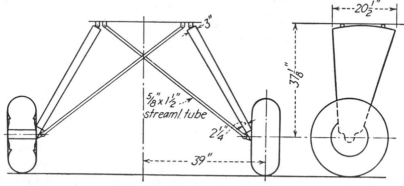

FIG. 65.—Landing gear.

A considerable contribution to the total parasite drag of a normal airplane is supplied by the *landing gear*. The drag coefficient for a wheel without fairing was found to be 0.70, with the usual fairing from hub to rim 0.43 and complete fairing from hub to edge of tread 0.23. The axle, struts, wires, shock absorber, and all joints must be counted, with their total projected area, *i.e.*, corresponding to $C_p = 1$. This would also include the effect of the mutual interference of these parts. Figure

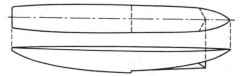

FIG. 66.—Model of a seaplane float.

65 shows a landing gear with low-pressure wheels and faired struts and wires, designed for an airplane of 3000 lb. weight. This undercarriage has a parasite area of 1.75 sq. ft. Under average conditions the total drag of the landing gear may amount to 15 to 20 per cent of the total drag of the complete plane (except the wing drag).[1] This is the reason why modern airplanes designed for high speed use retractable landing gear. The drag is then present only immediately after starting and during the

[1] *Cf.* WIESELSBERGER, C., in W. F. Durand, "Aerodynamic Theory," Vol. 4, p. 146.

landing maneuver.   In the latter instance an increase of air resistance is rather desirable.

In the case of seaplanes, *floats* take the place of the landing gear. The usual forms of float show a drag coefficient of 0.16 to 0.20.   This figure can probably be improved by better streamlining (Fig. 66).

All values given in this section must be considered as rough estimates only and may be used to determine the order of magnitude rather than the actual amount of air resistance.   In practical design work the chief reliance must be on experimental results originating either in wind-tunnel tests performed on models of the body shapes actually used or, better, in full-scale experiments with the bodies themselves.

Many problems connected with parasite drag and important in aircraft design work cannot be discussed here.   One of them concerns the aerodynamics of *cooling*.   All aircraft engines have arrangements for the cooling of their cylinders.   Whether air cooling or water cooling is applied, in either case the forced heat transfer from the engine to the surrounding air is achieved by exposing certain body surfaces to the air stream.   That means that a certain amount of drag is utilized for a useful purpose and thus must not be considered as "parasite."   To find the right balance between the thermal effect and the drag employed is a problem that could be solved only by combining aero- and thermodynamic theories.

Problem 12.   An airplane has a well-streamlined fuselage of circular cross section of diameter $d = 7$ ft.   Assume that the fuselage contributes 55 per cent to the total parasite area.   What is the amount of total parasite drag and power it requires to an altitude of 25,000 ft. and speed of 280 ft./sec.?

Part Two

THE AIRPLANE WING

CHAPTER VI

FUNDAMENTAL NOTIONS. GEOMETRY OF WINGS

1. The Three Coefficients. The wing of an airplane can roughly be described as a flat or slightly cambered plate, symmetric with respect to a median plane. As is indicated by the term "plate" one of the dimensions of the wing, the *thickness t*, is much smaller than the others. Moreover, in the wing plane, the dimension perpendicular to the median

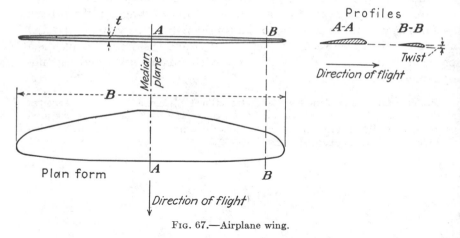

Fig. 67.—Airplane wing.

plane, the *span B*, is usually considerably greater than the width parallel to the median plane (Fig. 67). The contour of the plate can be considered as a plane curve, called the *plan-form* of the wing. In the early days of flying, forms imitating bird wings were frequently used. Later, a rectangular plan-form slightly rounded off at the wing tips was the rule. Today there is a marked preference for tapered plan-forms. The cross sections of the wing in planes parallel to the median plane are called *profiles*. The profiles of the wing of an airplane usually vary in shape, size, and orientation as one proceeds from the median plane toward the wing tips. For example, for a tapered wing (Fig. 67) the chord length of the profile decreases toward the tips, and the thickness is reduced at the same time. Furthermore, compared with the orientation of the central profile, a profile near the wing ends usually has its "nose" turned slightly downward. This is described as the *twist* of the wing. In most cases a certain profile shape is maintained throughout the span or, at least, throughout the major part. This is then called the *principal profile* of the wing. If the wing is not interrupted by the fuselage, the principal profile will be found as the shape of the cross section in the median plane.

As is clear from these remarks, the geometry of the wing depends on a considerable number of parameters, the influence of all of which on the aerodynamic characteristics of the wing must be studied. For conventional wing forms the influence of the shape of the principal profile is more

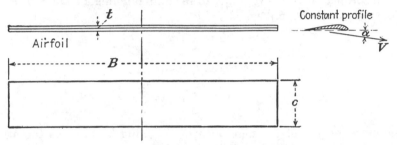

FIG. 68.—Airfoil.

important than that of most other parameters. In order to study this influence separately from that of the plan-form, twist, etc., a rectangular, untwisted wing with invariable profile is generally used, in experimental investigations. Such a wing is a prismatic body whose shape and size are completely determined by the shape of the profile and the two sides of the rectangular plan-form (Fig. 68). Throughout this book the term *airfoil*, often applied to any kind of wing, will be used exclusively to designate this prismatic wing of rectangular plan-form. One of the sides of the rectangle is the span B of the wing; the other, the *chord length* of the profile, will be denoted by c. The ratio B/c is called the *aspect ratio*. If $S = Bc$ denotes the area of the airfoil, the aspect ratio can be written in any one of the following forms:

$$\mathcal{R} = \frac{B}{c} = \frac{B^2}{S} = \frac{S}{c^2} \qquad (1)$$

Consider an airfoil moving through a bulk of air at rest with a constant velocity V parallel to the median plane of the airfoil. According to Sec. IV.2 the forces exerted by the air on the airfoil will then depend on its shape, size, and orientation, on the velocity of the motion, and on the density and viscosity of the air. The shape of the airfoil can be defined by giving the shape of the profile and the aspect ratio $\mathcal{R}$. The size can be defined by giving the chord length c. Finally, the orientation can be determined by giving the so-called "angle of attack," *i.e.*, the angle α between the velocity V and an arbitrarily chosen direction in the median plane of the airfoil.

The resultant force F that the air exerts on the wing will have its line of action situated in the median plane (Fig. 69). The magnitude, direction, and line of action of this force are completely determined when its components D and L, parallel and perpendicular to the velocity V, and its moment M with respect to an arbitrarily chosen point O in the median plane are known. The component L perpendicular to V is called the *lift*, the component D parallel to V the *drag*, and M the *pitching moment* of the wing.

According to Sec. IV.2, lift, drag, and pitching moment of a wing of any form can be written as

$$L = \frac{\rho}{2} V^2 S C_L$$

$$D = \frac{\rho}{2} V^2 S C_D$$

$$M = \frac{\rho}{2} V^2 c S C_M$$

where the area S of the plan-form of the wing is taken as the area of reference (in the case of an airfoil, $S = Bc$) and C_L, C_D, C_M are dimensionless coefficients, known as the coefficients of lift, drag, and moment; the chord length c is chosen as reference length for the moment coefficient. For an airfoil of given profile these coefficients can depend only on the Reynolds number Re, the aspect ratio R, and the angle of attack α (Sec. IV.2).

In wind-tunnel experiments airfoils of the aspect ratio $R = 6$ are generally used. For a given Reynolds number the three coefficients C_L, C_D, C_M are then found as functions of the angle of attack α. Experimental results will be discussed in Chaps. VII and X, while Chaps. VIII and IX will be devoted to the fundamentals of the aerodynamic wing theory, the principal aim of which is to predict the values of the three coefficients for a wing of given shape flying under given conditions.

Problem 1. The following values of the three coefficients were found for an airfoil with Clark Y section and aspect ratio 6:

α, deg.	C_L	C_D	C_M
0	0.384	0.0172	−0.070
3	0.602	0.0288	−0.073
6	0.819	0.0464	−0.079
9	1.034	0.0700	−0.054

The angle of attack refers to a middle line (chord) of the profile. The moment coefficient refers to a point on this straight line whose distance from the leading end is $c/4$ and is counted positive if it tends to increase the angle of attack. The chord length was 5 in., the air speed 110 m.p.h., the pressure 32 in. Hg, and the temperature 76°F. Compute L, D, and M, and find the line of action of the resulting force for each of the four α-values.

2. Geometry of Airfoil Profiles.   Sets of Profiles.   A common characteristic of all airfoil profiles is the pointed trailing end (B in Fig. 69). Some profiles have a *cusp* at the trailing end; *i.e.*, the upper and lower

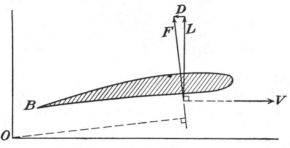

FIG. 69.

part of the contour have a common tangent there.   As this means that at some distance from the trailing end the profile is still very thin, profiles with a *corner* at the trailing end are preferred for structural reasons.   In this case, the upper and lower contour meet at the trailing end under a certain small angle.   Whether cusp or corner, the point B is a "singular point" in the sense of differential geometry.   It will be explained in Chap. VIII that a pointed trailing end (or at least a trailing end which is not rounded off too much) is essential in producing the lift.

The definition of the three coefficients C_L, C_D, C_M requires a precise definition of the chord length c, necessary to evaluate the airfoil area $S = Bc$, and the choice of a clearly defined reference line from which the angle of attack α is measured. The earlier profiles are rather thin and strongly cambered, such that one of the tangents of the lower

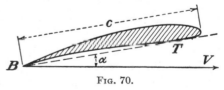

FIG. 70.

contour passes through the singular point B (BT in Fig. 70).   This tangent, called the *chord*, can then be used as the line of reference in measuring the angle of attack.   The chord length c can be defined as the distance of B from that tangent of the profile which is perpendicular to the chord.   These definitions obviously do not apply to more general profiles.   Definitions of chord and chord length valid for any profile would be useful, but so far no general definitions have been adopted. One possible procedure would be to use the circle C with the center at the singular point B and tangent to the nose of the profile (Fig. 71). Its radius would be the chord length, and the line joining the point of contact to the singular point B would be used as reference line for measuring the angle of attack.   This definition would fail only if the circle C has more than one point of contact with the profile, a case that is very unusual.

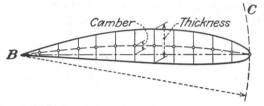

Fig. 71.—Camber and thickness of a profile. The mean camber line is the locus of the centers of the straight lines perpendicular to the chord.

Other notions that should be defined in a way valid for any profile are the *thickness* and the *camber*.   One way of doing this is to consider the segments cut out by the upper and lower contour of the profile on straight lines perpendicular to the chord.   The longest of these segments then gives the thickness of the profile, and the locus of their centers is defined as the *mean camber line* (Fig. 71).   Finally, the greatest distance of the mean camber line from the chord would give the camber of the

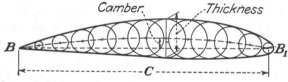

Fig. 72.—Camber and thickness of a profile. The mean camber line is the locus of the centers of circles inscribed to the profile.

profile.   Another possibility is to define the mean camber line as the locus of the centers of the circles inscribed to the profile.   This mean camber line is well defined through the main part of the profile; it starts at B and, in general, ends at the center B_1 of a circle that has a four-point contact with the nose of the profile (Fig. 72).   The diameter of the largest circle among those defining the mean camber line can then be defined as the thickness of the profile, and the greatest distance between the line BB_1 and a point of the mean camber line as the camber of the profile.   If these definitions were adopted, it would seem advisable to define the length $\overline{BB_1}$ as chord length and to use the line BB_1 as line of reference for measuring the angle of attack.

All these difficulties encountered in defining the various concepts related to the geometry of profiles are avoided when the profile is described not only by its graph or by the coordinates of a certain number of its points but by precise instructions that make it possible to determine any desired number of points, once the values of certain parameters like chord, camber, or thickness are given. This is the case for the profiles of certain sets any particular profile of which is completely defined by a small number of constants. As an example we shall discuss two such sets that have been developed by the NACA and are known as the *four-digit series* and the *five-digit series*.

The starting point for the construction of these profile series was the observation that for not too thick profiles the form of the mean camber

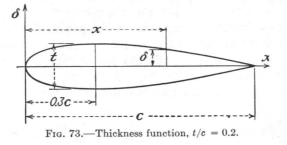

Fig. 73.—Thickness function, $t/c = 0.2$.

line and the manner in which the thickness of the profile varies along the chord affect the aerodynamic characteristics of the profile almost independently of one another. The thickness variation can be explained most easily for a symmetrical profile, where the mean camber line is straight (Fig. 73). Based on two successful profiles, the Clark Y and the Göttingen 398, the NACA recommends the following *thickness function:* Denoting the chord length by c and the (maximum) thickness of the profile by t, the ordinates δ of the contour of the "normal" symmetrical profile are given by[1]

$$\delta = \pm t \left[1.4845 \sqrt{\frac{x}{c}} - 0.6300 \frac{x}{c} - 1.7580 \left(\frac{x}{c}\right)^2 + 1.4215 \left(\frac{x}{c}\right)^3 - 0.5075 \left(\frac{x}{c}\right)^4 \right] \quad (2)$$

At $x = 0$ the derivative $d\delta/dx$ is infinite, corresponding to the leading end. For $x = c$, the value of δ is $\pm 0.0105t$, which means a practically sharp trailing end. For $x = 0.3c$, the derivative $d\delta/dx$ vanishes, and δ assumes its maximum value $0.5t$. The radius of curvature at the leading edge $x = 0$ is found to equal $1.10 \, t^2/c$.

The values of δ/c for $t = 0.2c$ are given in Table 6.

[1] See *NACA Tech. Rept.* 460 (1933).

TABLE 6.—NACA THICKNESS FUNCTION

$\frac{x}{c}$	0	.0125	.025	.050	.075	.100	.15	.20	.30
$\frac{\delta}{c}$	0	.0316	.0436	.0592	.0700	.0780	.0891	.0956	.1000
$\frac{x}{c}$	.40	.50	.60	.70	.80	.90	.95	1.000	
$\frac{\delta}{c}$	.0967	.0882	.0761	.0611	.0437	.0241	.0134	.0021	

Any profile of the so-called *four-digit series* is completely defined by four digits, the first two digits determining the form of the mean camber line, the last two the relative thickness of the profile, *i.e.*, the thickness t expressed in per cent of the chord length c. The mean camber line of this family consists of two parabolic arcs with a common vertex A whose axes are parallel to the y-axis (Fig. 74). If the coordinates of the point A are denoted by x_A and y_A, respectively, the equations of these parabolic arcs are

$$y = \frac{y_A}{x_A^2} x(2x_A - x) \qquad \text{for } 0 \leqq x \leqq x_A$$

$$y = \frac{y_A}{(c - x_A)^2} (c - x)(c + x - 2x_A) \qquad \text{for } x_A \leqq x \leqq c$$

(3)

The first digit of the profile number indicates the maximum camber y_A in per cent of the chord c or $100y_A/c$, which is supposed to be a single-

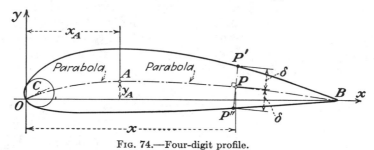

FIG. 74.—Four-digit profile.

digit integer. The second digit is the abscissa of the point A in tenths of the chord length, that is, $10x_A/c$, also supposed to be a single-digit integer. The last two digits give the relative thickness of the airfoil in per cent, $100t/c$. This is supposed to be an integer of not more than two digits and is always written with two digits. Thus the profile number 2409 indicates a profile with $y_A = 0.02c$, $x_A = 0.4c$, and $t = 0.09c$.

Given the chord length c and the thickness t of a profile, the value of δ corresponding to any abscissa x can be computed from (2). On the mean camber line given by (3), locate the point P of the abscissa x, and determine on the normal of the mean camber line in P the profile points P' and P'' in such a manner that $\overline{PP'} = \overline{PP''} = \delta$ (Fig. 74). The drawing of the contour is simplified by the fact that the radius of curvature at the origin O of the coordinates is independent of the camber. The center of curvature C lies on the tangent of the mean camber line at O, and the radius of curvature $\overline{OC}$ equals $1.10t^2/c$. Figure 75 gives a small selection of profiles of the four-digit series.

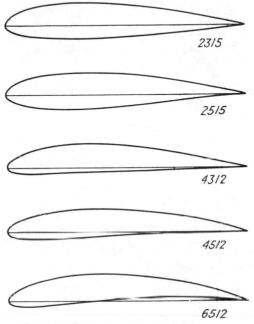

FIG. 75.—Examples of the four-digit series.

The two parabolic arcs of which the mean camber line of any profile of the four-digit series consists have a common tangent in A but different radii of curvature. Now it is a matter of general experience that the aerodynamic performance of an airfoil may be impaired even by very small irregularities of its shape. It seems advisable, therefore, to avoid the sudden change of the curvature at A. This condition is fulfilled for the airfoils of the *five-digit series* of the NACA.[1]

In the five-digit series the same thickness function as in the four-digit series is used. The relative thickness $100t/c$ is expressed by the last two digits of the profile number in the same way as before. The mean camber line consists either of an arc of a cubic parabola and a portion of a straight line or of two cubic parabolas. In the first case, the mean camber line is "simple," in the second S-shaped, or "reflexed."

[1] *NACA Tech. Rept.* **537** (1935), **610** (1937).

In either case, the nose and the tail portion of the mean camber line are fitted together at a point where both have the curvature zero. This point B lies slightly behind the position A of the maximum camber

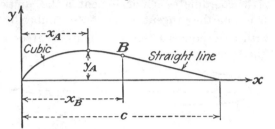

Fig. 76.—Mean camber line of the simple five-digit profiles.

(Fig. 76). The first two digits of the profile number again define the maximum camber y_A and its position x_A, but the correspondence between the values of digits and those of y_A/c, x_A/c is now different and is given

TABLE 7.—CAMBER OF THE NACA FIVE-DIGIT PROFILES

Second digit	1	2	3	4	5
$100x_A/c$	5	10	15	20	25
$100y_A/c$	1.1	1.5	1.8	2.1	2.3 for first digit = 2
$100y_A/c$		2.3	2.8	3.1	for first digit = 3
$100y_A/c$		3.1	3.7	4.2	for first digit = 4
$100y_A/c$		4.6	5.5	6.2	for first digit = 6

in Table 7. The third digit is 0 for the simple and 1 for the reflexed mean camber lines. Figure 77 shows some of the five-digit airfoils.[1]

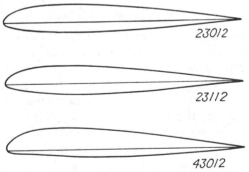

Fig. 77.—Examples of the five-digit series.

Another way of avoiding the sudden change of curvature of the mean camber line of the four-digit airfoils has recently been suggested.[2]

[1] According to *NACA Rept.* 610 (1937).

[2] MUNK, M., On the Geometry of Streamlining, "Th. v. Kármán Anniversary Volume," p. 8, 1941.

It consists in using a single conic arc as mean camber line. In order to obtain a shape similar to that of a given four-digit profile the tangents of this conic arc at the leading and trailing end of the chord OB (Fig. 74) are made to coincide with the tangents of the given mean camber line at O and B, and the maximum cambers of both are made equal. If the maximum camber of the four-digit profile lies at 30 per cent of the chord, that of the modified profile will lie at 40 per cent.

Problem 2. Draw the NACA 2312 profile, choosing the chord length as 10 in., and compare it with that five-digit profile which comes closest to it.

*Problem 3. Compute the jump of curvature on the upper contour of the profile 4210 that is due to the transition between the two parabolas of the mean camber line. How large is the sudden change of centripetal acceleration for an air particle moving at the speed of 300 m.p.h.?

3. Theoretically Developed Airfoil Sections. The two-dimensional irrotational flow around a profile of given shape can be determined theoretically by means of a powerful mathematical method known as the method of *conformal mapping*. From the knowledge of the flow pattern, lift and moment coefficients can then be found. A discussion of this method is beyond the scope of this book, but the main results of the theory will be derived in an elementary way in Chap. VIII. The objective of the present section is a different one.

In airfoil theory as well as in many other branches of mathematical physics, a very successful procedure is to study the so-called "inverse problem." In the present case the direct problem would be to find the flow around a given profile from the hydrodynamic equations. The inverse problem means that a flow determined by some simple formula is assumed and then the profile sought for which this flow is the solution satisfying the hydrodynamic equations. Some sets of profiles that have been developed on this basis will be discussed here.

Such a theoretically developed profile presents the following advantages: (1) The contour of the profile is not fitted together artificially from arcs of various curves but is given by a single analytical expression with continuous derivatives of any order or by the equivalent prescription for its geometrical construction. The slope of the tangent, the curvature, and all derivatives of the curvature with respect to the arc length of the contour thus vary in a continuous manner at all points of the contour (except, of course, at the trailing end). (2) For these profiles theoretical values of the coefficients C_L and C_M for infinite aspect ratio are known in advance; they agree, for moderate angles of attack, with the values obtained by experiment and can be used in a way that will be explained later in the case of finite aspect ratio, too. (3) It is even possible to construct profiles according to given aerodynamic specifications. For example, one can develop a profile for which the line of action of the resultant air force passes through one and the same point for all (moderate) angles of attack.

The simplest set of such theoretical profiles is due to N. Joukowski.[1] This set is important particularly from the historical point of view since it was the starting point of the theory. The *Joukowski profiles* are obtained in the following way: Consider two circles C_1 and C_2 that are tangent to one another at the point B' situated on the negative x-axis (Fig. 78) and have centers M_1 and M_2 such that the y-axis bisects the angle M_1OM_2. Let P_1 be an arbitrary point of C_1 and P_2 a correspond-

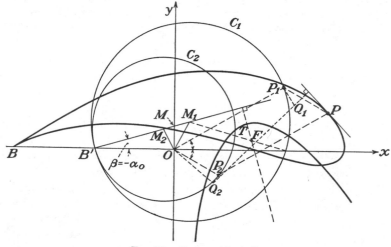

Fig. 78.—Joukowski profile.

ing point on C_2 such that the x-axis bisects the angle P_1OP_2. Determine the point P so that the vector $\overrightarrow{OP}$ is the sum of the vectors $\overrightarrow{OP_1}$ and $\overrightarrow{OP_2}$. As P_1 describes the circle C_1 and, simultaneously, P_2 the circle C_2 in the opposite sense, the point P will describe a Joukowski profile. In particular, when P_1 and P_2 coincide at B', the point P takes the position B on the x-axis such that $\overline{OB} = 2\overline{OB'}$. The point B is a cusp of the profile; the angle that the common tangent of the upper and lower parts of the contour at B forms with the x-axis is twice as big as the angle $M_1B'O$. The tangent of the contour at any position of P can be found in the following way: Determine the point Q_1 and Q_2 so that $P_1Q_1 \perp OP_1$, $P_2Q_2 \perp OQ_2$ and $OQ_1 \parallel M_1P_1$, $OQ_2 \parallel M_2P_2$; then Q_1Q_2 is perpendicular to the tangent of the contour at P.

The shape of the profile obviously depends on two parameters, the angle M_1OM_2 and the angle $OB'M_1$. The first of these parameters determines the relative thickness, the second the relative mean camber

---

[1] *Z. Flugtech. u. Motorluftschiffahrt*, 1910, p. 281; 1912, p. 81.

of the profile. In particular, if the angle M_1OM_2 is zero, M_1 and M_2 coincide on the y-axis (Fig. 79) and the profile has zero thickness and consists of a circular arc through B, having its center on the axis of y.

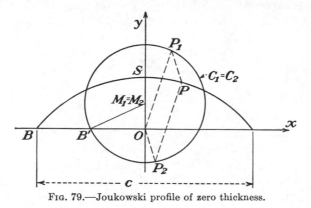

FIG. 79.—Joukowski profile of zero thickness.

This circular arc meets the y-axis at a point S such that $\overline{OS} = 2\overline{OM}_1$. The relative mean camber of this profile is

$$\frac{\overline{OS}}{2\overline{OB}} = \frac{\overline{OM}_1}{\overline{OB}} = \frac{\overline{OM}_1}{2\overline{OB'}} = \tfrac{1}{2}\tan\beta,$$

where β is the angle $OB'M_1$. If, on the other hand, the angle $OB'M_1$ is zero, M_1 and M_2 lie on the x-axis so that O is the center of $\overline{M_1M_2}$ and

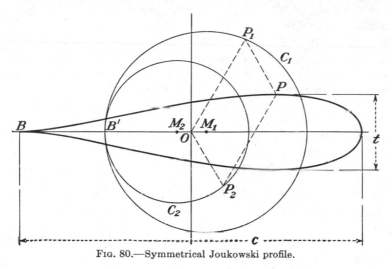

FIG. 80.—Symmetrical Joukowski profile.

the profile has zero camber and is symmetrical with respect to the x-axis (Fig. 80). For moderately thick, symmetrical Joukowski profiles, $\overline{OM}_1$

is small compared with $\overline{OB'}$. The relative thickness t/c of the profile is then found to be approximately 1.3 $\overline{OM_1}/\overline{OB'}$.

The theory of airfoils with infinite aspect ratio furnishes the following aerodynamic characteristics of the Joukowski profiles: If the airfoil moves parallel to $B'M_1$ through a bulk of air at rest, the lift is zero. The line $B'M_1$, called the *first axis* of the profile, thus indicates the so-called "zero lift direction" (see Sec. VII.1). If the direction of flight makes the angle α' with the zero lift direction, the lift coefficient is

$$C_L = 8\pi \frac{a}{c} \sin \alpha' \qquad (4)$$

where a denotes the radius $B'M_1$ and c the chord length to which C_L refers. Furthermore, the *aerodynamic center* (a.c.) F (see Secs. VII.1 and VIII.3) lies on the line joining M_1 to the point of intersection of the circle C_1 and the positive x-axis. If $\overline{B'O}$ is denoted by b, the distance $\overline{M_1F}$ equals b^2/a (Fig. 78). The coefficient of the moment with respect to the point F is

$$C_{M_0} = -4\pi \frac{b^2}{c^2} \sin 2\beta \qquad (5)$$

where β is again the angle $OB'M_1$. This moment is independent of the angle of incidence α' and is considered positive if acting counterclockwise. The vertex T of the *metacentric parabola* (see Sec. VIII.4) lies on the perpendicular drawn from F to the first axis $B'M$ and halves the distance of F from $B'M_1$. The tangent of the metacentric parabola at T is parallel to the first axis.

The Joukowski profile is not suitable for practical applications, for several reasons. First the leading portion is too massive in comparison with the thin tail, and second the maximum camber lies rather close to the center of the chord, whereas a position within the forward third of the chord is preferred. To avoid these disadvantages, more general prescriptions for theoretical profiles have been sought. Such prescriptions are most easily formulated if use is made of a particular vector notation known as the *complex number notation of vectors*.

The position of a point P with respect to a rectangular system of coordinates O, x, y can be described by giving either the vector $\overrightarrow{OP}$ or the length $\overline{OP}$ and the angle that OP makes with the x-axis. It is convenient to denote the vector $\overrightarrow{OP}$ by a single symbol z, adopting the following two rules: (1) If z_1 and z_2 denote the vectors $\overrightarrow{OP_1}$ and $\overrightarrow{OP_2}$, then $z_1 + z_2$ represents the vector $\overrightarrow{OP}$, which is commonly called the sum of the vectors $\overrightarrow{OP}_1$ and $\overrightarrow{OP}_2$, that is, the point P is the fourth corner of the parallelogram determined by the points O, P_1, P_2 (Fig. 81).   (2) Departing from the customary vector notation, the product $z = z_1 z_2$ will be used to denote the vector $\overrightarrow{OQ}$, where $\overline{OQ} = \overline{OP}_1 \times \overline{OP}_2$ and $\angle QOx = \angle P_1 Ox + \angle P_2 Ox$ (Fig. 82).   The length of the vector z thus equals the product of the lengths of the vectors z_1 and z_2.   The so-called

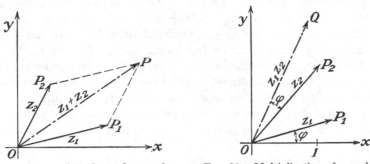

Fig. 81.—Summation of complex numbers.	Fig. 82.—Multiplication of complex. numbers.

"argument of z," that is, the angle that the vector z forms with the x-axis, equals the sum of the arguments of z_1 and z_2.   It can easily be seen that the rules (1) and (2) are identical with the rules of addition and multiplication of complex numbers, if a complex number $x + iy$ is represented by a vector z with the components x and y.   Consequently, within this system of notation, ordinary (real) numbers are represented by vectors directed along the positive or negative x-axis, according to their sign.

The following examples will make the reader familiar with the complex number notation of vectors.   Let z denote the vector $\overrightarrow{OP}$ of the length $\overline{OP} = r$ and the argument $\angle POx = \varphi$

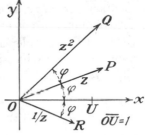

Fig. 83.—Square and reciprocal value.

(Fig. 83).   Then z^2 denotes the vector $\overrightarrow{OQ}$ of the length r^2 and the argument 2φ. $1/z$ accordingly

denotes a vector $\overrightarrow{OR}$ such that the product of z and $1/z$ represents the unit vector $\overrightarrow{OU}$ directed along the positive x-axis.   The vector $1/z$ therefore has the length $1/r$ and the argument $-\varphi$.   Generally, the vector z^k, where k is a real number, has the length r^k and the argument $k\varphi$.   The vector az, where a is a real number, has the length ar and the argument φ.   It is easily seen that the familiar algebraic rules concerning the combination of addition and multiplication, arranging and reducing algebraic expressions, can be applied to such vector formulas.

In airfoil theory the following use is made of this notation: Let C_1 be a circle in the x-y-plane, and denote by z_1 the vector from O to an arbitrary point P_1 of the circle C_1. Then, by a certain algebraic formula, a vector $\overrightarrow{OP} = z$ may be made to correspond to the vector z_1. When the end point P_1 of z_1 describes the circle C_1, the end point P of z will describe a closed curve C. By an appropriate choice of the formula connecting z with z_1, the curve C can be given the shape of a suitable profile. This seemingly artificial procedure has the following advantage: The same equation between z and z_1 that relates the circle C_1 to the contour of the profile relates the well-known streamlines of the flow around the circle to the streamlines of the flow around the profile. In addition, the lift and moment coefficients for any angle of attack and any point of reference are determined in a simple way by the radius and center of C_1 and the parameters appearing in the transformation formula.

A Joukowski profile can be defined by such a formula in the following way: In Fig. 78, denote the length $\overline{OB'}$ by b. Then b^2 represents a vector of the length $\overline{OB'^2}$ directed along the positive x-axis. According to the description given in connection with Fig. 78 the vector $\overrightarrow{OP} = z$ is the sum of the vectors $\overrightarrow{OP_1} = z_1$ and $\overrightarrow{OP_2} = z_2$. Furthermore, the arguments of the vectors z_1 and z_2 are equal and opposite, so that the product $z_1 z_2$ will be directed along the positive x-axis. Now, by elementary geometry, it can be shown that for such two circles as C_1 and C_2 (common tangent in B', $\angle M_1 O y = \angle M_2 O y$) the product of the lengths $\overline{OP_1}$ and $\overline{OP_2}$ is independent of the position of P_1 on C_1. The constant value of this product is found when the point P_1 is made to coincide with B'. Since P_1 and P_2 then coincide, $\overline{OP_1} \times \overline{OP_2} = \overline{OB'^2} = b^2$.

For an arbitrary position of P_1 on C_1, the vector $\overrightarrow{OP_2}$, therefore, is represented by b^2/z_1 and the formula defining the Joukowski profile is

$$z = z_1 + \frac{b^2}{z_1} \tag{6}$$

Here b^2 is a positive real number; the end point P_1 of the vector z_1 must be made to describe a circle C_1, having its center M_1 to the right of the y-axis and passing through the point B' with the coordinates $-b$, 0.

In order to avoid the cusp at the trailing end, Kármán and E. Trefftz[1] have proposed the following formula, to be applied to a circle C_1 of the

[1] *Ibid.*, **9**, 111 (1918).

type indicated above:

$$z = nb \frac{(z_1 + b)^n + (z_1 - b)^n}{(z_1 + b)^n - (z_1 - b)^n} \tag{7}$$

Here n is a positive real number, usually only slightly smaller than 2. The trailing edge B of the profile corresponds again to the point B'; with $z_1 = -b$ formula (7) furnishes $z = -nb$ for the trailing edge B.

For $n = 2$ formula (7) coincides with the Joukowski formula (6), since the numerator of the fraction in (7) becomes $2(z^2 + b^2)$ and the

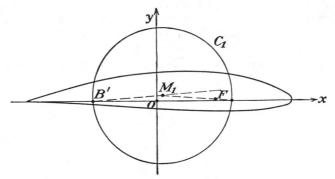

FIG. 84.—Kármán-Trefftz profile.

denominator $4bz$. If n is smaller than 2, say, $n = 2 - (\epsilon/\pi)$, the upper and lower parts of the contour meet at the trailing edge under the angle ϵ, the so-called "tail angle." Figure 84 shows a Kármán-Trefftz profile with $n = 1.9$; the tail angle ϵ equals $\pi/10$ rad. or 18°. This profile has been traced by applying formula (7) to a number of points on the circle C_1 whose center M_1 has the coordinates $x = y = 0.0707b$. The computations are rather cumbersome and are best arranged in form of a table giving the lengths and arguments of z_1, $z_1 + b$, $z_1 - b$, $(z_1 + b)^n$, etc.

The aerodynamic characteristics of the Kármán-Trefftz profiles are as follows: The zero lift direction (first axis) is given by the line $B'M_1$, joining the center of the circle M_1 to the point B', which is transformed into the trailing edge of the profile. The lift coefficient corresponding to the angle of incidence α' is again given by formula (4). The aerodynamic center F lies on the line joining the center M_1 to the point of intersection of the circle C_1 and the positive x-axis so that

$$\overline{M_1F} = \tfrac{1}{3} \frac{(n^2 - 1)b^2}{a}$$

The coefficient of the moment with respect to the aerodynamic center F is

$$C_{M_0} = -4\pi \frac{n^2 - 1}{3} \frac{b^2}{c^2} \sin 2\beta \tag{8}$$

where β is the angle $OB'M_1$.

The Kármán-Trefftz formula (7) is only a slight modification of the Joukowski formula (6), avoiding the inconveniently thin tail end of the Joukowski profiles. A much more general formula furnishing practically all kinds of profiles was proposed by R. v. Mises (1920).[1] This formula is

$$z = z_1 + \frac{a_1}{z_1} + \frac{a_2}{z_1^2} + \cdots + \frac{a_n}{z_1^n} \tag{9}$$

where in most cases the number n of the terms following z_1 need not be taken greater than 2 or 3. Here again, the point z_1 describes a circle C_1 passing through the point $B'(-b,0)$ and containing the origin O in its interior. The constants $a_1, a_2, \ldots, a_n$ in (9) cannot be taken quite arbitrarily, but must fulfill the condition that the $n+1$ roots of the equation $dz/dz_1 = 0$, or

$$z_1^{n+1} - a_1 z_1^{n-1} - 2a_2 z_1^{n-2} - \cdots - na_n = 0 \tag{10}$$

represent n points inside the circle C_1 and the point B'. In order to obtain such values of the coefficients, one may choose, inside the circle C_1, n points $U_1, U_2, \ldots, U_n$ represented by $z_1 = u_1, z_1 = u_2, \ldots, z_1 = u_n$ and write Eq. (10) as

$$(z_1 + b)(z_1 - u_1)(z_1 - u_2) \cdots (z_1 - u_n) = 0 \tag{11}$$

Considering that according to (10) the coefficient of z_1^n must vanish, we see that $u_1, u_2, \ldots, u_n$ are subject to the condition

$$u_1 + u_2 + \cdots + u_n = b \tag{12}$$

This means that one has to choose the points $U_1, U_2, \ldots, U_n$, inside the circle C_1, such that the sum of the vectors $\overrightarrow{OU}_1, \overrightarrow{OU}_2, \ldots, \overrightarrow{OU}_n$ is equal and opposite to the vector $\overrightarrow{OB'}$ or that O is the centroid of the points $B', U_1, U_2, \ldots, U_n$. Working out (11) for $n = 3$ and comparing the coefficients of the various powers of z_1 with those of (10), we obtain

$$a_1 = b(u_1 + u_2 + u_3) - u_1 u_2 - u_2 u_3 - u_3 u_1 = b^2 - u_1 u_2 - u_2 u_3 - u_3 u_1$$
$$a_2 = \tfrac{1}{2}[u_1 u_2 u_3 - b(u_1 u_2 + u_2 u_3 + u_3 u_1)] \tag{13}$$
$$a_3 = \tfrac{1}{3} b u_1 u_2 u_3$$

In this way the coefficients in Eq. (9) can be determined. From this equation successive points z of the profile corresponding to the points z_1 on the circle C_1 are found.

The aerodynamic characteristics of the Mises airfoils are the following: The zero lift direction is again given by the line $B'M_1$, joining the

[1] *Ibid.*, **8**, 1, 68 (1917); **11**, 68, 87 (1920).

center of the circle M_1 to the point B' where this circle intersects with the negative x-axis. The lift coefficient corresponding to the angle of attack is given by formula (4). The vector $\overrightarrow{M_1F}$ from the center of the circle C_1 to the aerodynamic center F is found in the following way: Applying the rules of complex vector notation, divide the vector a_1 by the vector $\overrightarrow{B'M_1}$. The vector represented by this quotient is parallel and equal to the vector $\overrightarrow{M_1F}$. The coefficient of the moment with respect to the aerodynamic center F is

$$C_{M_0} = -4\pi \frac{b^2}{c^2} \sin 2 \, (\beta - \gamma) \tag{14}$$

where β denotes the angle $OB'M_1$ and 2γ is the argument of a_1. The line through M_1 that forms the angle γ with the x-axis is called the *second axis* of the profile.

The Joukowski profiles are included in the class of profiles defined by (9). Comparison of (6) and (9) shows that, for a Joukowski profile, $n = 1$ and $a_1 = b^2$. Equation (10) accordingly takes the form

$$z_1^2 - b^2 = 0$$

There is only one point U, and $u_1 = b$. On the other hand, practically any given profile can be represented by (9) to a sufficient degree of approximation if the number of terms on the right side of (9) is sufficiently increased. The rules already given for the determination of the aerodynamic characteristics hold good for any number of terms.

It is possible to choose a profile so as to fulfill the condition $C_{M_0} = 0$, the importance of which will be explained later in Sec. VIII.4. All that is necessary is to assume the points $U_1, U_2, \ldots, U_n$ so that the argument of a_1 equals 2β, the double of the angle $OB'M_1$. To do this, choose the point B' on the negative x-axis and some points $U_1, U_2, \ldots, U_n$, compute the coefficient a_1 according to (13), and determine the argument 2γ of a_1. Through B' then draw a straight line that forms the angle γ with the x-axis, and assume on this line the point M_1 so that the circle with the center M_1 and the radius M_1B' contains all points $U_1, U_2, \ldots, U_n$ in its interior. For the profiles thus obtained the line of action of the resultant air force passes through the point F for all (moderate) angles of attack.

Profiles of this type are shown in Figs. 85 and 86. The transformation formula (9) for the first profile is:

$$z = z_1 + \frac{b^2}{z_1} (1 + \kappa i) + \frac{\kappa b^3 i}{2z_1^2} \tag{15}$$

with $\kappa = 0.4$. Here $n = 2$, and the coefficients are $a_1 = b^2(1 + \kappa i)$,

$a_2 = \kappa b^3 i/2$. The expression (10), that is, the derivative dz/dz_1 multiplied by z_1^3, is

$$z_1^3 - b^2 z_1(1 + \kappa i) - \kappa b^3 i = (z_1 + b)(z_1^2 - b z_1 - \kappa b^2 i)$$

Thus the roots, except for $z_1 = -b$, are $b(1 \pm \sqrt{1 + 4\kappa i})/2$, and this leads to

$$u_1 = (\ \ 1.10 + 0.33i)b \quad \text{or} \quad \overline{OU}_1 = 1.15b, \quad \angle U_1 Ox = 17°$$
$$u_2 = (-0.10 - 0.33i)b \quad \text{or} \quad \overline{OU}_2 = 0.35b, \quad \angle U_2 Ox = 253°$$

From $a_1 = b^2(1 + 0.4i)$ we have $2\gamma = \arg a_1 = \text{arc tan } 0.4 = 21.8°$.

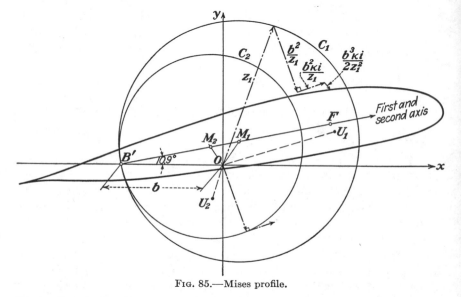

FIG. 85.—Mises profile.

Hence, to obtain $C_{M_0} = 0$, the point M_1 must be so chosen that $\angle M_1 B'O = 10.9°$.

The radius $\overline{M_1 B'}$ must be so large as to make the circle C_1 enclose the points U_1 and U_2. This is fulfilled if we set $\overline{M_1 B'} = 1.18b$. The circle C_2, obtained in exactly the same way as in Fig. 78, serves to construct the "Joukowski term" b^2/z_1. To construct the term $b^2 \kappa i/z_1$, one has only to turn the vector b^2/z_1 through 90° and to change its length in the ratio κ.

Figure 86 shows a profile constructed according to the transformation formula

$$z = z_1 + \frac{b^2}{z_1}(1 + \mu^2 i) - \frac{b^4 \mu^2 i}{3z_1^3} \tag{16}$$

for $\mu = 0.4$ (see Prob. 7).

Starting with one definite profile of the general theoretical type and gradually changing the parameter values in an appropriate way, one can derive sets of profiles with their camber, thickness, etc., varying according to practical requirements. In such sets, which would replace the rather artificially constructed profiles of the four- and five-digit series, the discontinuities mentioned in the foregoing would be avoided and a sound basis for comparison of experimental and theoretical coefficient values obtained.

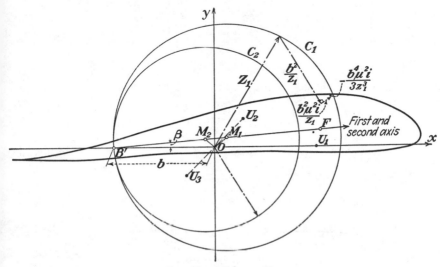

Fig. 86.—Mises profile.

Problem 4. Draw the Joukowski profile corresponding to a circle C_1 with coordinates of center M_1: $x = 0.15b$, $y = 0.10b$.

Problem 5. Determine the aerodynamic center, the moment coefficient with respect to it, and the metacentric parabola for the circular arc profile of Fig. 79.

Problem 6. Draw the Kármán-Trefftz profile for $n = 1.94$ corresponding to the same circle as the Joukowski profile in Prob. 4, and determine the aerodynamic characteristics.

Problem 7. Determine the roots of Eq. (10) for the profile of Fig. 86, based on Eq. (16). Compute the coordinates of the center M_1 for $\overline{B'M_1} = 1.12b$ under the condition of vanishing C_{M_0}, and find the distance $\overline{M_1F}$ of the aerodynamic center. How does the position of F vary with the value of μ while b and $\overline{B'M_1}$ remain constant?

Problem 8. Draw the Mises profile with $n = 2$ and the following values of u_1, u_2:

$$u_1 = b(1.01 + 0.16i) \qquad u_2 = b(-0.01 - 0.16i)$$

The coordinates of the center M_1 are $x = y = 0.10b$. Find F and C_{M_0}.

Problem 9. Draw the Mises profile with $n = 2$, the vector u_1 having the length $b/2$ and forming the angle $240°$ with the x-axis. Determine u_2 from (12), and set up the equation for the transformation. Find out a suitable circle to which to apply this transformation.

4. Geometry of Airplane Wings. While the shape of an airfoil is completely determined by the profile and the aspect ratio, the shape of an actual wing cannot be described so easily. By definition, the plan-form of an airfoil is a rectangle; the plan-form of actual wings is never strictly rectangular. Even if the chord length is constant over almost the entire

Fig. 87.—Dihedral angle.

span, the wing tips will be rounded off. Modern designers prefer tapered plan-forms for which the chord length gradually decreases toward the wing tips. As the chord length varies, the profile, too, must vary. Geometrically, the easiest way would be to conserve the shape of the profile by letting all its dimensions decrease in the same proportion as the chord length, but this is very rarely feasible. Another important

Fig. 88.—Plan-form of the Taube.

difference between the shape of an airfoil and an actual wing is introduced by the necessity of providing smooth transitions between the wing surface and such interruptions as the fuselage or engine nacelles. Finally, the port and starboard halves of the wing may be set so that the tips are raised with respect to the roots. Seen in the direction of flight, such a

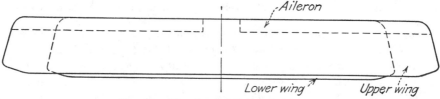

Fig. 89.—Plan-form of Bréguet's biplane.

wing appears as a flat V. The angle with which each limb of the V rises from the horizontal is called the *dihedral angle* (Fig. 87); its influence on the lateral stability of the plane is considerable (see Chap. XX). The following is a short review of some aspects of a complete description of the shape of a wing.

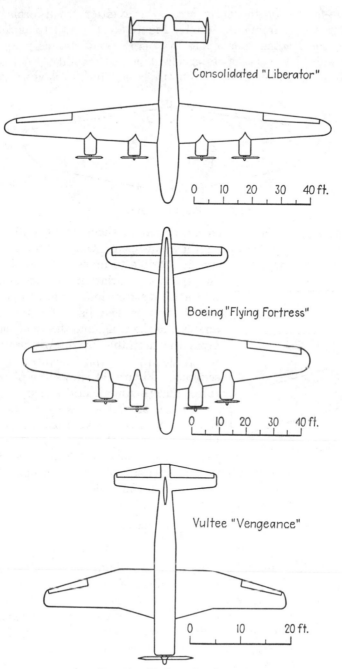

Consolidated "Liberator"

0   10   20   30   40 ft.

Boeing "Flying Fortress"

0   10   20   30   40 ft.

Vultee "Vengeance"

0   10   20 ft.

FIG. 90.—Modern plan-forms.

Plan-form. In the early stages of aviation, plan-forms imitating bird wings were frequently used. Figure 88 shows the plan-form of the Taube, the leading monoplane in Austria and Germany up to 1914, highly valued because of its excellent lateral stability. A less fanciful French design of the same period (Bréguet, biplane) is given in Fig. 89.

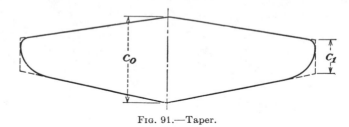

FIG. 91.—Taper.

At that time there was no aerodynamic theory that could give any guidance in the selection of the plan-form. Only later did Prandtl's wing theory (Chap. IX) make it possible to investigate the influence of

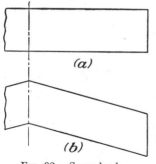

(a)

(b)

FIG. 92.—Sweepback.

the spanwise variation of the chord length on the lift distribution. Knowledge of the lift distribution is the basis for the structural analysis of the wing, and this analysis, in turn, gives the desirable height of the wing spars, which determines the thickness of the wing sections. Figure 90 shows the plan-forms of three modern American designs.

Theory as well as experimental evidence shows that for the conventional plan-forms and for moderate angles of attack the spanwise variation of the chord length has but little influence on the aerodynamic characteristics of a wing as long as the ratio of the span to the average chord is not modified. It is usual to define the *aspect ratio of a wing of arbitrary plan form* by

$$\mathcal{R} = \frac{B^2}{S} \qquad (17)$$

where B is the span and S the area of the wing. This is a generalization of formula (1) defining the aspect ratio of an airfoil. From the aerodynamic point of view the aspect ratio should be made as great as possible, but obvious limits are set by structural considerations.

The plan-form of most modern designs consists essentially of two equal trapezoids (Fig. 91); *i.e.*, the chord length c varies linearly with the spanwise coordinate. The ratio c_0/c_1 is known as the taper ratio of the trapezoidal wing. Taper ratios up to 2.5 are not uncommon. The plan-form, however, is by no means completely determined by the span-wise distribution of the chord length. For example, the rectangular plan-form of Fig. 92*a* and the *sweepback* form of Fig. 92*b* have constant chord lengths. For the trapezoidal form of Fig. 91 the amount of sweep-back is usually specified in the following manner: With the exception

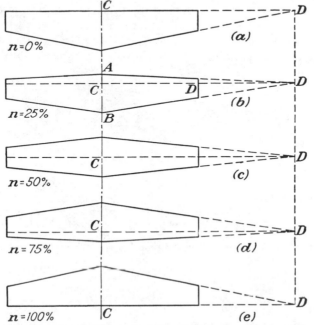

Fig. 93.—Plan-forms with constant taper ratio and varying sweepback.

of the wing of the taper ratio 1 (wing of constant chord length) there exists, for any trapezoidal wing, one straight line CD (Fig. 93) that is perpendicular to the median plane and divides all chords in the same ratio. The value of the ratio $\overline{AC}/\overline{AB}$ (see Fig. 93*b*) together with that of the taper ratio completely defines the trapezoid. With

$$100\,\frac{\overline{AC}}{\overline{AB}} = n$$

the line CD is called the n per cent line and the wing is described as a wing with straight n per cent line. Figure 93 shows trapezoidal wings with taper ratio 2.0 and varying sweepback. The main aerodynamical influence of sweepback is to shift the line of action of the lift backward. The sweepback is thus connected with the longitudinal stability of the plane (see Chap. XVII).

Profile Variation. In order to obtain a smooth flow at the wing tips, it seems desirable to let the thickness of the profiles decrease toward the wing tips. Since for the aerodynamically satisfactory profiles the relative thickness (*i.e.*, the ratio of thickness and chord length) varies within a narrow range only, the decrease of thickness toward the wing tips implies a decrease in chord length. This is one of the reasons that recommend tapered plan-forms. The relative thickness and consequently the shape of the profile can then be kept constant over the central portion of the span. Near the wing tips, structural considerations often enforce a variation of the relative thickness.

When the wing sections and their relative attitudes at two or more stations along the span have been chosen according to aerodynamic

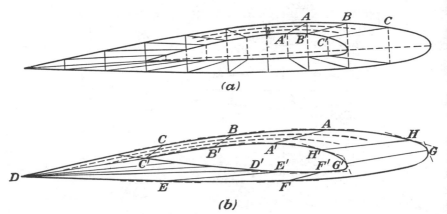

Fig. 94.—Insertion of intermediate profiles.

and structural considerations, intermediate sections must be determined so as to provide for a satisfactorily smooth surface. If it is desired to insert n intermediate profiles between two given ones (Fig. 94), corresponding points of the two profiles may be connected by straight lines AA', BB', . . . and the distances $\overline{AA'}\ \overline{BB'}$, . . . divided into $n + 1$ equal parts. The correspondence between the points of two profiles can be established by taking points with parallel tangents as in Fig. 94*b* or points whose abscissas on the chord have the same value if referred to the chord length (Fig. 94*a*). With reflexed profiles, and in certain other cases in the neighborhood of the trailing edge, a modified procedure may be devised.

If the n intermediate profiles are constructed in the foregoing way and placed at equal distances along the span, the wing surface will be a ruled surface and, if the method of Fig. 94*b* is used, a developable surface.

Twist. Wings are generally given a twist so that the angle of attack is not constant all over the span. A decrease of the angle of attack toward the wing tips (called *washout*) helps to concentrate the lift in much the same way as the spanwise decrease of chord length: a comparatively larger part of the total lift is contributed by the central portion of the wing, and therefore the bending moment at the wing roots will be smaller for the same total lift.

Apart from this structural advantage, washout is also applied for reasons of lateral stability. The dangerous condition of wing-tip stall— a stalling that starts at one of the wing tips and may easily upset the lateral balance of the airplane—must be avoided by keeping the angle of attack at the ends of the wing a good deal lower than in the central part. This is even more necessary with modern trapezoidal plan-forms than with elliptical or rectangular ones. The twist necessary to avoid wing-tip stall during landing operations may be so great that in level flight the wing tips operate under an angle of attack which is zero or even negative.

Interruptions. Engine nacelles can be built into the wing structure in such a way as to become a part of the wing not only from the structural but also from the aerodynamic point of view. At and near the location of the nacelles the shape of the wing sections then differs considerably from that of the principal profile; but these sections may still be considered as airfoil sections possessing definite aerodynamic coefficients C_L, C_D, C_M that depend on the angle of attack. Prandtl's wing theory has been applied to such wings and has been found to furnish useful results.

The influence of a fuselage interrupting the wing is less clear. If a wing model is tested, combined with a vertical plate in its median plane, the flow around the wing is essentially the same except for the increased friction. This would seem to indicate that a very narrow fuselage with comparatively high vertical side walls interrupting the wing will not modify essentially the flow past the two half wings. The situation changes when width and height of the fuselage become of the same order of magnitude. Under certain simplifying assumptions it has been found[1] that an infinitely long cylindrical fuselage of circular cross section of the diameter d contributes to the lift of the wing in the same way as an equivalent central wing portion of the spanwise extension $d(1 - d/B)$, where B is the span of the wing. Experiments indicate that with properly designed fillets between wing and fuselage the lift of the fuselage portion between the two half wings may almost equal the lift of an equivalent central wing portion whose spanwise extension is equal to the width of the fuselage. In all cases, the drag of the combination of fuselage and wing exceeds considerably the sum of the drags of the two component parts considered independently of each other (see Sec. V.6).

Biplanes. In the case of biplanes, further geometric parameters must be introduced that define the arrangement of the two wings. Figure 95 shows a cross section through the wings of a biplane. For the

[1] LENNERTZ, J., *Z. angew. Mathematik u. Mechanik*, **7**, 249 (1927).

upper profile as well as for the lower one, a well-defined chord line according to Sec. 2 must be chosen. The angle δ between the two chords is called their *decalage*. The distances of two corresponding points of the wings, *e.g.*, of the forward end points of the chords, measured parallel and normal to the bisectrix of the angle between the chords are defined as the *stagger* and the *gap*, respectively. For twisted wings and nonrectangular plan-forms all these parameters may vary along the span. The span of the lower wing is usually smaller than that of the upper one. The tips of the lower wing are thus less likely to hit the ground when the

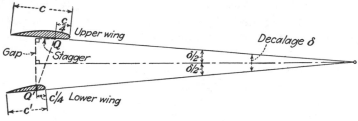

Fig. 95.—Geometry of biplane.

airplane tilts somewhat about the longitudinal axis while taxiing on the ground. Usually, in accordance with the reduced span, the chord length of the lower wing is also smaller than that of the upper wing. Aerodynamic points of view for the choice of the area ratio will be given in Chap. IX. The gap, as a rule, approximately equals the greatest chord length of the two wings.

From the aerodynamic point of view a biplane and a monoplane of equal total wing area can be made equally satisfactory. However, the smaller span made possible by the use of two wings may become important where the available space is restricted, as on the deck of an aircraft carrier. In the past, the fact that the two wings may be connected by struts and wires so as to form a structural unit of considerable stiffness caused designers to prefer biplanes. Modern design places all structural elements in the interior of the wing. In fact, with modern speeds the drag of outside struts and wires becomes so important that structures of this kind must be avoided at all costs. Thus the biplane loses its structural superiority, and this is the reason why monoplanes are almost exclusively used today.

CHAPTER VII

EMPIRICAL AIRFOIL DATA

1. The Three Main Results. In the course of the last thirty-five years a great number of experiments have been carried out in many countries with the aim of exploring the characteristic properties of profiles. Such experiments show how the values of the coefficients

$$C_L = \frac{2L}{\rho V^2 S}, \qquad C_D = \frac{2D}{\rho V^2 S}, \qquad C_M = \frac{2M}{\rho V^2 Sc} \tag{1}$$

depend on the shape of the profile, the angle of attack α, the aspect ratio $\mathcal{R}$ of the airfoil, and the Reynolds number Re. The influence of Re is comparatively small; it will be discussed in Sec. 5 of this chapter. The curves showing the coefficients C_L, C_D, C_M as functions of the angle of attack α are known as the *characteristic curves* of an airfoil. The forms of these curves depend on the shape of the profile, on the aspect ratio of the airfoil, and, to a certain extent, on the Reynolds number. Before going into

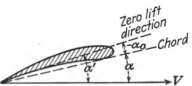

Fig. 95a.—Angle of attack and angle of incidence.

the details, the main results concerning the general form of these characteristic curves may be stated, which are of basic importance in all airplane computations.

a. Lift Coefficient. For each profile there exists a certain angle of attack α_0 for which the lift and, accordingly, the lift coefficient are zero. The corresponding direction of the velocity vector is called the *zero lift direction* or the *direction of the first axis of the profile*. If the angle of attack is measured from the chord as explained in Sec. VI.2, the angle α_0 will be negative for the usual profiles.

The difference

$$\alpha' = \alpha - \alpha_0 \tag{2}$$

measures the angle between the direction of the velocity V and the zero lift direction (Fig. 95a); it is often called the effective angle of attack. However, in the theory of the wing of finite span, the same term is frequently used to denote a differently defined angle of attack. On the other hand, the term angle of incidence is sometimes used as synonymous with the term angle of attack. In order to avoid confusion the following

139

terminology will be applied consistently throughout this book: The angle α, measured from a geometrically defined reference line, will be referred to as the *angle of attack;* the angle α', measured from the zero lift direction, will be called the *angle of incidence.* The term *effective angle of attack* will be reserved for use in the three-dimensional wing theory; it will be defined in Sec. IX.3.

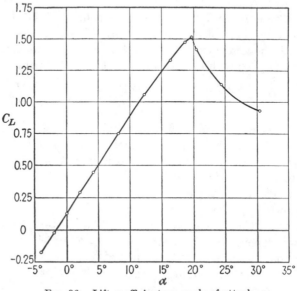

Fig. 96.—Lift coefficient vs. angle of attack.

The main result concerning the coefficient C_L of the usual wing profiles can be stated as follows: *For moderate values of the angle of incidence α' the lift coefficient is proportional to α'.*

$$C_L = k(\alpha - \alpha_0) = k\alpha' \tag{3}$$

the factor of proportionality k depending mainly on the aspect ratio but being almost independent of the shape of the profile.

If α' is expressed in degrees, the value of the *lift factor k* lies between 0.07 and 0.08 (for the usual aspect ratios between 5 and 7). If α' is expressed in radians, k accordingly varies from about 4.0 to 4.5. Figure 96 gives a typical C_L vs. α-curve; the profile (NACA 2409) is shown at the bottom of Fig. 103, the aspect ratio equals 6, and the Reynolds number, referring to the chord length, about 3×10^6. The angle of attack α is varied from -4 to $30°$. Since $C_L = 0$ for $\alpha = \alpha_0 = -2°$, the angle of incidence varies from -2 to $32°$. For angles of incidence up to about $\alpha' = 21°$ the curve is almost a straight line, the value of k being about 0.075. For higher angles of incidence the character of the curve changes abruptly.

The range of values of C_L within which formula (3) is applicable depends on the shape of the profile and on the value of the Reynolds number. The upper boundary of this range usually corresponds to C_L-values between 1.0 and 1.5 and the lower boundary to C_L-values between -0.4 and -0.6. The upper end point of the approximately straight portion of the C_L vs. α-curve is called the *stalling point*. The phenomena occurring beyond the stalling point will be discussed in Sec. X.1.

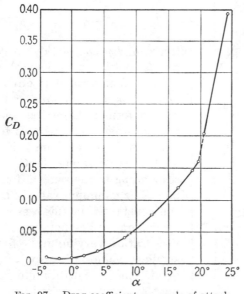

Fig. 97.—Drag coefficient vs. angle of attack.

More precise experiments show that the C_L vs. α-curve is slightly bent, so that the formula

$$C_L = k \sin (\alpha - \alpha_0) = k \sin \alpha' \qquad (4)$$

may be considered a slight improvement on (3). Since, for $\alpha' = 15°$, the difference between $\sin \alpha'$ and α' (expressed in radians) is less than 1.2 per cent, the difference between formulas (3) and (4) is unimportant. Both formulas were originally suggested by the two-dimensional wing theory (see Chap. VIII) and have been found to agree well with the experimental evidence.

b. Drag Coefficient. Within the range interesting in most aeronautical problems the C_L vs. α-curve can be approximated by a straight line; the C_D vs. α-curve, however, is of parabolic character (Fig. 97). It has a minimum close to $\alpha' = 0$. This minimum value of the drag coefficient depends on the shape of the profile and on the value of the

Reynolds number. (For ordinary wings the influence of surface roughness need not be considered.) For values of α' different from zero, the drag coefficient increases proportional to α'^2 at a rate that depends mainly on the aspect ratio. Since within the range of linearly increasing C_L the lift coefficient is proportional to the angle of incidence α', the main character of the C_D vs. α-curve can be stated as follows: *Within the range of linearly increasing C_L the drag coefficient follows the parabolic law*

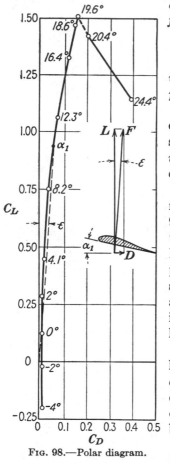

$$C_D = a + \frac{C_L^2}{b} = a + \frac{k^2}{b}\alpha'^2 \qquad (5)$$

where a depends mainly on the shape of the profile and b mainly on the aspect ratio.

The first term a is often called the coefficient of *parasite drag* or *profile drag;* the second term is known as the coefficient of *induced drag.* The reasons for this terminology will be explained in Sec. IX.3.

Figure 97 shows the C_D vs. α-curve corresponding to the same airfoil as that of Fig. 96. The minimum value of C_D is $a = 0.007$; the constant b in this case equals about 16. In general, b lies between 15 and 20 for aspect ratios varying from 5 to 7, and the smallest values of a so far reached equal about 0.006. Further information concerning these coefficients will be given in the following sections of this chapter.

A comprehensive representation of the relations between C_L, C_D, and α' is the *polar diagram* introduced by G. Eiffel. In this diagram, C_L is plotted vs. C_D (Fig. 98). According to (5) the curve thus obtained should be a parabola whose axis coincides with the C_D-axis and whose vertex has the abscissa

Fig. 98.—Polar diagram.

$C_D = a$. On this curve the points corresponding to certain round values (or to the actual experimental values) of α (or α') are marked. By using the α-scale thus obtained as well as the C_D- and C_L-scales on the coordinate axes, it is easy to read off the values of any two of the quantities C_D, C_L, α that correspond to a given value of the third quantity. Moreover, if the same unit is used for plotting C_D and C_L, the line joining the origin to the point corresponding to $\alpha = \alpha_1$ indicates the direction of the resultant force F experienced by an airfoil moving horizontally from right to left under the angle of attack α_1 (see Fig. 98, inset). Because the C_D-values are much smaller than the C_L-values, the unit for C_D in the standard graphs is taken five times as large as the unit for C_L. Figure 99 shows the polar diagram obtained in this manner. (The diagrams of Figs. 96 to 99 and 103 refer to the same airfoil.)

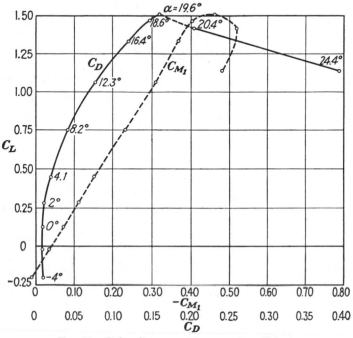

FIG. 99.—Polar diagram with moment coefficient.

The angle ϵ between the lift L and the resultant force F has the tangent $D/L = C_D/C_L$. Since for the usual angles of attack ϵ is small, this tangent practically equals the angle ϵ expressed in radians:

$$\epsilon \sim \tan \epsilon = \frac{D}{L} = \frac{C_D}{C_L} = \frac{a}{C_L} + \frac{C_L}{b} \tag{6}$$

This angle has the following significance: Suppose that by an appropriate device (tail surface) the airfoil is kept at a constant angle of attack but otherwise left free to glide down under the influence of its weight W. If the airfoil moves at constant velocity, the forces W, L, and D must be in equilibrium; i.e., the resultant, F, of L and D, must be equal and opposite to W (Fig. 100). Since the lift L is perpendicular to the velocity V of the airfoil and the weight W is perpendicular to the horizontal plane, the angle between V and the horizontal plane equals the angle ϵ between L and F. For this reason, ϵ is called the *gliding angle* of the airfoil.

The minimum value of the gliding angle, $(C_D/C_L)_{min}$, is found by drawing the tangent from the origin to the polar curve and forming the quotient C_D/C_L for the point of contact. In order to find ϵ_{min} analytically, the derivative $d\epsilon/dC_L$ computed from (6) must be set equal to zero:

$$\frac{d\epsilon}{dC_L} = -\frac{a}{C_L^2} + \frac{1}{b} = 0 \quad \text{or} \quad C_L = \sqrt{ab} \tag{7}$$

Introducing this value of C_L into (6), we find

$$\epsilon_{min} = \frac{a}{\sqrt{ab}} + \frac{\sqrt{ab}}{b} = 2\sqrt{\frac{a}{b}} \tag{8}$$

In the case represented by Fig. 98, ϵ_{min} has the value 0.0426 equivalent to 2.45° and its reciprocal $(C_L/C_D)_{max}$ equals 23.4. The highest values of this quotient that have been reached so far are about 24 to 26 for aspect ratios between 5 and 7.

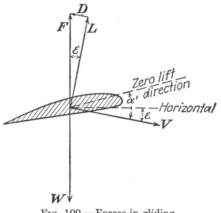

FIG. 100.—Forces in gliding.

c. *Moment Coefficient.* The moment (pitching moment) and the moment coefficient depend, of course, on the position of the point with respect to which the moment is taken. If the coefficients of lift and drag and the moment coefficient with respect to one certain point of reference are known, the moment coefficient with respect to any other point can be computed by the rules of elementary statics. The outstanding fact concerning the variation of the moment coefficient with the angle of attack can be stated as follows: *For any given profile there exists a certain point F with respect to which the moment M and the moment coefficient have practically constant values as long as the angle of attack does not exceed the range of linearly varying C_L. The constant value of the moment coefficient depends on the shape of the profile but is practically independent of the aspect ratio.*

The point F is known as the *aerodynamic center* (a.c.) or the *focus* of the profile. For the usual profiles it lies near the chord line, about a quarter chord length aft the leading end, in accordance with the two-dimensional wing theory. The moment with respect to the aerodynamic center will be denoted by M_0. The moment coefficient C_{M_0} has small negative values for simple (*i.e.*, unreflexed) profiles according to the terminology introduced in Sec. VI.2, the absolute value of C_{M_0} decreasing with decreasing camber. (Note that the moment is counted positive

if it tends to raise the leading end of the profile.)   The coefficient C_{M_0}
is zero for symmetrical and slightly reflexed profiles and positive for
strongly reflexed profiles.   For the modern slightly cambered simple
profiles, C_{M_0} varies between 0 and -0.1.

Experimental results concerning the pitching moments of airfoils
can be represented in various ways.   If C_{M_0} is plotted against α or α' or
against C_L, the diagram will consist of a straight line parallel to the axis
of abscissas.   But if an airfoil is to be investigated experimentally, the
position of the aerodynamic center is not known beforehand.   The
customary procedure is to measure the moment with respect to some
well-defined point of reference and then to compute the moment coefficient with respect to the quarter chord point, $i.e.$, the point on the chord

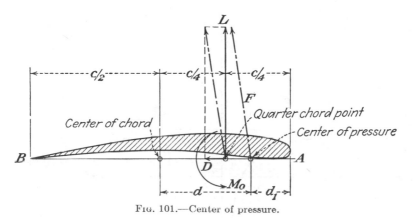

Fig. 101.—Center of pressure.

at one-quarter the chord length aft the leading end.   In most cases, this
moment coefficient will be fairly independent of the angle of attack, an
indication that an aerodynamic center exists and that it lies near the
quarter chord point.

In order to obtain an approximate relation between the moment
coefficients with respect to various points of reference, we may assume
that the aerodynamic center coincides with the quarter chord point and
that, for the usual small angles of attack, the lift is practically perpendicular and the drag practically parallel to, and coinciding with, the
chord AB (Fig. 101).   The thrusts that the air exerts on the profile are
statically equivalent to the forces L and D applied to the aerodynamic
center and a couple of the moment M_0.   Under the assumptions here
introduced the moment of this system with respect to the leading end A
of the chord is

$$M_1 = M_0 - \frac{c}{4} L \qquad (9)$$

The corresponding moment coefficient therefore equals

$$C_{M_1} = C_{M_0} - \tfrac{1}{4}C_L \tag{10}$$

The point where the line of action of the resultant force intersects the chord is sometimes called the *center of pressure;* if it has the distances d and d_1 from the center of the chord AB and from the leading end A, respectively (Fig. 101), the ratios d/c and d_1/c are given by

$$\frac{d}{c} = \frac{C_{M_0}}{C_L} + \tfrac{1}{4}, \qquad \frac{d_1}{c} = -\frac{C_{M_0}}{C_L} + \tfrac{1}{4} \tag{11}$$

Usually the graphs showing d/c or d_1/c in dependence on the angle of attack are used instead of the C_M vs. α-curve as a means of specifying

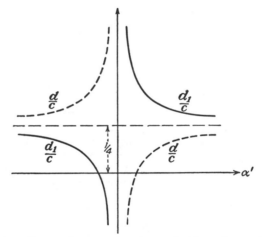

Fig. 102.—Center of pressure location vs. incidence (approximately).

the position of the line of action of the resultant force. Since C_{M_0} is constant and C_L proportional to α', Eqs. (11) represent equilateral hyperbolas with a vertical asymptote at $\alpha' = 0$ and a horizontal asymptote at the ordinate $\tfrac{1}{4}$. Figure 102 shows these hyperbolas for d/c and d_1/c. In comparing these curves with those showing experimental values of d/c and d_1/c, it must be kept in mind that Eqs. (11) have been derived under assumptions which may fail considerably at points where D is not small as compared with L.

Figure 103 shows the complete diagram for the NACA airfoil 2409. Besides the curves of Figs. 96 and 97, Fig. 103 gives the empirical values of d_1/c (expressed in per cent) and the values of L/D. The scales are those used in the NACA reports. Note that the scale for d_1/c is chosen in such a manner that the line $d_1/c = 100$ per cent coincides with the α-axis.

The reports of the Aerodynamic Institute at Göttingen generally use $-C_{M_1}$, the negative of the moment coefficient with respect to the leading end of the chord, as abscissa in a graph whose ordinates represent C_L. Figure 99 shows such a graph in addition to the polar diagram.

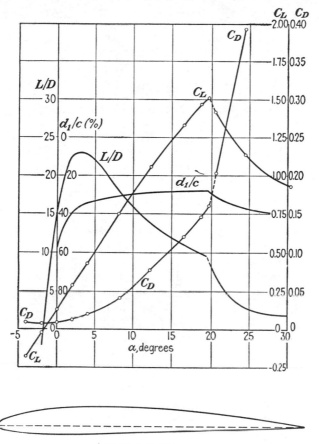

NACA 2409

Fig. 103.—Characteristics of the profile NACA 2409.

According to (10) this graph should be a straight line of the slope $\frac{1}{4}$ intersecting the axis of abscissas at the point with the abscissa $-C_{M_0}$. Most empirical curves show this behavior except, sometimes, for a narrow region at small angles of incidence, where Eq. (10) fails to give a good approximation.

The existence of the aerodynamic center, *i.e.*, of a point with constant moment, has been predicted by the two-dimensional wing theory (see Chap. VIII). This result has subsequently been confirmed by practically all experiments.

Problem 1. The following table gives the coefficients of lift, drag, and moment with respect to the leading end of the chord as functions of the angle of attack α. Plot the curves C_L vs. α, C_D vs. α, and d_1/c vs. α and the polar diagram including the moment curve C_{M_1} vs. C_L.

$\alpha =$	-4.5	-1.6	1.3	4.2
$C_L =$	0.124	0.327	0.549	0.738
$C_D =$	0.0185	0.0222	0.0331	0.0495
$C_{M_1} =$	-0.110	-0.156	-0.208	-0.257

$\alpha =$	8.6	11.6	14.5	17.5
$C_L =$	1.033	1.192	1.314	1.369
$C_D =$	0.0860	0.118	0.152	0.203
$C_{M_1} =$	-0.335	-0.368	-0.388	-0.430

Problem 2. Formula (11) gives the position of the center of pressure with sufficient accuracy only if the angle of attack is so small that the lift is practically perpendicular and the drag practically parallel to, and coinciding with, the chord. Indicate the modifications necessary for greater angles of attack.

*Problem 3. If the existence of the aerodynamic center of a profile is taken for granted, for how many angles of attack must the coefficients C_L, C_D, C_M be measured in order to determine the position of the aerodynamic center? Try to locate the aerodynamic center of the profile of Prob. 1 (Göttingen 404).

2. Influence of Aspect Ratio. Within the range of linearly varying C_L the dependence of the coefficients of lift, drag, and pitching moment on the angle of incidence α' is defined by the parameters k, a, b, and C_{M_0}. The lift factor k introduced in (3) can be described as the slope of the C_L vs. α-diagram. The parameters a and b are defined by (5); here a is the coefficient of profile drag of the airfoil, $b/2$ is the parameter of the parabola of the polar diagram,[1] and C_{M_0} denotes the coefficient of the pitching moment with respect to the aerodynamic center. The parameters k, a, b, C_{M_0} depend on the shape of the profile, the aspect ratio, and the Reynolds number. The influence of the aspect ratio will be discussed in this section, that of the shape of the profile in Sec. 4, and the influence of the Reynolds number in Sec. 5.

The dependence of k and b on the aspect ratio is fairly well represented by formulas furnished by the three-dimensional wing theory (see Chap. IX). Although this theory uses a series of simplifying assumptions that cannot be expected to be rigorously fulfilled under actual conditions, the theoretical formulas for k and b give a good account of the experimental results. The parameters a and C_{M_0} are practically independent of the aspect ratio.

a. Slope of the C_L vs. α-diagram. The experimental values of the slope of the C_L vs. α-diagram for the usual wing profiles are fairly well represented

---

[1] The parabola $y^2 = 2px$ has the "parameter" p.

by

$$k = \frac{0.1}{1 + 2/R} \qquad (12a)$$

if α is expressed in degrees or by

$$k = \frac{5.7}{1 + 2/R} \qquad (12b)$$

if α is expressed in radians.   According to these formulas the lift coefficient $C_L = k\alpha'$ assumes the value 1.0 at the angles of incidence

$$\begin{aligned}
\alpha' &= 10° &&\text{for } R = \infty \\
\alpha' &= 12\tfrac{1}{2}° &&\text{for } R = 8 \\
\alpha' &= 13\tfrac{1}{3}° &&\text{for } R = 6 \\
\alpha' &= 14° &&\text{for } R = 5
\end{aligned}$$

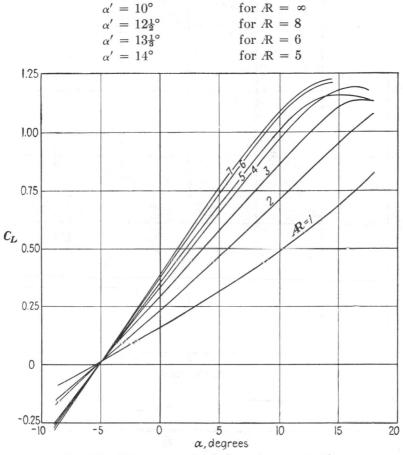

Fig. 104.—Lift curves of airfoils for various aspect ratios.

(How an infinite aspect ratio can be realized experimentally will be explained in Sec. 3.)   Formulas (12) apply to aspect ratios as small as 2 and to all forms of profiles used today for wings and tail surfaces.   They can also be used in a certain way in the case of propeller blades with

profiles that do not deviate too much from the usual wing sections
(see Sec. XI.4).

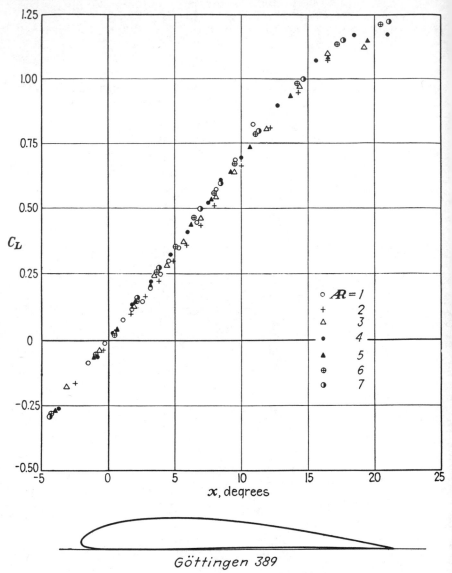

Göttingen 389

FIG. 105.—Lift curves of Fig. 104, reduced to the aspect ratio 5.

Formulas (12) can be checked in the following manner: Figure 104
shows the experimental C_L vs. α-curves obtained with airfoils of the
same profile but of different aspect ratios. The linear portions of these
curves clearly have different slopes. In Fig. 105 the same C_L-values are

plotted against the variable

$$x = \frac{1 + \frac{2}{5}}{1 + 2/\mathcal{R}} \alpha' = \frac{1.4\alpha'}{1 + 2/\mathcal{R}} \tag{13}$$

According to (12) the linear portions of the curves showing C_L vs. x should coincide on the straight line that joins the origin to the point $x = 14°$, $C_L = 1.0$. The transformation of the C_L vs. α-curve into the

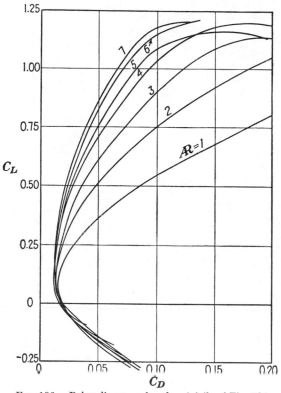

FIG. 106.—Polar diagrams for the airfoils of Fig. 104.

corresponding C_L vs. x-curve is called the "reduction of the C_L vs. α-curve to the aspect ratio 5." Figure 105 shows that this reduction of the curves of Fig. 104 confirms the prediction of the theory. The experiments represented in Fig. 104 were performed at the Aerodynamic Institute of Göttingen;[1] the profile used is shown at the bottom of Fig. 105.

    b. Parameter of the Polar Parabola. The three-dimensional wing theory supplies a coefficient of "induced" drag of the form C_L^2/b where b

[1] *Göttinger Ergebnisse*, **1** (1921).

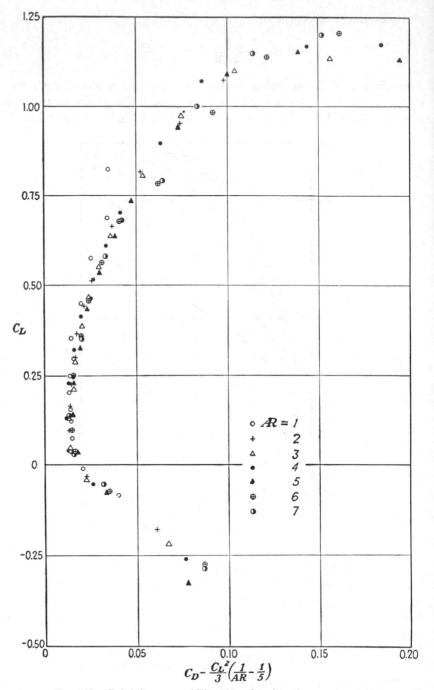

Fig. 107.—Polar diagrams of Fig. 106 reduced to the aspect ratio 5.

depends on the aspect ratio.   No account is given of the parasite drag
coefficient a introduced in (5).   Comparison of the experimental values
of C_D with the formula suggested by the theory leads to the following
result: *The experimental values for the drag coefficient C_D are fairly well
represented by formula* (5), *if the parameter b is taken as*

$$b \sim 3\!R \tag{14}$$

Again this statement applies only to the range of linearly varying C_L.
Furthermore, the airfoil is supposed to have a profile of the usual type
and an aspect ratio of 2 or more.

Figure 106 shows the polar diagrams for the airfoils of Fig. 104.
While the minimum value a of C_D is practically the same for all curves,
the curves corresponding to larger aspect ratios open up more than those
corresponding to smaller aspect ratios.   Figure 107 shows the same
experimental results reduced to the aspect ratio 5.   Here each point
has the same ordinate C_L as in Fig. 12, but the abscissa is changed from
C_D to

$$x = C_D - \frac{C_L^2}{3}\left(\frac{1}{\!R} - \tfrac{1}{5}\right) \tag{15}$$

If Eq. (5) is valid with the value (14) of b, the right-hand side of (15)
equals $a + C_L^2/15$ and consequently is independent of $\!R$.   In fact, the
points marked in Fig. 107 lie fairly well on a single curve.   Moreover,
in the range of practical use this curve is nearly a parabola with the
parameter $\tfrac{1.5}{2}$, thus corresponding (except for the value a) to the theoretical polar diagram for an airfoil of the aspect ratio 5.

Introduction of (14) into expression (8) furnishes the minimum
gliding angle

$$\epsilon_{\min} = 2\sqrt{\frac{a}{3\!R}} \tag{16a}$$

and its reciprocal, the maximum of the ratio C_L/C_D,

$$\left(\frac{C_L}{C_D}\right)_{\max} = \tfrac{1}{2}\sqrt{\frac{3\!R}{a}} \tag{16b}$$

c. Moment coefficient.   In its present form the three-dimensional
wing theory does not furnish any indication for assuming an influence
of the aspect ratio on the position of the aerodynamic center F and on
the moment coefficient C_{M_0}.   Experiments seem to indicate that, if
such an influence exists at all, it must be extremely small.

The situation is not quite so simple when the moment coefficient
with respect to some other point of reference is considered.   Take, for

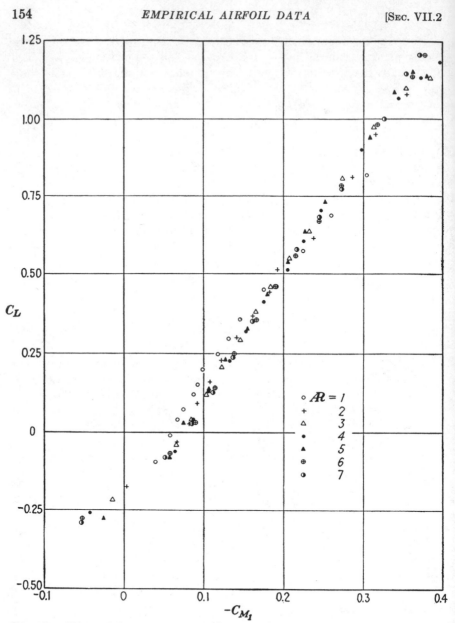

FIG. 108.—Lift coefficient vs. moment coefficient with respect to the leading end for the airfoils of Fig. 104.

example, the coefficient C_{M_1} with respect to the leading edge. According to (10) this coefficient is expressed in terms of C_L and of C_{M_0}. Now C_L is related to the angle of incidence α' in a manner that depends on the aspect ratio [see (3) and (12)]. The relation between C_{M_1} and α' consequently is not independent of the aspect ratio. On the other hand, if C_{M_1} is plotted against C_L rather than against α', the result is

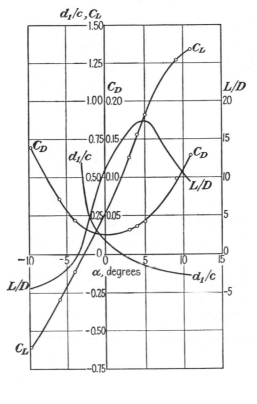

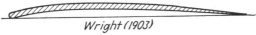

Wright (1903)

FIG. 109.

the same for all aspect ratios. The same applies to the moment coefficient with respect to any other point as well as to the relative distance d_1/c of the line of action of the resultant force from the leading end or from any other point. *When considered as functions of C_L, the moment coefficient and the relative distance of the line of action of the resultant force with respect to any fixed point are independent of the aspect ratio.*

In Fig. 108 the abscissas and ordinates represent $-C_{M_1}$ and C_L, respectively. The points indicate experimental values obtained with the group of airfoils represented in Fig. 104; they are clearly lying on the same straight line of the slope 4:1.

Problem 4. For an airfoil of the aspect ratio 5 the following values of C_L and C_D have been observed. The profile is to be used for the wing of a glider plane of the aspect ratio 12. Reduce the curves C_L vs. α and C_D vs. α to this aspect ratio.

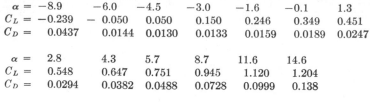

$\alpha =$	-8.9	-6.0	-4.5	-3.0	-1.6	-0.1	1.3
$C_L =$	-0.239	-0.050	0.050	0.150	0.246	0.349	0.451
$C_D =$	0.0437	0.0144	0.0130	0.0133	0.0159	0.0189	0.0247

$\alpha =$	2.8	4.3	5.7	8.7	11.6	14.6
$C_L =$	0.548	0.647	0.751	0.945	1.120	1.204
$C_D =$	0.0294	0.0382	0.0488	0.0728	0.0999	0.138

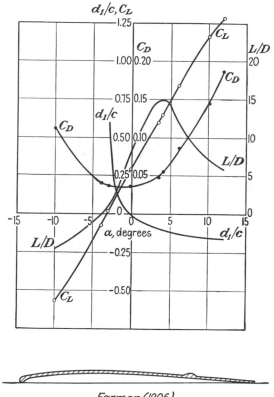

Farman (1906)

Fig. 110.

Problem 5. Determine the minimum gliding angle of the wing of Prob. 4 (Göttingen 436). How is the minimum gliding angle modified by the presence of an additional constant parasite drag that increases the value of a by 0.02?

3. Historical Development of Wing Profiles. Before entering into a study of the comparatively small variations of the aerodynamic characteristics encountered in modern wings, some properties of earlier profiles may be discussed.

Otto Lilienthal, the pioneer of gliding, was the first to realize the importance of a carefully shaped wing section. As the result of experiments carried out in the years 1871 to 1890, he found that camber and an appropriate thickness distribution improved the aerodynamic qualities

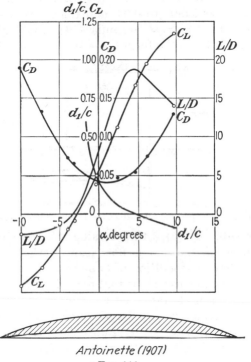

Antoinette (1907)

Fig. 111.

of a wing as compared with a plane plate. The brothers Wilbur and Orville Wright, who performed, in 1903, the first successful motor flight, used the profile that is shown in Fig. 109 together with its aerodynamic characteristics.[1] In the first decade of this century, French aviators developed wing sections that today appear somewhat strange. Figures 110 and 111 show such sections used by Farman and Antoinette, and the respective diagrams. The diagrams of Figs. 109 to 111 represent the results of experiments carried out in 1910 in the Aerodynamic Laboratory in Moscow under the direction of N. Joukowski.[2] Airfoils of an extremely

---

[1] In the Fig. 109 to 111 there should be read $-10d/c$ instead of d_1/c.

[2] LOUKIANOFF, G. S., *Z. Flugtech. u. Motorluftschiffahrt*, **3**, 153 (1912).

small aspect ratio (about 0.6) were mounted between parallel glass panes, leaving a clearance of about $\frac{1}{12}$ in. only (Fig. 112). In this way it has been possible to enforce a practically two-dimensional flow and thus to realize the condition of infinite aspect ratio. The air speed used in these tests was about 55 ft./sec. In all three diagrams the curve C_L vs. α exhibits the characteristic fairly straight portion with a slope corresponding to $C_L = 1$ for $\alpha' = 9$ to $10°$. This agrees well with the foregoing equations (12a) and (12b) and with the theory that gives $C_L = 1$ for $\alpha' = 1/(2\pi)$ rad. $= 9.1°$ (see Sec. VIII.2). The minimum values of C_D are about 0.03 to 0.04, and the C_D vs. α-curves are of the parabolic type. For $Æ = \infty$, formula (14) would give $b = \infty$; and, with this value, (5) takes the form $C_D = a = $ const. The actual curves

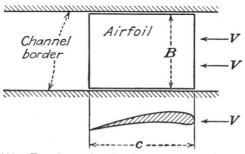

Fig. 112.—Experimental realization of two-dimensional flow.

could be represented by (5) with a b-value of about 100, corresponding to an aspect ratio of 33.3. It is not clear whether this discrepancy is due to the fact that the condition of two-dimensional flow was not completely realized. It may as well be that formula (14) is an approximation valid for moderate aspects ratios only. The maximum value of C_L/C_D (about 17 to 20) cannot be compared immediately with the results of modern experiments carried out with airfoils of finite aspect ratio. The position of the line of action of the resultant force is given in terms of d_1/c; the curves have the expected character according to Fig. 102. At the time when these experiments were published, the existence of the aerodynamic center and, consequently, relation (11) were not yet known.

The next phase of the development of airfoil sections, covering about the second decade of this century, may be illustrated by the profiles and diagrams of Figs. 113 and 114. The first is the profile Göttingen 360, one of a series published in 1920; the second was designed by the author[1] for a 600-hp. airplane built in 1915. The tendency at that time was toward smooth, well-rounded forms. The diagrams of Fig. 113 and Fig. 111 refer to airfoils of the aspect ratio 5 and 6, respectively. The

---

[1] R. v. Mises, Ein 600-P.S. Grossflugzeug vom Jahre 1916, *Beiträge zur Flugtechnik,* her. R. Katzmayr, Wien, 1937, p. 17.

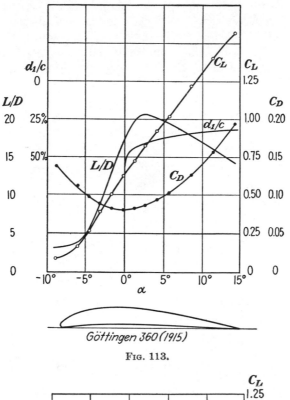

Fig. 113.

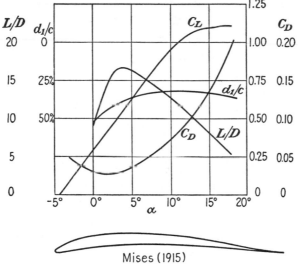

Fig. 114.

slope of the straight portion of the C_L vs. α-curves corresponds to $C_L = 1$ for $\alpha' = 13.6°$ in the first and $\alpha' = 14.5°$ in the second case. The maximum values of C_L/C_D are 15.7 and 16.7. According to (16b), this corresponds to the profile-drag coefficients 0.015 and 0.016, respectively, and to $(C_L/C_D)_{max} = 17.3$ for the Göttingen 360 profile when reduced to

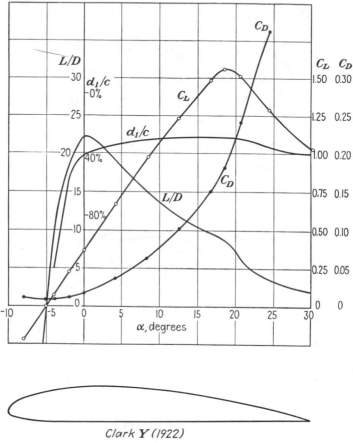

Clark **Y** (1922)

Fig. 115.

aspect ratio 6. The moment coefficient with respect to the quarter chord point is about -0.08 in the first and -0.06 in the second case.

A successful American profile of 1922 is the Clark Y-profile shown in Fig. 115. Tested at the aspect ratio 6 it gives $C_L = 1$ for $\alpha' = 13.7°$ and $(C_L/C_D)_{max} = 22$, corresponding to the profile drag coefficient $a = 0.010$.

The latest development is illustrated by the profiles of Figs. 116 and 117, taken from the NACA four- and five-digit series (see Sec. VI.2).

The camber of these modern profiles is smaller than that of the older ones, the lower contour is nearly straight, and the thickness comparatively large. The diagrams for NACA 2409 and NACA 23009 are given for $R = 6$. The slope of the straight portion of the curves C_L vs. α corresponds to $C_L = 1$ for $\alpha' = 13.7°$ in the first and $\alpha' = 13.2°$ in the second case. The maximum value of C_L/C_D is about 24, corresponding to the profile-drag coefficient $a = 0.0078$. The decrease of the value

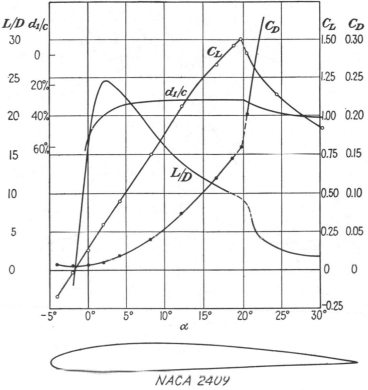

NACA 2409

Fig. 116.

of $1000a$ from 40 to 30 in the first period, through values of 13 to 10 in the second, down to 7.8 in recent designs shows the progress of the development.

4. Influence of the Shape of the Profile. Systematic experimental investigations of the influence of the shape of the profile on the aerodynamic characteristics of an airfoil were not undertaken before about 1920. The four- and five-digit series of the NACA (Sec. VI.2) were developed for this purpose. With the exception of the coefficient of profile drag a, the aerodynamic parameters of an airfoil depend only to a very limited extent on the shape of the profile.   On the other hand, the two-dimensional airfoil theory furnishes definite statements concerning the variation of C_L and C_{M_0} (see Chap. VIII).   The main experimental results are given below.

  a. Lift Coefficient.   The experiments on the four-digit family of airfoils show that *the slope k of the straight portion of the curve C_L vs. α*

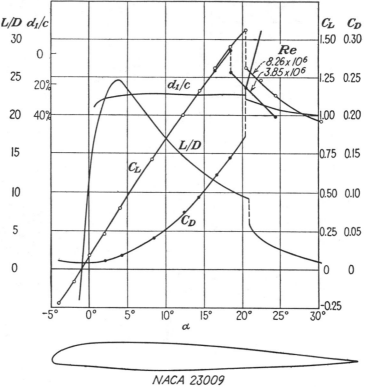

NACA 23009
Fig. 117.

decreases slightly with increasing thickness and is almost independent of the camber.   In Sec. VII.2 the slope k was given as 0.10 for the infinite aspect ratio if α' is expressed in degrees.   Reduced to $\mathcal{R} = \infty$, the experimental results show a variation of k from $k = 0.103$ for the relative thickness $t/c = 0.06$ to $k = 0.097$ for $t/c = 0.21$.   These are values for an airfoil with the camber designation 24, that is, an airfoil with a maximum camber of 2 per cent of the chord length located at 40 per cent of the chord aft the leading end.   For an airfoil of the camber designation 64 the corresponding k-values are 0.104 and 0.096.

  The two-dimensional airfoil theory (which does not take into account any viscosity effects) does not predict a decrease of k with increasing thickness. In discussing the theoretically developed airfoil sections in Sec. VI.3, it was mentioned that the theoretical lift coefficient equals $8\pi(a/c) \sin \alpha'$, where a is the radius of the circle C_1 from which the airfoil is developed. If $\sin \alpha'$ is replaced by the small angle α' expressed in radians, this gives $k = 8\pi a/c$. To increase the thickness of a Joukowski profile without changing the camber means to let the point M_1 move to the right without changing the slope of the line $B'M_1$ (Fig. 78). In this case the radius $a = \overline{B'M_1}$ and the chord length c increase with the thickness in such a manner that the ratio a/c remains essentially unchanged (it increases only very slightly). The quotient a/c is always greater than $\frac{1}{4}$, and thus the theoretical value of k is greater than $8\pi/4 = 2\pi$ if α' is expressed in radians, or greater than $2\pi(\pi/180) = 0.11$ if α' is expressed in degrees. Now, the profiles of the four-digit series of the NACA are not Joukowski profiles. However, experiments made with Joukowski profiles also give k-values, which, reduced to $R = \infty$, are about 5 per cent smaller than those predicted by the two-dimensional theory.[1] This discrepancy is probably due to the effects of viscosity (see also Sec. VII.5). It seems that viscosity is the principal cause for the influence of the thickness on the aerodynamic characteristics of airfoils.

The straight portion of the C_L vs. α-line is completely determined by the slope k and the zero lift angle, *i.e.*, the angle of attack α_0 for which $C_L = 0$. In the case of the Joukowski airfoils discussed in Sec. VI.3 the zero lift direction is given by $B'M$, the camber of the corresponding mean camber line equals $2\overline{OM}$, and the chord length of this line, which may be considered as the length c, is $4\overline{B'O}$ (Fig. 78). Consequently,

$$-\alpha_0 \sim -\tan \alpha_0 = \frac{\overline{MO}}{\overline{B'O}} = \frac{\frac{1}{2}y_{max}}{\frac{1}{4}c} = 2\frac{y_{max}}{c} \qquad (17)$$

If α_0 is expressed in degrees and the relative camber y_{max}/c in per cent, this gives

$$-\frac{\alpha_0}{\text{relative camber}} = 1.15 \qquad (18)$$

The following values of this quotient have been observed with profiles whose maximum camber was located at the center of the chord (second digit 5):

Relative thickness = 6	9	12	15	18	21 per cent
$-\alpha_0$/rel. camber = 1.0	1.0	1.0	1.0	1.0	0.9 for rel. camber = 2%
$-\alpha_0$/rel. camber = 1.1	1.0	1.0	1.0	1.0	0.9 for rel. camber = 4%
$-\alpha_0$/rel. camber = 1.05	1.05	1.05	1.0	0.95	0.9 for rel. camber = 6%

[1] Betz, A., *Z. Flugtech. u. Motorluftschiffahrt*, **10**, 173 (1919).

Modern profiles have their maximum camber closer to the leading end. In these cases the observed values of the quotient considered in the foregoing are slightly reduced, in some cases to as much as 90 per cent of those given in the table. The two-dimensional theory of thin airfoils (Sec. VIII.6) predicts a similar influence of the mean camber position. The vagueness in the definition of the shape of the "thin" airfoil to be substituted for the actual airfoil on the one hand and the difficulties of obtaining reliable experimental values concerning the rather small influence of the maximum camber position on the other hand prevent a closer checking of these details. The following statement sums up the situation: *Experiments with the profiles of the four-digit series show, in agreement with the two-dimensional airfoil theory, that the angle between chord and zero lift direction expressed in degrees equals about 0.9 to 1.0 times the relative camber expressed in per cent, the smaller values corresponding to maximum camber positions nearer to the leading end.*

b. *Moment Coefficient.* Equation (5) of Sec. VI.3 gives the moment coefficient for a Joukowski profile. If $\overline{4B'O}$ is taken for c and y_{max} equal to $2\overline{OM}$ and, on the other hand, the angle β is considered as small, $\sin 2\beta$ approximately equals $2\overline{OM}/\overline{OB'} = 4y_{max}/c$ and (5) reduces to

$$-C_{M_0} = \frac{\pi y_{max}}{c} \tag{19}$$

The moment coefficient with respect to the aerodynamic center should therefore be proportional to the relative camber. The theory of thin airfoils (Sec. VIII.6) shows that the factor which is π for the Joukowski profile should decrease slightly as the maximum camber approaches the leading end. The experiments on the four-digit airfoils confirm the proportionality with the relative maximum camber and the decrease of the factor with changing camber position but give much smaller values than the theoretical ones. The empirical results show a large dispersion and do not lead to a convincing conclusion. The highest factor, equal to 2.7, was found for the profile 4506 possessing a relative thickness of 6 per cent and a relative camber of 4 per cent located at the center of the chord. Definitely smaller factors are found for higher values of the relative thickness. This fact is another indication that the viscosity effects, neglected in the theory, determine the influence of the thickness on the aerodynamic characteristics.

In additional experiments the mean camber line was changed so as to transform the original simple profiles into reflexed ones. In accordance with the prediction of the two-dimensional airfoil theory the (negative) C_{M_0} was found to decrease substantially in absolute value. Also, positive C_{M_0}-values were observed with reflexed profiles.

As regards the position of the aerodynamic center, it seems that for most profiles it lies slightly in front of the quarter chord point, so that 0.24c is a better approximation for the distance from the leading end than the usually adopted value of one-quarter chord length.

The following statement summarizes the situation as far as the moment coefficient is concerned: *Experiments on simple profiles confirm the prediction of the two-dimensional airfoil theory that the moment coefficient C_{M_0} with respect to the aerodynamic center is proportional to the relative*

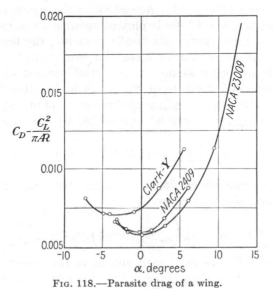

$$C_D - \frac{C_L^2}{\pi \mathcal{R}}$$

Fig. 118.—Parasite drag of a wing.

camber.   The empirical value of the factor of proportionality is smaller than the theoretical one.   For reflexed profiles the increase of the negative C_{M_0} and the occurrence of positive C_{M_0} values are confirmed.

c. Drag Coefficient. The two-dimensional wing theory in its present form does not furnish any drag and consequently does not give any guidance for the discussion of the experimental results concerning the drag coefficient. The three-dimensional wing theory gives a drag coefficient of the form C_L^2/b, where b equals about $3\mathcal{R}$ [the theoretical value for elliptic lift distribution being $\pi\mathcal{R}$ according to Eq. (19) of Sec. IX.5]. However, no profile-drag coefficient a is furnished by the three-dimensional wing theory. Therefore, the formula

$$C_D = a + \frac{C_L^2}{b} \tag{20}$$

must be considered as half empirical.   There are two ways of comparing the values furnished by this formula with those found by experiments. The first is to plot the polar diagram with the observed values of C_L and

C_D and to determine the constants a and b for the parabola that best fits this empirical curve. The other way is to use the theoretical value for b and to study the behavior of the difference $C_D - C_L^2/b$. This method, used more frequently than the first one, is not well justified, since the theoretical value $b = \pi R$ is derived under the assumption of an elliptic lift distribution which is only approximately fulfilled in the case of a rectangular airfoil (see Sec. IX.5).

Figure 118 shows the result of such an investigation carried out for the airfoils of Figs. 115 to 117. According to the aspect ratio 6 used in these experiments, $C_D - C_L^2/6\pi$ is plotted against the angle of attack α. The second and third curves practically coincide; the first curve lies a little higher. All three curves have the same parabolic shape, the minimum ordinate being located in the neighborhood of $\alpha = 0$. If a denotes the minimum ordinate of the curve and C_L' the lift coefficient that belongs to the corresponding α, the results obtained with the airfoils of the four-digit series can be represented by the formula[1]

$$C_D - \frac{C_L^2}{6\pi} = a + 0.0062(C_L - C_L')^2$$

If C_L' is taken as zero, which would mean that the drag minimum occurs at incidence zero, this formula gives

$$C_D = a + C_L^2 \left(\frac{1}{6\pi} + 0.0062 \right) = a + \frac{C_L^2}{5.37\pi}$$

i.e., the values of C_D observed with airfoils of the aspect ratio 6 would correspond to the theoretical values for airfoils with the aspect ratio 5.37.

The C_D vs. C_L-curve may then be represented by

$$C_D = a + \frac{C_L^2}{\pi M R} \tag{21}$$

where M is about 0.9 for ordinary profiles. This correction factor M, often called the Munk factor, will be discussed later in the performance computation (Sec. XV.1).

As regards the influence of the relative thickness t/c on the profile drag a, the results of the NACA experiments with the airfoils of the four-digit series can be represented by the formula

$$a = a' + 0.0056 + 0.01 \frac{t}{c} + 0.1 \left(\frac{t}{c} \right)^2$$

where a' depends on the relative camber according to the following:

Relative camber =	2	4	6%
a' =	0.0005	0.001	0.002

[1] *NACA Tech. Rept.* 460 (1933).

The maximum camber is supposed to be located at a distance between $0.2c$ and $0.5c$ aft the leading edge. These values may be cautiously applied to similar profiles, but as long as no proper theoretical basis is available *it is impossible to make general statements concerning the influence of the shape of the profile on the drag coefficient.*

Problem 6.   Compute the lift coefficient for an airfoil of the four-digit series at $\alpha = 6°$, aspect ratio 7, relative thickness 12 per cent, relative camber 4 per cent.

Problem 7.   What is the value of the moment with respect to the a.c. for a wing of 200 ft.² area and 5.5 ft. chord length at a dynamic pressure of 100 lb./ft.² if the profile has a relative maximum camber of 3 per cent and the proportionality factor is known to be one-half that for a Joukowski profile?

Problem 8.   Determine the minimum gliding angle for an airfoil for which

$$C_D = 0.011 + 0.006(C_L - 0.3)^2 + \frac{C_L^2}{\pi R}$$

for the aspect ratios 6, 8, and 10.

5. Influence of the Reynolds Number.   Degree of Turbulence.   For airfoils the chord length is usually taken as the reference length in defining the Reynolds number.   Thus, in a wind tunnel with an air speed of 70 ft./sec. for the usual size of airfoil models (5 by 30 in.) and with the standard value of the kinematic viscosity $\nu = 1.57 \times 10^{-4}$ ft.²/sec.,

$$Re = \frac{Vc}{\nu} = \frac{70 \times \frac{5}{12}}{1.57 \times 10^{-4}} = 1.86 \times 10^5$$

For a full-scale wing of the chord length $c = 6$ ft., flying at 200 m.p.h. = 293 ft./sec. at an altitude of 10,000 ft., where the ν-value is about 27 per cent higher than the standard value used in the foregoing (see Sec. IV.1 and Table 1 in Chap. I), $Re = 8.8 \times 10^6$.   With modern high-speed aircraft, values of Re of 30 million and more occur.   Obviously, it is very difficult to reach such values of Re in wind-tunnel experiments by increasing the size of the model and the air speed of the tunnel.   It is easier to use a wind tunnel in which the air is kept under high pressure (20 to 25 atm.).[1] At a pressure of 20 atm. the kinematic viscosity is reduced to about one-sixteenth to one-twentieth of its value at standard pressure (depending on the temperature), so that the Reynolds number in the case of the above experiment considered would be raised to about $Re = 3 \times 10^6$.

The general experience with viscous fluids shows that the resistance coefficients decrease with increasing Re, but at a very small rate once a certain value of Re is exceeded.   In the present context this applies to the coefficient of profile drag.   Experiments made in the full-scale wind tunnel of the NACA with a special profile (NACA 23012) gave values of the profile-drag coefficient a decreasing from 0.009 at $Re = 1.7 \times 10^6$

---

[1] A description of the variable-density tunnel of the NACA is found in *NACA Tech. Rept.* 416 (1932).

to 0.0075 at $Re = 7.5 \times 10^6$. This means that within this range the parasite-drag coefficient a is reduced by about 17 per cent for Re increasing to 4.5 times its original value.   Probably a further 4.5-fold increase of Re to about 34×10^6 will be accompanied by a considerably smaller reduction of a.

The slope k of the straight portion of the C_L vs. α-diagram also increases slightly with Re. For the profile mentioned the value of k (corresponding to α expressed in degrees) increased from 0.095 at $Re = 1.6 \times 10^6$ to 0.10 at $Re = 4.5 \times 10^6$, both values being reduced to infinite aspect ratio according to $(12a)$.   This suggests again that the failure to reach the theoretical value $k = 0.11$ must be attributed to viscosity effects.   Such effects decrease with increasing Re, since $Re = \infty$ characterizes the perfect fluid.

Even if the Reynolds number of the wind-tunnel experiment equals that corresponding to the full-scale conditions, a certain caution is necessary in applying the experimental results to an airplane in flight.   In fact, the results of the experiment would rigorously apply to the flying airplane only if (1) the model in the wind tunnel were exposed to a strictly steady and uniform stream and (2) the airplane were flying through a bulk of air perfectly at rest.   If these two conditions were fulfilled, the relative motions of the air with respect to the model and the airplane, respectively, would be the same (at the same Re) and the aerodynamic coefficients obtained in the experiment could be applied without hesitation to the flying airplane.   But air under atmospheric conditions is never completely at rest, even when there is no major current (wind), and the stream of air in a wind tunnel is never strictly steady and uniform, even if all efforts are made to render the stream as uniform as possible.   In either case the air is turbulent; i.e., small velocity fluctuations, rapidly changing in magnitude and direction, are superimposed on the state of rest or of steady uniform motion.   Since these fluctuations cannot be expected to follow the same pattern or to assume the same proportions in the wind tunnel and under the actual conditions of flight, they may seriously interfere with the applicability of the results of wind-tunnel tests.   The same difficulty arises when results obtained in different wind tunnels are to be compared; scale and pattern of the velocity fluctuations may vary considerably from one wind tunnel to another.

To make the results of different wind-tunnel tests or of a wind-tunnel test and the free-air conditions comparable, two assumptions are usually introduced which so far have proved helpful.   According to the first assumption the fluctuation can be sufficiently described by a single parameter, called the *degree of turbulence*, which under definite conditions (*e.g.*, for a given wind tunnel) depends only on the Reynolds number. Second, it is assumed that for two different sets of conditions, *e.g.*, for two wind tunnels or for one wind tunnel and the free air, there exists a factor λ such that all phenomena observed at the Reynolds number Re in the first case occur at λRe in the second. In order to find this λ, one characteristic experiment may be used, for example, the sudden drop of the drag coefficient of a sphere (Sec. V.3) or the maximum value of the lift coefficient of an airfoil (see Sec. X.1). If with a certain airfoil the maximum is reached at $Re_1 = 1.3 \times 10^6$ in one wind tunnel and at $Re_2 = 3.1 \times 10^6$ in another wind tunnel, the ratio $Re_2/Re_1 = 2.38$ is taken for λ. If the same phenomenon is observed at $Re_0 = 3.4 \times 10^6$ in free air, the factor $\lambda = Re_0/Re_1 = 2.62$ is called the turbulence factor of the first wind tunnel. For each test carried out in this wind tunnel, $\lambda \times$ actual Reynolds number is called the *effective Reynolds number* of the test.

Tentative values of the turbulence factor for the full-scale wind tunnel and for the (smaller) variable-density tunnel at Langley Field are 1.1 and 2.64, respectively. These values are based on the observations of maximum lift coefficients. Accordingly, the C_L- and C_D-values for an airfoil measured at $Re = 3 \times 10^6$ in the variable-density tunnel are considered to be correct at $Re = 8 \times 10^6$ in free air.

Problem 9. If it is known that the parasite-drag coefficient a of an airfoil decreases by 15 per cent with an increase of Re to four times its value, what is the value of a to be expected in actual flight at sea level under standard conditions at a velocity of 170 m.p.h. for a chord length of 5 ft. if $a = 0.009$ was observed in a wind-tunnel test at $Re = 1,250,000$ and the turbulence factor is known as $\lambda = 2.9$?

CHAPTER VIII

THE WING OF INFINITE SPAN

1. The Momentum Equation for Irrotational Flow. The term "infinite span" is used to indicate that the flow round an airfoil is assumed to be two-dimensional, the streamlines lying in planes parallel to the median plane of the airfoil and the stream pattern being the same in all those planes. For the usual aspect ratios this type of flow may be considered as a first approximation to the actual flow. Moreover, the theory of the wing of infinite span is a necessary first step for the theory of the wing of finite aspect ratio. If in a wind tunnel with two vertical walls the airfoil model is made to extend close to the walls, the flow around the airfoil will be nearly two-dimensional so that the results of the theory of the wing of infinite span can in this way be checked experimentally (see Sec. VII.3).

In this chapter the theory of the wing of infinite span will be developed under the assumption that the air flow past the wing is an *irrotational flow of an incompressible perfect fluid*. Under these conditions the form of the momentum equation, Eq. (7), Sec. III.2 can be modified. As the flow under consideration is irrotational, the constant H of Bernoulli's equation has the same value, not only along each streamline, but throughout the entire field of flow. From Bernoulli's equation, Eq. (4), Sec. II.2, the pressure is found as[1]

$$p = \gamma H - \gamma h - \frac{\rho}{2} V^2 \tag{1}$$

In the vertical plane of flow, choose the x-axis horizontal and the y-axis vertical and directed upward, so as to have $y = h$. Denoting by C' the contour of the profile and by C'' an arbitrary curve enclosing the profile (Fig. 119), apply Eq. (7), Sec. III.2, to the fluid between the contour C' of the airfoil and the curve C''. The resultant of the thrusts that the fluid outside C' transmits across the curve C'' on the fluid within C'', if a layer of unit thickness is considered, has the components

$$P_x = -\int p \cos (n,x) \, dl, \qquad P_y = -\int p \sin (n,x) \, dl \tag{2}$$

where n denotes the outward normal and dl the length of an element of C''

---

[1] Note that here, and up to Eq. (5), V is used for the velocity amount at an arbitrary point of the plane; later, V will again be reserved for the (constant) velocity at infinity.

($dl \times 1$ being the area element).   If, in computing these integrals, the curve C'' is described in the positive (counterclockwise) sense,

$$dl \cos (n,x) = dy \qquad \text{and} \qquad dl \cos (n,y) = -dx,$$

where dx is positive on the lower part CDA of C'' and negative on the upper part ABC, while dy is positive on the right-hand part DAB of C'' and negative on the left-hand part BCD.

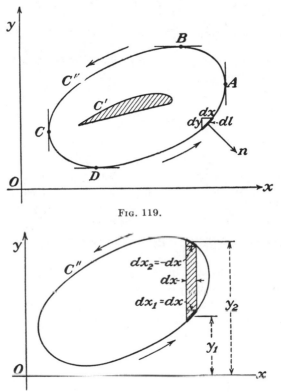

Fig. 119.

Fig. 120.

Figs. 119–120.—Application of momentum theorem to the two-dimensional flow around an airfoil.

Substituting the value of p obtained from Bernoulli's equation (1) in expressions (2) for P_x and P_y and taking account of the fact that for the present system of coordinates $W_x = 0$ and $W_y = -W$, we find

$$P_x + W_x = -\gamma \int H \cos (n,x) \, dl + \gamma \int h \cos (n,x) \, dl$$

$$+ \frac{\rho}{2} \int V^2 \cos (n,x) \, dl \quad (3)$$

$$P_y + W_y = -\gamma \int H \sin (n,x) \, dl + \gamma \int h \sin (n,x) \, dl - W$$

$$+ \frac{\rho}{2} \int V^2 \sin (n,x) \, dl \quad (4)$$

Now, $\gamma \int H \cos (n,x) \, dl = \gamma H \int dy = 0$, since after making the round of C'' we return to the y-value from which we started. Similarly,

$$\gamma \int H \sin (n,x) \, dl = -\gamma H \int dx = 0.$$

Furthermore, substituting y for h,

$$\gamma \int h \cos (n,x) \, dl = \gamma \int y \, dy = \tfrac{1}{2}\gamma \int d(y^2) = 0$$

for the same reason. The contribution to the integral

$$\gamma \int h \sin (n,x) \, dl = -\gamma \int y \, dx$$

made by two elements lying on the upper and lower parts of C'' between two parallels to the y-axis equals $\gamma(y_2 - y_1)|dx|$, where $|dx|$ is the absolute value of the distance between the two parallels to the axis of y (Fig. 120). Now, $\gamma(y_2 - y_1)|dx|$ is the weight of a fluid mass enclosed between the two elements of C'' of width $|dx|$. The integral $\gamma \int h \sin (n,x) \, dl$ thus is seen to equal the weight $\bar{W}$ of a fluid mass that would completely fill the space inside C''. Because of the presence of the airfoil the weight W of the fluid actually contained within C'' is less than $\bar{W}$. The difference $\bar{W} - W$ (appearing in the expression for $P_y + W_y$) equals the weight of the fluid needed to fill the space occupied by the airfoil.

Taking account of the various reductions discussed in the foregoing, we finally obtain from Eq. (7), Sec. III.2, *the components X and Y of the resultant thrust* that the fluid exerts on a portion of unit width of the airfoil as

$$X = \frac{\rho}{2} \int [V^2 \cos (n,x) - 2V_x V_n] \, dl$$

$$Y = \bar{W} - W + \frac{\rho}{2} \int [V^2 \sin (n,x) - 2V_y V_n] \, dl \quad (5)$$

The difference $\bar{W} - W$ is known as the buoyancy and, in accordance with Archimedes' principle, is the only air reaction on the body if the velocity is everywhere zero. In the case of the flow past a wing, the velocity terms have large values as compared with the buoyancy. Accordingly, the buoyancy will be neglected in the following discussion. With

$$V^2 = V_x^2 + V_y^2, \qquad V_n = V_x \cos (n,x) + V_y \sin (n,x)$$

the expressions for X and Y take the form

$$X = \frac{\rho}{2} \int \left[(V_y^2 - V_x^2) \cos (n,x) - 2V_x V_y \sin (n,x) \right] dl$$

$$Y = \frac{\rho}{2} \int \left[(V_x^2 - V_y^2) \sin (n,x) - 2V_x V_y \cos (n,x) \right] dl \qquad (6)$$

These equations constitute the basis for the computation of the force that the fluid exerts on an airfoil of infinite span.

The *resultant moment* (with respect to the origin of the coordinates) of the thrusts that the fluid exerts on the airfoil can be found in a similar way from Eq. (11), Sec. III.3. The general case of a control surface S'' of arbitrary form may be left as an exercise to the reader (see Prob. 1); here C'' is assumed as a circle with the origin O as the center. In this case the moment M_z'' of the thrusts that the fluid outside S'' transmits through this surface is zero, since the lines of action of these thrusts are normal to the circle C'' and therefore pass through the point of reference O. If the entire interior of S'' were occupied by fluid, the moment of the weight of this portion of the fluid with respect to O would also vanish since O would be the center of gravity (c.g.) of this portion of the fluid. The moment of the weight of the fluid between C'' and the contour C' of the airfoil therefore equals the negative moment of the weight of the fluid needed to fill the volume occupied by the airfoil. This may be called the moment of the buoyancy. In our case of the flow past an airfoil this moment is very small as compared with the moment of the dynamic thrusts exerted on the airfoil and will be disregarded. Thus Eq. (11), Sec. III.3, furnishes as the moment M_z' of the thrusts acting on the airfoil

$$M_z' = -\rho \int V_n (x V_y - y V_x)\, dl$$

Now $x V_y - y V_x$ is the moment of the velocity vector with respect to O. This vector can be decomposed into two components one of which, V_n, is normal to the circle C'', while the other, V_t, is tangential to this circle, pointing in the counterclockwise direction. Since the moment of the component V_n with respect to O vanishes, the moment of the velocity vector with respect to O equals $R V_t$, where R is the radius of the circle C''. Thus

$$M_z' = -\rho R \int V_n V_t\, dl \qquad (7)$$

From Eqs. (5) and (7) the forces X, Y and the moment M_z' can be computed as soon as the velocity distribution along the control surface C'' is known.

*Problem 1. Prove that in the case of a cylindrical control surface of arbitrary cross section C''' the moment of the thrusts acting on the surface of the airfoil is given by

$$M'_z = \frac{\rho}{2} \int \left\{ x[(V_x^2 - V_y^2) \sin (n,x) - 2V_x V_y \cos (n,x)] \right.$$

$$\left. - y[(V_y^2 - V_z^2) \cos (n,x) - 2V_z V_y \sin (n,x)] \right\} dl$$

if the moment of the buoyancy is neglected. Note that the differentials under the integral signs in (6) are not the force components dX and dY acting at a respective point.

*Problem 2. Prove that in the case of two-dimensional irrotational flow the force and moment acting on a body within the control surface C'' can be expressed in the form

$$\overrightarrow{F} = -\frac{\rho}{2} \int V^2 \overrightarrow{n}' \, dl, \qquad \overrightarrow{M} = -\frac{\rho}{2} \int V^2 (\overrightarrow{r} \times \overrightarrow{n}') \, dl$$

where $\overrightarrow{r}$ is the radius vector of a point of C'' and $\overrightarrow{n}'$ the vector of unit length, symmetrical to the normal of C'' with respect to the velocity vector.

2. The Lift of an Airfoil of Infinite Span. One way to compute the components X and Y of the resultant of the thrusts that the fluid exerts on the airfoil would be to determine the velocity distribution along the sur-

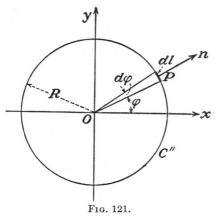

face of the airfoil, to derive from this the pressure distribution by means of Bernoulli's equation, and then to sum (integrate) the contributions of the thrusts acting on the elements of the airfoil surface. Equations (6), however, make it possible to find X and Y in a different manner, using the velocity distribution along a suitably chosen curve C'' surrounding the profile. In the following, this curve will be taken as a circle of infinite radius. In this

FIG. 121.

way it will not be necessary to determine the velocity distribution in the immediate neighborhood of the profile; it will be sufficient *to know the behavior of the velocity distribution at a great distance from the profile.* When this procedure is used, the case of a combination of airfoils (biplane, wing, and tail) is also covered.

In the preceding section the x-axis was taken horizontal so as to make the x-component of the weight of the fluid vanish. Since, ultimately, the influence of the weight of the fluid (buoyancy) has been neglected, Eqs. (6) and (7) remain valid, whatever the orientation of the axes of x and y may be. Let us now give to the x-axis the direction of the velocity V of the uniform rectilinear motion of the airfoil. For the relative motion of the fluid with respect to the airfoil (inverse flow) the velocity components V_x and V_y at a point P then will tend toward the values $-V$ and 0, respectively, as the distance of P from the airfoil increases indefinitely.

For a circle of the finite radius R we have $dl = R \, d\varphi$ (Fig. 121), where φ is the angle between the radius vector OP and the x-axis. If on this circle the velocity components had the constant values $V_x = -V$, $V_y = 0$, Eqs. (6) would furnish

$$X = -\frac{\rho}{2} V^2 R \int_0^{2\pi} \cos \varphi \, d\varphi = 0, \qquad Y = \frac{\rho}{2} V^2 R \int_0^{2\pi} \sin \varphi \, d\varphi = 0 \quad (8)$$

Actually, the relations $V_x = -V$, $V_y = 0$ are not exactly fulfilled on any circle of finite radius R, but the respective differences between V_x, V_y and $-V$, 0 tend to zero as R increases indefinitely. The actual flow may therefore be considered as the result of the superposition of the uniform flow with the velocity V in the negative x-direction and flow patterns for which the velocity tends to zero as R increases indefinitely. Two such stream patterns have been discussed in Sec. II.6, the circulating motion and the bicirculating motion. For these flow patterns the velocity is inversely proportional to R

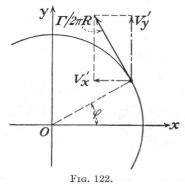

Fig. 122.

and R^2, respectively, and therefore tends to zero with increasing R. As has already been mentioned in Sec. II.6, these are the only two-dimensional irrotational velocity distributions that have this property. It can be shown that velocity distributions for which the velocity is inversely proportional to a fractional power of R, such as $\sqrt{R}$, are incompatible with the conditions for irrotational motion. Thus, at a great distance R from the profile the velocity distribution can be considered as the result of the superposition of (1) a constant velocity with the components $V_x = -V$, $V_y = 0$; (2) a distribution for which the velocity is proportional to $1/R$, that is, a circulating motion; (3) a distribution for which the velocity is proportional to $1/R^2$, that is, a bicirculating motion; (4) distributions for which the velocity is proportional to $1/R^3$, $1/R^4$, etc.

In the evaluation of X and Y we need not consider the contributions from (3) and (4). In fact, if $dl = R \, d\varphi$ is substituted in (6), terms of the order of $1/R$ in the expressions for V_x^2, V_y^2, $V_x V_y$ may supply a finite contribution, while terms of the order $1/R^2$ or higher would furnish products containing the factor $1/R$. If the circulation of the circulating motion is denoted by Γ, the corresponding velocity at a point of the circle of radius R has the magnitude $\Gamma/2\pi R$. Its components are (Fig. 122)

$$V_x' = -\frac{\Gamma}{2\pi R}\sin\varphi, \qquad V_y' = \frac{\Gamma}{2\pi R}\cos\varphi \qquad (9)$$

The velocity components due to the contributions from (1) and (2) are therefore

$$V_x = -V - \frac{\Gamma}{2\pi R}\sin\varphi, \qquad V_y = \frac{\Gamma}{2\pi R}\cos\varphi \qquad (10)$$

In order to evaluate X and Y from Eqs. (6) we need the expressions for V_x^2, V_y^2, and V_xV_y following from (10). In forming these expressions, terms containing the factor $1/R^2$ may be omitted, since they do not furnish a contribution to X or Y. So we obtain from (10)

$$V_x^2 = V^2 + \frac{\Gamma V}{\pi R}\sin\varphi, \qquad V_y^2 = 0, \qquad V_xV_y = -\frac{\Gamma V}{2\pi R}\cos\varphi \qquad (11)$$

Introducing this in Eqs. (6) with $dl = R\,d\varphi$, it follows that

$$X = \frac{\rho}{2}\int_0^{2\pi}\left[-\left(V^2 + \frac{\Gamma V}{\pi R}\sin\varphi\right)\cos\varphi + \frac{\Gamma V}{\pi R}\sin\varphi\cos\varphi\right]R\,d\varphi$$

$$= -\frac{\rho}{2}RV^2\int_0^{2\pi}\cos\varphi\,d\varphi = 0 \qquad (12)$$

$$Y = \frac{\rho}{2}\int_0^{2\pi}\left[\left(V^2 + \frac{\Gamma V}{\pi R}\sin\varphi\right)\sin\varphi + \frac{\Gamma V}{\pi R}\cos^2\varphi\right]R\,d\varphi$$

$$= \frac{\rho}{2}RV^2\int_0^{2\pi}\sin\varphi\,d\varphi + \rho\frac{\Gamma V}{2\pi}\int_0^{2\pi}(\sin^2\varphi + \cos^2\varphi)\,d\varphi = \rho\Gamma V$$

because the integrals of $\sin\varphi$ and $\cos\varphi$ over the period 0 to 2π vanish.

Since the x-axis has been given the direction of the motion of the airfoil and the y-axis is directed upward, $-X$ and Y represent drag and lift. As dl (that is, $dl \times 1$) was written for the element of area, the results obtained can be stated as follows: *In a two-dimensional irrotational flow an airfoil has no drag but experiences a lift that equals per unit of span length the product of density, velocity, and circulation:*

$$L' = \rho\Gamma V \qquad (13)$$

This result is known as the *Kutta-Joukowski* theorem.

Since the circulation Γ is not known, the Kutta-Joukowski theorem does not permit an immediate determination of the lift on an airfoil of a given shape; but important conclusions can be reached by exploiting the following two facts: (1) Any velocity distribution obtained by the superposition of the velocities of two irrotational flows around a profile represents another irrotational flow around the same profile. In fact, it has been seen in Chap. II that the superimposing of irrotational stream patterns always leads to patterns which also are irrotational. Moreover,

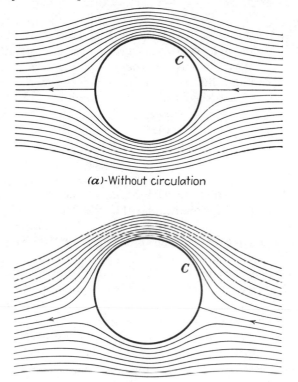

(a)-Without circulation

(b)-With positive circulation

Fig. 123.—Irrotational flow around a cylinder.

if each of the two original velocity distributions gives a tangential flow at the points of the contour, the sum will be tangential, too. (2) Experimental evidence shows that the flow around an airfoil of a given shape is completely determined by the magnitude of the velocity of the undisturbed stream and by the angle of attack. This is not the case for cylindrical bodies of arbitrary cross sections, *e.g.*, for a circular cylinder. It is the principal merit of Joukowski's airfoil theory to have shown that in the case of a profile with a pointed trailing edge there exists only one irrotational stream pattern which fulfills all geometrical and physical conditions and that this flow is completely determined by the shape of the profile and the velocity of the undisturbed stream.

In order to make this clear, let us consider again the two-dimensional irrotational flow around a circular cylinder of infinite length. A circle C then takes the place of the wing profile. Choose the center of this circle as the origin, and lay the x-axis opposite to the velocity V of the undisturbed flow. Any flow around the cylinder must then fulfill the following conditions: At an infinite distance from the circle C, the velocity components are $V_x = -V$, $V_y = 0$; at any point on C, the velocity is tangential to C. Now, one possible flow pattern will be symmetrical with respect to the axis of x (Fig. 123a). In this flow the cylinder obviously does not experience any lift since the magnitude of the velocity and therefore the pressure are the same at any two symmetrically situated points. Another flow around the cylinder, with the same values of V_x and V_y in the undisturbed stream (*i.e.*, at infinity), can be obtained

by superimposing on this symmetrical flow a circulating motion with the origin as the center (Fig. 123b). At infinite distance from the circle C the velocity of the circulating motion vanishes, and at any point on C

B

FIG. 124.

it is tangential to C. Thus, to each value of the circulation Γ there corresponds an irrotational flow around the cylinder, the velocity of the undisturbed stream being the same for all these flow patterns. That is, the flow around a circular cylinder is in no way determined by the velocity of the undisturbed stream.

The situation is essentially the same for a cylinder whose cross section differs slightly from a circle. If, however, the contour of the cross section contains an arc of a very small radius of curvature (B in Fig. 124), most of these stream patterns would lead to extremely high velocity values at B so that, according to Bernoulli's equation, the pressure there would become negative. Such flow patterns have to be discarded. One might expect that in the case of a sharp trailing edge there exists no continuous flow pattern for which the pressure is everywhere positive. At least, it has been seen in Sec. IV.4 that past a body with two sharp edges a discontinuous flow pattern develops which shows a region of dead water behind the body. Yet Joukowski observed that in the case of stream-lined bodies with one sharp edge the infinity of stream patterns corresponding to Γ-values from $-\infty$ to $+\infty$ contains one and only one pattern for which the velocity at the edge remains finite. While the theory of perfect fluids leaves the value of the circulation Γ completely undetermined in the case of a well-rounded body, this value is thus seen to be uniquely determined in the case of a body with a sharp trailing edge. This theoretical result, found by Joukowski as early as 1906, agrees completely with the experimental evidence. It had been known from the very beginning of flight that wings with a sharp trailing edge must be used in order to obtain a well-defined lift.

Let us now turn to the above announced conclusions from Eq. (13). As the circulation, by its definition, depends linearly on the velocity values, the circulation Γ of the sum of two flows each of which has the circulation Γ', Γ'', respectively, must equal $\Gamma' + \Gamma''$. Choose the x-axis parallel to the chord of the profile, and let it point from the trailing to the leading end (Fig. 125). Consider first the flow where the velocity of the undisturbed stream has unit magnitude and the direction opposite to that of the x-axis. In this flow the circulation around the profile will have a definite value Γ_1. If the velocity of the undisturbed stream has the same direction as above but the magnitude V_1, the circulation will have the value $V_1\Gamma_1$ since the new stream pattern can be obtained by

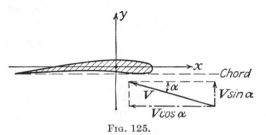

Fig. 125.

superimposing V_1 times the original pattern. (Using a sufficiently small velocity unit, one can always assume that V_1 is an integer.)

Consider next the case where the velocity of the undisturbed stream has the y-direction and unit magnitude. The circulation around the profile then will have a definite value Γ_2. If the velocity of the undisturbed stream has the y-direction and the magnitude V_2, the circulation will be $V_2\Gamma_2$.

Finally, consider the case where the velocity of the undisturbed stream has the magnitude V and forms the angle α with the negative x-direction. Since this velocity has the components $V_x = -V \cos \alpha$ and $V_y = V \sin \alpha$, this case can be considered as the result of the superposition of the two previous cases with $V_1 = V \cos \alpha$ and $V_2 = V \sin \alpha$. Consequently, the circulation around the profile will have the value

$$\Gamma = V(\Gamma_1 \cos \alpha + \Gamma_2 \sin \alpha)$$

With

$$C = 2 \sqrt{\Gamma_1^2 + \Gamma_2^2}, \qquad \sin \alpha_0 = - \frac{\Gamma_1}{\sqrt{\Gamma_1^2 + \Gamma_2^2}}, \qquad \cos \alpha_0 = \frac{\Gamma_2}{\sqrt{\Gamma_1^2 + \Gamma_2^2}}$$

the circulation can be written as

$$\Gamma = \tfrac{1}{2}CV(\sin \alpha \cos \alpha_0 - \cos \alpha \sin \alpha_0) = \tfrac{1}{2}CV \sin (\alpha - \alpha_0) \qquad (13a)$$

The lift per unit length of the airfoil is obtained by introducing this expression in (3). Thus the Kutta-Joukowski theorem yields

$$L' = \frac{\rho}{2} V^2 C \sin (\alpha - \alpha_0) \tag{14}$$

The wing area S to which this value of the lift refers is a rectangle whose sides are the chord length c and the unit of length, respectively. The lift coefficient as defined at the beginning of Chap. VI is the quotient $L'/\frac{1}{2}\rho V^2 S$ where now $S = c \times 1$. Therefore, Eq. (14) is equivalent to

$$C_L = \frac{C}{c} \sin (\alpha - \alpha_0) \tag{15}$$

This formula is identical with the empirical formula, Eq. (4), Sec. VII.1, where k appeared instead of C/c. Since α is the angle of attack (Fig. 95a) and $C_L = 0$ for $\alpha = \alpha_0$, the angle α_0 defines the zero lift direction. Thus $\alpha - \alpha_0$ is the angle of incidence α'.

For small angles of incidence $\sin \alpha'$ can be replaced by α'. This leads to the simplified formula

$$C_L = \frac{C}{c} \alpha' \tag{15'}$$

in accordance with Eq. (3), Sec. VII.1. *The two-dimensional airfoil theory gives the curve C_L vs. α, in good agreement with the experimental evidence for moderate angles of incidence, as an inclined straight line.*

Relations (14) and (15) have been obtained without referring to the particular shape of the profile except for the existence of a sharp trailing end. To go beyond these results, which are an immediate consequence of the Kutta-Joukowski formula (13), *i.e.*, to compute the constants C and α_0, requires a deeper analysis. The values of the factor C and of the zero lift angle α_0 depend on the shape of the profile. Both can be determined theoretically if the (conformal) transformation is known that maps the exterior of the profile on the exterior of a circle.[1] It can be shown that C must equal $8\pi a$, that is, four times the circumference of the circle C_1, and that the zero lift direction is given by the straight line joining M_1 to that point B' of the circle C_1 which corresponds to the trailing end B of the profile. In the cases discussed in Sec. VI.3, where profile forms were derived by applying conformal transformations to a given circle, the values of C and α_0 are thus known in advance. For all ordinary profiles the radius a of the circle C_1 is found to be only slightly

---

[1] For those familiar with the theory of conformal mapping it may be added that the transformation under consideration must leave the points at infinity unchanged. This condition determines the radius a and the center M_1 of the circle C_1 into which the profile is transformed.

greater than $c/4$. In Sec. 6 of the present chapter an approximate solution of the complete problem will be given for thin and slightly cambered profiles. There the value $c/4$ for a will be found as the first approximation, sufficiently exact for most computations; that is, the constant C/c in (14) and (15) equals 2π, and Eq. (15′) gives

$$C_L = 2\pi\alpha' \qquad \text{if } \alpha' \text{ is expressed in radians}$$

$$C_L = \frac{2\pi^2}{180}\,\alpha' = 0.11\alpha' \qquad \text{if } \alpha' \text{ is expressed in degrees}$$

This means that C_L equals 1 for an angle of incidence of about 9°. As has been mentioned in Sec. VII.3, the early Moscow experiments made under the condition of infinite span gave a slope of the curve C_L vs. α' that practically corresponds to this value. Most later experimenters found a slightly smaller value of the slope, giving $C_L = 1$ for an angle of incidence of about 10°. As regards the theoretical values of the zero lift angle α_0, these, too, are found to agree well with the experimental evidence (see Sec. VII.4 and Sec. 6 of the present chapter). *The two-dimensional wing theory seems to overestimate the slope of the C_L vs. α-line by about 10 per cent as compared with the experimental value found for the same angle of incidence. The zero lift angle α_0 actually is 1 to 2° greater than the theoretical value.*

Problem 3. In a wind-tunnel experiment carried out under the conditions of infinite span the lift force on an airfoil of 1.08 ft.² area was found to be 34.7 lb. at an angle of attack 3° and 48.1 lb. at an angle of attack 6°. Compute the zero lift angle and the slope of the C_L vs. α-line if the dynamic pressure of the wind was $q = 40.5$ lb./ft.²

Problem 4. The weight of an airplane flying at 220 ft./sec. at sea level under an angle of incidence of 3° is 4000 lb. Compute the required area S of the wing if the actual lift coefficient is assumed to be three-fourths the theoretical value for a thin airfoil of infinite span. How will the angle of incidence change if the airplane flies at an altitude of 25,000 ft. with 85 per cent of the sea-level velocity?

Problem 5. What is the range in which the angle of incidence α' varies if the velocity of an airplane varies between 280 and 330 ft./sec., the load per unit wing area W/S being 32 lb./ft.³, ρ lying between 0.0022 and 0.0018 slug/ft.³.? Assume the slope of the C_L vs. α-curve as two-thirds the theoretical value for a thin airfoil with infinite aspect ratio.

3. The Pitching Moment of an Airfoil of Infinite Span. A relation for the pitching moment of an airfoil of infinite span can be obtained from the moment of momentum formula (7). Denote by M' the pitching moment exerted on a portion of unit length of the airfoil. With $dl = R\,d\varphi$, Eq. (7) gives

$$M' = -\rho R^2 \int_0^{2\pi} V_n V_t\,d\varphi \tag{16}$$

Here, again, we try to find the limit of this expression for $R = \infty$; *i.e.*, we want to compute the moment M' from our knowledge about the velocity distribution at infinity. If a limit exists, the integral must be of the order $1/R^2$; that is, the contributions that the finite terms and the terms of the order $1/R$ in $V_n V_t$ furnish to the integral must vanish. The contributions of the terms of higher order than $1/R^2$ will have no significance, as $R \to \infty$. Thus, for the computation of the integral (16) only the first three contributions mentioned in the beginning of Sec. 2 of the present chapter need be taken into account.

In Sec. 2 the velocity components on the control circle were given as

$$V_x = -V - \frac{\Gamma}{2\pi R} \sin \varphi, \qquad V_y = \frac{\Gamma}{2\pi R} \cos \varphi \qquad (10)$$

In these expressions the constant term $-V$ represents the velocity of the undisturbed stream, and the terms containing the factor $1/R$ correspond

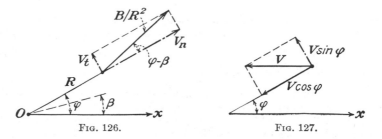

Fig. 126.   Fig. 127.

to a circulating motion with the circulation Γ; terms of an order higher than $1/R$ did not contribute to the lift. Now, however, terms with the factor $1/R^2$ must not be omitted. In other words, a bicirculating motion as discussed in Sec. II.6 must be added to the two component motions considered in the foregoing, the uniform flow and the circulating motion. While the circulating motion is determined by one scalar parameter, *viz.*, the circulation Γ, the bicirculating motion was seen to be determined either by two scalar parameters, the magnitude B and the angle β, or by the bicirculation vector $\overrightarrow{B}$. As has been shown in Sec. II.6, the velocity at a point with the polar coordinates R, φ has the magnitude B/R^2 and forms the angle $\varphi - \beta$ with the radius vector (Fig. 126). Accordingly, the contributions that the bicirculating motion furnishes to the velocity components V_n and V_t are $(B/R^2) \cos (\varphi - \beta)$ and $(B/R^2) \sin (\varphi - \beta)$, respectively. The velocity of the undisturbed stream furnishes the contributions $-V \cos \varphi$ and $V \sin \varphi$ to V_n and V_t, respectively (Fig. 127), while the circulating motion has an entirely tangential velocity $\Gamma/2\pi R$. Thus, the complete values for V_n, V_t to be used in (16) are

$$V_n = -V \cos \varphi + \frac{B}{R^2} \cos (\varphi - \beta)$$

$$V_t = V \sin \varphi + \frac{\Gamma}{2\pi R} + \frac{B}{R^2} \sin (\varphi - \beta) \tag{17}$$

Forming the product $V_n V_t$ and disregarding terms of higher order than $1/R^2$,

$$V_n V_t = -V^2 \sin \varphi \cos \varphi - \frac{\Gamma V}{2\pi R} \cos \varphi$$

$$+ \frac{BV}{R^2} [\sin \varphi \cos (\varphi - \beta) - \cos \varphi \sin (\varphi - \beta)]$$

$$= -\tfrac{1}{2} V^2 \sin 2\varphi - \frac{\Gamma V}{2\pi R} \cos \varphi + \frac{BV}{R^2} \sin \beta$$

Equation (16) then gives

$$M' = -\rho R^2 \int_0^{2\pi} \frac{BV}{R^2} \sin \beta \, d\varphi = -2\pi \rho B \, V \sin \beta \tag{18}$$

since, again, the integrals of $\sin \varphi$, $\cos \varphi$, etc , over a period vanish. Now, if the bicirculation vector B is decomposed into two components parallel and normal to the direction of the undisturbed flow, the normal component is $B \sin \beta$. In analogy to the Kutta-Joukowski theorem, $L' = \rho \Gamma V$, we therefore have the theorem: *The pitching moment (with respect to the origin of the coordinates) exerted on a portion of unit width of the airfoil equals -2π times the product of the density, velocity, and the component B_n of the bicirculation vector normal to the direction of the undisturbed flow:*[1]

$$M' = -2\pi \rho V B_n \tag{18'}$$

The situation here is much the same as with the Kutta-Joukowski formula (13). Since the bicirculation vector is not known, the present theorem does not permit an immediate evaluation of the pitching moment. However, it is possible to proceed in the same manner as in the previous section, *i.e.*, to exploit the facts that the superposition of irrotational flows around a profile leads to other flows of the same kind and that the actual flow is uniquely determined by the magnitudes V and α.

Let us take the x-axis parallel to the zero lift direction of the profile, pointing from the trailing to the leading end. Consider first the flow

---

[1] It should be noted that the bicirculation vector is not independent of the position of the origin with respect to which the motion is analyzed into component parts with velocities proportional to the successive powers of $1/R$.

where the velocity of the undisturbed stream has unit magnitude and the direction opposite to that of the x-axis. Let $\overrightarrow{B_1}$ be the bicirculation vector for this flow. The magnitude of this vector and the angle that it forms with the x-axis will be denoted by B_1 and β_1 respectively (Fig. 128). For the flow where the velocity of the undisturbed stream has unit magnitude and the direction of the y-axis, the corresponding quantities will be denoted by $\overrightarrow{B_2}$, B_2, and β_2. The general case, where the undisturbed stream has a velocity of the magnitude V, forming the angle of incidence α' with the x-axis, can be considered as the result of the superposition of the two previous cases where the velocities at a great distance from the profiles have the magnitude $V \cos \alpha'$ and $V \sin \alpha'$, respectively, instead

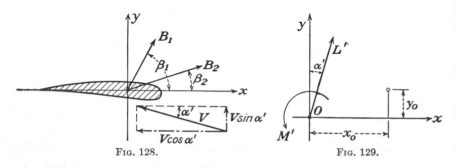

FIG. 128.   FIG. 129.

of the unit magnitudes considered in the foregoing. The bicirculation vector thus can be obtained as the sum of the vectors $(V \cos \alpha')\overrightarrow{B_1}$ and $(V \sin \alpha')\overrightarrow{B_2}$. The component B_n of the bicirculation vector normal to the direction of V equals the sum of the corresponding components of these two vectors. Since the first vector makes the angle $\alpha' + \beta_1$ with the direction of V and the second the angle $\alpha' + \beta_2$, the component of the bicirculation vector normal to the direction to V has the value

$$B_n = V[B_1 \cos \alpha' \sin (\alpha' + \beta_1) + B_2 \sin \alpha' \sin (\alpha' + \beta_2)]$$

Equation (18) therefore furnishes the pitching moment with respect to the origin of the coordinates as

$$M' = -2\pi\rho V^2[B_1 \cos \alpha' \sin (\alpha' + \beta_1) + B_2 \sin \alpha' \sin (\alpha' + \beta_2)] \quad (19)$$

Let us now compute the moment M'_0 with respect to the point with the coordinates x_0, y_0. The thrusts that the fluid exerts on the airfoil are statically equivalent to the force L' applied to the origin of coordinates, together with a couple of moment M'. The force L', given by Eq. (14), is perpendicular to the direction of the undisturbed flow (Fig. 129) and consequently has the components $L'_x = L' \sin \alpha'$ and $L'_y = L' \cos \alpha'$.

With respect to the point with the coordinates x_0, y_0, the thrusts exerted on the airfoil therefore have the moment

$$M_0' = M' - x_0 L_y' + y_0 L_x' = M' - L'(x_0 \cos \alpha' - y_0 \sin \alpha')$$

Substituting the value of L' from (14) and of M' from (19), one has

$$M_0' = -2\pi\rho V^2[B_1 \cos \alpha' \sin (\alpha' + \beta_1) + B_2 \sin \alpha' \sin (\alpha' + \beta_2)$$
$$+ \frac{C}{4\pi} \sin \alpha'(x_0 \cos \alpha' - y_0 \sin \alpha')]$$

Making use of $\sin (\alpha' + \beta_1) = \sin \alpha' \cos \beta_1 + \cos \alpha' \sin \beta_1$ and of the corresponding relation for $\sin (\alpha' + \beta_2)$, we finally obtain M_0' in the form

$$M_0' = -2\pi\rho V^2(a \sin \alpha' \cos \alpha' + b \sin^2 \alpha' + b' \cos^2 \alpha')$$

where

$$a = B_1 \cos \beta_1 + B_2 \sin \beta_2 + \frac{C}{4\pi} x_0$$

$$b = B_2 \cos \beta_2 - \frac{C}{4\pi} y_0, \qquad b' = B_1 \sin \beta_1$$

This gives the moment with respect to any arbitrary point x_0, y_0. We now select a particular point determined by the conditions that the coefficient a vanishes and b and b' are equal. In this case the expression in the parentheses reduces to $0 + b' (\sin^2 \alpha' + \cos^2 \alpha') = b'$. It is easily seen that the point with the coordinates

$$x_0 = -\frac{4\pi}{C} (B_1 \cos \beta_1 + B_2 \sin \beta_2), \qquad y_0 = -\frac{4\pi}{C} (B_1 \sin \beta_1 - B_2 \cos \beta_2)$$

fulfills the conditions $a = 0$ and $b = b' = B_1 \sin \beta_1$. The pitching moment for this particular point of reference therefore equals

$$M_0' = -2\pi\rho V^2 B_1 \sin \beta_1 \tag{20}$$

This formula does not include the angle of incidence α'. The corresponding moment coefficient, which is by definition the quotient $M'/\frac{1}{2}\rho V^2 Sc$, equals therefore, since $S = c \times 1$,

$$C_{M_0} = -4\pi \frac{B_1}{c^2} \sin \beta_1 \tag{20'}$$

This result can be stated as follows: *For any given profile there exists a certain point with respect to which the moment coefficient is independent of the angle of incidence.* It has already been stated in Sec. VII.1 that this is confirmed by the experimental evidence.

The existence of the aerodynamic center, *i.e.*, of a point with respect to which the pitching moment has a constant value, was found as a theoretical result by R. v. Mises in 1920. He also showed that for the usual profiles the aerodynamic center lies near to the forward quarter chord point. This fact suggests that it is desirable to choose this quarter chord point as the point of reference for the pitching moment, a procedure that is often adopted today.

The exact position of the aerodynamic center and the constant value M_0' of the pitching moment can be determined theoretically, if the conformal transformation is known that maps the exterior of the profile on the exterior of a circle and leaves infinity unchanged. In the case of the profiles discussed in Sec. VI.3, which are derived by a given conformal transformation of a circle, the position of the aerodynamic center and the value M_0' are immediately determined. In particular, it is possible to derive profiles for which $M_0' = 0$ (see the last group discussed in Sec. VI.3). For these profiles the line of action of the lift force passes through the aerodynamic center for all values of the angle of attack. The analysis also shows in accordance with the experiments (see Sec. VII.4) that, for all ordinary profiles with simple mean camber lines, C_{M_0} has a small negative value, about -0.05 to -0.10, and increases to zero or to small positive values when reflexed camber lines are considered.

Problem 6. If the aerodynamic center coincides with the one-quarter point and lift and moment coefficients for the angle of attack $\alpha = 7°$ have been found as $C_L = 0.770$ and $C_M = 0.127$ with respect to the center of the chord, compute the constant C_{M_0}.

Problem 7. The lift coefficients C_L of a certain airfoil for the angles of attack $\alpha = 5°$ and $\alpha = 8°$ have been found to be 0.724 and 0.926, respectively. The corresponding moment coefficients with respect to the leading end were -0.225 and -0.275. If it is known that the aerodynamic center lies on the chord, find its position.

4. The Metacentric Parabola. It was seen that the lift L' is proportional to the sine of the angle of incidence α' (14) and that the moment M_0' with respect to a certain point F, the aerodynamic center of the profile, is constant. Thus

$$L' = L_0' \sin \alpha', \qquad M_0' = hL_0' \qquad (21)$$

where L_0' and h are independent of α'. The factor L_0' has the dimension of a force per unit of length, and h is a length.

If for a certain profile the aerodynamic center F, the zero lift direction FA, and the values of L_0' and h are known, the line of action for a given angle of incidence can be found in the following way (Fig. 130): The resultant force is perpendicular to the velocity V; it has the magnitude L' and the moment M_0' with respect to the aerodynamic center F. The distance of its line of action from F is therefore M_0'/L'; according to (21) this equals $h/\sin \alpha'$. In other words, if the foot of the perpendicular drawn from F to the line of action of the lift L' is denoted by G, the distance $\overline{FG}$ must equal $h/\sin \alpha'$. The distance of the point G from the line FA is $\overline{FG}$ sin α'. This product has accordingly the value h, which is independent of the angle of incidence. This solves the problem: In order to find the required line of action, draw the parallel to the velocity V through F, and determine its point of intersection G with the straight line plotted parallel to the zero lift direction at a distance h from F (see Fig. 130). The line of action passes through G and is perpendicular to the velocity V.

This result can also be expressed in the following terms: The feet of the perpendiculars drawn from the aerodynamic center on the lines of action corresponding to different angles of attack lie on a straight line GB which has the direction of zero lift and the distance h from F. Now, a well-known theorem concerning the ordinary parabola states: The feet of the perpendiculars dropped from the focus on the tangents of the

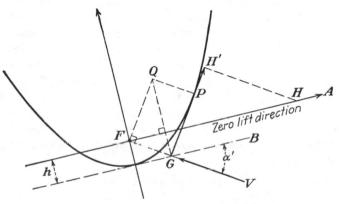

FIG. 130.—Metacentric parabola ($M_0' > 0$).

parabola lie on the straight line that is tangent to the parabola at its vertex. Thus it is seen that *all possible lines of action of the lift L' are tangent to a parabola whose focus is the aerodynamic center F and whose axis is perpendicular to the zero lift direction F*. This parabola is called the *metacentric parabola* of the profile. The meaning of this term will be discussed in Chap. XVII.

If H is a point on the axis FA such that $\overline{FH}$ represents the magnitude L_0', then the projection $\overline{GH'}$ of $\overline{FH}$ on the line of action of the resultant force represents the magnitude L' of this force.

For the purpose of drawing the metacentric parabola it may be useful to determine its point of contact with a particular line of action. Through F draw the perpendicular to the velocity V (Fig. 130) and through G the perpendicular to FA. Through the point of intersection Q of these two perpendiculars draw the parallel to the velocity V; its point of intersection with the line of action is the point of contact P. In the theory of stability (Chap. XVII) the point P is called the *metacenter* of the profile corresponding to the angle of incidence α'.

In drawing the diagram of Fig. 130 the quantity M_0' was assumed to be positive. However, for most airfoils, M_0' is negative. The metacentric parabola and the line of action of the lift L' then have the position indicated in Fig. 131. The significance of the position of the metacentric parabola for the stability of the profile will also be discussed in Chap. XVII.

In the particular case where $h = 0$ and therefore $M_0' = 0$, all possible lines of action, of the lift L' pass through the aerodynamic center F.

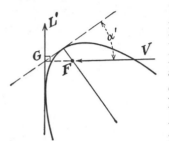

The metacentric parabola reduces then to the focus F. Since the ratio of h to the chord length c is determined by the shape of the profile, the condition $h = 0$ can be considered as an equation between the parameters that determine the shape of the profile. The result here obtained may thus be stated as follows: *If the parameters that determine the shape of the profile are chosen so as to satisfy one certain*

Fig. 131.—Metacentric parabola ($M_0' < 0$).

equation, the line of action of the lift L' passes through the focus F of the profile for all angles of incidence. It would be logical to restrict to this case alone the use of the expression "aerodynamic center" and to speak of the "focus" in the general case.

Problem 8. For a thin profile the moment coefficient with respect to the aerodynamic center is given as -0.10; the zero lift direction forms the angle $\alpha_0 = -4°$ with the chord that is taken as the x-axis. The aerodynamic center has the coordinates $-0.24c$ and $0.03c$ with respect to an origin at the leading edge. Plot the metacentric parabola, find its equation, and find the line of action and the metacenter for the angle of attack $\alpha = 4°$.

Problem 9. In a wind-tunnel experiment the lift coefficient C_L has been measured for two α-values so as to determine the C_L vs. α-line. For how many α-values must the moment coefficient with respect to a given point be measured in order to determine the metacentric parabola?

5. Vortex Sheets, Another Approach. In the preceding sections the two-dimensional flow around an airfoil has been studied as the result of the superposition of a uniform motion, a circulating motion, a bicirculating motion, etc. For the nth of these components of the flow, the velocity at any point P is proportional to $r^{-(n-1)}$, where r is the distance of P from a fixed point O. This approach has proved useful in computing the lift L' and the moment M', since there only the first three components had to be considered, the others having no influence on L' and M'. This approach, however, is not the only possible one. In this section another theory, useful for many purposes, will be discussed.

The two-dimensional flow around an airfoil can also be considered as the result of the superposition of a uniform motion (whose velocity V is equal and opposite to the velocity of the airfoil) and an *infinite number of circulating motions,* each of which has a separate center.   This kind of analysis will lead us to a determination not only of the total lift but also of the lift distribution over the chord of the profile, in the case of infinite aspect ratio.   Furthermore, it will be generalized in the next chapter so as to apply to the three-dimensional problem of a wing of finite span.

Let us first take up again the simple circulating motion introduced in Sec. II.5.   It will be explained now why this type of motion is often referred to as a *vortex motion* although it is actually irrotational.   Accord-

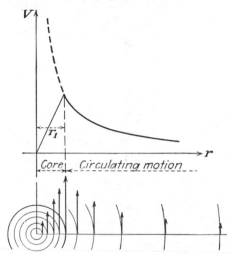

Fig. 132.—Circulating motion with solid rotating core.

ing to Eq. (23), Sec. II.5, the velocity at the center of a circulating motion would be infinite, which, of course, is physically impossible.   The two-dimensional flow in circular streamlines for which the velocity distribution is given by the solid curve in Fig. 132 has the advantage of being physically possible while still agreeing with the equations of motion of perfect fluids.   Outside the small circle $r = r_1$ this modified stream pattern coincides with that of a circulating motion, the velocity distribution being given by

$$V = \frac{\Gamma}{2\pi r} \qquad\qquad r \geqq r_1 \quad (a)$$

Inside this circle $r = r_1$ the fluid rotates as a rigid body with the angular velocity

$$\omega = \frac{V_1}{r_1} = \frac{\Gamma}{2\pi r_1^2} \qquad\qquad r \leqq r_1 \quad (b)$$

i.e., here the velocity is distributed according to

$$V = \omega r = \frac{\Gamma}{2\pi r_1^2} r \qquad\qquad r \leqq r_1 \quad (c)$$

Condition (b) assures the continuity of the velocity across the circle $r = r_1$.

Seen in space, the rotating core forms a thin circular cylinder of the radius r_1. Such a rotating cylinder of fluid is called a *vortex filament*, and the double product of its angular velocity ω and the area of its cross section is called its *vorticity*. According to (b) the vorticity in our case equals the circulation of the (irrotational) circulating motion prevailing outside the vortex filament. Thus a circulating motion may be conceived as attached to a straight vortex filament of the vorticity Γ. If

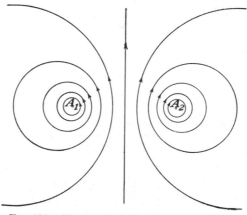

Fig. 133.—Vortex pair with opposite vorticities.

r_1 is made to tend toward zero, this vortex filament reduces to a straight *vortex line* of the vorticity Γ. It is usual to say that the vortex filament or the vortex line of the vorticity Γ *induces* the circulating motion or *induces*, at any point P, a velocity which is perpendicular to the plane determined by P and the vortex line and has the magnitude $V = \Gamma/2\pi r$, where r denotes the distance of P from the vortex line. The expression "induce" is used rather than "produce" in order to avoid representing the velocity as being caused by the vortex line, while still stating that there is a definite relation between the two. If the strictly two-dimensional point of view is adopted, one may speak of a *vortex point inducing a circulating motion* in the plane of flow.

Two or more irrotational flow patterns can be superimposed, as was seen in Sec. II.5, furnishing, then, a new irrotational flow. Figure 133, for example, shows the streamlines of the flow induced by two vortex points (A_1 and A_2) of equal vorticity and opposite sense of rotation, *i.e.*, the sum of two opposite circulating motions of equal circulation having their centers in A_1 and A_2, respectively. Kármán's vortex street (see Sec. IV.4) is an example of a flow induced by an infinite number of vortex points, superimposed on a uniform velocity field.   In each such case the motion can be described as a two-dimensional flow that is irrotational everywhere except at certain *singular points* which coincide with the vortex points inducing the flow.   The circulation as defined in Sec. II.5 is zero for any circuit that does not enclose singular points.   For a circuit enclosing vortex points the circulation equals the sum of the vorticities of the vortex points that lie inside the circuit.

The field of flow around an airfoil must be assumed to be free of singularities since in the neighborhood of a singular point the velocity would tend toward an infinite value and thus the pressure would become

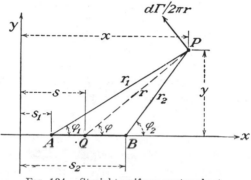

Fig. 134.—Straight uniform vortex sheet.

negative.   Accordingly, the flow cannot be considered as induced by *discrete* vortex points situated either in the region outside the profile or on its boundary.   Nevertheless, the concept of a velocity distribution induced by vortex points or vortex lines is of the greatest importance in airfoil theory, owing to the following fact: By considering a continuous distribution of vortex lines, each of infinitesimal vorticity, over a closed cylindrical surface (or a continuous distribution of vortex points along the closed cross-sectional curve), one arrives at an induced velocity distribution that is free of singularities.   This may be proved in the following manner.

Consider first (Fig. 134) the velocity induced at the point P with the coordinates x, y by a vortex line of the infinitesimal vorticity $d\Gamma$, which is perpendicular to the x-y-plane and intersects this plane at the point Q with the coordinates s, 0.   This velocity is perpendicular to QP and has the magnitude $d\Gamma/2\pi r$, where r denotes the distance $\overline{QP}$.   The components of this velocity with respect to the coordinate axes are

$$-\frac{d\Gamma}{2\pi r}\sin\varphi, \qquad \frac{d\Gamma}{2\pi r}\cos\varphi$$

respectively, where φ is the angle between the x-axis and the line QP.

Now let us assume that we have a *uniformly distributed vorticity* along the portion AB of the x-axis between the points $x = s_1$ and $x = s_2$. If Γ' denotes the vorticity per unit of length, or the *vortex density*, an element of the length ds carries the vorticity $d\Gamma = \Gamma' ds$. Situated at Q, such an element will induce at P the velocity components just found. The entire vortex sheet extending from $x = s_1$ to $x = s_2$ therefore induces at P the velocity components

$$v_x = -\int \frac{\sin\varphi}{2\pi r}\, d\Gamma = -\frac{\Gamma'}{2\pi}\int_{s_1}^{s_2}\frac{\sin\varphi}{r}\, ds$$

$$v_y = \int \frac{\cos\varphi}{2\pi r}\, d\Gamma = \frac{\Gamma'}{2\pi}\int_{s_1}^{s_2}\frac{\cos\varphi}{r}\, ds$$

Now, $r^2 = (x - s)^2 + y^2$. Since x, y is a fixed point, differentiation of this relation with respect to s furnishes $2r\, dr = -2(x - s)\, ds$, and consequently

$$\cos\varphi\, ds = \frac{x - s}{r}\, ds = -dr$$

Similarly, differentiation of $\tan\varphi = y/(x - s)$ furnishes

$$\frac{d\varphi}{\cos^2\varphi} = \frac{y\, ds}{(x - s)^2}.$$

Hence,

$$\sin\varphi\, ds = \frac{y}{r}\, ds = \frac{(x - s)^2}{r\cos^2\varphi}\, d\varphi = r\, d\varphi$$

Therefore,

$$v_x = -\frac{\Gamma'}{2\pi}\int_{\varphi_1}^{\varphi_2} d\varphi = \frac{\Gamma'}{2\pi}(\varphi_1 - \varphi_2), \qquad v_y = -\frac{\Gamma'}{2\pi}\int_{r_1}^{r_2}\frac{dr}{r} = \frac{\Gamma'}{2\pi}\log\frac{r_1}{r_2} \qquad (22)$$

where φ_1 and φ_2 denote the angles xAP and xBP and r_1 and r_2 the distances $\overline{AP}$ and $\overline{BP}$, respectively (Fig. 134).

The velocity induced by a uniform vortex sheet AB at an arbitrary point P is completely determined by (22). The component v_x is finite everywhere, while v_y becomes infinite at the points $A(r_1 = 0)$ and $B(r_2 = 0)$ but remains finite at all other points. For points P on the x-axis to the right of the vortex sheet we have $\varphi_1 = \varphi_2 = 0$ and therefore

$v_x = 0$. For points P to the left we have $\varphi_1 = \varphi_2 = \pi$ and therefore again $v_x = 0$.

Consider now points P lying in the upper half plane where y is positive. If P is very close to the vortex sheet, *i.e.*, if its x is between s_1 and s_2 and its y is a very small positive quantity, φ_1 tends toward zero, φ_2 toward π, and consequently v_x toward $-\Gamma'/2$ (Fig. 135a). On the other hand, if a point P situated in the lower half plane approaches the vortex sheet, φ_1 tends toward 2π, φ_2 toward π, and accordingly v_x toward $+\Gamma'/2$ (Fig.

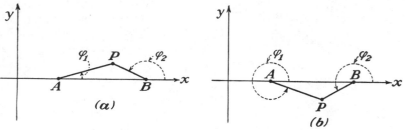

Fig. 135.—Discontinuity at the vortex sheet.

135b). In other words, v_x has the constant value $-\Gamma'/2$ on the upper side and the constant value $+\Gamma'/2$ on the lower side of AB. Thus v_x increases discontinuously by the amount Γ' when the point P crosses the vortex sheet, passing from the side of positive y to that of negative y. It could be guessed that there must occur some discontinuity of v_x at the vortex sheet, for all vortex lines of the sheet induce a negative v_x at the upper side and a positive v_x at the lower side of the sheet, as can be

Fig. 136.

learned from the small circular streamlines shown in Fig. 136. There it is also seen that the vertical velocity components as induced by immediately neighboring points cancel each other. It is noteworthy that the amount Γ' of the jump of v_x does not depend on the length $s_2 - s_1$ of the vortex sheet.

Finally, consider a cylindrical vortex sheet whose cross section is a closed curve C with continuously varying tangent (Fig. 137). Choosing a point O of this curve as the origin of the arc length, we can define any point Q of the curve by giving the length s of the arc OQ measured in the counterclockwise sense. The distribution of the vorticity along C is supposed to be given by a continuous density function $\Gamma'(s)$, such that an element of the length ds situated at s carries the vorticity $\Gamma'(s)\,ds$. Considering the curve as consisting of an infinite number of infinitesimal rectilinear elements, each carrying a uniformly distributed vorticity, and applying formulas (22), we can determine the velocity distribution induced by the vortex sheet.

Since the curve C is closed, there is no singularity of the type encountered at the edges of a straight vortex sheet (points A and B in Fig. 134). In fact, consider two consecutive elements of the curve C with the common point Q (Fig. 138), carrying the vorticities $\Gamma_1'\,ds$ and $\Gamma_2'\,ds$, respectively. Each of these elements separately would induce a velocity in Q with a denominator of the order of magnitude ds. Accordingly, from formulas (22) it can be seen that the velocity which the two elements combined induce at Q remains finite if the difference $\Gamma_1' - \Gamma_2'$ and the angle ϵ formed by the two elements are infinitesimally small. These conditions are fulfilled when the vorticity distribution $\Gamma'(s)$ is a continuous function of s and the curve C has a continuously turning tangent. In this case the velocities induced by the vortex sheet under consideration are finite everywhere.

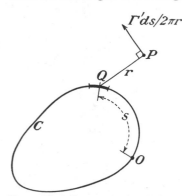

Fig. 137.—Cross section of a cylindrical vortex sheet.

For a straight vortex sheet, the velocity component parallel to the sheet induced at a point P was seen to change discontinuously if the point P crosses the sheet and the amount Γ' of this jump was independent of the length of the vortex sheet. That a discontinuity of the same kind exists for the velocity distribution induced by a curved vortex sheet can be seen as follows: Consider the cross section C of the vortex sheet (Fig. 137) as consisting of an element of the length ds, containing the point Q, and the rest C^* of C, and let the point P cross the curve C at Q. The

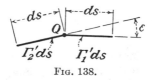

Fig. 138.

velocity components induced by the vorticity spread over C^* vary continuously as P crosses C. The velocity distribution induced by the element ds containing the point Q is of the type already studied since an infinitesimal element of a curve can be considered as straight. Accordingly, the tangential velocity component induced by ds increases discontinuously by $\Gamma'(s)$ as P crosses the curve C. The velocity distribution induced by the entire vortex sheet therefore possesses the same discontinuity. We thus have the following result:

A cylindrical vortex sheet with a continuous distribution of vorticity over its closed cross-sectional curve C induces finite velocities in the whole space. The velocity distribution is continuous everywhere except on the sheet itself, where the tangential velocity component jumps by an amount equal to the vortex density $\Gamma'(s)$ at the point s.

If the curve C does not extend to infinity and if, moreover, the vortex density Γ' is finite everywhere on C, the magnitude of the velocity induced at infinity is zero. In fact, the velocity that an element of the vorticity $\Gamma'\,ds$ induces at the distance r has the magnitude $\Gamma'\,ds/2\pi r$. The magnitude of the resultant velocity induced by the entire vortex sheet cannot be greater than the sum of the magnitudes of the velocities induced by the elements, *viz.*, $\int(\Gamma'/2\pi r)\,ds$. This certainly is not greater than $\Gamma'_{\max}\int(1/2\pi r)\,ds$, where $\Gamma'_{\max}$ denotes the maximum absolute value of Γ'. If the distances r of P from the points of C increase without bounds while the length $\int ds$ of the curve C is finite, the integral $\int ds/r$ tends toward zero.

The value of the circulation along an arbitrary circuit equals the total vorticity carried by that portion of the curve C which lies inside the circuit. If, in particular, the circuit does not include any portion of C, the circulation is zero. For example, the circulation is zero for the circuits 1 and 2 of Fig. 139 and equals $\displaystyle\int_A^B \Gamma'\,ds$ for circuit 3 and $\displaystyle\oint \Gamma'\,ds$ for circuit 4, the latter integral extending over the entire vortex sheet.

Let us now return to the problem of determining the two-dimensional flow around an obstacle of a given shape, for example, a wing profile C, when the velocity at infinity has a given magnitude and direction. Consider the uniform flow the velocity of which has this magnitude and direction, and superimpose on this the flow induced by a vortex sheet which has the contour C of the profile as cross section. The resulting flow evidently fulfills the conditions at infinity (since the induced velocities vanish there) but, in general, will not be tangential to C. An

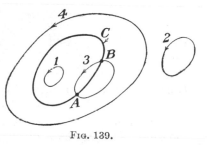

Fig. 139.

attempt may be made to determine the distribution of the vorticity along C, that is, the function $\Gamma'(s)$, in such a way as to fulfill this last condition. An approximate result can be obtained by dividing the contour C into a large number n of equal segments and assuming discrete vortex points $A_1, A_2, \ldots, A_n$ of the vorticities $\Gamma_1, \Gamma_2, \ldots, \Gamma_n$ in the centers of these segments (Fig. 140). Compute the velocity component, normal to the contour, that is induced in each of the points B_1, B_2, . . . , B_n separating two consecutive segments, and express the condition that in each of these points the sum of the induced normal velocity components plus the normal velocity component of the uniform flow is zero. Thus, n linear equations for the n vorticities Γ_1, Γ_2, . . . Γ_n are obtained, and at first glance it would appear that the unknown vorticities may be found from these equations.

This, however, is not the case, for the equations are not independent. In fact, for the uniform flow as well as for the flow induced by the vortices the flux across the curve C is identically zero. Since the elements into which C has been divided are of equal length, this means that the sum of the normal velocity components at the n points B_1, B_2, . . . , B_r is zero for each of the two flows, whatever the values of Γ_1, Γ_2, . . . , Γ_n

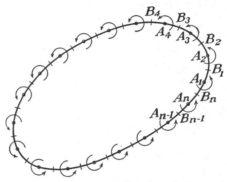

FIG. 140.—Approximation of a vortex sheet by a finite number of vortices.

are. Thus, not more than $n - 1$ of the n linear equations can be independent, and the problem of finding the n vorticities Γ_1, Γ_2, . . . Γ_n is undetermined. In order to arrive at a unique solution, an additional condition must be used. For example, the circulation Γ along a circuit that encloses the entire curve C may be prescribed. Since the value of Γ equals the sum $\Gamma_1 + \Gamma_2 + . . . + \Gamma_n$, this condition furnishes another equation for the vorticities Γ_1, Γ_2, . . . , Γ_n, which thus can be determined. This is in accordance with Sec. VIII.2, where it was shown that the two-dimensional flow around a well-rounded body is not uniquely determined by its shape and by the velocity of the undisturbed stream. There exists, in general, a different flow pattern for each arbitrarily chosen value of the total circulation Γ. Only in the case of an obstacle with a sharp trailing edge can the value of Γ be determined from the condition that the velocity at the trailing edge must remain finite.

If the proper value of Γ is known and the linear equations for Γ_1, Γ_2, . . . , Γ_n are solved, a vorticity distribution over the contour C is found which fulfills approximately the condition that the induced flow will be tangential to C. If the number n is more and more increased and the length of the elements $\overline{B_1B_2}$, $\overline{B_2B_3}$, etc., becomes smaller and smaller, the approximate solution will tend toward an exact one. The vortex distribution, obtained by spreading each of the vorticities Γ_1, Γ_2, . . . , Γ_n uniformly over the length of the corresponding segment, will tend toward a certain limit as the segments are taken smaller and smaller. The result, which is proved rigorously in potential theory, may be stated as follows: *The two-dimensional flow around a closed contour C with a given velocity at infinity and a given circulation Γ for a circuit enclosing C can be considered as the result of the superposition of a uniform flow, the velocity of which equals the given velocity at infinity, and the flow induced by a cylindrical vortex sheet, the cross section of which coincides with C. The vortex distribution along C is uniquely determined when the total amount Γ of the vorticity is given. The velocity is everywhere finite except, possibly, at points of C where the direction of the tangent changes suddenly.*

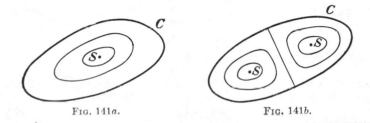

Fig. 141a.   Fig. 141b.

The problem of determining the flow around a given profile thus is reduced to the problem of finding a vortex distribution along the contour C of this profile. An approximate solution can be obtained in the manner already outlined by subdividing C into a finite member of segments, etc. This method has been applied to many practical problems. Without entering into details here, only one general fact concerning the vortex density Γ' need be mentioned.

Consider the streamlines of the field resulting from the superposition of the uniform flow and the field induced by the vortex sheet on C. The distribution of the vorticity along C is determined from the condition that there is no flow across C. Consequently, the streamlines of the resulting flow must not cross C, each streamline remaining entirely either inside or outside C. Now, it is easily seen that the two-dimensional irrotational field of flow of an incompressible fluid which cannot leave the finite region enclosed by a curve C must possess some singularity. In fact, the curve C itself is a closed streamline, and the neighboring inner streamlines also must be closed curves. If, progressing toward the interior of C, we continue to plot the streamlines as in Figs. 141a or 141b, we necessarily obtain streamlines that contract more and more around certain points S.

At such a point the direction of the velocity is not defined. On the other hand, the velocity cannot be zero there, for the condition for an irrotational flow, $dV/dn - V/R = 0$ (see Sec. II.4), shows that the magnitude V of the velocity (which is a positive quantity) increases toward the concave side of the streamlines. But the existence of a finite velocity of undefined direction would be inconsistent with the fact, already proved, that the field of flow induced by a vortex sheet is regular and continuous everywhere but on the sheet. It follows that the field obtained by the superposition of the uniform flow and the field induced by the vortex sheet on C *must have the velocity zero throughout the interior of* C.

On the other hand, it has been seen that at any point of C there is a difference of the amount Γ' between the tangential velocity components at the exterior and interior of C. Inside, all velocities are zero. Since the normal velocity of the resulting flow outside is also zero, it follows that *at any point P of the curve C the velocity V of the outside flow, tangential to C, equals the vortex density* Γ' *at this point.* Positive values of the velocity V correspond to the counterclockwise sense of rotation around C. The determination of the vortex density Γ' therefore furnishes immediately the velocity along C, and the pressure exerted on C can then be found from Bernoulli's equation.

Problem 10. Study the manner in which the stream pattern of Fig. 133 changes when the vorticity of the right-hand vortex point is doubled, tripled, etc.

Problem 11. Determine the flow induced by a cylindrical vortex sheet of circular cross section and constant vortex density Γ'.

*Problem 12. Show that the symmetrical flow past a circle of radius a (Fig. 123a) can be obtained by superimposing a uniform flow of the velocity $\overrightarrow{V}$ and a bicirculating motion whose bicirculation vector equals $-a^2\overrightarrow{V}$. Using this result, show that the same flow can also be considered as the sum of the uniform flow and the flow induced by a vortex distribution over the circle with the vortex density $\Gamma' = 2V \sin \vartheta$ at the point P, where ϑ is the angle of OP with the direction of the uniform flow.

*Problem 13. Try to find approximate values for the vortex density Γ' at the points of the circle by subdividing the periphery into 16 equal parts and applying the method described in the preceding section. Compare the values found in this way with the exact solution $\Gamma' = 2V \sin \vartheta$ as given in Prob. 12.

6. Theory of Thin Airfoils. It has been mentioned that the so-called *direct problem, i.e.,* the theoretical determination of the aerodynamic coefficients of a given profile, is by no means easy to solve. In view of the fact that, within the range of the usual profiles, these coefficients vary but little, it may even be doubtful whether it is worth while to attempt a rigorous solution of this difficult problem. However, a method that furnishes approximate information without involving too complicated analysis may often prove useful. In the following, such a method, known as the *theory of thin airfoils,* will be discussed on the basis of the

approach outlined in the preceding section.   The method was developed
for the first time by M. Munk.

Modern wing sections have a maximum thickness of about 10 to 12
per cent of the chord and a maximum mean camber of about 2 to 1 per
cent of the mean chord.   For such almost flat, thin profiles an approximate theory would replace the airfoil by its mean camber line C and
regard C as deviating but little from a straight line that is supposed to

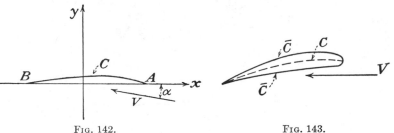

Fig. 142.   Fig. 143.
Figs. 142–143.—Replacement of the thin airfoil by its mean camber line.

form a small angle with the direction of the undisturbed stream (Fig. 142).
For the infinitely thin profile thus obtained to produce a lift, the pressure
must have different values on the upper and lower side of C.   According
to Bernoulli's equation, the magnitude of the (tangential) velocity will
also have different values on the two sides of C.   Since the velocity
distribution induced by a vortex sheet with C as the cross section has a
discontinuity of just this kind, an attempt may be made to build up the
flow around the infinitely thin profile by
superimposing on a uniform field of flow the
field induced by a vortex sheet with the cross
section C.   The vortex density $\Gamma'(s)$ of this
sheet must be determined in such a way that
there is no resulting flow across the line C.

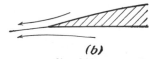

(a)

This approach is in accordance with the
results of the preceding section.   Indeed, the
two-dimensional irrotational flow round any
given profile $\bar{C}$ can be considered as the result
of the superposition of a uniform flow and the

(b)
Fig. 144.

flow induced by a certain vorticity distribution on $\bar{C}$.   In the present case
the upper and lower part of $\bar{C}$ are considered as coinciding on C (Fig. 143).
This leads to the assumption of a single vortex layer on C.

There arises, however, the following difficulty: The leading end of
the actual profile is rounded, and the trailing end is pointed.   It has been
seen in Sec. VIII.2 that in this case, for any given angle of attack, there
exists only one pattern of continuous flow, which does not involve an

infinite velocity at the trailing end.   According to this pattern the air
does not flow round the pointed trailing edge (Fig. 144a), since this would
lead to an infinite velocity, but leaves the trailing end B in a direction
near to that of the mean camber line (Fig. 144b).   When the profile is
replaced by its mean camber line, the leading end A becomes pointed, too.

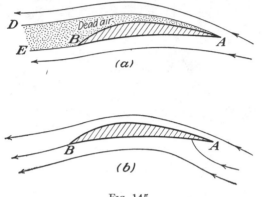

FIG. 145.

For an arbitrarily given angle of attack there exists, then, no continuous
flow pattern with finite velocities at both points A and B.   Actually, the
flow past a profile with two sharp edges A and B (Fig. 145) is of the
discontinuous type (Helmholtz flow) discussed in Sec. IV.4, with a narrow
dead-air region bounded by the two lines AD and BE seen in Fig. 145a.

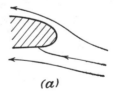

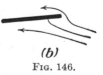

FIG. 146.

If the sharp edge at A is gradually replaced by
a rounded nose of increasing radius of curvature, the
upper boundary of the dead-air region approaches
more and more the upper side of the profile and the
wake line BE.   At a certain amount of curvature the
dead-air region disappears, the flow has a smooth
wake line as shown in Fig. 144b, and the situation at
the leading end is that shown in Fig. 146a.   This is
the case with an ordinary thin airfoil.   On the other
hand, if we disregard, in the case of the double-
pointed profile (Fig. 145), the condition of finite
velocity at the leading end, we find a flow as shown
in Fig. 145b.   Here the situation in the wake is
practically the same as in Fig. 144b; and the situation at the front,
except for the point A itself, is approximately that of Fig. 146a.
Thus we come to the conclusion that *an approximate solution for a
thin airfoil with two sharp edges, under a small angle of attack, can
be found by considering the irrotational flow with finite velocity at the
trailing end but infinite velocity at the leading end.*   This idea of dealing

with a thin profile with pointed leading and trailing ends is due to W. M. Kutta (1902), who thus anticipated to a certain extent Joukowski's exact solution for a profile with rounded leading end and pointed trailing edge. The Antoinette profile, mentioned in Sec. VII.3, was suggested by Kutta's theory.

Returning now to the case of an infinitely thin airfoil, *i.e.*, to a profile replaced by its mean camber line (*AB* in Fig. 147), we see that the solution we have to look for is the irrotational flow pattern which fulfills the condition of finite velocity at the point *B*, but not at *A*.

In the preceding section the flow induced by a plane vortex sheet with a uniformly distributed vorticity has been studied, and the velocity at the ends of the sheet was seen to be infinite. It can be easily proved also, that when the cross section of the sheet is curved and if the vorticity is not uniformly distributed, the velocity at the ends becomes infinite unless the vortex density Γ' here is zero. (Since Γ' equals the difference

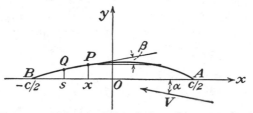

FIG. 147.—Velocities induced along the vortex sheet.

between the velocity values on the two sides of the sheet, it is obvious that the velocity distribution must possess some singularity when $\Gamma' \neq 0$ up to the end of the sheet.) Kutta's condition stipulating a finite velocity at the trailing end therefore implies that the vortex density Γ' is zero here. In addition, the distribution of vorticity Γ' over the sheet is subject to the condition that there is no resulting flow across the line *C*.

To derive the mathematical expression for these conditions, refer the mean camber line *C* to a system of rectangular coordinates whose origin *O* coincides with the center of the chord *AB* and whose positive x-axis is directed along *OA* (Fig. 147). The ordinates of the mean camber line are then given by a function $y(x)$ that is defined for $-c/2 \leqq x \leqq c/2$. These ordinates as well as the slope of the mean camber line, $\beta = dy/dx$, are supposed to be small. Let V denote the magnitude of the velocity of the undisturbed stream and α the angle of attack, which is also supposed to be small. At any point of the mean camber line *C* the uniform flow of the velocity V then has a velocity component, normal to *C*, of the magnitude $V \sin (\alpha + \beta) \sim V(\alpha + \beta)$. The vortex density Γ' must be determined so that for any point on *C*, this normal velocity component is canceled by the corresponding component of the induced flow.

If terms of higher order are neglected, the distance of the points P and Q on C, with the abscissas x and s, respectively (see Fig. 147), can be considered as equal to $x - s$. The line PQ has nearly the direction of the tangent of C at P; consequently, the velocity induced in P by $\Gamma'(s)\,ds$ at Q has nearly the direction of the normal of C at P. The magnitude of this induced velocity equals the quotient $\Gamma'(s)\,ds$ divided by 2π times the distance $\overline{PQ} = x - s$; and the total velocity induced by the entire sheet is $\displaystyle\int \frac{\Gamma'(s)\,ds}{2\pi(x - s)}$, the integral extended from the point B to the point A.

We thus arrive at the following problem: *For a given function $\beta(x)$, defined in the interval $-c/2 \leq x \leq c/2$, determine $\Gamma'(x)$ so that*

$$\Gamma'\left(-\frac{c}{2}\right) = 0 \quad \text{and} \quad \frac{1}{2\pi} \int_{-c/2}^{c/2} \frac{\Gamma'(s)\,ds}{x - s} = -V[\alpha + \beta(x)] \quad (23)$$

for any x in the interval $-c/2 \leq x \leq c/2$, where V and α are given constants.

If this problem is solved, the circulation along any circuit surrounding the airfoil is found as

$$\Gamma = \int_{-c/2}^{c/2} \Gamma'(s)\,ds \qquad (24)$$

According to the Kutta-Joukowski theorem (13) the lift then equals

$$L = \rho V \int_{-c/2}^{c/2} \Gamma'(s)\,ds \qquad (24a)$$

and the lift coefficient

$$C_L = \frac{2}{cV} \int_{-c/2}^{c/2} \Gamma'(s)\,ds \qquad (24b)$$

The moment and the moment coefficient can be found in the following way: According to the preceding section, the tangential velocity components on the two sides of a vortex sheet differ by an amount equal to the vortex density Γ'. In the present case, Γ' has been determined so that the normal velocity components vanish all along the sheet. Consequently, Γ' represents the difference of the values of the velocity at the two sides of the profile. If the arithmetical mean of these values is denoted by V_m, the difference Δp of the values of the pressure below and above the profile is found from Bernoulli's equation as

$$\Delta p = \frac{\rho}{2}\left[\left(V_m + \frac{\Gamma'}{2}\right)^2 - \left(V_m - \frac{\Gamma'}{2}\right)^2\right] = \rho V_m \Gamma'$$

Now, the flow around a nearly straight thin airfoil flying at a small angle of attack differs but very little from the undisturbed flow. Within the accuracy of the present approximation it is therefore justified to take $V_m = V$. This gives the lift

$$L = \int_{-c/2}^{c/2} \Delta p \, ds = \rho V \int_{-c/2}^{c/2} \Gamma' \, ds$$

in accordance with (24a) and the moment with respect to the center of the chord

$$M = \int_{-c/2}^{c/2} s \, \Delta p \, ds = \rho V \int_{-c/2}^{c/2} s \Gamma'(s) \, ds \qquad (25a)$$

The moment coefficient with respect to the center of the chord equals, according to the definition of C_M given in Sec. VI.1,

$$C_M = \frac{2}{c^2 V} \int_{-c/2}^{c/2} s\Gamma\,'(s) \, ds \qquad (25b)$$

The problem of determining the coefficients of lift and moment of a thin airfoil thus is reduced to that of finding the vortex density Γ' which satisfies the conditions (23).

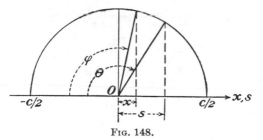

FIG. 148.

In dealing with Eqs. (23) to (25) it is helpful to replace the variables x and s by new variables φ and ϑ defined by

$$\frac{2x}{c} = -\cos \varphi, \qquad \frac{2s}{c} = -\cos \vartheta \qquad (26)$$

As x and s vary from $-c/2$ to $c/2$, the new variables, φ and ϑ, as seen in Fig. 148, vary from 0 to π. With $ds = c \sin \theta \, d\vartheta/2$, Eqs. (23), (24b), and (25b) take the form

$$\Gamma_\vartheta' = 0 \qquad \text{for} \qquad \vartheta = 0$$

$$\frac{1}{2\pi} \int_0^\pi \frac{\Gamma_\vartheta' \sin \vartheta \, d\vartheta}{\cos \vartheta - \cos \varphi} = -V(\alpha + \beta_\varphi) \qquad (23')$$

$$C_L = \frac{1}{V} \int_0^\pi \Gamma_{\vartheta}' \sin \vartheta \, d\vartheta \tag{24c}$$

$$C_M = -\frac{1}{2V} \int_0^\pi \Gamma_{\vartheta}' \cos \vartheta \sin \vartheta \, d\vartheta \tag{25c}$$

where Γ_{ϑ}' and β_{φ} are now functions of the new variables ϑ and φ.

Consider first the case of a *symmetrical thin airfoil*. The mean camber line then coincides with the chord AB, and we have $\beta = dy/dx = 0$. In this case Eqs. (23') are satisfied by the following two equivalent expressions

$$\Gamma_{\vartheta}' = 2\alpha V \frac{1 - \cos \vartheta}{\sin \vartheta} \quad \text{or} \quad \Gamma_{\vartheta}' = 2\alpha V \tan \frac{\vartheta}{2} \tag{27}$$

In fact, upon substituting the first expression for the vortex density Γ_{ϑ}' in the second equation (23') the factor $\sin \vartheta$ cancels out, and the left-hand side becomes

$$\frac{\alpha V}{\pi} \int_0^\pi \frac{d\vartheta}{\cos \vartheta - \cos \varphi} - \frac{\alpha V}{\pi} \int_0^\pi \frac{\cos \vartheta \, d\vartheta}{\cos \vartheta - \cos \varphi} \tag{28}$$

Now, the following integral formula will be proved at the end of this section:

$$\frac{1}{\pi} \int_0^\pi \frac{\cos n\vartheta \sin \varphi}{\cos \vartheta - \cos \varphi} \, d\vartheta = \sin n\varphi \quad \text{for } n = 0, 1, 2, 3, \cdots \tag{29}$$

With $n = 0$ this gives

$$\int_0^\pi \frac{d\vartheta}{\cos \vartheta - \cos \varphi} = 0 \tag{29'}$$

and, with $n = 1$,

$$\int_0^\pi \frac{\cos \vartheta \, d\vartheta}{\cos \vartheta - \cos \varphi} = \pi \tag{29''}$$

Expression (28) is thus seen to equal $-\alpha V$, as it should, according to the second equation (23'), since $\beta = 0$ in the present case. *Thus (27) is proved to be the solution of our problem in the case of a straight mean camber line.*

Upon introducing (27) into the formulas (24c) and (25c), the lift and drag coefficients are found:

$$C_L = 2\alpha \int_0^\pi (1 - \cos \vartheta) \, d\vartheta = 2\pi\alpha \tag{30}$$

$$C_M = -\alpha \int_0^\pi (1 - \cos \vartheta) \cos \vartheta \, d\vartheta = \int_0^\pi \cos^2 \vartheta \, d\vartheta = \frac{\pi}{2} \alpha \tag{31}$$

Equation (30) shows that the slope of the C_L vs. α-line for the infinitely thin straight airfoil is 2π. From (31) the distance d of the line of action from the center of the chord can be found since $Ld = M$ and therefore $C_L d = C_M c$.

$$d = \frac{C_M}{C_L} c = \frac{c}{4}$$

The center of pressure therefore lies a quarter of the chord aft the leading edge, irrespective of the value of the (small) angle of attack. The moment of the aerodynamic forces with respect to the forward quarter chord point thus being zero for all angles, this point is seen to be the aerodynamic center of the thin symmetrical profile. The moment coefficient with respect to it is zero, as was to be expected for a symmetrical profile. These results bear out the statements made in Sec. VII.1c.

The case of a *curved mean camber line* can be treated by assuming that the vortex density Γ' can be represented as a sum of the form

$$\Gamma_\vartheta' = 2V \left(\Gamma_0' \frac{1 - \cos \vartheta}{\sin \vartheta} + \Gamma_1' \sin \vartheta + \Gamma_2' \sin 2\vartheta + \cdots + \Gamma_n' \sin n\vartheta \right) \tag{32}$$

where the constants Γ_0', Γ_1', . . . , Γ_n' should be determined by the shape of the mean camber line. Expression (32) obviously satisfies the first condition (23'), that is, $\Gamma_\vartheta' = 0$ for $\vartheta = 0$. Furthermore, the first term on the right-hand side of (32) is essentially the expression (27). The evaluation of the left-hand side of the second equation (23') is rendered possible by the following integral formula, which will also be proved at the end of this section:

$$\frac{1}{\pi} \int_0^\pi \frac{\sin n\vartheta \sin \vartheta}{\cos \vartheta - \cos \varphi} \, d\vartheta = -\cos n\varphi \tag{33}$$

Upon substituting (32) in the second equation (23') and making use of (28) and (33), the second condition (23') takes the form

$$\Gamma_0' + \Gamma_1' \cos \varphi + \Gamma_2' \cos 2\varphi + \cdots + \Gamma_n' \cos n\varphi = \alpha + \beta_\varphi \tag{34}$$

Without restricting the generality of the discussion, the function β_φ, which determines the shape of the profile, can be represented by a sum of the form

$$\beta = \beta_0 + \beta_1 \cos \varphi + \beta_2 \cos 2\varphi + \cdots + \beta_n \cos n\varphi \qquad (35)$$

where the constants $\beta_0, \beta_1, \ldots, \beta_n$ depend on the shape of the mean camber line. Comparing (34) and (35), we see that the following conditions must be fulfilled:

$$\Gamma_0' = \alpha + \beta_0, \qquad \Gamma_1' = \beta_1, \qquad \Gamma_2' = \beta_2, \qquad \cdots, \qquad \Gamma_n' = \beta_n \qquad (36)$$

This solves the problem. The coefficients of expression (32) for the vortex density Γ' are determined in (36) by the coefficients of expression (35) given for the slope β of the mean camber line.

Since

$$\int_0^\pi \sin p\vartheta \sin q\vartheta \, d\vartheta = 0 \qquad (37)$$

if p and q are integers different from each other, and since

$$\int_0^\pi \sin^2 p\vartheta \, d\vartheta = \frac{\pi}{2} \qquad (37')$$

formulas (24c) and (25c) give the following values of the coefficients of lift and moment:

$$C_L = 2\pi(\alpha + \beta_0)$$
$$\qquad + 2 \int_0^\pi (\beta_1 \sin \vartheta + \beta_2 \sin 2\vartheta + \cdots + \beta_n \sin n\vartheta) \sin \vartheta \, d\vartheta$$
$$= 2\pi \left(\alpha + \beta_0 + \frac{\beta_1}{2} \right) \qquad (38)$$

and

$$C_M = \tfrac{1}{2}\pi(\alpha + \beta_0)$$
$$\qquad - \tfrac{1}{2} \int_0^\pi (\beta_1 \sin \vartheta + \beta_2 \sin 2\vartheta + \cdots + \beta_n \sin n\vartheta) \sin 2\vartheta \, d\vartheta$$
$$= \tfrac{1}{2}\pi \left(\alpha + \beta_0 - \frac{\beta_2}{2} \right) \qquad (39)$$

These equations, for the first time given by M. Munk (1922), include the principal results of the thin-wing theory. From (38) it is seen that the slope of the C_L vs. α-curve is 2π, independently of the shape of the profile, while the zero lift angle $\alpha_0 = -\beta_0 - \beta_1/2$ depends on the shape.

From (39) the moment coefficient C_{M_0} can be derived with respect to the one-quarter point on the chord. Since, according to the elementary laws of statics, $M_0 = M - (c/4)L$, it follows that

$$C_{M_0} = C_M - \tfrac{1}{4}C_L = -\frac{\pi}{4}(\beta_1 + \beta_2) \tag{40}$$

As this is independent of α, the quarter point is seen to be the aerodynamic center.

The outcome, which is in best accordance with what was seen in the preceding sections of this chapter, can be summarized as follows: *For a sufficiently thin and flat profile the slope of the C_L-curve is 2π, and the aerodynamic center coincides with the quarter point on the chord. The zero lift angle and the moment coefficient for the a.c. depend on the shape and are determined by the first three coefficients β_0, β_1, β_2, in the development (35) of dy/dx.*

The question remains to determine β_0, β_1, and β_2 from the given form of the profile. Suffice it to say that, if the ordinate y of the mean camber line is given as a polynomial in x, all β-values can be computed by means of the well-known trigonometric formulas that express each power of $\cos y$ in terms of $\cos y$, $\cos 2y$, $\cos 3y$, etc. The first of these expressions are

$$\cos^2 \varphi = \tfrac{1}{2}(1 + \cos 2\varphi), \qquad \cos^3 \varphi = \tfrac{1}{4}(3 \cos \varphi + \cos 3\varphi), \\ \cos^4 \varphi = \tfrac{1}{8}(3 + 4 \cos 2\varphi + \cos 4\varphi) \text{ etc.} \tag{41}$$

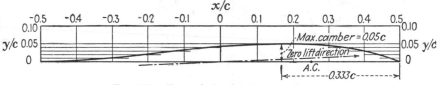

Fig. 149.—Example for the thin wing theory.

Take, for example, the mean camber line shown in Fig. 149 corresponding to the equation

$$y = \kappa c \left(1 - 4\frac{x^2}{c^2}\right)\left(\lambda + \frac{x}{c}\right) \tag{42}$$

with $\kappa = 0.0422$ and $\lambda = 0.5$. By differentiation we find that

$$\beta = \frac{dy}{dx} = -8\kappa \frac{x}{c}\left(\lambda + \frac{x}{c}\right) + \kappa\left(1 - 4\frac{x^2}{c^2}\right) = \kappa\left(1 - 8\lambda\frac{x}{c} - 12\frac{x^2}{c^2}\right)$$

Substituting $-\tfrac{1}{2}\cos \varphi$ for x/c according to (26) and using (41),

$$\beta = \kappa(1 + 4\lambda \cos \varphi - 3\cos^2 \varphi) = \kappa(-\tfrac{1}{2} + 4\lambda \cos \varphi - \tfrac{3}{2}\cos 2\varphi) \tag{43}$$

Thus $\beta_0 = -\kappa/2$, $\beta_1 = 4\kappa\lambda$, $\beta_2 = -3\kappa/2$ (and all further $\beta_n = 0$). This supplies the zero lift angle $\alpha_0 = -(4\lambda - 1)\kappa/2$ and the moment coefficient with respect to the aerodynamic center $C_{M_0} = -\pi(4\lambda - \frac{3}{2})\kappa/4$. The detailed discussion of these results for various values of κ and λ as well as the application of the general method to other examples is left to the reader (see Probs. 14 to 16).

APPENDIX

Let us now proceed to prove the integration formulas (29) and (33). As a first step, we establish the equation

$$I_n = \int_{-\pi}^{\pi} \cos nx \cot \frac{x}{2} \, dx = 0 \qquad \text{for } n = 1, 2, 3, \cdots \quad (44)$$

Indeed, $\cos nx$ is an even function and $\cot x/2$ an odd function of x (Fig. 150). The contributions that two symmetrically situated intervals like

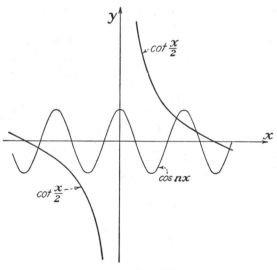

FIG. 150.

$-a \leqq x \leqq -b$ and $b \leqq x \leqq a$ furnish to the integral I_n will therefore cancel. It is true that immediately to the left and right of $x = 0$ the negative and positive contributions both become infinite. But if the integral is taken from $-\pi$ to $-\epsilon$ and from ϵ to π, the sum of these two expressions vanishes, however small ϵ is. The so-called "principal value" of I_n is therefore zero. That this value has to be taken in all integrals occurring in the present argument follows from what they stand for in the vortex theory.

Next, we prove the formula

$$J_n = \int_{-\pi}^{\pi} \sin nx \cot \frac{x}{2} dx = 2\pi \qquad \text{for } n = 1, 2, 3, \cdots \qquad (45)$$

For $n = 1$ this is easily seen to be correct. In fact,

$$J_1 = \int_{-\pi}^{\pi} \sin x \cot \frac{x}{2} dx = 2 \int_{-\pi}^{\pi} \cos^2 \frac{x}{2} dx = \int_{-\pi}^{\pi} (1 + \cos x) \, dx = 2\pi \qquad (46)$$

For an integer $n > 1$ we compute the difference $J_n - J_{n-1}$ using the transformation

$$[\sin nx - \sin (n-1)x] \cot \frac{x}{2} = 2 \cos \left(n - \frac{1}{2} \right) x \sin \frac{x}{2} \cot \frac{x}{2}$$

$$= 2 \cos \left(n - \frac{1}{2} \right) x \cos \frac{x}{2} = \cos nx + \cos (n-1)x \qquad (47)$$

It follows that

$$J_n - J_{n-1} = \int_{-\pi}^{\pi} \cos nx \, dx + \int_{-\pi}^{\pi} \cos (n-1)x \, dx = 0 \qquad (48)$$

since each of these two integrals is zero for any integer n greater than 1. Thus J_n is the same for all n. As the functions under the integral signs in I_n and J_n are periodic with the period 2π, the interval of integration, $-\pi$ to π, can be replaced by the interval from $\varphi - \pi$ to $\varphi + \pi$ without changing the value of the integrals.

Consider now the integral (29). Using the familiar formulas for the sine and cosine of the sum and the difference of two angles,

$$\cot \frac{\varphi - \vartheta}{2} + \cot \frac{\varphi + \vartheta}{2} = \frac{2 \sin \varphi}{\cos \vartheta - \cos \varphi} \qquad (49)$$

The integral (29) therefore equals

$$\frac{1}{2\pi} \int_{0}^{\pi} \left(\cot \frac{\varphi - \vartheta}{2} + \cot \frac{\varphi + \vartheta}{2} \right) \cos n\vartheta \, d\vartheta = \frac{1}{2\pi} \int_{-\pi}^{\pi} \cot \frac{\varphi + \vartheta}{2} \cos n\vartheta \, d\vartheta$$

With $x = \varphi + \vartheta$ this can be written as

$$\frac{1}{2\pi} \int_{\varphi - \pi}^{\varphi + \pi} \cot \frac{x}{2} (\cos nx \cos n\varphi + \sin nx \sin n\varphi) \, dx$$

$$= \frac{I_n}{2\pi} \cos n\varphi + \frac{J_n}{2\pi} \sin n\varphi = \sin n\varphi \qquad (50)$$

Thus (29) is proved for $n = 1$, and since $I_0 = J_0 = 0$, it holds also for $n = 0$.

Finally, consider the integral (33). The numerator of the fraction appearing under the integral sign can be written as

$$\sin n\vartheta \sin \vartheta = \tfrac{1}{2} \cos (n - 1)\vartheta - \tfrac{1}{2} \cos (n + 1)\vartheta$$

Upon making use of (29), the integral (33) is thus seen to equal

$$\tfrac{1}{2} \frac{\sin (n - 1)\varphi}{\sin \varphi} - \tfrac{1}{2} \frac{\sin (n + 1)\varphi}{\sin \varphi} = - \cos n\varphi \tag{51}$$

which completes the proof.

Problem 14. What condition must be fulfilled for the class of thin airfoils represented by (42) so that all lines of action of the lift forces intersect at one point for any (small) angle of incidence (existence of a center of pressure)?

Problem 15. For the thin airfoil (42), express maximum camber and maximum camber position in terms of κ and λ. Discuss how the zero lift angle and C_{M_0} change with those parameters. Show that C_{M_0} increases if the shape becomes more and more reflexed.

Problem 16. The mean camber line of a thin airfoil is represented by

$$y = \kappa c \left(1 - 4 \frac{x^2}{c^2} \right) \left(1 + \mu \frac{x}{c} + \nu \frac{x^2}{c^2} \right)$$

Find C_L and C_{M_0} in terms of κ, μ, and ν, and determine the condition for the existence of a center of pressure. With this condition fulfilled, find the useful range of the parameter μ by discussing the zero lift angle α_0 in terms of the maximum camber position. Plot the metacentric parabola for $\kappa = 0.05$, $\mu = 1.5$, and $\nu = 0.2$.

CHAPTER IX

THE WING OF FINITE SPAN

1. Curved Vortex Lines. The role of straight vortex lines in the theory of two dimensional fluid motion has been explained in Secs. VIII.5 and 6. The two-dimensional irrotational flow of an incompressible fluid around a cylindrical airfoil was represented as the result of the superposition of a uniform flow and the field induced by a system of vortex lines coinciding with the generatrices of the cylinder. In this section the notion of the vortex line will be extended so as to become useful in the analysis of general three-dimensional stream patterns. For this purpose *curved vortex* lines must be considered.

As a preparatory step the velocity distribution induced by a finite portion $\overline{S_1S_2}$ of a straight line will be defined in the following way: Consider a point P with the distance R from the line S_1S_2. Denote by α and β the angles that S_1P and S_2P make with the direction S_1S_2 (Fig. 151). The velocity induced at P will then be a vector of the magnitude

$$V = \frac{\Gamma}{4\pi R} (\cos \alpha - \cos \beta) \tag{1}$$

with a constant Γ. The direction of this vector is normal to the plane S_1S_2P and pointing counterclockwise when seen in the direction S_2S_1. The direction S_1S_2 is called the orientation of the vortex line and Γ its vorticity. This definition is in accordance with that of the velocity induced by an infinite vortex line as given in the two dimensional theory. In fact, if both points S_1 and S_2 go in opposite directions toward infinity, we shall have $\alpha = 0$, $\beta = \pi$. Thus $\cos \alpha - \cos \beta = 2$ and the magnitude V becomes $\Gamma/2\pi R$; as to the direction, the agreement is obvious.

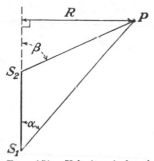

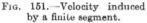

Fig. 151.—Velocity induced by a finite segment.

By definition the streamlines of the induced velocity field are circles whose planes are normal to the line S_1S_2. Along each circle the velocity is constant, but in the case of a finite S_1S_2 the velocity on circles of the same radius R will vary according to the position of the center with respect to S_1S_2.

211

In Sec. II.4 the mean rotation of a fluid element in a two-dimensional flow was defined as the arithmetical mean of the angular velocities of two line segments, one parallel and one normal to the direction of flow. A general and, in a certain way, a more satisfactory *definition of mean rotation* will now be given for the three-dimensional case.

Consider a fluid element around the point O that includes the neighboring point P. As the element moves along, OP will in general change its direction because of the fact that the velocity vectors of O and P can be different. If V_x, V_y, V_z are the velocity components of O and if $V_x + dV_x$, $V_y + dV_y$, $V_z + dV_z$ are those of P, the rotation of P can depend only on the relative velocities dV_x, dV_y, dV_z. In Fig. 152a the point P' is the projection of P on the x-y-plane passing through O. Then the rotation of OP' about the z-axis through O is usually called the z-component of the rotation of OP. It is seen in Fig. 152b, which shows the x-y-plane, how

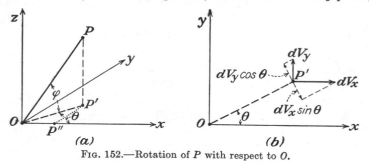

(a)   (b)

FIG. 152.—Rotation of P with respect to O.

the rotation of OP' is determined by dV_x and dV_y. In fact, $dV_y \cos \vartheta$ is one component of the relative velocity of P' normal to OP'. This means that dV_y produces a positive rotation with angular velocity $dV_y \cos \vartheta / \overline{OP'}$; likewise, dV_x causes OP' to rotate in the negative sense with angular velocity $dV_x \sin \vartheta / \overline{OP'}$. Thus the total angular velocity of OP' is

$$\frac{dV_y}{\overline{OP'}} \cos \vartheta - \frac{dV_x}{\overline{OP'}} \sin \vartheta \tag{2}$$

We now define the *z-component of the mean rotation* of the fluid element around O as *the average value of expression* (2) *for all points P in the neighborhood of O*. Note that dV_y is the change of V_y for the displacement OP. Now this displacement OP can be resolved (Fig. 152a) into the displacements OP'', $P''P'$, $P'P$ in the directions of x, y, z, respectively. Therefore,

$$dV_y = \frac{\partial V_y}{\partial x} \overline{OP''} + \frac{\partial V_y}{\partial y} \overline{P''P'} + \frac{\partial V_y}{\partial z} \overline{P'P}$$

$$= \left(\frac{\partial V_y}{\partial x} \cos \vartheta + \frac{\partial V_y}{\partial y} \sin \vartheta + \frac{\partial V_y}{\partial z} \tan \varphi \right) \overline{OP'}$$

where φ is the angle between OP and the x-y-plane. The same relation holds for dV_x if V_y is replaced by V_x, and thus expression (2) equals

$$\frac{\partial V_y}{\partial x} \cos^2 \vartheta + \frac{\partial V_y}{\partial y} \sin \vartheta \cos \vartheta + \frac{\partial V_y}{\partial z} \tan \varphi \cos \vartheta - \frac{\partial V_x}{\partial x} \sin \vartheta \cos \vartheta$$

$$- \frac{\partial V_x}{\partial y} \sin^2 \vartheta - \frac{\partial V_x}{\partial z} \tan \varphi \sin \vartheta$$

As was to be expected, the rotation depends, not on the distance $\overline{OP}$, but only on the orientation of OP determined by ϑ and φ. If ϑ passes through all values from 0 to 2π (at any constant value of φ) the mean values of $\sin \vartheta$, $\cos \vartheta$ and of $\sin \vartheta \cdot \cos \vartheta$ are zero, while the mean values of $\sin^2 \vartheta$ and $\cos^2 \vartheta$ are known to be $\frac{1}{2}$. It follows that the z-component of the mean rotation equals

$$\omega_z = \tfrac{1}{2}\left(\frac{\partial V_y}{\partial x} - \frac{\partial V_x}{\partial y}\right) \tag{3}$$

The first term in the parentheses is obviously the rotation about the z-axis of a segment OP when P lies on the x-axis through O, and the second term (including the minus sign) is the same for P lying on the y-axis. Thus *the z-component of the mean rotation of a fluid element is half the difference of the crosswise derivatives of V_x and V_y or the arithmetical mean of the rotations of two segments in the x- and y-directions.* Here z is an arbitrary direction and x, y may be any pair of directions orthogonal to z and to each other, with an appropriate definition of their positive sides. It is left to the reader (see Prob. 1) to prove that (3) agrees with the definition given in the two-dimensional case.

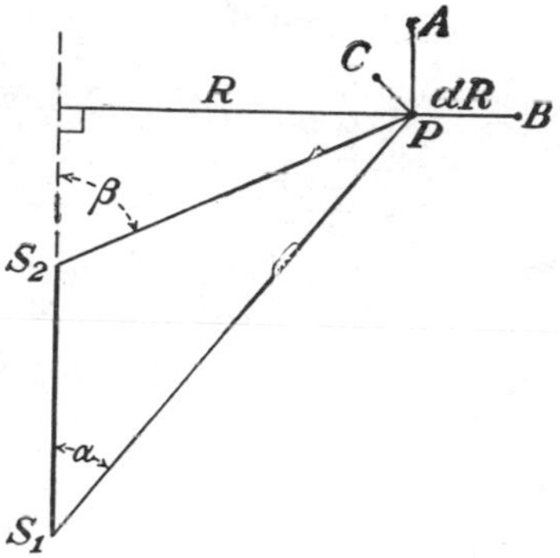

It can easily be seen that, if ω_x and ω_y (which are defined in an analogous way) and ω_z are zero for a point O, the component of mean rotation vanishes for all directions through O. If this is the case in all points of a fluid mass, the motion is called irrotational.

Fig. 153.

To return to the flow induced by a finite segment S_1S_2, it will be proved that this flow is not irrotational. In Fig. 153 the line PA is drawn parallel to S_1S_2, and PB and PC are normal and tangential to the circular streamline through P. The velocity of P is given by Eq. (1). In determining the mean rotation at P it suffices to carry out the computation for the velocity distribution given by

$$V_1 = \frac{\cos \alpha}{R} \tag{3'}$$

then to add the analogous contribution due to the term $V_2 = -\cos \beta/R$, and finally to multiply by $\Gamma/4\pi$.

Consider first the component ω_a of the mean rotation at P for the direction $PA\|S_1S_2$. In the plane PBC normal to PA we have a two-dimensional motion in concentric circles. According to Eq. (20), Chap. II, the mean rotation in this plane has the magnitude

$$\omega_a = \tfrac{1}{2}\left(\frac{V_1}{R} - \frac{dV_1}{dn}\right)$$

where R is the radius of curvature of the streamline and dV_1/dn denotes the derivative of V_1 in the direction normal to the streamline. Here $dn = -dR$, as has been pointed out in the discussion of the circulating

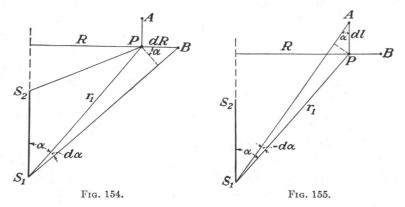

FIG. 154.   FIG. 155.

motion in Sec. II.5. As regards V_1, the numerator as well as the denominator of the right side of (3') depends on R. With the notations of the Fig. 154, $d\alpha = dR \cos \alpha/r_1$ and $R = r_1 \sin \alpha$. Consequently,

$$\frac{dV_1}{dn} = -\frac{dV_1}{dR} = \frac{\cos \alpha}{R^2} + \frac{\sin \alpha}{R}\frac{d\alpha}{dR} = \frac{\cos \alpha}{R^2} + \frac{\cos \alpha}{r_1^2}$$

and

$$\omega_a = \tfrac{1}{2}\left(\frac{V_1}{R} - \frac{dV_1}{dn}\right) = -\frac{\cos \alpha}{2r_1^2} \tag{4}$$

When $\omega_a > 0$, the vector $\overrightarrow{\omega_a}$ is directed from P toward A.

Next consider the component ω_b, in the direction PB, of the mean rotation at P due to the field V_1. The line element PC remaining tangential to the circular streamline of P rotates in the plane PBC (Fig. 153). The angular velocity of this element around PB therefore is zero. The line element PA of the length dl possesses an angular velocity around PB due to the fact that the points P and A have different velocities V_1 and $V_1 + dV_1$. Since R has the same value for P and A, the

increment dV_1 corresponds to the change of α only. With the notations
of Fig. 155, $d\alpha = -dl \sin \alpha / r_1$, and again $R = r_1 \sin \alpha$. Consequently,

$$dV_1 = -\frac{\sin \alpha}{R} d\alpha = \frac{dl}{Rr_1} \sin^2 \alpha = \frac{dl}{r_1^2} \sin \alpha$$

The angular velocity of PA around PB is $-dV_1/dl$. The mean rotation
in the plane PAC is thus seen to have the magnitude

$$\omega_b = \tfrac{1}{2}\left(0 - \frac{\sin \alpha}{r_1^2}\right) = -\frac{\sin \alpha}{2r_1^2}\,. \tag{5}$$

When $\omega_b > 0$, the vector $\overrightarrow{\omega_b}$ is directed from P toward B. Finally, the
component ω_c of the mean rotation at P vanishes since PA rotates in the

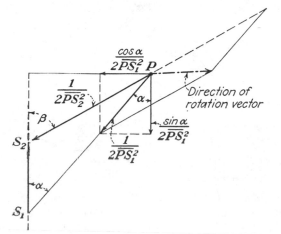

FIG. 156.—Rotation vector at P due to the velocity field induced by the segment S_1S_2.

plane PAC and PB in the plane PBC. The result is that the mean
rotation at P due to the field V_1 is a vector with the components (4) and
(5), *i.e.*, of the magnitude $1/2r_1^2$ directed from P toward S_1 (Fig. 156).

Similarly, the mean rotation at P due to the field V_2 is a vector of the
magnitude $1/2r_2^2$ in the direction S_2P. Taking account of the factor $\Gamma/4\pi$,
we thus have the theorem: The flow induced by a finite segment S_1S_2
is rotational. The rotation vector at a point P is $\Gamma/4\pi$ times the difference of a vector of the magnitude $1/2\overline{PS_1^2}$ in the direction PS_1 and a
vector of the magnitude $1/2\overline{PS_2^2}$ in direction PS_2.

Consider now the field of flow induced by a *polygonal vortex line*
consisting of the straight portions S_1S_2, S_2S_3, . . . $S_{n-1}S_n$ (Fig. 157),
all of which have the same vorticity Γ. The velocity that this polygonal
vortex line induces at a point P is by definition the vector sum of the
velocities induced by the portions S_1S_2, S_2S_3, etc. The rotation vector

at P for the resulting flow is the sum of the rotation vectors corresponding to the flow patterns induced by each of the segments S_1S_2, S_2S_3, etc. It is seen that this sum, except for the factor $\Gamma/4\pi$, consists of two terms only, *viz.*, the vector $1/2\overline{PS}_1^2$ in the direction PS_1 and the vector $1/2\overline{PS}_n^2$ in the direction S_nP, since for each intermediate corner the two terms cancel each other. If the polygon is closed, S_n coincides with S and the rotation is zero. *The flow induced by a closed vortex polygon is irrotational.*

This result, being independent of the number and the length of the sides of the polygon, applies also to a *curved vortex line*, which can be considered as the limit of a polygon with an indefinitely increasing number of sides. In this case the velocity induced at a point P can be found by integrating the contributions due to the elements ds of the curve. For a line element ds (Fig. 158) the angle β becomes $\beta = \alpha + d\alpha$; consequently, $\cos \alpha - \cos \beta = \cos \alpha - \cos (\alpha + d\alpha) \sim \sin \alpha \, d\alpha$. Fur-

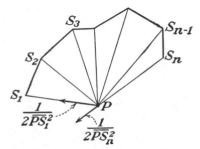

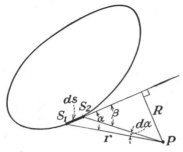

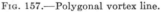

FIG. 157.—Polygonal vortex line.   FIG. 158.—Closed vortex line.

ther, we have $r \, d\alpha = ds \sin \alpha$, $R = r \sin \alpha$. According to (1) the contribution of the element ds to the velocity at P is an infinitesimal vector that is normal to the plane determined by ds and P and has the magnitude

$$dV = \frac{\Gamma}{4\pi} \frac{\sin \alpha \, d\alpha}{R} = \frac{\Gamma}{4\pi} \frac{R}{r^3} \, ds \qquad (6)$$

This can be interpreted as the velocity due to a rotation about the line ds as axis with the angular velocity $\Gamma \, ds/4\pi$, divided by r^3. To obtain the total velocity induced at P, the components of this vector with respect to the coordinate axes must be considered separately. The x-component of the total velocity at P is then obtained by integrating the x-components induced by the elements of the curved vortex line, etc. We are not interested in the details of this computation but note that *the velocity induced at a point P by an element of the length ds at the point S of a vortex line can be obtained by dividing by $(PS)^3$ the velocity which the point P would assume in a rotation with the tangent in S as the axis and with the angular velocity $\Gamma \, ds/4\pi$.* In electrodynamics the magnetic force induced at P by the element ds of an electric current of the intensity

$\Gamma/4\pi$ is given by the same formula, known as the *Biot-Savart formula*. It is partly because of this analogy that the term "induce" is used in describing the relation between a vortex line and the associated velocity distribution.

The velocity that a closed vortex line induces in accordance with the Biot-Savart formula is finite and continuous everywhere except for the points of the vortex line. The velocity induced at infinity is zero if the vortex line remains within a finite region.

In Sec. II.5 the *circulation* along an arbitrary plane circuit has been defined in the case of a two-dimensional flow. This definition can be immediately extended to a plane or skew curve in a three-dimensional field of flow. Consider a closed curve L which is entirely situated in the fluid and on which a sense of progression has been defined. From a point P of L, proceed in this sense to a neighboring point P' of L. Denoting by dl the infinitesimal distance $\overline{PP'}$ and by V_l the projection of the velocity at P on the directed line PP', we define as the *circulation along the circuit L* the value of the integral

$$\oint V_l \, dl$$

extended over the closed line L.

Equation (25), Sec. II.5, establishes a simple relation between the circulation along a plane circuit in a two-dimensional field of flow and the integral of the rotation extended over the area enclosed by this circuit. An analogous formula exists in the three-dimensional case. This is obvious for a plane circuit, for, by definition, the circulation is independent of the velocity components normal to the circuit. If the plane is taken as the x-y-plane, the circulation will be twice the area integral of the mean rotation as computed from the x- and y-components of the velocity. But, according to (3) this is the z-component of the mean rotation of the three-dimensional motion. Thus, for a circuit in the x-y-plane, Γ equals $2\int\omega_z \, dA$, independent of whether or not the motion has any component in the z-direction.

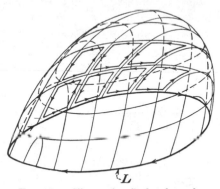

Fig. 159.—Illustrating Stokes formula.

Consider now a skew closed curve L lying on a surface S (Fig. 159). Denote by A the part of S that is bounded by L. By a network of two families of curves drawn on S, divide A into infinitesimal meshes. In the same way as in the two-dimensional case, the circulation along L is seen to equal the sum of the circulations along the contours of all meshes. Now, an infinitesimal mesh can be considered as plane. The circulation along the contour of such a mesh therefore equals the double product of the area dA of the mesh and the mean rotation of the motion parallel to its plane. This mean rotation, according to (3), is the component of

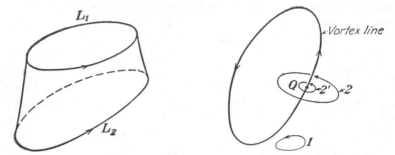

Fig. 160.—Surface subtended by two circuits.   Fig. 161.—Circuits in the velocity field of a vortex line.

the rotation in the direction normal to dA and will be designated by ω_n. Consequently,

$$\oint V_l \, dl = 2 \int \omega_n \, dA \qquad (7)$$

where the first integral is extended over the boundary of the area for which the second integral is taken. This relation (7) is known as the *Stokes formula*.

It follows from (7) that, if the field of flow is regular and irrotational at all points of the surface A enclosed by L, the circulation along L is zero. Moreover, the circulation has the same value for any two circuits L_1 and L_2 between which a surface A can be extended at all points of which the field of flow is regular and irrotational (Fig. 160), since (7) applies to this A, with L_2 and the reversed L_1 as its boundaries. These results enable us to compute the circulation for the various types of circuit in the field of flow induced by a closed vortex line. For a circuit, like circuit 1 in Fig. 161, which does not encircle the vortex line, the circulation is zero. For a circuit, like circuit 2, which encircles the vortex line, the circulation does not change when the circuit is contracted around the vortex line. For an infinitesimal circle 2′ surrounding the vortex line

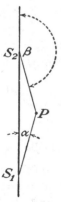

Fig. 162.

at some point Q, finite contributions to V_l have no influence since the circumference of 2′ is infinitesimal. Thus only that infinitesimal element of the vortex line which contains the point Q can furnish a contribution to the circulation along 2′. But such an element can be considered as straight (Fig. 162). According to Eq. (1), the velocity induced at a point P by a straight portion S_1S_2 of a vortex line has a constant value along the circular streamline through P, and the circulation along this circle of radius R equals, according to (1),

$$2\pi R V = \frac{\Gamma}{2}(\cos\alpha - \cos\beta)$$

When P comes closer and closer to the segment S_1S_2, the angles α and β tend toward 0 and π, respectively, and thus the circulation toward Γ. We have thus established the following result: *In the irrotational field of flow induced by a closed vortex line of the vorticity Γ, the circulation is zero for any circuit that does not encircle the vortex line and equals Γ for all circuits encircling (in a single turn) the vortex line.* The sense of progression on the circuit is here supposed to be counterclockwise when seen from the side toward which the vortex line points.

Problem 1. Explain that in the case of a flow in the x-y plane Eq. (20), Chap. II,

$$\omega = \tfrac{1}{2}\left(\frac{V}{R} - \frac{dV}{dn}\right)$$

is only a special case of formula (3),

$$\omega_z = \tfrac{1}{2}\left(\frac{\partial V_y}{\partial x} - \frac{\partial V_x}{\partial y}\right)$$

Problem 2. Compute the components of the mean rotation according to (3) for a fluid mass rotating about the z-axis as a rigid body with the angular velocity ω.

Problem 3. Compute the components of the mean rotation at a point x, y, z in a flow pattern with the velocity components.

$$V_x = (y - z)A, \qquad V_y = (z - x)A, \qquad V_z = (x - y)A$$

with

$$A = x^2 + y^2 + z^2 - xy - yz - zx + a^2, \qquad a = \text{const.}$$

Show that the rotation vector has everywhere the same direction and that its magnitude on the quadrics $A = \text{const.}$ is constant.

Problem 4. Compute the velocity induced by a circular vortex line at a point on its axis.

Problem 5. Considering the field of flow induced by a finite straight segment S_1S_2 of a vortex line of the vorticity Γ, compute the circulation along a circle that has its center on S_1S_2 and its plane normal to this segment. Verify that half the difference of the circulations for two such circles in the same plane equals the integral of the normal component ω_n of the rotation, extended over the circular ring.

Problem 6. A vortex line has the form of a rectangle in the x-y-plane, with the corners $(0, -b/2)$, $(0, b/2)$, $(-a, b/2)$, $(-a, -b/2)$. Compute the induced velocity at any point P of the x-y-plane, in particular for P lying on the y-axis. Give also the values for infinite a (horseshoe vortex).

2. Vortex Sheet and Discontinuity Surface.

In Sec. VIII.5 a cylindrical vortex sheet has been defined as a continuous distribution of parallel straight vortex lines of infinitesimal vorticity over a cylindrical surface. The theory of wings of finite span requires the discussion of curved vortex sheets of more general character.

Consider a closed or open surface S which is completely covered by a family of closed curves such that through any point of S passes

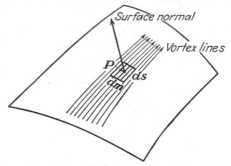

Fig. 163a.—Piece of a vortex sheet with distributed vorticity, built up of vortex lines with infinitesimal vorticity $d\Gamma$.

one and only one of these curves. Any two indefinitely near curves of this family form a strip of infinitesimal width on the surface S. Considering the curves as vortex lines, attribute to each such strip a certain infinitesimal vorticity $d\Gamma$ and a certain sense of progression. The velocity that an element of the strip induces at a point P can be found from the Biot-Savart formula (6) if there $d\Gamma$ is substituted for Γ. The surface S built up from such strips is called a *curved vortex sheet*. The velocity that the total sheet induces at the point P is found by forming the sum (integral) of the infinitesimal contributions furnished by the strips of the surface S. Thus, to find any velocity component in P it is necessary to compute a double integral, extended over the arc elements ds of each vortex line according to the Biot-Savart formula and over all elements dm across these vortex lines where $d\Gamma = \Gamma'\, dm$. At any point P that does not lie on the surface S the velocity distribution found in this way will be continuous.

Let us study the velocities at points in the neighborhood of S. If P is a point of S (Fig. 163a), the double integral over $ds\, dm$ cannot be used for computing the velocity induced at P since the denominator r^3 that comes in according to the Biot-Savart formula becomes zero for the element $ds\, dm$ that surrounds P. This difficulty also arises when the surface S is a cylinder and the vortex lines on it are straight lines. For this case it was shown in Sec. VIII.5 what kind of irregularity in the velocity distribution occurs when the point under consideration approaches the vortex sheet: the velocity becomes discontinuous, and the velocity vector changes abruptly from one side of the sheet to the other, the change being equal to a vector of magnitude Γ', tangential to the sheet and perpendicular to the vortex lines.   This discontinuity originates only in the influence of the small element $ds\,dm$ surrounding P.   But an infinitesimal surface element can always be considered as plane and the curve elements on it as elements of parallel straight lines (Fig. 163a). Thus it is obvious that, whatever the shape of the vortex sheet, the vector of induced velocity will change suddenly as one passes from one

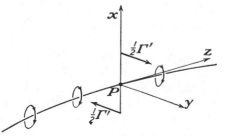

Fig. 163b.—Velocity discontinuity at a point of the vortex sheet.

side of the sheet to the other; the change is parallel to the sheet and normal to the vortex line through P; if the direction on the normal to the surface in which one proceeds is taken as the positive x-axis and the direction of the vortex line as the z-axis, the increase $\Gamma' = d\Gamma/dm$ has the positive y-direction.   This is indicated in Fig. 163b, where the small circles are streamlines of the circulating motion induced by the vortex line through P.   The sense of the circulation is counterclockwise when seen from the positive side of the vortex line.   The abrupt change of velocity corresponds to the opposite directions of the circulating velocities on both sides of the surface.

The field of flow induced by a curved vortex sheet is irrotational since it is defined as the superposition of an infinity of irrotational flow patterns, each induced by a closed vortex line.   Accordingly, the circu-

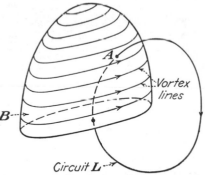

Fig. 164.—Circulation along a circuit piercing the vortex sheet.

lation will be zero on each circuit which has no points in common with the vortex sheet as long as, between such circuits, a surface can be extended on which the velocity is everywhere regular.   On the other hand, it was learned in Sec. 1 of this chapter that, for a circuit surrounding a single vortex line in the right sense, the circulation equals the vorticity of this line.   If several vortex lines or a system of vortex lines are present, the circulation on a circuit surrounding all or any part of them will equal the sum of all vorticities enclosed in the circuit.   This applies

to the curved vortex sheet also.   If a circuit L pierces the vortex sheet
at a point A as shown in Fig. 164, *its circulation must equal the vorticity*
$\int d\Gamma$ *of all the vortex lines between A and the boundary B of the sheet*, provided the sense of progression on the circuit is taken counterclockwise as
seen from the side toward which the vortex lines point.   (Note that
vorticity is a positive quantity, like mass or magnitude of velocity, while
circulation on a circuit can be positive or negative.)

As in the preceding chapters of this book, the present discussion
of the flow around a finite wing will be restricted to the case of permanent,
or *steady, motion*.   That means that all variables have values that do not
vary with time.   If a vortex sheet exists in a steady flow, the sheet and
the value of the vorticity at any point of the sheet are invariable.   It
follows that any particle on the sheet must move in a direction tangential
to the sheet.   In fact, if a fluid particle moved from one side of the sheet
to the other, its velocity would change in a discontinuous manner.   Such
a discontinuous variation of the velocity would imply infinite accelerations, which are physically impossible.   Note that this refers to the
resultant velocities of the particle, not to the velocity as induced just
by the respective vortex sheet.   For example, a stream pattern may
consist of the velocities induced by two different vortex sheets S_1 and S_2
and of a superimposed uniform flow.   Then, at each point of S_1, the
velocity resulting from the three contributions must be parallel to S_1.
It is one of the most important facts, which must not be overlooked, that
in a steady flow any vortex sheet must be tangential to the resultant stream.

It has been learned that a vortex sheet is a surface along which the
otherwise irrotational velocity distribution shows a discontinuity.
Inversely, *any discontinuity surface in a steady flow can be considered as a
vortex sheet* in the following sense: At each point of such a surface S
there is a certain vector difference $\overrightarrow{d}$ between the velocities on both sides
of the surface.   These vectors $\overrightarrow{d}$ are everywhere parallel to S.   If the
lines on S that cross the direction of $\overrightarrow{d}$ at right angles are taken as vortex
lines with a vortex density Γ' equal to the amount of d, the vortex sheet
thus constructed would induce a velocity field with just the actual
discontinuities along S.   Thus the existing flow pattern can be considered
as the sum of a flow induced by the vortex sheet on S and some irrotational flow that is continuous at all points of S.   This applies, for example, to the discontinuous flow past an obstacle studied in Sec. IV.4
(Helmholtz flow).   Here the discontinuity surface separates the wake
from the proper flow.   If the flow is supposed to be parallel to the
x-y-plane, the separation surface can be considered as a vortex sheet
with the vortex lines parallel to z and of a vortex density Γ' equal to the

flow velocity along the boundary of the wake. Another example is supplied by the thin-wing theory (Sec. VIII.6). Here, the cylindrical wing is a discontinuity surface for the velocity distribution (thus providing the pressure difference on both sides) and, on the other hand, is carrying the Γ-distribution, which was computed in the preceding chapter. This is not the place to elaborate this idea, which is one of the finest achievements of Helmholtz's theory of perfect fluid flow. But the reader is advised to bear in mind that in the case of a steady irrotational flow of a perfect fluid the terms *discontinuity surface, separation surface, and vortex sheet mean one and the same.*

The principal application of the concept of vortex sheet, in the theory of flight, is indicated by the use made of it in Sec. VIII.5, where a wing of infinite span was dealt with. There it was seen that the two-dimensional flow past such a wing can be considered as the result of superposition of a uniform field of flow and the field induced by a cylindrical vortex sheet whose cross section coincides with the contour of the wing profile. A similar result holds in the case of the three-dimensional flow past a body of simple form. To formulate the theorem correctly, the meaning of "simple form" must be defined. Spheres, cylinders of finite length, etc., are simple, and so are all bodies produced by continuous transformation of one of them. But a torus, a ring, or a pretzel are not simply connected. In these cases, there exist closed curves (loops) in the space, linked to the body in such a way that they cannot be separated from it without penetrating through the body. Any surface subtended by these curves would intersect the body. When such curves do not exist, the body is called simple or simply connected. Moreover, a body will be said to be regular if it has a continuously varying tangential plane, *i.e.*, no sharp edges, corners, etc. Then the following two theorems can be proved:

1. *The continuous irrotational flow past a finite, simply connected body of regular form is completely determined by the magnitude and direction of the velocity vector $\overrightarrow{V}$ at infinity.*

2. *This flow can be considered as the result of the superposition of the uniform flow with the velocity $\overrightarrow{V}$ and the flow induced by a certain vortex sheet that is spread over the surface of the body.*

No proof of these theorems will be attempted here. A demonstration of rather heuristic value can be given along the lines indicated in Sec. VIII.5 for the corresponding statement in the two-dimensional case. But there is one decisive difference between the two cases. In Chap. VIII it was found that the shape of the obstacle and the velocity vector at infinity do not fully determine the flow. One parameter, the circulation along a curve encircling the body, could still be chosen arbitrarily.

Nothing of this kind happens in the three-dimensional space. Here, in fact, the circulation must be zero for any curve outside the simply connected body since a surface can be subtended by this curve that does not interfere with the body. Note that, if the two-dimensional flow is conceived as a problem of space, the body involved is a cylinder of infinite length, thus not meeting the specifications of the foregoing theorems.

By "flow past a body" is meant a velocity distribution over the whole space outside the body, fulfilling the conditions that the velocity vector equals at infinity the given $\overrightarrow{V}$ and is tangential to the body surface at all points on this surface S. On the other hand, the vortex sheet on S induces a velocity at all points outside and inside S, and it is known from what has already been said that the resultant velocity distribution has a discontinuity along S with the jump of magnitude Γ' at all points on S. Now, it can be seen in the same way as in the two-dimensional case that the velocity in the interior of S can only be zero: no continuous irrotational flow is possible in the interior of a simply connected closed surface. It follows that the jump on S is identical with the outside velocity. One thus can add the following to the two theorems already stated:

3. *The vortex sheet on the surface of the body determines immediately the flow velocity at its points; this velocity has the magnitude Γ' and is directed normal to the vortex line, pointing to the right as one goes along in the direction of the vortex line.*

It may be added that the induced contribution to the velocity field has a physical significance in itself. If the air is originally at rest, disturbed only by the body that moves at a velocity V (opposite to the afore-mentioned vector $\overrightarrow{V}$), the actual velocities in the whole space are just those induced by the vortex sheet on S.

Problem 7. If a body of revolution moves parallel to its axis, what are the vortex lines on its surface that induce the velocity field?

*Problem 8. It is known that, if a sphere of radius a moves with a velocity V in the x-direction, the velocity of a fluid element at a surface point P with $\angle POx = \vartheta$ has a component of magnitude $\frac{1}{2}V \sin \vartheta$ in the direction perpendicular to OP in the plane OPx. Compute for the inverse motion the circulation along a closed circuit that pierces the sphere at two points P_1 and P_2 when the angles $P_1Ox = \vartheta_1$ and $P_2Ox = \vartheta_2$ are given.

3. The Flow Past a Wing of Finite Span. The mathematical problem indicated in the three theorems at the end of the preceding section has been solved for various bodies of particularly simple form, like a sphere, an ellipsoid, etc. The common feature of all these solutions is the following: If from the computed velocities on the surface of the body the pressure values are derived by means of Bernoulli's equation and properly integrated, it is found that, in general, a resultant moment (couple)

exists, but in no case a resultant force.   This fact, which can be proved
to hold true for all kinds of simply connected finite bodies, is known as
D'Alembert's paradox.   The term originally referred to the conviction
that a simple body like a sphere should meet a resistance (drag) when
moving uniformly through the air.   Today, one is not so much surprised
that the perfect fluid theory does not furnish a drag force; one rather
misses the lift that a body like an airplane wing doubtless experiences in
uniform flight.

Another kind of "paradox" was encountered in the two-dimensional
problem.   It has already been mentioned in Chap. IV and then was
repeatedly elaborated that the two-dimensional motion past a cylinder
of regular cross section, *e.g.*, past a circular cylinder, is not uniquely
determined in the theory of perfect fluids.   Such a cylinder, according
to this theory, never experiences a drag force, but it can (Sec. VIII.2)
experience a lift (*i.e.*, a force normal to the velocity) depending on the
amount of circulation in the outside stream.   But this amount is left
completely undetermined by the theory.   The answer to this paradox, at
least in the case of a cross section of the form of the usual wing profile, was
found to lie in the existence of the sharp trailing end of the profile.
Joukowski pointed out (see Sec. VIII.2) that the condition of finite
velocity at the sharp end removes the indetermination and leads to a
definite value of lift, in best accordance with the observations.

Now, in the three-dimensional case, again, the *sharp trailing edge*
of the wing supplies the clue for solving the paradox, though in a different
and more complicated way.   If the problem stated in theorems 1 and 2
of the preceding section is solved for a body whose surface includes
points with a very strong curvature, it will necessarily be found that
at such points the velocity increases to considerable amounts.   This
implies that the pressure at those points must be low and, under certain
circumstances, will drop below zero.   At sharp edges, *i.e.*, infinite curvature, the velocity supplied by the theory would be infinite and therefore
the pressure at any rate negative.   This is why the foregoing statements
have been restricted to bodies of regular shape.   Theorems 1 to 3 do
not apply to a wing with a sharp edge; or if they were used for computing
the field of flow, in this case they would furnish a stream pattern including infinite velocities and negative pressures.   Here, one is reminded
of Helmholtz's discovery regarding the two-dimensional flow around a
sharp-edged obstacle: In cases where no continuous flow with everywhere
positive pressure value exists, a certain discontinuous flow takes place.
In Fig. 47c (page 87) the actual flow, when conceived as a phenomenon
in space, has two discontinuity surfaces, each of them starting at one
of the two sharp edges of the obstacle.   In the case of an airplane wing
the situation is a little different since, essentially, only one sharp edge is

present.   It may be surmised that *in the real flow past a wing a discontinuity surface exists which sets in along the sharp trailing edge.*

There is another way of reasoning that supports this hypóthesis. If it is assumed that the theory of a wing of infinite span is correct—and there is no reason to doubt it—then the flow around a wing of large aspect ratio must show, at least in the region near the center, some

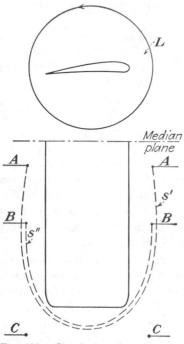

resemblance to what was found for the infinite wing.   Now, it was seen that in the latter case the circulation along a curve surrounding the wing profile (L in Fig. 165) has a definite positive value on which the lift depends.   It can hardly be expected that, if in the case of a finite wing the curve is drawn in the plane AA near the longitudinal symmetry plane of the wing, the circulation should be zero. But a finite circulation along L is incompatible with the assumption that the flow outside the wing is everywhere continuous, as was learned in Sec. 2 of this chapter.   A surface subtended by L (like s' in Fig. 165) must cross vortex lines somewhere if the circulation is to be different from zero.   It seems a reasonable assumption that the circulation value for a curve L decreases as its plane shifts from AA to BB and drops finally to zero if its plane no longer intersects

Fig. 165.—Circulation along the circuit L for various locations.

the wing, as in CC.   The consequence is that some vortex lines must extend between the planes AA and BB as well as between BB and CC.

Based on such ideas in a rather vague form and on certain observations on moving wings, F. W. Lanchester, as early as 1897, laid the foundation of a theory of wings with finite aspect ratios.   His first attempts to come to practical conclusions were outdone by L. Prandtl, who, between 1911 and 1917, developed an elaborate theory, susceptible of quantitative results, in best agreement with the experimental evidence. The main idea of the wing theory of Lanchester and Prandtl can be stated in the following way: *The flow of a perfect fluid past a wing of finite span exhibits a discontinuity surface that extends downstream from the trailing edge of the wing.   With increasing aspect ratio the discontinuity becomes less and less pronounced, so as to disappear completely for infinite aspect ratio.*   It is obvious that the afore-mentioned difficulty is resolved in this way.   For infinite span the velocity distribution becomes continuous and Joukowski's theory can be retained.   On the other hand, for the wing of finite span D'Alembert's paradox based on the assumption of a continuous flow outside the wing is invalidated.   Thus the way for a satisfactory explanation of the facts is opened.

Any velocity distribution with a discontinuity surface U can be considered, as was learned in Sec. 1 of this chapter, as the result of the superposition of a velocity distribution continuous on U and the field induced by a vortex sheet along U.   It has been stated in the foregoing that the continuous flow past a simply connected body can be obtained by superimposing on the uniform flow with the velocity V the flow induced by a vortex sheet which coincides with the surface of the body.   Consequently, the discontinuous flow postulated by the three-dimensional wing theory can be represented as the result of the superposition of (1) the uniform flow with the velocity V; (2) the flow induced by a vortex sheet extending over the surface W of the wing; and (3) the flow induced by a vortex sheet U of an as yet unknown shape which extends downstream from the trailing edge of the wing.

It follows from D'Alembert's paradox that the vortex sheet U must extend to infinity since otherwise the fluid still would not exert any resultant force on the wing.   In fact, the introduction of the vortex sheet U is kinematically equivalent to extending the wing by adding an infinitely thin, rigid sheet U.   If the sheet U would not extend to infinity, the surfaces W and U would form a finite body to which D'Alembert's principle applies.   The total force on

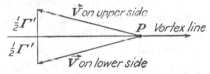
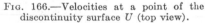

FIG. 166.—Velocities at a point of the discontinuity surface U (top view).

U and W would thus be zero.   But U cannot experience any lift, either positive or negative, since the pressure in the free stream must be continuous; i.e., its value at any point of U must be the same on the upper and lower sides of U.

The equality of pressure on both sides of the free vortex sheet U restricts in a definite way the discontinuity of the velocity occurring at its points.   In fact, from Bernoulli's equation it follows that under this condition the velocity must be the same on both sides.   The two velocity vectors on the upper and lower side, at any point P on U, are therefore symmetrical (Fig. 166) to the straight line through P that is perpendicular to the vector that represents the difference of the two vectors.   It was learned in Sec. 2 of this chapter that this perpendicular has the direction of the vortex line through P.   As the amount of the difference equals the vortex density Γ', the situation at each point on U is like that shown in Fig. 166.   The velocity has on both sides the same component parallel to the vortex line, while the components normal to it are $+\Gamma'/2$ and $-\Gamma'/2$, respectively. On the surface of the wing, we know that the velocity vector is everywhere perpendicular to the vortex line (theorem 3 in the preceding section).

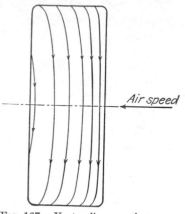

These remarks make it possible to form at least a rough picture of the vortex-line distribution on W and on U.

On the upper surface of the wing the flow can be supposed to have essentially the direction of the undisturbed flow. In Fig. 167 this flow is assumed to be directed from the right toward the left. Hence, the vortex lines on the upper side of the wing have essentially the spanwise direction. The sense of progression of these vortex lines is from port to starboard (from the left to the right of the pilot), according to the rule already given in theorem 3.

Fig. 167.—Vortex lines on the upper wing surface (top view).

On the lower surface of the wing the situation is less simple. For the usual values of the angle of attack the two-dimensional flow past an airfoil has a stagnation point near the leading edge on the lower contour of the profile (Fig. 168). The three-dimensional flow past a wing will have a stagnation line on the lower surface of the wing. In front of this stagnation line the flow along the lower wing surface is directed toward the leading edge; at the rear, toward the trailing edge. Seen from above, the vortex lines on the lower wing surface therefore must have the form indicated in Fig. 169. Near the stagnation line AB the vortex lines are closed curves that lie entirely on the lower wing surface and enclose the stagnation line AB. All vortex lines that are encountered beneath between the stagnation line and the leading edge are of this type. On the other hand, since all vortex lines must

Fig. 168.—Stagnation line.

be closed, the lines farther toward the trailing edge must join the lines on the upper wing surface (Fig. 167) so as to form closed curves. The result is shown in Fig. 170, where the dotted lines indicate the portions of the vortex lines on the lower surface of the wing. In general, a vortex line consists of an upper part directed port to starboard and a lower part lying farther toward the trailing edge and having opposite direction (see 2-2' or 3-3' in Fig. 170). Only certain lines that pass beneath between the stagnation line and the leading edge lie completely on the lower surface of the wing and surround the stagnation line AB, as already noted (see 1-1' in Fig. 170).

In outlining this picture it was necessary to assume that the vortex lines on the wing surface W deviate to a certain degree from the spanwise direction. Near the wing tips the vortex lines must be curved in the sense

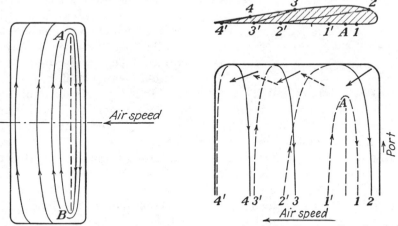

FIG. 169.—Vortex lines on the lower wing surface (top view).   FIG. 170.—Vortex lines on the wing (solid parts on upper surface).

indicated in Fig. 170. Since along W the velocity is normal to the vortex lines, the velocity vector must be directed inward on the upper side (solid vector in Fig. 170) and outward on the lower side (dotted vectors). Thus, there exists a spanwise velocity component toward the median plane of the wing on the upper surface and away from the median plane on the lower surface of the wing. Figure 171 shows these spanwise velocity components as they appear in a plane normal to the direction of flight. The existence of this encircling flow can be understood easily if the distribution of pressure over the wing surface is taken into account. A lift is produced only if there exists overpressure on the lower side and underpressure on the upper side of the wing. This difference of

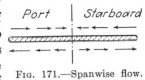

FIG. 171.—Spanwise flow.

pressure must be larger in the median plane of the wing and must vanish at the wing tips since outside the wing a pressure jump cannot exist. On the lower surface of the wing we therefore have a decrease of pressure from the median plane toward the wing tips and on the upper side an increase. Since the fluid has in a certain way the tendency to move in the direction of decreasing pressure, the spanwise flow indicated in Fig. 171 is understandable.

We are now in a position to make some statements concerning the free vortex sheet U. With the exception of the vortex lines surrounding the stagnation line AB, the vortex lines on W can be described approximately as oval curves lying in planes sloping toward the trailing edge of the wing. The last line shown in Fig. 170 is the line 4-4', which extends downside to the trailing edge. If the following line 5-5' lay entirely on the upper wing surface, it would surround a stagnation line there, similar to the stagnation line AB on the lower wing surface. This case would correspond to the two-dimensional flow with the stagnation points A and C shown in the Fig. 172; the fluid would flow around the trailing edge and, if this is sharp or has too strong a curvature, will assume an infinite velocity. As this is physically impossible, the vortex line 5 cannot be closed on the wing surface W.

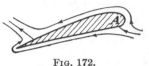

Fig. 172.

In order to see what happens, let us assume for the moment that the discontinuity surface U is replaced by a body of very small but finite thickness. Then 5', the second part of vortex line 5, would lie on the lower side of this body; and, for reasons of continuity, more lines of the

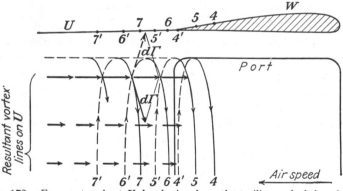

Fig. 173.—Free vortex sheet U developing from the trailing end of the wing.

same shape would follow, as 6-6', 7-7', etc., with one part on the upper side (solid line) and another half on the lower side (dotted lines), the latter shifted to the left. In reality, the two sides of U are not distinguishable, and both the solid and the dotted parts of the vortex lines lie on the same surface. Therefore, what presents itself at each point P on U is a vortex vector that is the sum of the two vectors of magnitude $d\Gamma$, one in the direction of the solid line passing through P, the other in the direction of the dotted line. As seen in Fig. 173 the resultant vector will have approximately the direction of the forward velocity on the port half of the wing and the opposite direction on the starboard half.   Combining this result with the foregoing, we come to the following conclusion: *It must be expected that the discontinuity surface or free vortex sheet U has the form of a long ribbon whose width is the wing span, with the vortex lines running almost parallel to the direction of flight, pointing forward on the port and backward on the starboard side.   The vortex lines on the wing are essentially parallel to the span direction, at the tips slightly curved backward on the upper side and forward on the lower side.*

4. Prandtl's Wing Theory.   The preceding qualitative analysis of the flow around a wing can be supplemented by a quantitative theory that furnishes numerical data concerning the influence of the aspect ratio on the forces acting on an airfoil.   Prandtl has worked out such a theory, which is in good agreement with a large body of experimental evidence and is generally accepted today.

An attempt at building a rigorous theory on the fundamental ideas developed in the foregoing meets with considerable difficulties.   The shape of the free vortex sheet U and the distribution of the vortex lines and of the vortex density on U and on the wing surface W are not known but must be determined so as to fulfill the following two conditions: (1) The resultant flow obtained by superimposing the uniform flow with the velocity V and the flow induced by the vortex sheets U and W must be tangential to the surfaces U and W.   (2) At the two sides of U the velocity of the resultant flow must have the same magnitude, since the pressure on the two sides of the free vortex sheet must have the same value.   Mathematically speaking, this constitutes a so-called "boundary-value" problem, and one of a most complicated nature.

In Prandtl's wing theory the problem is enormously simplified by a number of assumptions that have been found to give a satisfactory first approximation.   (1) The trailing edge of the wing is supposed to be a straight line, and the free vortex sheet U is assumed to have the form of a plane ribbon that is parallel to the velocity $\overrightarrow{V}$ at infinity and extends from the trailing edge downstream to infinity.   (2) The resulting vortex lines on U are assumed to be straight lines parallel to the velocity $\overrightarrow{V}$. (3) The wing is supposed to be infinitely thin so that the vortex lines on the upper and lower surfaces can be replaced by resultant vortex lines on a simple sheet.   These vortex lines would have to supply the connecting link between the forward- and backward-running vortex lines on U and would accordingly be of the form indicated in Fig. 174a. Since this pattern of vortex lines is still too difficult to handle, a further simplifying assumption is introduced: (4) Each vortex line from its beginning to its end is supposed to consist of three straight parts as shown in Fig. 174b, two of these parts extending to infinity.   The spanwise parts of these so-called "horseshoe vortices" are moreover supposed to lie on the one straight segment AB, which thus represents the total wing.

In the case of Fig. 174a an infinitesimal vorticity $d\Gamma$ is attributed to each strip between two infinitely near vortex lines. If the width of such a strip is denoted by dm, the vortex density $\Gamma' = d\Gamma/dm$ is finite. However, under the assumptions of Fig. 174b, the spanwise parts of these strips have zero width. The vortex density therefore becomes infinite

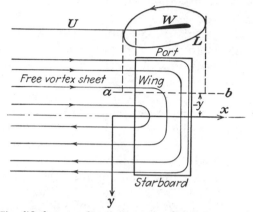

Fig. 174a.—Simplified vortex distribution according to assumptions (1) to (3).

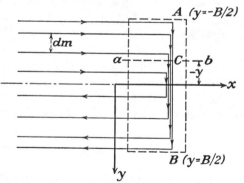

Fig. 174b.—Further simplification according to assumption (4).

along the line AB, which must be considered as a vortex line of finite vorticity. This vorticity is not a constant. At some point C between A and B it has the value $\int \Gamma' \, dm$, the integral extended over the segment AC across the vortex sheet. Thus it is zero at A and increasing toward the middle plane. From the center on toward the starboard side, the value decreases, since here vortex strips are leaving, and it finally drops to zero at the end point B. The line AB, which replaces the whole wing and carries a finite, variable vorticity, is known as the *lifting line* or *supporting line*.

In the following a system of rectangular coordinates will be used whose x-y-plane coincides with the plane of the free vortex sheet U and whose x-z-plane is the longitudinal symmetry plane of the wing, the x-axis having the direction of the velocity of flight, the y- and z-axes pointing to starboard and downward, respectively. In the case of Fig. 174a, consider a plane section a-b of the wing, normal to the y-axis and at the abscissa y with respect to the median plane of the wing. In the plane of this section, draw a circuit L surrounding the wing. The circulation $\Gamma(y)$ along the circuit L equals the integral of the (infinitesimal) vorticities of those vortex strips which pierce the plane of L. These strips can also be described as the vortex strips which, on the free vortex sheet U, lie to the port (or to the starboard) side of the plane of the circuit L. As y increases from $-B/2$ to $B/2$ the value of $\Gamma(y)$ increases from zero to a positive maximum at $y = 0$ and then drops again to zero at $y = B/2$. The same holds true in the case of Fig. 174b, where the span-

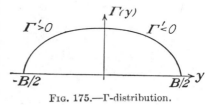

FIG. 175.—Γ-distribution.

wise parts of all vortex lines are concentrated along AB. It is seen that in this case the function $\Gamma(y)$ represents also the aforementioned concentrated vorticity for any point y of the lifting line.

In Fig. 175 a graph of $\Gamma(y)$ is shown. On the left-hand side (port) the derivative $d\Gamma/dy$ is positive and coincides exactly with what was called Γ' before. On the right half of the figure (starboard) $d\Gamma/dy$ has negative values and corresponds to the former $-\Gamma'$, since according to our definitions dm points in the negative y-direction when the vortex lines are directed in the negative x-direction. The function $\Gamma(y)$ is called the Γ-*distribution* or the *lift distribution over the span*.

If $\Gamma(y)$ is known, the field of flow induced by the vortex system, which consists of the lifting line and the vortex sheet, can be computed. In Prandtl's wing theory the z-component of the velocity induced at points near the lifting line plays an important role. This velocity component is called the *down-wash* velocity; its value will be denoted by $w(y)$. The relation between the functions w and Γ will be established later in this section.

So far, the shape of the wing (plan-form, profile) has not been taken into account. Prandtl has proposed the following manner of connecting the as yet unknown function Γ with the shape of the wing: Suppose Γ to be given and the corresponding down-wash velocity to be computed. Any value of y between $-B/2$ and $B/2$ determines a certain cross section or profile of the wing. Let O be the point where the lifting line pierces the plane of the cross section (Fig. 176). Within the framework of

Prandtl's theory the velocity at O is obtained by the superposition of the induced velocity at O and the velocity V in the negative x-direction. As in the two-dimensional theory of thin airfoils, it is assumed that the difference between the actual velocity at any point and the velocity V is small as compared with V. The magnitude of the resulting velocity then practically equals V. Considering, however, the velocity induced by the vortex sheet, the resulting velocity is no longer parallel to the x-axis; it has the small downward component w and a small spanwise component. The effect of the first is to lower the angle of incidence. Denote by $\overrightarrow{V'}$ the velocity vector obtained by adding the vectors of

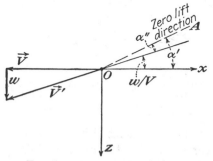

the magnitudes V and w in the negative x-direction and in the z-direction, respectively. Considering w to be small as compared with V and neglecting higher order terms, we can state that the velocity vector V' has the magnitude V but forms the small angle w/V with the x-axis. In Fig. 176 let OA indicate the zero lift direction of the profile and α' the angle of incidence, *i.e.*, the angle between the velocity V of the undis-

Fig. 176.—Down-wash velocity.

turbed stream and the zero lift axis. The resulting velocity V' then forms the so-called *effective angle of incidence* α''

$$\alpha'' = \alpha' - \frac{w}{V} \tag{8}$$

with the zero lift direction.

Now, the position of the lifting line with respect to the wing is in no way specified. The pattern of the flow in the close neighborhood of the profile is rather complicated, depending on the profile as well as on the plan-form of the wing. The vector $\overrightarrow{V'}$ therefore cannot be the velocity vector for all points in the vicinity of the wing. At best, the vector $\overrightarrow{V'}$ may represent a kind of average velocity vector near the wing in the plane section under consideration. On the other hand, in the two-dimensional flow around an airfoil the velocity vector of the undisturbed stream (the velocity at infinity) constitutes such an average velocity vector in the vicinity of the airfoil. It therefore seems reasonable to introduce the following hypothesis: *In each cross section of a wing of finite span the velocity distribution in the immediate neighborhood of the wing is assumed to be identical with the velocity distribution of the two-*

dimensional flow round this cross section, with $\overrightarrow{V}'$ *(instead of* $\overrightarrow{V}$*) as the velocity vector at infinity.* This hypothesis expresses the fundamental idea of Prandtl's quantitative approach; it has been found to lead to results that, to a remarkable extent, agree with the experimental evidence.

It was learned in Sec. VIII.2 that in the two-dimensional flow the circulation around a profile can be written as $KV\alpha'$, where K is a constant [called $C/2$ in Eq. (13a), Chap. VIII] depending on the form and size of the profile, V the velocity at infinity, and α' the angle of incidence. The only change necessary in the present context is the replacement of the angle of incidence α' by the effective angle of incidence α''. Thus,

$$\Gamma = KV\alpha'' = KV\left(\alpha' - \frac{w}{V}\right)$$

In general, the values of K and α' vary with y since the wing is not exactly prismatic. Γ and w also are functions of y. The preceding relation therefore can be written more clearly as

$$\Gamma(y) = K(y)V\left[\alpha'(y) - \frac{w(y)}{V}\right] \tag{9}$$

Since it is possible to express the down-wash velocity $w(y)$ in terms of $\Gamma(y)$, Eq. (9) furnishes the means for determining the unknown function

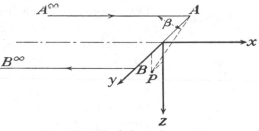

Fig. 177.—Single horseshoe vortex.

$\Gamma(y)$ when $K(y)$ and $\alpha'(y)$ representing the influence of the shape of the wing are given.

We now proceed to derive the formula giving $w(y)$ in terms of $\Gamma(y)$. Consider a single horseshoe vortex $A^\infty ABB^\infty$ (Fig. 177) of the infinitesimal vorticity $d\Gamma(y)$. Let P be an arbitrary point in the y-z-plane. The velocity induced at P may be decomposed in two components normal and parallel to the y-z-plane. The component parallel to the y-z-plane will be called the velocity induced in this plane. The velocity induced at P by any element of the segment AB of the horseshoe vortex is perpendicular to the plane containing this element and the point P. The segment AB therefore does not make any contribution to the velocities induced in the y-z-plane. On the other hand, the velocity induced at P by any element of the infinite segments $A^\infty A$ or BB^∞ of the horseshoe vortex lies completely in the y-z-plane. According to Eq. (1) the velocity induced by $A^\infty A$ has the magnitude

$$\frac{d\Gamma(y)}{4\pi \overline{AP}} (\cos \alpha - \cos \beta)$$

where α is the angle between $A^\infty A$ and $A^\infty P$ at the infinitely distant point A^∞ and β is the angle $A^\infty AP$. Hence $\cos \alpha = 1$, $\cos \beta = 0$, and the induced velocity has the magnitude

$$\frac{1}{4\pi} \frac{d\Gamma(y)}{\overline{AP}}.$$

This velocity is perpendicular to AP as indicated in Fig. 178. Seen in the direction of flight, the sense of this velocity vector corresponds to a clockwise rotation around A. The velocity distribution induced in the y-z-plane by the segment $A^\infty A$ is thus seen to be that of a circulating motion with the center A and the circulation $-d\Gamma/2$. That is, the velocity component parallel to the y-z-plane induced at any point P by the semi-infinite vortex line $A^\infty A$ equals one-half the velocity induced at P by a vortex line of the vorticity $-d\Gamma$ coinciding with A but extending both ways to infinity (two-dimensional motion). The same statement holds true for the velocity distribution induced in the y-z-plane by the semi-infinite vortex line BB^∞, but the vorticity of the equivalent infinite vortex line must be taken as $d\Gamma$, since the sense of progression on BB^∞ is opposite to that on $A^\infty A$. Taking into account the contributions made by all horseshoe vortices, we come to the following result: *The velocity component parallel to the y-z-plane induced at P equals one-half the velocity induced at P by a plane vortex sheet consisting of straight vortex lines that are parallel to the x-axis and extend both ways to infinity. This sheet extends spanwise from $y = -B/2$ to $y = B/2$; and if $\Gamma(y)$ denotes the distribution of circulation around the wing, the vortex density of the sheet equals the negative derivative $-d\Gamma/dy$.*

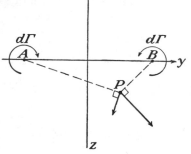

FIG. 178.—Contributions of the semi-infinite legs of the horseshoe vortex.

From this theorem the down-wash velocity $w(y)$ can be found when $\Gamma(y)$ is given. Consider the vortex strip with the abscissa η and the infinitesimal width $d\eta$. Now, upon writing $\Gamma'(y)$ as an abbreviation for

$d\Gamma/dy$, the vorticity of this strip equals $-\Gamma'(\eta)\,d\eta$ and the down-wash velocity induced at a point P of the y-axis (Fig. 179) equals

$$\frac{-\Gamma'(\eta)\,d\eta}{4\pi(\eta - y)} = \frac{\Gamma'(\eta)\,d\eta}{4\pi(y - \eta)}$$

It is easily seen that this holds true irrespective of whether the point P lies on the right or the left side of the vortex strip under consideration. The down-wash velocity produced at P by the entire vortex sheet therefore equals

$$w(y) = \frac{1}{4\pi} \int_{-B/2}^{B/2} \frac{\Gamma'(\eta)\,d\eta}{y - \eta} \tag{10}$$

Together with Eq. (8), this equation renders possible the determination of the function $\Gamma(y)$.

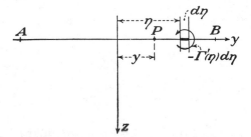

FIG. 179.—Contribution of a vortex element at η to the down-wash at y.

According to Prandtl's hypothesis the velocity distribution and consequently the pressure distribution along the contour of any cross section of the wing determined by a certain value of y are the same as for the two-dimensional flow around this cross section with $\overrightarrow{V'}$ as the velocity vector at infinity and with the value of $\Gamma(y)$ computed from Eqs. (9) and (10). According to the Kutta-Joukowski theorem the resultant pressure force exerted on a slice of unit length in span direction has the value $L' = \rho V'\Gamma(y)$ and, therefore, for a slice of the wing of the thickness dy has the magnitude $dF = \rho V'\Gamma(y)\,dy$. The force is perpendicular to the vector $\overrightarrow{V'}$, but the magnitude V' in the formula can be replaced by V. If the force dF is decomposed into two components parallel and normal to the direction of flight, the magnitudes of these components are obtained by multiplying dF with $\sin \epsilon$ and $\cos \epsilon$, respectively, where ϵ is the small angle between $\overrightarrow{V}$ and $\overrightarrow{V'}$. The tangent of this small angle is w/V. Neglecting higher order terms, one has $\sin \epsilon = w/V$, $\cos \epsilon = 1$.

The force perpendicular to $\overrightarrow{V}$ thus equals

$$dL = dF \cos \epsilon = dF = \rho V\Gamma(y)\,dy \tag{11'}$$

and that parallel to V,

$$dD = dF \sin \epsilon = dF \frac{w}{V} = \rho w \Gamma(y) \, dy \qquad (12')$$

The infinitesimal forces dL combine to the *lift*

$$L = \rho V \int_{-B/2}^{B/2} \Gamma(y) \, dy \qquad (11)$$

and the infinitesimal forces dD to the *drag*

$$D = \rho \int_{-B/2}^{B/2} w(y) \Gamma(y) \, dy \qquad (12)$$

When $\Gamma(y)$ is found from (9) and (10), lift and drag can be determined from (11) and (12).

The drag value in this argument is usually called the *induced drag* in order to distinguish it from the parasite drag, which must be added to keep in accordance with the experimental evidence (see Sec. VII.1 and the next section of this chapter). It may be remarked that the induced drag is a pure perfect-fluid effect and has nothing to do with viscosity. It is not correct to think that in a perfect fluid drag cannot exist because there is no dissipation of energy. If the steady motion with the body at rest is considered, no work is done against the drag. In the direct motion with the body moving through a bulk of air originally at rest, new kinetic energy in the air can be permanently produced, which accounts for the work measured by the product drag times velocity. This is also true in the case of the Helmholtz motion past a disk or any bluff body with a dead-air region downstream. Here, too, the computed drag is essentially a perfect-fluid phenomenon.

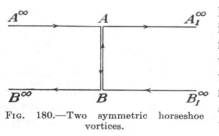

FIG. 180.—Two symmetric horseshoe vortices.

The result stated in the foregoing theorem that the velocities induced in the y-z-plane by the horseshoe vortex sheet are one-half the velocities induced by both sides infinite vortex lines can also be made plausible in the following way: Consider two adjacent horseshoe vortex lines as shown in Fig. 180. They are obviously equivalent to the two straight vortex lines $A^\infty A_1^\infty$ and $B_1^\infty B^\infty$ since the contributions of the elements along AB cancel each other. On the other hand, each of the two horseshoes will have the same effect at any point of the y-z-plane, and thus the actual velocities induced in such points are half those induced by $A^\infty A_1$ and $B_1^\infty B^\infty$. Another result that can be deduced in this way or by simple direct computation is that the velocities induced at a large (infinite) distance downstream by $A^\infty ABB^\infty$ alone are just the same as would be induced by the vortex lines $A^\infty A_1^\infty$ and $B_1^\infty B^\infty$ and thus the double of those induced in the y-z-plane.

Problem 9. Compute and discuss the velocities induced in the x-z-plane by a horseshoe vortex of vorticity Γ whose legs coincide with $y = \pm b$ from $x = -\infty$ to $x = 0$.

Problem 10. Using the result of Prob. 9, compute the velocity induced in the plane $x = -\infty$ by the horseshoe vortex sheet, and show that it is twice the velocity induced in the corresponding point $x = 0$.

Problem 11. As an extreme case of Γ-distribution one may assume $\Gamma = $ const. over the whole span. This means that the vortex sheet is replaced by one single horseshoe vortex of vorticity Γ with its corners at the wing tips. Discuss for this case the down-wash velocity and the drag in terms of lift and the drag coefficient in terms of lift coefficient.

5. Elliptic Lift Distribution. The mathematical problem of determining the vorticity distribution $\Gamma(y)$ of a given wing from Eqs. (8) and (9) will be discussed in a general way in the last section of this chapter. Here, a particular case in which the mathematical work is comparatively

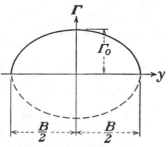

FIG. 181.—Elliptic Γ-distribution.

simple and which possesses a great practical importance will be dealt with.

As has been observed by Prandtl, particularly simple relations are obtained for the so-called "elliptic lift distribution" given by

$$\Gamma(y) = \Gamma_0 \sqrt{1 - \left(\frac{2y}{B}\right)^2} \tag{13}$$

where Γ_0 is a constant and B denotes the span. The curve representing the lift distribution $\Gamma(y)$ obviously is the upper half of the ellipse shown in Fig. 181; the equation of this ellipse is

$$\left(\frac{\Gamma}{\Gamma_0}\right)^2 + \left(\frac{2y}{B}\right)^2 = 1$$

Γ has the maximum value Γ_0 in the median plane of the wing ($y = 0$) and drops to zero at the wing tips ($y = \pm B/2$). The elliptic lift distribution thus follows the general trend discussed in connection with Fig. 175.

Differentiating (13) with respect to y, one obtains

$$\Gamma'(y) = \frac{1}{2}\Gamma_0 \left[1 - \left(\frac{2y}{B}\right)^2\right]^{-\frac{1}{2}} 2\left(-\frac{2y}{B}\right)\frac{2}{B} = -\frac{4\Gamma_0 y}{B^2}\left[1 - \left(\frac{2y}{B}\right)^2\right]^{-\frac{1}{2}}$$

From (10) the down-wash velocity is then obtained as

$$w(y) = -\frac{\Gamma_0}{\pi B^2} \int_{-B/2}^{B/2} \left[1 - \left(\frac{2\eta}{B}\right)^2 \right]^{-\frac{1}{2}} \frac{\eta \, d\eta}{y - \eta} \tag{14}$$

This integral can be reduced to one that was evaluated in Sec. VIII.6. Since the values of y and η are restricted to the interval from $-B/2$ to $B/2$, both $2y/B$ and $2\eta/B$ range from -1 to $+1$. Accordingly, the substitution

$$-\frac{2y}{B} = \cos \varphi, \qquad -\frac{2\eta}{B} = \cos \vartheta \tag{14'}$$

is allowed. With this substitution, the expression $[1 - (2\eta/B)^2]^{-\frac{1}{2}}$ occurring in (14) can be written as

$$(1 - \cos^2 \vartheta)^{-\frac{1}{2}} = \frac{1}{\sin \vartheta}$$

Furthermore, from the definition of ϑ,

$$d\eta = \frac{B}{2} \sin \vartheta \, d\vartheta$$

Expression (14) for the down-wash velocity then takes the form

$$w_\varphi = -\frac{\Gamma_0}{\pi B^2} \int_0^\pi \frac{1}{\sin \vartheta} \frac{-\dfrac{B}{2} \cos \vartheta \, \dfrac{B}{2} \sin \vartheta \, d\vartheta}{-\dfrac{B}{2} (\cos \varphi - \cos \vartheta)} = -\frac{\Gamma_0}{2\pi B} \int_0^\pi \frac{\cos \vartheta \, d\vartheta}{\cos \varphi - \cos \vartheta} \tag{15}$$

where the limits $\eta = -B/2$ and $\eta = B/2$ of the integral (14) have been replaced by the corresponding values $\vartheta = 0$ and $\vartheta = \pi$, respectively. The integral occurring in (15) has been evaluated in the appendix to Chap. VIII; it has the value $-\pi$. The down-wash velocity corresponding to the elliptic lift distribution is thus seen to have the constant value

$$w = \frac{\Gamma_0}{2B} \tag{16}$$

For a constant down-wash velocity Eqs. (11) and (12) reduce to

$$L = \rho V \int_{-B/2}^{B/2} \Gamma(y) \, dy \tag{11''}$$

and

$$D = \rho w \int_{-B/2}^{B/2} \Gamma(y) \, dy = \frac{w}{V} L \tag{12''}$$

The definite integral occurring in these formulas equals the area of the semi-ellipse shown in Fig. 181.   Now, the axes of this ellipse have the lengths B and $2\Gamma_0$, and the area of the ellipse equals $\pi/4$ times the product of these two lengths.   Consequently,

$$L = \rho V \frac{1}{2} \frac{\pi}{4} B \, 2\Gamma_0 = \frac{\pi}{4} \rho V \Gamma_0 B \tag{17}$$

and

$$D = \frac{w}{V} L = \frac{\Gamma_0}{2BV} L = \frac{\pi}{8} \rho \Gamma_0^2 \tag{18}$$

The lift and drag coefficients defined in Sec. VI.1 equal

$$C_L = \frac{L}{(\rho/2) V^2 S} = \frac{\pi B}{2S} \frac{\Gamma_0}{V} \tag{17'}$$

$$C_D = \frac{D}{(\rho/2) V^2 S} = \frac{\pi}{4S} \left(\frac{\Gamma_0}{V} \right)^2 \tag{18'}$$

Eliminating the ratio Γ_0/V by dividing the square of C_L by C_D, we obtain

$$\frac{C_L^2}{C_D} = \pi \frac{B^2}{S} = \pi \!R \tag{19}$$

where, according to the definition given in Eq. (1), Chap. VI, B^2/S has been replaced by the aspect ratio $\!R$.   The curve showing C_L as a function of C_D has been introduced in Sec. VII.1 as the polar diagram. The result expressed by Eq. (18) therefore can be stated as follows: *The polar diagram of a wing with elliptic lift distribution is a parabola.   The vertex of this parabola coincides with the origin and its axis with the axis of C_D; the parameter equals $\pi\!R/2$.*[*]

It has been stated in Sec. VII.1 that the polar diagram of a wing of conventional design is fairly well represented by an expression of the form

$$C_D = \frac{C_L^2}{3\!R} + a$$

where a is a small positive constant.   The three-dimensional wing theory is thus seen to supply an expression for the relation between C_D and C_L that is adequate to within a small additional amount of drag.   This additional drag is due to viscosity effects that have been neglected in the theory given in the foregoing.   It is remarkable that the result derived under the assumption of an elliptic lift distribution seems to be in substantial agreement with the observations covering a wide range of conventional wing designs.   This point will be taken up later.

[*] It is usual to call p the "parameter" of the parabola $y^2 = 2px$.

A further assumption concerning the shape of the wing will now be introduced. In the theory developed in the preceding section the profile of the wing was permitted to vary along the span. In the following, it will be assumed that all cross sections of the wing have the same shape, only the size determined by the chord length being allowed to vary. Moreover, the wing is assumed to be without twist; *i.e.*, the zero lift direction is the same for all profiles. According to these assumptions, Eq. (9) takes a simpler form: the angle of incidence no longer depends on y but has a constant value α', and the factor $K(y)$ is proportional to the chord length, $K(y) = \frac{1}{2}kc(y)$, where k is a constant depending only on the common shape of the profiles of the wing. From the discussion at the end of Sec. VIII.2, the constant k is known to be approximately 2π for the usual profiles. Upon introducing the value w from (16), Eq. (9) can then be written as

$$\Gamma(y) = \tfrac{1}{2}kc(y)V\left(\alpha' - \frac{\Gamma_0}{2BV}\right) \qquad (20)$$

Since on the right side $c(y)$ is the only quantity depending on y, the chord length $c(y)$ must be proportional to $\Gamma(y)$:

$$c(y) = c_0\sqrt{1 - \left(\frac{2y}{B}\right)^2} \qquad (21)$$

If we introduce (21) in (20), we find that, according to (13), the constant c_0, that is the chord length at $y = 0$ is connected with Γ_0 by the relation

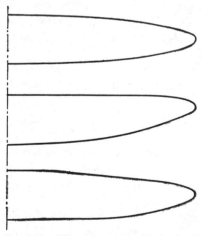

$$\Gamma_0 = \tfrac{1}{2}kc_0V\left(\alpha' - \frac{\Gamma_0}{2BV}\right) \qquad (22)$$

A wing the chord length of which varies according to (21) will be called an *elliptic wing*. The preceding argument shows that *an elliptic wing with invariable profile without twist experiences an elliptic lift distribution.* The chord length being given by (21), the plan-form of an elliptic wing is determined as soon as the maximum chord length c_0 and the shape of either the leading or the trailing edge is given.

Fig. 182.—Plan-forms with elliptic chord distribution.

Figure 182 shows the plan-forms of some elliptic wings of the aspect ratio 6. Actual plan-forms do not deviate too much from that of an elliptic wing.

Equation (22) can be solved for Γ_0:

$$\Gamma_0 = \frac{2BV\alpha'}{1 + \dfrac{4B}{kc_0}}$$

Introducing this in the expression for C_L obtained from (17), we find that

$$C_L = \frac{\pi B \Gamma_0}{2SV} = \pi \frac{B^2}{S} \frac{\alpha'}{1 + \dfrac{4B}{kc_0}} = \frac{\pi R \alpha'}{1 + \dfrac{4B}{kc_0}}$$

Here the ratio B/c_0 can be expressed in terms of the aspect ratio of the wing. In fact, the area S of an elliptic wing of the span B and the maximum chord c_0 equals the area of a half ellipse with the axes B and $2c_0$. From $S = \pi c_0 B/4$ the aspect ratio is obtained as

$$R = \frac{B^2}{S} = \frac{4B^2}{\pi c_0 B} = \frac{4B}{\pi c_0}$$

Consequently, $2B/c_0 = \pi R/2$. Introducing this in the expression for the lift coefficient already obtained, we find that

$$C_L = \frac{\pi R \alpha'}{1 + \dfrac{\pi}{k} R} = \frac{2\pi \alpha'}{\dfrac{2}{R} + \dfrac{2\pi}{k}} \tag{23}$$

Since k equals approximately 2π, this relation is near to

$$C_L = \frac{2\pi \alpha'}{1 + 2/R} \sim \frac{6\alpha'}{1 + 2/R} \tag{23'}$$

which has been given in Sec. VII.1 as a fair representation of empirical facts. The results obtained for the wing of elliptical plan-form hold essentially also for rectangular and trapezoidal wing shapes. The small corrections that can be applied in these cases will be developed in Sec. 7 of this chapter.

It seems appropriate to stress the fact that the results expressed by (19) and (23), *i.e.*, the parabolic form of the polar diagram and the dependence of this form on the aspect ratio, and the relation between lift coefficient, angle of attack, and aspect ratio, were not known as empirical facts before the wing theory was developed. These facts as well as the existence of the aerodynamic center in the two-dimensional case have been predicted by the theory. Experiments carried out a posteriori have confirmed these theoretical predictions to a degree that is remarkable in view of the numerous idealizations of the theory.

Problem 12. A wing of $S = 280$ ft.[2] and aspect ratio 7, flying at 10,000 ft. altitude with a velocity of 220 ft./sec., supports a given weight of 8500 lb. Compute the value of circulation in the middle plane Γ_0, the down-wash velocity w, the induced drag, the geometrical angle of incidence α', and the effective angle of incidence α''. The Γ-distribution is assumed to be elliptic.

Problem 13. An elliptic wing of aspect ratio 6 and profile-drag coefficient $a = 0.02$ works under conditions where the lift to drag ratio is 15. What angle of incidence is used? Compute also the circulation Γ_0 if the air speed is 260 ft./sec. and the span $B = 40$ ft.

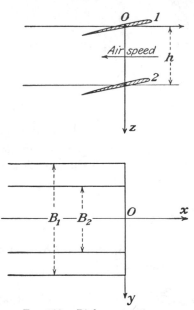

FIG. 183.—Biplane notation.

6. Biplane Theory. Another achievement of Prandtl's wing theory is a satisfactory account of the mutual influence of the wings of a biplane. In the following an outline of this biplane theory will be given. In order to simplify the argument as much as possible, elliptic lift distribution will be assumed for both wings.

Figure 183 shows a cross section of the two wings of a biplane without stagger. In this case the lifting lines replacing the two wings may be assumed to lie in a plane that is perpendicular to the velocity V of the undisturbed stream, the distance h of these lifting lines corresponding to the average distance between the two wings. Each lifting line is supposed to trail behind it a horseshoe vortex sheet of a width equal to the span of the wing under consideration. These vortex sheets are parallel to the velocity V of the undisturbed stream.

In the following all quantities concerning the upper wing will be given the subscript 1 and those concerning the lower wing the subscript 2. For each point P_1 of the upper lifting line (Fig. 184) and for each point

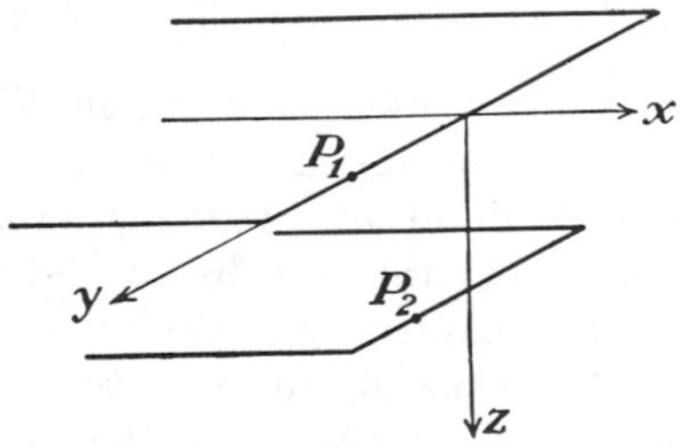

FIG. 184.—Biplane represented by two vortex sheets.

P_2 of the lower lifting line the total down-wash velocity induced by both vortex sheets must be computed. If w_1 is the total down-wash velocity at P_1 and α' the angle of incidence of the corresponding cross section of the upper wing, the circulation around this cross section is then assumed to equal

$$\Gamma_1 = KV \left(\alpha' - \frac{w_1}{V} \right)$$

in accordance with Prandtl's hypothesis (page 234). The factor K depends on the shape of the cross section and is furnished by the two-dimensional theory. According to (11) and (12) the lift and the drag of the upper wing are given by

$$L_1 = \rho V \int_{-B_1/2}^{B_1/2} \Gamma_1(y_1) \, dy_1 \quad \text{and} \quad D_1 = \rho \int_{-B_1/2}^{B_1/2} \Gamma_1(y_1) w_1(y_1) \, dy_1$$

In the last formula, $w_1(y_1)$ can be written as the sum of the down-wash velocity w_{11} induced by the upper vortex sheet and the down-wash velocity w_{12} induced by the lower vortex sheet at the point y_1 of the upper wing. Accordingly,

$$D_1 = D_{11} + D_{12} = \rho \int \Gamma_1(y_1) w_{11}(y_1) \, dy_1 + \rho \int \Gamma_1(y_1) w_{12}(y_1) \, dy_1$$

where the first term represents the drag that the upper wing would experience if used as a monoplane wing.

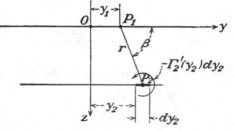

FIG. 185.—Down-wash at the upper wing due to the lower wing.

In order to evaluate D_{12}, let us determine the down-wash velocity w_{12}. As has been shown in Sec. 3 of this chapter, the velocity component parallel to the y-z-plane that a sheet of horseshoe vortices induces at a point P of this plane equals one-half the velocity induced at P by a sheet of straight vortex lines that are parallel to the x-axis and extend both ways to infinity, provided that this sheet has the same Γ-distribution as the original horseshoe sheet. Accordingly, the element of the vorticity $-\Gamma'(y_2) \, dy_2$ at P_2 (Fig. 185) induces at P_1 a velocity of the magnitude

$$- \frac{1}{4\pi} \frac{\Gamma_2'(y_2)}{r} \, dy_2$$

where r denotes the distance P_1P_2. Since this velocity is perpendicular to P_1P_2, its down-wash component equals

$$- \frac{1}{4\pi} \frac{\Gamma_2'(y_2)}{r} \cos \beta \, dy_2$$

where β denotes the angle between P_1P_2 and the positive y-axis. Now, $\cos \beta = (y_2 - y_1)/r$ and $r^2 = h^2 + (y_2 - y_1)^2$. The down-wash velocity induced at P_1 thus can be written as

$$w_{12}(y_1) = -\frac{1}{4\pi} \int_{-B_2/2}^{B_2/2} \frac{y_2 - y_1}{h^2 + (y_2 - y_1)^2} \Gamma_2'(y_2)\, dy_2$$

The integral can be transformed by means of an integration by parts according to $\int u\, dv = uv - \int v\, du$, where

$$u = \frac{y_2 - y_1}{h^2 + (y_2 - y_1)^2} \qquad dv = \Gamma_2'(y_2)\, dy_2$$

and consequently

$$du = \frac{h^2 + (y_2 - y_1)^2 - 2(y_2 - y_1)^2}{[h^2 + (y_2 - y_1)^2]^2}\, dy_2 = \frac{h^2 - (y_2 - y_1)^2}{[h^2 + (y_2 - y_1)^2]^2}\, dy_2$$

$$v = \Gamma_2(y_2)$$

It follows, since the function $\Gamma_2(y_2)$ vanishes for the limits $y_2 = \pm B_2/2$, that

$$w_{12} = \frac{1}{4\pi} \int_{-B_2/2}^{B_2/2} \frac{h^2 - (y_2 - y_1)^2}{[h^2 + (y_2 - y_1)^2]^2} \Gamma_2(y_2)\, dy_2$$

With this value of w_{12} the expression for D_{12} furnishes

$$D_{12} = \frac{\rho}{4\pi} \int_{-B_1/2}^{B_1/2} \int_{-B_2/2}^{B_2/2} \Gamma_1(y_1)\Gamma_2(y_2) \frac{h^2 - (y_2 - y_1)^2}{[h^2 + (y_2 - y_1)^2]^2}\, dy_1\, dy_2 \qquad (24)$$

The value of this expression is not altered if y_1 and y_2 are made to change roles. Consequently, the drag D_{12} induced on wing 1 by vortex sheet 2 and the drag D_{21} induced on wing 2 by vortex sheet 1 have the same value. The total drag of the biplane thus equals

$$D = D_1 + D_2 = (D_{11} + D_{12}) + (D_{21} + D_{22}) = D_{11} + 2D_{12} + D_{22}$$

where D_{11} and D_{22} are the drags that the wings would experience if used separately as monoplane wings, while $2D_{12}$ is the additional drag due to the mutual influence of the wings, found in (24).

If both wings have elliptic lift distributions, the relations already obtained can be evaluated without much difficulty. In the case of a monoplane wing with elliptic lift distribution, Eqs. (13) and (17) give

$$\Gamma(y) = \frac{4L}{\pi\rho VB} \sqrt{1 - \left(\frac{2y}{B}\right)^2} \qquad (25)$$

Furthermore, from (17) and (18) the drag is seen to equal

$$D = \frac{\pi}{8} \rho \left(\frac{4L}{\pi \rho V B}\right)^2 = \frac{2}{\pi \rho V^2} \frac{L^2}{B^2} \tag{26}$$

Relation (25) holds true, if either the subscript 1 or the subscript 2 is attached to Γ, y, L, B, and relation (26) furnishes D_{11} or D_{22} according to whether the subscript 1 or 2 is attached to L and B on the right-hand side. The drag D_{12} is then found from (24) as

$$D_{12} = \frac{4L_1L_2}{\pi^3 \rho V^2 B_1 B_2} \int_{-B_1/2}^{B_1/2} \int_{-B_2/2}^{B_2/2} \sqrt{\left[1 - \left(\frac{2y_1}{B_1}\right)^2\right]\left[1 - \left(\frac{2y_2}{B_2}\right)^2\right]}$$

$$\times \frac{h^2 - (y_2 - y_1)^2}{[h^2 + (y_2 - y_1)^2]^2} \, dy_1 \, dy_2$$

Upon introducing $\eta_1 = 2y_1/B_1$ and $\eta_2 = 2y_2/B_2$ as new variables of integration, the limits of both integrals become -1 and $+1$, and D_{12} can be written in the form

$$D_{12} = \frac{2L_1L_2}{\pi \rho V^2 B_1 B_2} \sigma \tag{27}$$

where

$$\sigma = \frac{B_1 B_2}{2\pi^2 h^2} \int_{-1}^{1} \int_{-1}^{1} \sqrt{(1 - \eta_1^2)(1 - \eta_2^2)}$$

$$\times \frac{1 - \left(\frac{B_2}{2h}\eta_2 - \frac{B_1}{2h}\eta_1\right)^2}{\left[1 + \left(\frac{B_2}{2h}\eta_2 - \frac{B_1}{2h}\eta_1\right)^2\right]^2} \, d\eta_1 \, d\eta_2$$

Since the expression under the integral sign is a function of η_1 and η_2, depending on the parameters B_1/h and B_2/h, tho so-called "interference factor" σ can be evaluated for each combination of these ratios. The results of the rather laborious numerical computa-

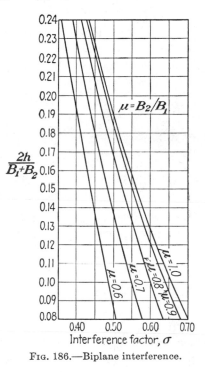

FIG. 186.—Biplane interference.

tions are shown in Fig. 186, where the span ratio B_2/B_1 and the ratio $2h/(B_1 + B_2)$ of the gap h to the average span have been used as the independent parameters. In preparing this figure it has been assumed that $B_2 \leqq B_1$. However, the fact that wing 1 is the upper one has not been used in the preceding argument. Figure 186 therefore furnishes the interference factor for all cases, provided that the wing of greater span is labeled 1.

Upon using the values of D_{11} and D_{22} as furnished by (26), the total drag of the biplane is obtained as

$$D = D_{11} + 2D_{12} + D_{22} = \frac{2}{\pi \rho V^2}\left(\frac{L_1^2}{B_1^2} + 2\sigma \frac{L_1 L_2}{B_1 B_2} + \frac{L_2^2}{B_2^2}\right) \qquad (28)$$

Upon introducing the coefficients

$$C_D = \frac{D}{\frac{\rho}{2} V^2 S}, \qquad C_{L_1} = \frac{L_1}{\frac{\rho}{2} V^2 S}, \qquad C_{L_2} = \frac{L_2}{\frac{\rho}{2} V^2 S}$$

where S denotes the sum of the areas of the two wings, Eq. (28) yields

$$C_D = \frac{1}{\pi}\left(\frac{S}{B_1^2} C_{L_1}{}^2 + 2\sigma \frac{S}{B_1 B_2} C_{L_1} C_{L_2} + \frac{S}{B_2^2} C_{L_2}{}^2\right)$$

This relation, which corresponds to (19), shows that the drag coefficient of a biplane is a quadratic form of C_{L_1} and C_{L_2}, the coefficients depending on the interference factor σ and the aspect ratios B_1^2/S, $B_1 B_2/S$, B_2^2/S. Introducing the ratios

$$\frac{B_2}{B_1} = \mu, \qquad \frac{L_2}{L_1} = \nu$$

it follows that $L_1 = L/(1 + \nu)$, $L_2 = \nu L/(1 + \nu)$ with $L = L_1 + L_2$. Then (28) can be written as

$$D = \frac{2L^2}{\pi \rho V^2 B_1^2} \frac{\mu^2 + 2\mu\nu\sigma + \nu^2}{\mu^2(1 + \nu)^2} \qquad (28')$$

The first term represents the drag of a monoplane with lift L and span B_1. Therefore, if we define the factor M by

$$M = \frac{(1 + \nu)\mu}{\sqrt{\mu^2 + 2\mu\nu\sigma + \nu^2}}$$

(28') takes the form

$$D = \frac{2L^2}{\pi \rho V^2 (MB_1)^2}$$

This M is often called *Munk's span factor*; MB_1 the equivalent monoplane span; and $(MB_1)^2/S$ the equivalent monoplane aspect ratio.

In designing a biplane with given values of B_2/B_1 and $2h/(B_1 + B_2)$, the ratio L_1/L_2 can be assigned an arbitrary value by disposing of the chord lengths and the angles of incidence of the two wings in a suitable manner. It is therefore reasonable to ask which value of L_1/L_2 gives the smallest possible drag D for a given value of the total lift $L_1 + L_2$.

According to well-known rules the extreme values of a function of L_1 and L_2 for a given value of $L_1 + L_2$ are reached where the partial derivatives of this function with respect to L_1 and L_2 are equal. Computing these derivatives from (28), we find that

$$\frac{L_1}{B_1^2} + \sigma \frac{L_2}{B_1 B_2} = \frac{L_2}{B_2^2} + \sigma \frac{L_1}{B_1 B_2}$$

or

$$\frac{L_1}{L_2} = \frac{B_1/B_2 - \sigma}{B_2/B_1 - \sigma}$$

and

$$\frac{L_1}{L_1 + L_2} = \frac{1 - \sigma B_2/B_1}{1 - 2\sigma B_2/B_1 + (B_2/B_1)^2}$$

Substitution of this in (28) furnishes, after some transformations,

$$D_{\min} = \frac{L^2}{\pi(\rho/2)V^2 B_1^2} \cdot \frac{1 - \sigma^2}{1 - 2\sigma B_2/B_1 + (B_2/B_1)^2} \tag{29}$$

The second factor thus gives the optimum value of Munk's span factor,

$$M_{\text{opt}} = \sqrt{\frac{1 - 2\sigma\mu + \mu^2}{1 - \sigma^2}}$$

which is reached when the lift ratio ν is connected with the span ratio μ by the above relation:

$$\nu = \frac{\mu - \sigma}{(1/\mu) - \sigma}$$

Note that σ depends on $2h/(B_1 + B_2)$ as well as on B_2/B_1.

All the formulas in this section are valid in the case of elliptic lift distribution on both wings. For actual wings with, in general, a somewhat different distribution, the formula

$$D = \frac{2L^2}{\pi\rho V^2 (MB_1)^2}, \qquad C_D = \frac{C_L^2}{\pi(MB_1^2)/S} = \frac{C_L^2}{\pi M^2 \mathcal{R}}$$

can be maintained if the span factor M is given a more general meaning. It then must be defined as an average value of

$$\frac{C_L}{\sqrt{\pi C_D \mathcal{R}}}$$

for various angles of attack. If this idea is applied to a monoplane, it turns out that M^2 is slightly smaller than 1, between 0.90 and 0.95, and in the case of biplanes 90 to 95 per cent of the above value

$$\frac{\mu(1 + \nu)}{\sqrt{\mu^2 + 2\mu\nu\sigma + \nu^2}}.$$

Problem 14.   Compute the equivalent monoplane span for a biplane whose upper and lower spans are 42 ft. and 36 ft., respectively, the gap being 6 ft., if 60 per cent of the total weight is supported by the upper and 40 per cent by the lower wing.

Problem 15.   If the two spans and the gap have the values used in the preceding problem, compute the optimum lift ratio and the corresponding span factor.

7. General Lift Distribution.   The expressions for lift and drag coefficients found in Sec. 4 of this chapter were derived under the assumption that the down-wash velocity w has a constant value over the span. It was also seen that $w = $ const. can be reached with a wing of elliptic plan-form and uniform profiles without twist.   But the general equations (9) and (10), which establish two relations between the distributions of vorticity $\Gamma(y)$ and of down-wash $w(y)$, allow us to deduce expressions for both functions in the case of an arbitrarily given wing shape.   Lift and drag coefficients can then be derived from Γ and w.   The mathematics to be used is much the same as that which served in the thin-wing theory in Chap. VIII.

The wing will be supposed to be given by its plan-form expressed in the form $c(y)$, that is, chord length as function of y, and by its twist expressed as $\alpha_0(y)$, that is, zero lift angle for each y.   Moreover, the two-dimensional problem for each profile will be supposed to be solved as far as the radius a of the mapping circle is concerned.   That is, for each profile the ratio a/c and thus the coefficient $K(y)$ in (9),

$$K(y) = 4\pi a = 4\pi \frac{a}{c} (y)c(y) \tag{30}$$

are supposed to be known.   In most cases, it will be sufficient to attribute the value $\frac{1}{4}$ to the ratio a/c so as to have

$$K(y) = \pi c(y) \tag{30'}$$

In both equations (9) and (10) we introduce the angular variables φ and ϑ instead of y and η as in (14') by

$$y = -\frac{B}{2} \cos \varphi, \qquad \eta = -\frac{B}{2} \cos \vartheta$$

They then can be written in the form

$$\Gamma(y) = \Gamma_\varphi = K_\varphi V \left(\alpha_{\varphi}' - \frac{w_\varphi}{V} \right), \qquad w(y) = w_\varphi = \frac{1}{4\pi} \frac{2}{B} \int_0^\pi \frac{d\Gamma_\vartheta}{\cos \vartheta - \cos \varphi} \tag{31}$$

where $d\Gamma_\vartheta$ stands for $\Gamma'(\eta)\, d\eta$ or $(d\Gamma_\vartheta/d\vartheta)\, d\vartheta$.   To solve these equations, suppose Γ_φ developed in a trigonometric series:

$$\Gamma_\varphi = 2BV(A_1 \sin \varphi + A_2 \sin 2\varphi + A_3 \sin 3\varphi + \cdots) \tag{32}$$

where the A_1, A_2, A_3, . . . are unknown (dimensionless) constants to be determined by Eqs. (31). Cosine terms are omitted since Γ_φ has to be assumed as vanishing at the points $y = \pm B/2$ or $\varphi = 0$, π. To introduce (32) into the second equation (31), compute first

$$d\Gamma_\varphi = 2BV(A_1 \cos \varphi + 2A_2 \cos 2\varphi + 3A_3 \cos 3\varphi + \cdots) \quad (33)$$

Upon replacing φ by ϑ in this expression and introducing it in the second equation (31), the nth term on the right-hand side of the expression for w_φ becomes

$$\frac{1}{2\pi B} 2BVnA_n \int_0^\pi \frac{\cos n\vartheta}{\cos \vartheta - \cos \varphi} d\vartheta = \frac{nA_nV}{\sin \varphi} \frac{1}{\pi} \int_0^\pi \frac{\cos n\vartheta \sin \varphi}{\cos \vartheta - \cos \varphi} d\vartheta \quad (34)$$

The last integral is exactly the same as that considered in Eq. (29), Chap. VIII. It was proved in the Appendix to Chap. VIII that its value is $\pi \sin n\varphi$. Thus,

$$w_\varphi = \frac{V}{\sin \varphi} (A_1 \sin \varphi + 2A_2 \sin 2\varphi + 3A_3 \sin 3\varphi + \cdots) \quad (35)$$

The first equation (31) can be written as

$$\Gamma_\varphi + K_\varphi w_\varphi = VK_\varphi \alpha_\varphi'$$

On the left-hand side, introduce Γ_φ from (32) and w_φ from (35). Then, dividing by $2BV$, we find

$$A_1 \left(\sin \varphi + \frac{K_\varphi}{2B} \right) + A_2 \left(\sin 2\varphi + \frac{2K_\varphi}{2B} \frac{\sin 2\varphi}{\sin \varphi} \right)$$
$$+ A_3 \left(\sin 3\varphi + \frac{3K_\varphi}{2B} \frac{\sin 3\varphi}{\sin \varphi} \right) + \cdots = \frac{K_\varphi}{2B} \alpha_\varphi' \quad (36)$$

This equation determines the coefficients A_1, A_2, A_3, . . . *if for each* φ *(i.e., for each* y) *the value of incidence* α' *and the value of* K, *i.e.,* 4π *times the radius of the mapping circle, or approximately* π *times the chord length* c, *are given.*

The type of Eq. (36) is an unusual one. There is a denumerably infinite number of unknowns subject to the condition that the equality holds for all values of the continuous variable φ from 0 to π. Without making any attempt at a discussion from an exact mathematical standpoint, the following method of obtaining an approximate solution may be recommended:

Let us assume that a finite number n of terms in (32) is sufficient to determine Γ_φ. In this case, only n unknowns $A_1, A_2, \ldots, A_n$ appear in Eq. (36). On the other hand, we are content to fulfill relation (36) only for n suitably chosen values of φ, say, $\varphi_1, \varphi_2, \ldots, \varphi_n$. Then, writing down Eq. (36) for each of these φ-values, we have simply a system of n linear equations for the n unknowns $A_1, A_2, \ldots, A_n$. This method was devised and applied to several cases by Glauert. How it works out will be seen in the subsequent examples. First we may assume that the coefficients A_1, A_2, etc., are found and then ask what conclusions they allow concerning lift and drag as given by Eqs. (10) and (11) in Sec. 3 of this chapter.

Using the transformation (14'), we find $dy = (B/2) \sin \varphi \, d\varphi$ and therefore, from (10) and (11),

$$L = \frac{\rho}{2} BV \int_0^\pi \Gamma_\varphi \sin \varphi \, d\varphi \tag{37}$$

$$D = \frac{\rho}{2} B \int_0^\pi w_\varphi \Gamma_\varphi \sin \varphi \, d\varphi \tag{38}$$

If Γ_φ is substituted from (32) in the first of these equations we have on the right side integrals over products $\sin \varphi \sin n\varphi$; these all vanish except the first with $n = 1$, and this has the value $\pi/2$, as we know:

$$\int_0^\pi \sin m\varphi \sin n\varphi \, d\varphi \quad \begin{array}{ll} = 0 & \text{for } n \neq m \\ = \dfrac{\pi}{2} & \text{for } n = m \end{array} \tag{39}$$

Hence, the lift depends on the first coefficient A_1 only:

$$L = \frac{\rho}{2} BV(2BV)A_1 \frac{\pi}{2} = \rho B^2 V^2 \frac{\pi}{2} A_1 \tag{40}$$

In the expression for D the product of the two series (32) and (35) appears, the factor $\sin \varphi$ in (38) canceling out against the denominator in (35). The product includes terms of the form $\sin m\varphi \sin n\varphi$; and, again, all those with $n \neq m$ vanish according to (39), while those with $n = m$ give the integral value $\pi/2$. Therefore,

$$D = \frac{\rho}{2} B(2BV)V \frac{\pi}{2} (A_1^2 + 2A_2^2 + 3A_3^2 + \cdots)$$

$$= \rho B^2 V^2 \frac{\pi}{2} (A_1^2 + 2A_2^2 + 3A_3^2 + \cdots). \tag{41}$$

This can be written as

$$D = \rho \frac{\pi}{2} B^2 V^2 A_1^2 (1 + \delta) \quad \text{with} \quad \delta = \frac{2A_2^2 + 3A_3^2 + \cdots}{A_1^2} \tag{41'}$$

On passing to the coefficients C_L and C_D, the Eqs. (40) and (41') supply

$$C_L = \pi \frac{B^2}{S} A_1 = \pi \mathcal{R} A_1, \qquad C_D = \pi \mathcal{R} A_1^2 (1 + \delta) \qquad (42)$$

and

$$\frac{C_L^2}{C_D} = \frac{\pi \mathcal{R}}{1 + \delta} \qquad (42')$$

These formulas replace (17'), (18'), and (19), which were found for the case of elliptic lift distribution in the preceding section.

Let us now see under what conditions the drag coefficient C_D will be a minimum for a given value of C_L and a given aspect ratio. It is clear from (42) that to give C_L and the aspect ratio is the same as to give A_1. On the other hand, it follows from the definition of δ in (41') that δ cannot be negative. Thus the smallest C_D will be obtained if $\delta = 0$; and this implies, as (41') shows, that all coefficients $A_2 = A_3 = \cdots = 0$. That is, the "best" wing in the sense mentioned will be the wing for which expression (32) for Γ_φ reduces to the first term. But, according to $y = -(B/2) \cos \varphi$, this means

$$\Gamma_\varphi = \Gamma(y) = 2\pi V A_1 \sin \varphi = 2BV A_1 \sqrt{1 - \frac{4y^2}{B^2}} \qquad (43)$$

This is exactly the formula for the elliptic lift distribution with

$$\Gamma_0 = 2BV A_1.$$

The very interesting result can be stated as follows: *For a given aspect ratio the wing that has the smallest drag for a given lift is the wing with elliptic Γ-distribution.*

In the general case the relations (42) and (42') hold. The latter must not be misunderstood. It does not mean that the polar curve, relating C_L and C_D, is always a parabola. For the expression δ depends on the coefficients A_2, A_3, . . . , and these are determined by Eq. (36), which includes α'. Thus δ depends on the angle of incidence and, in general, is not a constant if the C_L- and C_D-values for variable α' are considered.

Let us now proceed to the computation of the coefficients A_1, A_2, etc., for a given wing. As already stated, we choose a number n of φ-values and ask that (36) shall be satisfied for these φ by an equal number of A-values. It will be suitable to use an odd number n and to take one φ equal to $\pi/2$ and the others symmetrical and equidistant. For example, with $n = 7$,

$$\varphi_1 = 22.5°, \quad \varphi_2 = 45°, \quad \varphi_3 = 67.5°, \quad \varphi_4 = 90°, \cdot \cdot \cdot, \quad \varphi_7 = 157.5°$$

With this kind of choice the computation work is considerably reduced owing to the symmetry. If one chosen value is φ_1 and another $\pi - \varphi_1$, the corresponding points on the wing are symmetrical and K_φ and α_φ' will have the same value in both cases, say, K_1 and α_1'. The two equations, for φ_1 and $\pi - \varphi_1$, then are

$$A_1\left(\sin \varphi_1 + \frac{K_1}{2B}\right) + A_2\left(\sin 2\varphi_1 + \frac{2K_1}{2B}\frac{\sin 2\varphi_1}{\sin \varphi_1}\right)$$
$$+ A_3\left(\sin 3\varphi_1 + \frac{3K_1}{2B}\frac{\sin 3\varphi_1}{\sin \varphi_1}\right) + \cdots = \frac{K_1}{2B}\alpha_1'$$

$$A_1\left(\sin \varphi_1 + \frac{K_1}{2B}\right) - A_2\left(\sin 2\varphi_1 + \frac{2K_1}{2B}\frac{\sin 2\varphi_1}{\sin \varphi_1}\right)$$
$$+ A_3\left(\sin 3\varphi_1 + \frac{3K_1}{2B}\frac{\sin 3\varphi_1}{\sin \varphi_1}\right) - \cdots = \frac{K_1}{2B}\alpha_1'$$

The terms 1, 3, 5, . . . on the left-hand side remained unchanged, while the others change their signs. Now, the two equations can be replaced by their sum and their difference. If the difference is taken, only the terms including A_2, A_4, A_6, . . . remain and on the right-hand side we have zeros. Thus the equations can be satisfied by setting

$$A_2 = A_4 = A_6 = \cdots = 0.$$

If the sum is taken and divided by 2, only the first equation without the terms of even order remains. Thus, with $n = 7$, only four equations with the unknowns A_1, A_3, A_5, A_7 have to be solved. With the foregoing given values of φ_1, φ_2, . . . and writing a_1 for $K_1/2B$, a_2 for $K_2/2B$, . . . , b_1 for $K_1\alpha_1'/2B$, b_2 for $K_2\alpha_2'/2B$, etc., the equations are

$$A_1(0.383 + a_1) + A_3(0.924 + 7.24a_1)$$
$$+ A_5(0.924 + 12.06a_1) + A_7(0.383 + 7.00a_1) = b_1$$

$$A_1(0.707 + a_2) + A_3(0.707 + 3.00a_2)$$
$$+ A_5(-0.707 - 5.00a_2) + A_7(-0.707 - 7.00a_2) = b_2$$

$$A_1(0.924 + a_3) + A_3(-0.383 - 1.244a_3)$$
$$+ A_5(-0.383 - 2.072a_3) + A_7(0.924 + 7.00a_3) = b_3 \qquad (44)$$

$$A_1(1.000 + a_4) + A_3(-1.000 - 3.00a_4)$$
$$+ A_5(1.000 + 5.00a_4) + A_7(-1.000 - 7.00a_4) = b_4$$

The following solutions may be indicated:

a. Rectangular Wing without Twist, Aspect Ratio 7. If K_φ is taken as πc, it follows that $a_1 = a_2 = a_3 = a_4 = \pi/2R = 0.224$. On the right-hand side all b have the same value $0.224\alpha'$ and the solutions for the A are proportional to it. Thus we solve for $b_1 = b_2 = b_3 = b_4 = 1$, and the final result is found in the form

$$\frac{A_1}{0.224\alpha'} = 0.9520, \quad \frac{A_3}{0.224\alpha'} = 0.1225, \quad \frac{A_5}{0.224\alpha'} = 0.0258, \quad \frac{A_7}{0.224\alpha'} = 0.0050$$

Therefore the lift coefficient, according to (42), is

$$C_L = \pi R A_1 = \frac{\pi^2}{2} \times 0.952\alpha' = 4.70\alpha'$$

as compared with the value $(14\pi/9)\alpha' = 4.89\alpha'$ for elliptic lift distribution. To find the drag coefficient we compute

$$\delta = \frac{3A_3^2 + 5A_5^2 + 7A_7^2}{A_1^2} = 0.0557$$

and this gives, according to the second equation (42),

$$C_D = \pi R A_1^2(1 + \delta) = C_L A_1(1 + \delta) = 4.7\alpha' \times 0.952 \times 0.224\alpha' \times 1.0557$$
$$= 1.058\alpha'^2$$

as compared with $C_D = C_L^2/7\pi = 1.085\alpha'^2$ in the case of the elliptic wing.

b. *Rectangular Twisted Wing, $R = 7$.* The values of $a_1, a_2, \ldots$ are the same as in the first example. The twist may be assumed as a linear decrease of α' along the span, from a value α_0' in the middle plane to $\alpha_0'/2$ at the wing tips. As y is proportional to $\cos \varphi$, this means that, for the left half wing,

$$\alpha' = \alpha_0'(1 - \tfrac{1}{2} \cos \varphi)$$

and this gives the four values of b, except for the factor $0.224\alpha_0'$,

$$0.538, \qquad 0.646, \qquad 0.809, \qquad 1.000$$

If the equations (44) are solved for these values on the right-hand side, the values of A become, except for the factor $0.224\alpha_0'$,

$$0.7410, \qquad -0.00375, \qquad 0.03115, \qquad -0.00770$$

and $\delta = 0.0172$. This gives the lift and drag coefficients according to (42):

$$C_L = 3.65\alpha_0' = 4.86\frac{3\alpha_0'}{4}, \qquad C_D = 0.62\alpha_0'^2 = 1.10\left(\frac{3\alpha_0'}{4}\right)^2$$

Here $3\alpha_0'/4$ is introduced since this is the mean angle of incidence. The coefficients are then almost the same as in the former case. By combining this solution with the first, the solution for any other degree of linear twist can be found without solving any other system of linear equations.

c. *Tapered Wing without Twist, $R = 7$.* The chord length is assumed to follow the equation

$$c_\varphi = c_0(1 - \tfrac{1}{2} \cos \varphi)$$

The aspect ratio in this case is found to be $4B/3c_0$. Thus, in our case, $B/c_0 = 5.25$, and

$$\frac{K_\varphi}{2B} = \pi \frac{c_\varphi}{2B} = \pi \frac{c_0}{2B} \frac{c_\varphi}{c_0} = 0.299(1 - \tfrac{1}{2} \cos \varphi)$$

With the foregoing values of φ, we obtain

$$a_1 = 0.161, \qquad a_2 = 0.193, \qquad a_3 = 0.242, \qquad a_4 = 0.299$$

On the right-hand side we should have

$$b_1 = a_1\alpha', \qquad b_2 = a_2\alpha', \cdots$$

Thus, $b_1:b_2: \ldots = a_1:a_2: \ldots$. We solve for

$$b_1 = 0.538, \qquad b_2 = 0.646, \qquad b_3 = 0.809, \qquad b_4 = 1.000$$

and find the result in the form

$$\frac{A_1}{0.299\alpha'} = 0.737, \frac{A_3}{0.299\alpha'} = 0.0236, \frac{A_5}{0.299\alpha'} = 0.0318, \frac{A_7}{0.299\alpha'} = -0.0024$$

This allows us to compute the lift coefficient according to (42).

$$C_L = \pi \mathcal{R} A_1 = 4.84\alpha'$$

With the δ according to (41') the drag coefficient becomes

$$C_D = \pi \mathcal{R} A_1^2 \times 1.003 = 1.06\alpha'^2$$

These results compare with $C_L = 4.89\alpha'$ and $C_D = 1.085\alpha'^2$ in the case of the elliptic lift distribution. It is seen that the tapered wing approaches closer than the rectangular wing the conditions of elliptic distribution.

The taper ratio was here assumed as $1:2$ (maximum chord to chord at the wing tips). The same kind of computation leads to the required results for any other value of the taper ratio, also.

d. *Tapered Wing with Twist*, $\mathcal{R} = 7$. The same formula for the twist is applied as in case b. The values of $a_1, a_2, \ldots$ are the same as in case c, and the new values for the right-hand sides are

$$\frac{K_\varphi}{2B} \alpha_\varphi' = 0.299(1 - \tfrac{1}{2} \cos \varphi)^2 \alpha_0'$$

Thus we solve (44) for

$$b_1 = 0.289, \qquad b_2 = 0.417, \qquad b_3 = 0.654, \qquad b_4 = 1.000$$

and find the solutions for the quantities A_1 except for the factor $0.299\alpha_0'$.

$$0.585 \qquad -0.056 \qquad 0.037 \qquad -0.012$$

This gives, in the same way as before,

$$C_L = 3.84\alpha_0', \qquad \delta = 0.014, \qquad C_D = 0.68\alpha_0'^2$$

and by combination with the results of case c the values for different twist ratios can be found. When referred to the mean angle $3\alpha_0'/4$, the coefficients are again almost the same as in the earlier cases.

A problem which, in a certain sense, is inverse to that treated in this section, *viz.*, finding the twist of a wing for given plan-form and given lift distribution, will be briefly discussed in Sec. X.3.

Problem 16. Show that the polar curve for a rectangular untwisted wing has the same form $C_D = a + (C_L^2/b)$ (Chap. VII) as it has in the case of elliptic lift distribution. Compute the theoretical value of b for $R = 7$, 6, and 5.

Problem 17. Give the relation between the geometrical angle of incidence α' and the effective angle of incidence α'' at all points of the span for a rectangular untwisted wing of $R = 7$.

Problem 18. A rectangular wing of $R = 7$ is twisted in such a way that the angle of incidence decreases linearly from its value at the center α_0' to zero at the wing tips. Compute, by combining the solutions of Eqs. (43) given under cases a and b, the value of C_L and of δ.

*Problem 19. For a rectangular wing with a definite twist the relation between the incidence in the center and the incidence at $y = -B/2 \cos \varphi$ has the form

$$\alpha_\varphi' = \alpha_0' \pm c \cos \varphi.$$

Using the method of the preceding problem, discuss the relation between C_L and α_0' and between δ and α_0'. Take, in particular, the aspect ratio $R = 7$ and

$$c = 3° = 0.0524.$$

Show that the polar curve is no longer a parabola.

Problem 20. Use the approximation method established in the preceding section to compute lift and drag for an untwisted wing of elliptic plan-form ($R = 6$), and compare the result with the previously found exact solution.

Problem 21. A rectangular wing of aspect ratio 7 is twisted in such a way that in normal flight the incidence decreases from its value α_0' at the center to zero at the wing tip according to the parabolic law $\alpha_0' - \alpha' = \text{const.} \, y^2$. Compute lift and drag coefficient, in particular for $\alpha_0' = 7°$.

CHAPTER X

ADDITIONAL FACTS ABOUT WINGS

1. Stalling. In the discussion of the empirical airfoil data in Chap. VII it was stated that the lift coefficient increases proportionally to the angle of incidence only up to a certain point, which is called the *stalling point*. For the usual profiles stalling occurs at C_L-values between 1.2 and 1.5. At the stalling point the C_L vs. α-curve, as a rule, shows a sharp bend; in any case, there is no further increase at the same rate as before (see Figs. 115 to 117). This behavior differs strikingly from that predicted by the airfoil theory. If the range of the angles of incidence in the experiments is extended farther, it is seen that C_L drops

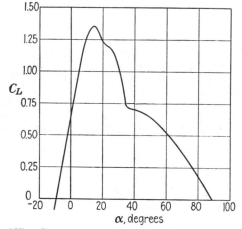

Fig. 187.—C_L vs. α-curve for an extended angular range.

to nearly zero for $\alpha = 90°$ (Fig. 187), while formula (14), Sec. VIII.2, gives the greatest value of C_L for $\alpha = 90°$. It is obvious that at $\alpha = 90°$ the problem is rather that of the motion past a rectangular plate opposed to the direction of flow. It was seen in Chap. IV that in this case a dead-air region occurs in the rear, with two discontinuity surfaces starting at the edges (Helmholtz flow). This flow pattern has evidently no resemblance to what was envisaged in the wing theory.

As will be shown in Sec. XIV.1, the velocity of an airplane of given weight in steady level flight is inversely proportional to the square root of C_L. The greatest available C_L therefore determines the smallest possible velocity. With a view to safety in landing the designer is interested in keeping this lower limit of the velocity of the airplane as low as possible, considering the other performance requirements. Hence the great importance attributed to the problem of stalling.

If the motion of air around an airfoil is made visible by smoke, the flow, which for small angles of incidence α' follows the contour of the profile, is seen to change into a discontinuous pattern when α' reaches the stalling value α'_s (Fig. 188). At a certain point near the nose the flow separates, and a region of dead air appears on the upper surface of the wing. Since the underpressure on the upper surface of the wing

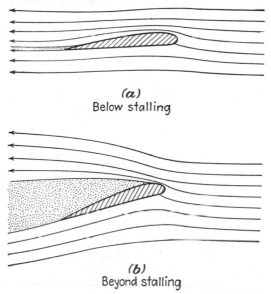

(a)
Below stalling

(b)
Beyond stalling

FIG. 188.—Flow pattern below and beyond stalling.

contributes to the lift to a much greater extent than the overpressure on the lower surface, the lift must decrease rapidly as the region of dead air develops. In fact, according to the theory of perfect fluids the pressure in the dead air would equal the pressure in the undisturbed stream at an infinite distance in the rear of the wing, i.e., the atmospheric pressure. The actual pressure behind the separation point will be slightly smaller. Up to angles of incidence of about 30° the observed shape of the dead-air region agrees roughly with the predictions that can be made on the basis of Helmholtz's theory of the discontinuous flow past an inclined flat plate.

The theory of perfect fluids can be applied only on the basis of the hydraulic hypothesis (Sec. IV.3), as follows: In the continuous flow pattern occurring below the stalling point the free stream is turbulent; i.e., the theoretical values of velocity and pressure materialize only as averages of fluctuating values. In the immediate neighborhood of the wing surface only, within an extremely narrow boundary layer, the velocity drops rapidly from its value in the free stream to zero. The turbulent fluctuations may be small, but there is no doubt that their intensity has a considerable influence on the occurrence of the separation that leads to stalling.

Even at angles of incidence far below the critical value α'_s a small region of dead air may be observed *behind the wing*, due to a separation very close to the trailing edge on the upper surface of the wing (see Fig. 188a). While the formation of such a region of dead air increases the profile drag, as was mentioned in Sec. V.5, it leaves the lift practically unchanged. This can be explained by the fact that in the con-

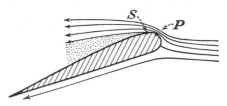

FIG. 189.—Separation at S.

tinuous flow around an airfoil the theoretical values of the velocity near the trailing edge do not differ much from the velocity V of the undisturbed stream. Accordingly, the pressure near the trailing edge does not differ much from the atmospheric pressure, and the formation of a small region of dead air near the trailing edge has only a very small influence on the lift. The influence on the moment coefficient with respect to the leading end may sometimes be more marked, because of the comparatively great distance by which the small changes of pressure near the trailing end have to be multiplied.

The phenomenon that develops when the stalling angle is reached and exceeded has quite a different character. The separation point S of this discontinuous flow pattern lies on the upper contour of the profile, near the leading end, a little behind the point P where the velocity of the continuous flow pattern assumes its greatest and accordingly the pressure its smallest value (Fig. 189). As the angle α' increases beyond α'_s, the stream withdraws more and more from the upper part of the contour, thus leaving more and more space for the dead-air region. The formation of this region can be understood in connection with the boundary-layer theory.

In the case of the continuous flow, the fluid particles in the boundary layer beyond P have to work their way against two kinds of resistance, (1) the shearing stress due to the friction at the surface of the airfoil and (2) the increase of pressure that corresponds to the decrease of the velocity of the free stream from its maximum value at P. (Note that the negative pressure gradient is the force per unit of volume acting on the fluid.) This pressure resistance increases with the angle of incidence since the difference between the pressure values at any two

points of the upper contour of the profile increases with this angle. At a
certain incidence $\alpha' = \alpha'_s$ the particles of the boundary layer are no
longer able to overcome the combined resistances and evade by separating
from the surface of the airfoil, thus starting the formation of the region
of dead air shown in Fig. 188b.   If the boundary layer is laminar, the fluid
in this layer moves in accordance with the Navier-Stokes equations
(see Sec. IV.1) and its behavior upstream of the separation point P is
accessible to mathematical investigation.   As has been stressed in Sec.

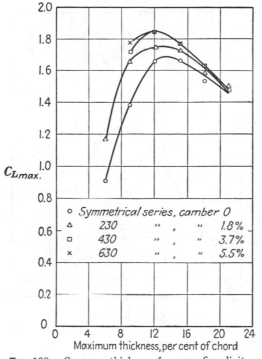

FIG. 190.— $C_{L_{max}}$ vs. thickness for some five-digit profiles.

IV.4, in this mathematical treatment the values of velocity and pressure
in the free stream are supposed to be known.   If these free-stream values
are taken from experiments, the boundary-layer theory furnishes the
position of the point of separation in good agreement with the observations.   Greater difficulties arise when, with increasing speed, the flow
in the boundary layer becomes turbulent.   This case will be considered
later.

Owing to the development of a region of dead air under stalling
conditions, the lift coefficient, in general, does not continue to increase
with the angle of attack but decreases after having reached a certain
maximum value $C_{L_{max}}$.   The question of how the value of $C_{L_{max}}$ is influ-

enced by the shape of the profile and by the Reynolds number has obviously great practical importance. The value of $C_{L_{max}}$ seems to be independent of the aspect ratio, which agrees with the fact that the separation is essentially determined by pressure differences. Accordingly, the angle of incidence α'_s at which stalling sets in will depend on the aspect ratio as required by the reduction formula (13) of Sec. VII.2. Suppose, for example, that for a certain profile $C_{L_{max}} = 1.2$. If, for infinite aspect ratio, the linear relation $C_L = 0.1\alpha'$ (α' expressed in

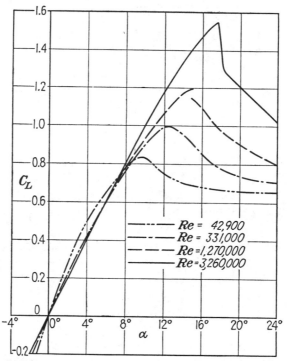

Fig. 191.—Increase of stalling C_L with Reynolds number (profile NACA 0015).

degrees) is assumed to be valid up to $C_{L_{max}}$, the stalling angle is $\alpha'_s = 12°$ for $\mathcal{R} = \infty$. On the other hand, as Eq. (13), Sec. VII.2, shows, the angle corresponding to $C_L = 1.2$ equals $\alpha'_s = 15°$ for $\mathcal{R} = 8$ and $\alpha'_s = 16.8°$ for $\mathcal{R} = 5$.

According to the best available experiments, the *influence of the shape* of the profile on the value of $C_{L_{max}}$ can be summed up by the following statements: *For the usual profiles there exists an optimum thickness (about 12 to 13 per cent of the chord length) that gives the highest value of $C_{L_{max}}$. The value of $C_{L_{max}}$ increases substantially with the camber of the airfoil and decreases when the radius of curvature at the leading end is reduced below its normal value.* Figure 190 represents the results of experiments on airfoils of the five-digit series of the NACA.[1] The relative maximum camber varied between 0 and 6 per cent, the relative thickness between 6 and 20 per cent, and the leading end had the normal curvature. Values of $C_{L_{max}}$ as high as 1.8 were observed. On the other hand, it must be kept in mind that the coefficient of profile drag a increases considerably with camber and thickness (see Sec. VII.4). Thus optimum

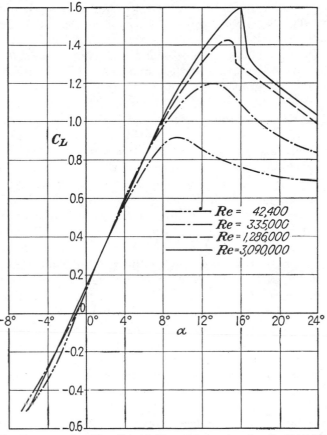

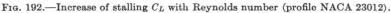

Fig. 192.—Increase of stalling C_L with Reynolds number (profile NACA 23012).

camber and thickness are determined by a compromise between the tendencies of raising $C_{L_{max}}$ and keeping a small.

As to the *influence of the Reynolds number* on $C_{L_{max}}$, the situation is much less clear. Within the range of values of Re usually encountered in experiments, the value of $C_{L_{max}}$ increases with Re. Figures 191 and 192 show typical curves C_L vs. α reduced to infinite aspect ratio, the first for a symmetrical profile, the second for the successful profile 23012.[2]

[1] *NACA Tech. Rept.* 610 (1937).

[2] *NACA Tech. Rept.* 586 (1937).

For this profile $C_{L_{max}}$ increases from 0.9 at Re = 42,400 up to 1.6 at Re = 3,090,000. These values of Re are computed according to the values of velocity, pressure, and temperature observed during the experiments in the variable-density wind tunnel at Langley Field. To be compared with values in the free atmosphere, these values of Re must be multiplied by the turbulence factor 2.64 of this tunnel (Sec. VII.5). The corresponding effective Reynolds numbers are thus found to be 112,000 and 8,200,000. The last value corresponds, for example, to a wing of 4 ft. chord length flying at an altitude of 3000 ft. with a velocity of 140 m.p.h.

From the standpoint of the boundary-layer theory the following explanation for the influence of the Reynolds number is usually offered. For moderate values of Re the boundary layer is laminar. At $\alpha' = \alpha'_s$ separation will then set in near to the leading end, as already explained. The position of the separation point S (Fig. 189) is determined by the following condition: As the particles of the boundary layer move beyond the point P where the velocity reaches its maximum value, their kinetic energy is gradually spent in overcoming the two kinds of resistance mentioned in the foregoing. Separation occurs at the point where the entire kinetic energy, which the particles possessed at P, has thus been spent. The presence of any source of energy on which the particles of the boundary layer can draw postpones separation. By a sufficiently great supply of energy to the boundary layer the occurrence of the separation may even be prevented entirely. Now, for high Reynolds numbers, the flow in the boundary layer tends to become turbulent. The fluid particles of the boundary layer and those of the free stream then do not keep strictly apart as in the laminar case; instead, a certain amount of mixing occurs. Fluid particles of the free stream entering the boundary layer import an amount of kinetic energy that is higher than the amount exported by the particles leaving the boundary layer. The turbulence in the flow of the boundary layer therefore tends to postpone the occurrence of the separation and to raise the value of $C_{L_{max}}$.

Modern airplanes flying at high speed reach values of Re considerably higher than those obtained in experiments. Whether the values of $C_{L_{max}}$ still improve in this range is uncertain. In any case it must be expected that the increase of $C_{L_{max}}$ with Re becomes less and less pronounced as Re increases.

What can be done to improve on stalling conditions will be discussed in the next section.

2. High-lift Devices. It has been seen in the preceding section that the value of $C_{L_{max}}$ can be raised by increasing the camber. But this method of improving $C_{L_{max}}$ can be applied within very narrow limits only, because of the increase of the profile drag a with the camber. Furthermore, the moment coefficient C_{M_0}, which is negative for the usual profiles, becomes unduly great in absolute value as the camber increases. Consequently, other methods of improving $C_{L_{max}}$ have been devised. According to the preceding section, the value of $C_{L_{max}}$ may be raised either by supplying additional energy to the boundary layer or by diminishing the resistance against which the particles in the boundary layer have to work their way. By either one of these two means the separation point S will be displaced downstream or, possibly, the occurrence of the separation prevented.

As already mentioned, the resistance against which the particles of the boundary layer progress consists in the friction along the solid surface of the wing and the pressure resistance.

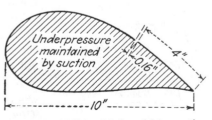

Fig. 193.—Hollow airfoil model for suction slit experiments.

An effective means of reducing the pressure resistance was proposed by L. Prandtl, as follows: A narrow, spanwise slit in the upper surface of the wing (Fig. 193) establishes a communication between the air flowing along the back of the wing and the air in the interior of the wing, where a suitable underpressure is maintained by means of a pump. At the location of the slit the pressure in the boundary layer is thus reduced, and the particles moving toward the slit encounter a smaller pressure resistance. Figure 194 shows the streamlines and the pressure distribution in the neighborhood of the slit. The particles of the boundary layer upstream of the slit are sucked into the interior of the wing, and downstream of the slit a new boundary layer is formed of particles that are deflected from the free stream toward the wing surface and that have not yet been exposed to the retarding influence of the friction at the wing surface. With extremely thick profiles experimental values of $C_{L_{max}}$ up to 5.0 have been obtained by this method, known as boundary-layer control by *suction*.[1] The best location of the suction slot seems to be at about 50 per cent of the chord.

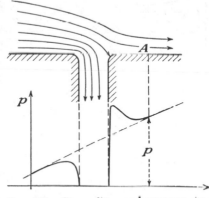

Fig. 194.—Streamlines and pressure distribution near a suction slit.

[1] Schrenk, O. *Z. Flugtech. u. Motorluftschiffahrt*, **22**, 259 (1931); *NACA Tech. Rept.* 385 (1931).

Another possibility of preventing the formation of a dead-air region is to supply kinetic energy to the boundary layer by ejecting a thin jet of air through a spanwise slit in the upper surface of the wing (Fig. 195). Although in a certain way this procedure, known as boundary-layer control *by pressure slots*, may appear to be the opposite of the suction method just described, it seems that the same effect can be produced. Experimental values of $C_{L_{max}}$ up to 3.3 have been observed.[1] In the form indicated in Fig. 195 this device has not been used in aeronautical design. However, these experiments may contribute to the understanding of another device, the *slotted wing*, introduced by G. Lachmann and Handley-Page and widely used in modern airplanes. Near the leading end a slot establishes a communication between the two sides of the wing (Fig. 196). Because of the difference in pressure between the upper and lower wing surfaces an upward current develops in this slot. The jet

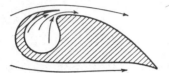

Fig. 195.—Boundary layer control by pressure slot.

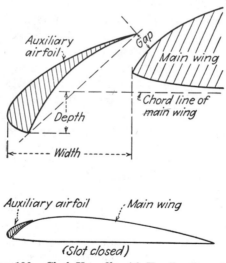

Fig. 196.—Clark Y-profile with Handley-Page slot.

of air emerging from the slot produces an effect similar to that described above. Figure 197 shows how the curve C_L vs. α of a profile is modified by such a slot. Values of $C_{L_{max}}$ up to 1.8 can easily be obtained in this way.

It is desirable that a high-lift device should permit of changing the aerodynamic characteristics of a wing *during flight*. In particular, for

[1] KATZMAYR, R. *Berichte der aeromechanischen Versuchsanstalt Wien*, **1**, 57 (1928); *NACA Tech. Mem.* 521 (1929).

landing operations, an increase in the value of $C_{L_{max}}$ is advantageous. The devices illustrated in Figs. 193 and 195 obviously fulfill this requirement since the pump necessary for the suction or the ejection of air may be operated at will.   In the case of a slotted wing this aim is less easily accomplished.   In normal flight the slot may be closed either by properly shaped sliding panels or by bringing the front part of the wing (auxiliary wing) in contact with the rear part (main wing), thus forming an uninterrupted wing.   In both cases a certain compromise must be

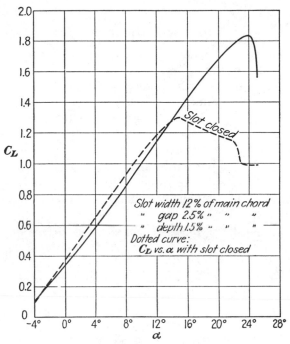

Fig. 197.—Lift curve of the profile of Fig. 196 ($\mathcal{R} = 6$).

worked out between the ideal aerodynamic design of the open slot and the kinematic requirements of the mechanism used for closing it.   The Handley-Page type of slot is operated automatically.   In normal flight the auxiliary wing is in contact with the main wing; at high angles of attack the underpressure on the upper side of the auxiliary wing becomes sufficiently great to separate it from the main wing, thus opening the slot.

Other possibilities of increasing the lift during landing operations are offered by *wings of variable shape, flaps, and spoilers*.   These devices may be used in combination with one another or with slots.

Soaring birds can modify the aerodynamic characteristics of their wings to a considerable extent: the camber and the area of the wing are changed by contracting or relaxing certain muscles.   (Earlier airplane types, like those of the brothers Wright and the Austrian Taube, used deformable wings instead of wings with ailerons.) Occasionally an experimental glider plane was built whose wing profile could be changed in a continuous manner during flight. However, the aerodynamic advantages of this design proved too slight. W. Schmeidler studied the

Fig. 198.—Plain split flap.

possibilities offered by a wing of variable area.[1]   Here, also, the weight and the complexity of the necessary mechanism were prohibitive.

All forms of flaps, the most widely used high-lift devices, constitute various compromises between the wish to modify the shape of the wing in an aerodynamically efficient manner and the necessity of keeping the mechanism by which this change of shape is effected sufficiently simple. Various arrangements are in use. Figure 198 shows the *split flap*, which usually extends over about 20 per cent of the chord length. In normal flight the flap is flush with the lower surface of the wing; in use, it is rotated about its leading edge by angles up to about 45°. Figure 199 shows the influence of a split flap on the aerodynamic characteristics of the NACA profile 23012 for $Re = 3,100,000$. The flap extends over 20 per cent of the chord length and over the entire span, the angle of deflection of the flap is 60°. The value of $C_{L_{max}}$, which equals 1.6 for the profile without flap, is raised to 2.45 when the flap is deflected. The additional increase in $C_{L_{max}}$ that would be obtained by using higher angles of deflection is very small. The maximum value of C_L/C_D, equaling 23 without flap,

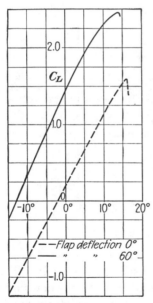

Fig. 199.—Effect of split flap.

is lowered to 5.   This considerable increase in drag is welcome when the split flap is used for reducing the landing speed but prevents the use of the split flap as a high-lift device for reducing the length of the take-off run.

Figure 200 shows the *trailing-end flap*, which can be rotated about an axis parallel to the span, similar to an aileron. If the wing is not interrupted by the fuselage, the flap extends over the central part of the wing, or else two symmetrical flaps are used. For sufficiently small angles of deflection of the flap, the theory of thin airfoils (Sec. VIII.6) may be applied to the wing with deflected flap. Within the framework of this theory, the coefficients of lift and moment depend linearly on the ordinates of the mean camber line. Consequently, the changes ΔC_L and ΔC_{M_0} of these coefficients, produced by deflecting the flap through the

FIG. 200.—NACA 23012 with trailing-end flap.

angle δ, can be found as the values of the coefficients of lift and moment for the broken camber line of Fig. 201. If λ denotes the ratio of the flap chord to the total wing chord and φ the angle defined by cos $\varphi = 1 - 2\lambda$, the result furnished by the thin-wing theory can be written as

$$\Delta C_L = 2\delta(\varphi + \sin \varphi - \lambda\pi)$$

and

$$\Delta C_{M_0} = -\tfrac{1}{2} \delta(\sin \varphi + \tfrac{1}{2} \sin 2\varphi)$$

In evaluating these formulas the angles φ and δ must be expressed in radians. The values of ΔC_L and ΔC_{M_0} obtained from these formulas refer, of course, to a wing of infinite aspect ratio. For a flap extending over 20 per cent of the chord length ($\lambda = 0.2$), the formulas give $\Delta C_L = 2.2\delta$ and $\Delta C_{M_0} = -0.64\delta$. The experimental evidence does not agree too well with these theoretical predictions. For a fixed angle of attack of the main wing, the curve ΔC_L vs. δ is not a straight line but has the character indicated in Fig. 202. Even for small angles δ, the theory leads to an overestimate of C_L. Furthermore, for a fixed value of δ

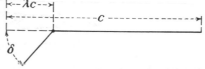

FIG. 201.—Trailing flap skeleton.

and varying angle of attack of the main wing the value of ΔC_L is not quite constant, as should be the case according to the theory. The value of $C_{L_{max}}$ obtainable with a given wing can be raised to about 1.0 by a trailing-end flap deflected through an angle of 30°. Values of the same order of magnitude can also be obtained with strongly cambered profiles. But the flap has the advantage of combining high values of the lift coefficient with the small values of the profile drag prevailing with undeflected flap.

While in the case of an ordinary trailing-end flap any leakage through a gap between main wing and flap seems to be highly detrimental,[1] a

[1] See *NACA Tech. Rept.* 554 (1936).

combination of *a flap and a properly shaped slot* (Fig. 203) proves to be very efficient. For the NACA profile 23021 and a flap length of about $0.25c$, the value of $C_{L_{\max}}$ was raised to 2.55 at $\delta = 30°$ and to 2.7 at $\delta = 60°$ as compared with $C_{L_{\max}} = 1.4$ for the airfoil without slot and flap.[1] It seems that there is a tendency toward separation near the leading end of the flap and that this tendency is counteracted by the thin jet of air

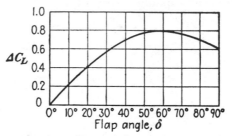

Fig. 202.—Lift increase due to trailing-end flap; angle of attack of main wing is 10°.

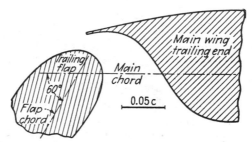

Fig. 203.—Flap and slot in combination; slot arrangement for best $C_{L_{\max}}$ at 60° flap angle.

Fig. 204.—Multiple flap arrangement.

flowing through the slot from the lower to the upper side of the wing. The *Venetian blind flap* (Fig. 204) is a combination of several slots and flaps, but it seems that such complicated devices do not pay, for the value of $C_{L_{\max}}$ is not increased beyond 2.8.

The *Fowler flap* (Fig. 205) combines the effects of increased camber and increased wing area. Referred to the original wing area, maximum lift coefficients of 2.45, 2.85, and 3.17 have been observed with Fowler flaps of 20, 30, and 40 per cent, respectively; the angles of deflection for which these values of $C_{L_{\max}}$ were obtained were 30, 40, and 40°, respectively.

[1] *NACA Tech. Rept.* **677** (1939).

Devices that aim at an increase of drag rather than of lift are *spoilers* and *brakes*. They are used in modern fighter planes for reducing the terminal speed of a dive. Figure 206 shows these flaplike devices, called spoilers if on the upper side of the wing and brakes if on the lower.

3. Pressure Distribution. For most performance computations it is sufficient to know the total lift and the total drag of a wing. For

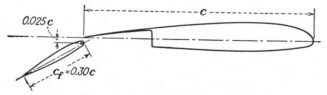

FIG. 205.—Fowler flap (experimental setup).

problems of structural analysis, however, information concerning the distribution of pressure over the wing surface is indispensable.

a. Chordwise Distribution. For given values of V and α the two-dimensional airfoil theory furnishes the velocity values along the contour of the profile. From these the pressure values can be computed by means of Bernoulli's equation. The curves of Fig. 207 show the theoretical distribution of the pressure over the two sides of the NACA profile 4412, as it flies at the angle of attack $\alpha = 8°$. The ordinates are the values of $P = (p - p_0)/q$, where p_0 is the pressure in the undisturbed flow and q the dynamic pressure corresponding to the speed of flight V. The area between the two curves, multiplied by q and chord length c, gives the lift per unit of span, that is, $L' = dL/dy$. The points above the horizontal axis correspond to the underpressure that occurs on the upper side of the airfoil; below are the points representing the

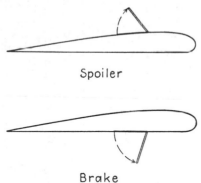

Spoiler

Brake

FIG. 206.—Spoiler and brake.

overpressure on the lower side. Experimental evidence confirms, in general, the shape of the theoretical curves. The points indicated in Fig 207 represent the observed distribution[1] of P. The absolute values of P found in the experiment are uniformly smaller than those predicted by the theory.

Since the upper surface of the wing contributes more to the lift than the lower one, protuberances and other irregularities in the form of the

[1] *NACA Tech. Rept.* 563 (1936).

wing surface on the upper side may be expected to make themselves more felt than on the lower side.

The theory of thin airfoils (Sec. VIII.6) does not describe the situation concerning the over- and underpressure correctly. Actually, with the notations of Sec. VIII.6 the velocity values at the two sides of the airfoils are $V \pm \frac{1}{2} \Gamma'$. Neglecting the square of the small vortex density Γ', the

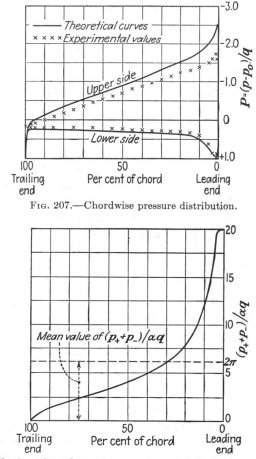

Fig. 207.—Chordwise pressure distribution.

Fig. 208.—Distribution of resultant pressure along the chord of an infinitely thin symmetrical airfoil.

squares of the velocity values on the two sides of the airfoil can be written as $V^2 \pm \Gamma' V$. According to Bernoulli's equation the underpressure $p_- = p_0 - p$ at the upper side would thus equal the overpressure $p_+ = p - p_0$ at the lower side. Although this does not coincide with the facts, the chordwise distribution of the sum $p_+ + p_- = \rho V \Gamma'$, as furnished by the theory, gives a fairly good approximation for the lift distribution.   Figure 208 shows the theoretical distribution for a thin symmetrical airfoil (straight mean camber line).

While the total lift is comparatively independent of the shape of the profile, as we know, the distribution of the underpressure on the upper side, particularly near the leading end, is very sensitive to the variation of the curvature.   A well-rounded leading end makes this distribution much more uniform and is therefore advantageous with respect to deferring the stalling point.

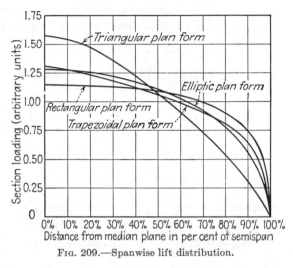

FIG. 209.—Spanwise lift distribution.

b. Spanwise Distribution.   The spanwise lift distribution can be determined by giving the value $L' = dL/dy$ for all values of y between 0 and $B/2$.   Instead, the *sectional lift coefficient*

$$C_L(y) = \frac{2\,dL}{\rho V^2\,dS} = \frac{2L'}{\rho V^2 c} \tag{1}$$

where c is the chord length at the distance y can be used.   It was seen in Eq. (11'), Chap. IX, that dL equals $\rho V\Gamma(y)\,dy$; thus

$$L' = \rho V\Gamma(y), \qquad C_L(y) = \frac{2\Gamma(y)}{Vc} \tag{1'}$$

Here $\Gamma(y)$ is the circulation in the section y.

In Sec. IX.7 a method has been developed that allows us to compute the Γ-distribution for any wing of given plan-form and twist.   If the wing, in particular, is untwisted, *i.e.*, if the angle of incidence α' is constant over the span, it turns out that, for each definite plan-form, Γ, and thus L', is proportional to α'.   In Fig. 209 the curves L' or Γ vs. y are plotted for several plan-forms.   The scale of the ordinates in each case is chosen in such a way as to give an equal value to the area that represents the total lift $L = \int L'\, dy$. One of the curves shows the elliptic wing distribution (Sec. IX.5).

The spanwise lift distribution observed in experiments generally follows closely the predictions of the theory, with the possible exception of the immediate vicinity of the wing tips, where the experimental values frequently are higher than the theoretical ones. Figure 210 shows a typical result.[1]

Sometimes, in wing design, a problem arises that is "inverse" to the problem discussed in Sec. IX.7 and is therefore easier to solve. Assume

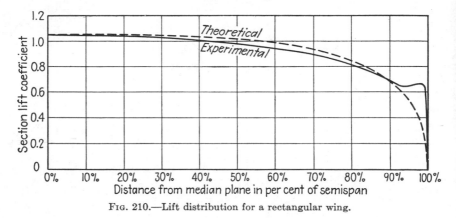

Fig. 210.—Lift distribution for a rectangular wing.

that a plan-form has been selected and that it is desired to realize a certain spanwise lift distribution (*e.g.*, with a view to keeping the bending moments on the wing within certain limits). This can be achieved by giving the wing an appropriate twist. The procedure to be followed in computing the twist may briefly be described as follows.

With the notation used in Chap. IX,

$$\cos \varphi = -\frac{2y}{B}$$

the intended load distribution may be given in the form of a trigonometric expression

$$\frac{L'}{2\rho V^2 B} = A_1 \sin \varphi + A_2 \sin 2\varphi + A_3 \sin 3\varphi + \cdots \qquad (2)$$

The coefficients $A_1, A_2, A_3, \ldots$ can easily be computed if L' is originally given as a polynomial in y or in some other way. Then, it follows from

[1] *NACA Tech. Rept.* 585 (1937).

Eq. (36), Chap. IX, that the angle of incidence $\alpha'(y)$ or α_φ' in the section determined by y or φ must have the value

$$\alpha_\varphi' = A_1 \left(\frac{2B}{K_\varphi} + \frac{1}{\sin \varphi} \right) \sin \varphi + A_2 \left(\frac{2B}{K_\varphi} + \frac{2}{\sin \varphi} \right) \sin 2\varphi$$

$$+ A_3 \left(\frac{2B}{K_\varphi} + \frac{3}{\sin \varphi} \right) \sin 3\varphi + \cdots \quad (3)$$

Here K_φ equals $\pi c_\varphi = \pi c(y)$ for thin profiles and, in general, is given by formula (30), Chap. IX. The differences of the α_φ-values for varying φ determine the twist of the wing. It is understood that the lift distribution (2) will occur only in one definite state of (normal) flight.

Problem 1. Compute the twist for a rectangular wing if it is required that the lift per unit of span shall decrease linearly from its value in the median plane to zero at the wing tips.

Problem 2. Find the twist for a wing of trapezoidal form under the assumption that the section load decreases from the median plane toward the wing tips proportional to y^2.

4. Influence of Compressibility. In the preceding discussion of theoretical and experimental results concerning the flow past a wing, the air has been considered as incompressible. This is certainly justified as long as the velocity does not exceed 250 to 300 ft./sec. For example, under standard conditions the atmospheric pressure at sea level is 2116.2 lb./ft.^{2}, while the dynamic pressure (see Table 2, page 32) corresponding to a velocity of 300 ft./sec. is found to equal 107 lb./ft.^{2} if the air is assumed to be incompressible. (The stagnation overpressure would equal 109 lb./ft.^{2} if the air is considered as a perfect gas flowing under adiabatic conditions; see Table 3, page 34.) In this case the pressure at the stagnation point is therefore only about 5 per cent greater than in the undisturbed stream, and the increase in density ρ is of the same order of magnitude. In most computations such density variations may safely be neglected. But at a velocity of 500 ft./sec. the value of the dynamic pressure is 297 lb./ft.^{2} for the incompressible fluid, and the increase of pressure and density at the stagnation point thus reaches about 14 per cent. Even then the theory of the incompressible fluid may be relied on to furnish an approximately correct over-all picture, but in the study of finer details the compressibility of the air must be taken into account.

The theory of compressible fluids is not developed far enough to answer satisfactorily the questions of practical importance. This is largely due to the tremendous mathematical difficulties caused by the fact that the problem is no longer linear. Because of this fact the method of superposition of flow patterns that is constantly used in the theory of incompressible fluids fails. For example, the flow around an airfoil has been studied in Secs. VIII.1 to 4 as the result of the super-position of a uniform flow and a circulating and a bicirculating motion and in Sec. VIII.5 as the sum of a uniform flow and the motion induced by certain vortex sheets. Since this powerful method of superposition can no longer be applied, the theory of compressible fluids must be built up along entirely different lines. So far, only a few first steps have been made, and a complete discussion of even these is beyond the scope of this book. Only a very elementary approach will be presented, which leads, however, to useful conclusions regarding the effect of compressibility on the aerodynamic characteristics of an airfoil.

The two-dimensional steady irrotational flow of an *incompressible* fluid is governed by two equations, the condition of continuity, $V\,dn =$ const. along each streamline (Sec. II.1), and the condition of vanishing rotation,

$$V/R - dV/dn = 0 \quad (\text{Sec. II.4}).$$

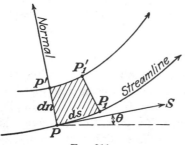

FIG. 211.

In applying these equations to the flow around an airfoil it is preferable to denote the velocity of the fluid particle under consideration by v and to reserve V for the velocity of the undisturbed stream. Moreover, since the rates of change of v in the directions tangential and normal to the streamline will now be studied simultaneously, the symbol dV/dn will be replaced by $\partial v/\partial n$; and $\partial v/\partial s$ will be used to denote the rate of change of v in the direction of the streamline. If ϑ denotes the angle that the velocity vector at a point P makes with a fixed direction (Fig. 211), the curvature of the streamline at P equals $1/R = \partial\vartheta/\partial s$. The condition of vanishing rotation thus takes the form

$$\frac{\partial v}{\partial n} - v\frac{\partial \vartheta}{\partial s} = 0 \qquad (4)$$

The condition of continuity can be written in a similar form. On the positive normal of the streamline through P (Fig. 211), take the point P' at the distance dn from P. The tangent of the streamline at P' forms the angle $\vartheta + (\partial\vartheta/\partial n)\,dn$ with the fixed direction. Consequently, the stream filament determined by P and P' has the width $dn + (\partial\vartheta/\partial n)\,dn\,ds$ at the distance ds from PP'. According to the condition of continuity the product of this width and the corresponding velocity $v + (dv/ds)\,ds$ must equal $v\,dn$. Multiplying the two expressions and retaining the terms of first order only, we obtain

$$\frac{\partial v}{\partial s} + v\frac{\partial \vartheta}{\partial n} = 0 \qquad (5)$$

The differential relations (4) and (5) for the magnitude v and the direction ϑ of the velocity vector embody all general conditions imposed on a two-dimensional steady irrotational flow of an incompressible fluid. Together with the boundary conditions of each particular problem, they determine the flow pattern.

In the case of a two-dimensional steady irrotational flow of a *compressible* fluid, nothing has to be changed in Eq. (4), which expresses only the kinematic condition of irrotationality. As regards (5), it must be kept in mind that the condition of continuity now reads $\rho v\, dn = $ const. (see Sec. II.1). Accordingly, v in (5) must be replaced by ρv. This leads to

$$\frac{\partial(\rho v)}{\partial s} + \rho v \frac{\partial \vartheta}{\partial n} = 0 \quad \text{or} \quad \frac{\partial v}{\partial s} + \frac{v}{\rho}\frac{\partial \rho}{\partial s} + v\frac{\partial \vartheta}{\partial n} = 0 \tag{6}$$

The differential relations (4) and (6) involving three variables, v, ϑ, and ρ are no longer sufficient for the determination of the stream pattern. This difficulty is not overcome by bringing in Bernoulli's equation, since this contains a new variable, the pressure p. In addition, an assumption already introduced and discussed in Sec. II.2 must be used. According to this assumption there exists a one-to-one correspondence between the values of pressure and density, for example, the isothermal relation $p/\rho = $ const. or the polytropic relation $p/\rho^\kappa = $ const. Whatever this relation may be, it is assumed that to each pressure value p corresponds a definite value of ρ as well as of the derivative $dp/d\rho$. Since, in any case, ρ increases with p, this derivative is positive and may be denoted by a^2:

$$\frac{dp}{d\rho} = a^2 \tag{7}$$

As is easily seen, a has the dimensions of a velocity. For reasons that will be given at the end of this section, the quantity a is called the *velocity of sound*. Its value depends on the value of the pressure p and may, therefore, vary from point to point. Under standard equilibrium conditions the velocity of sound in air at sea level is 1116 ft./sec. In the case of a compressible fluid in motion, where a varies with the position in space, it is better to refer to a as the *local velocity of sound*.

If the influence of gravity is neglected, Bernoulli's equation may be written in the differential form

$$\rho v \frac{\partial v}{\partial s} + \frac{\partial p}{\partial s} = 0$$

[see Eq. (3), Chap. II]. It follows from (7) that $dp = a^2\, d\rho$. Hence,

$$\rho v \frac{\partial v}{\partial s} + a^2 \frac{\partial \rho}{\partial s} = 0$$

or, multiplying by $v/\rho a^2$,

$$\frac{v}{\rho}\frac{\partial \rho}{\partial s} = -\frac{v^2}{a^2}\frac{\partial v}{\partial s} \tag{8}$$

The dimensionless ratio v/a is called the *Mach number* and is usually denoted by M:

$$\frac{v}{a} = v\sqrt{\frac{d\rho}{dp}} = M \tag{9}$$

For an incompressible fluid ($d\rho = 0$), the velocity of sound is infinite, and the Mach number everywhere zero.

Substituting from (8), we write (6) in the form

$$(1 - M^2)\frac{\partial v}{\partial s} + v\frac{\partial \vartheta}{\partial n} = 0 \tag{10}$$

For a compressible fluid this equation replaces (5), which applies only to an incompressible fluid, while (4) is valid in both cases. The com-

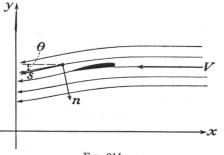

Fig. 211a.

parison of (5) and (10) shows that the factor $(1 - M^2)$ of the first term in (10) constitutes the only difference between the two sets of differential relations. Now, in general, M is not a constant but depends on the pressure p, which, itself, can be expressed in terms of v by means of Bernoulli's equation. Since M thus is a rather complicated function of v, the analogy between (5) and (10) is not so close as it may appear at first sight. However, there is a very important case where this correspondence between the compressible and incompressible fluid problem leads to useful conclusions.

Consider the flow round a slightly cambered thin airfoil flying at a small angle of attack (Fig. 211a). Taking the direction of the positive x-axis parallel but opposite to the direction of the velocity V of the undisturbed stream, we write the velocity components in the x- and y-direction as $-V + v_x$ and v_y, respectively. According to the nature of the flow, v_x and v_y will be small as compared with V. In the following, terms of higher order in v_x/V and v_y/V will be neglected consistently.

Since the tangent of the angle ϑ equals $-v_y/V$, this angle is small, too. The magnitude of the velocity, in this order of approximation, is

$$v = V - v_x,$$

and the derivatives occurring in (4) and (10) equal $-\partial v_x/\partial s$, and $-\partial v_x/\partial n$, respectively. The four derivatives occurring in (4) and (10) can therefore be considered as small quantities of first order. If quantities of higher order are to be neglected, the factors v and $(1. - M^2)$ occurring in these equations must be replaced by the values V and $(1 - M_0^2)$ that they assume in the undisturbed stream. In fact, the difference $V - v = v_x$ being of the first order, its products with the derivatives of ϑ are of the second order and have to be omitted. Similarly, the difference $M_0^2 - M^2$, which originates in the difference between V and v, will be small, of the first order at least. On the other hand, since ϑ is small, the directions of ds and dn nearly coincide with the directions of the negative x- and y-axes, respectively. Again neglecting higher order terms, we may replace $\partial/\partial s$ and $\partial/\partial n$ by $-\partial/\partial x$ and $-\partial/\partial y$, respectively. Equations (10) and (4) thus finally take the form

$$(1 - M_0^2) \frac{\partial v_x}{\partial x} - V \frac{\partial \vartheta}{\partial y} = 0$$

$$\frac{\partial v_x}{\partial y} + V \frac{\partial \vartheta}{\partial x} = 0 \tag{11}$$

In the case of an incompressible fluid, M_0 equals zero. Within the framework of the present approximation the equations of the compressible fluid therefore differ from those of the incompressible fluid only by the value of one constant, $(1 - M_0^2)$, instead of 1.

Equations (11) allow us to derive the stream pattern of a compressible fluid from that of an incompressible one. Assuming $M_0 < 1$, set

$$1 - M_0^2 = \lambda^2$$

and introduce in (11) new independent variables $x_1 = x$ and $y_1 = \lambda y$. Then, upon dividing the first equation by λ^2 and the second by λ, Eqs. (11) take the form

$$\frac{\partial v_x}{\partial x_1} - \frac{V}{\lambda^2} \frac{\partial(\lambda\vartheta)}{\partial y_1} = 0$$

$$\frac{\partial v_x}{\partial y_1} + \frac{V}{\lambda^2} \frac{\partial(\lambda\vartheta)}{\partial x_1} = 0 \tag{12}$$

These are exactly the equations for an incompressible flow, if v_x and $\lambda\vartheta$ are considered as the new unknowns and $V_1 = V/\lambda^2$ as the velocity at infinity. But when an x-y-plane is subject to the affine transformation $x_1 = x$, $y_1 = \lambda y$, all curves that had originally the slope ϑ transform into curves with the slope $\vartheta_1 = \lambda\vartheta$. Thus it is seen that the two sets of equations (11) and (12) define a one-to-one correspondence between certain stream patterns of a compressible and of an incompressible flow, which can be formulated in the following way: *If in an incompressible fluid flow with velocity V_1 at infinity the slope ϑ_1 of the streamlines and the ratio v_x/V_1 are everywhere small, there exists, approximately, a compressible fluid flow with the mean Mach number M_0 and the velocity $(1 - M_0^2)V_1$ at infinity of which the streamlines are the affine images*

$$x = x_1, \qquad y = \frac{y_1}{\sqrt{1 - M_0^2}},$$

of the original streamlines and for which the additional x-velocity v_x at each point is the same as in the original flow pattern. (See Fig. 212, where λ is assumed $1/\sqrt{2}$.)

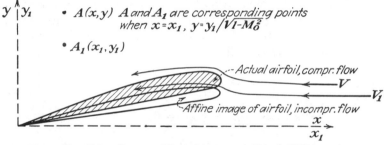

FIG. 212.—Compressible and incompressible fluid flow.

This correspondence allows us to estimate the influence of compressibility on the lift and the moment coefficients for an airfoil of moderate thickness and camber at small angles of attack. In fact, neglecting gravity, the Bernoulli equation in its differential form reads (Sec. II.2) for each streamline

$$v\frac{dv}{ds} + \frac{1}{\rho}\frac{dp}{ds} = 0, \qquad dp = -\rho v\, dv \tag{13}$$

In the framework of the present approximation v can be replaced by $-V + v_x$ where v_x is small. Then, calling ρ_0 and p_0 the values at infinity (which are the same on all streamlines) and omitting all terms of higher order, we find

$$dp = +\rho_0 V\, dv_x, \qquad p = p_0 + \rho_0 V \int dv_x \tag{14}$$

In the incompressible image of the flow the same relation holds, only with the V replaced by $V_1 = V/\lambda^2$. Thus it is seen that the pressure difference for any two points on the airfoil, e.g., one point on the lower side and one point on the upper side, is multiplied by λ^2 as one passes from the incompressible stream pattern to the corresponding compressible fluid flow. The lift and moment being determined by these pressure differences, one has $L = \lambda^2 L_1$ and $M = \lambda^2 M_1$. The lift and moment coefficients are reduced to the respective velocity squares, V^2 and V_1^2, thus:

$$\frac{C_L}{C_{L_1}} = \frac{L}{L_1}\frac{V_1^2}{V^2} = \lambda^2 \left(\frac{1}{\lambda^2}\right)^2 = \frac{1}{\lambda^2} \quad \text{and} \quad \frac{C_M}{C_{M_1}} = \frac{1}{\lambda^2} \tag{15}$$

In the affine transformation the angle of attack changes according to $\alpha_1 = \lambda\alpha$. Therefore the slope of the C_L vs. α-curve is

$$\frac{dC_L}{d\alpha} = \lambda\frac{1}{\lambda^2}\frac{dC_{L_1}}{d\alpha_1} = \frac{1}{\lambda}\frac{dC_{L_1}}{d\alpha_1} \tag{16}$$

The wing theory for incompressible flow leads to the result that the slope of the C_L-curve is constant with respect to α (below stalling) and only slightly dependent on the shape of the profile. Thus the main outcome of the present argument is as follows: *The slope of the C_L vs. α-curve, which is about 2π for infinite span in an incompressible fluid, is multiplied by the factor $1/\lambda = 1/\sqrt{1 - M_0^2}$ under the influence of compressibility.* This result is due to H. Glauert (1928). At an air speed $V = 600$ ft./sec. the increase is 18.5 per cent at sea level and 25 per cent at an altitude of 30,000 ft.

Moreover, the complete C_L vs. α-curve could be derived for each M_0-value if the curve were known for $M_0 = 0$ (incompressible case), for an airfoil whose thickness and camber are λ times as great as those of the actual profile. One would simply have. to multiply the abscissas by $1/\lambda$ and the ordinates by $1/\lambda^2$ (Fig. 213). A similar rule applies to the moment coefficient C_M. The position of the aerodynamic center is not changed.

It cannot be expected that this primitive attempt furnishes more than a first approximation. The assumptions that ϑ and v_x/V are small fail

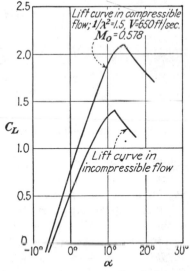

Fig. 213.—Lift at high velocities.

entirely near the stagnation point at the leading edge. At the stagnation point, v vanishes, and thus v_x/V equals 1; and somewhere at the nose of the profile ϑ must be 90°. However, for Mach numbers M_0 up to about 0.6 experimental evidence confirms to a large extent the theoretical result. Figure 214 shows the observed values of the slope ratio $(dC_L/d\alpha):(dC_{L_1}/d\alpha_1)$ for a thin propeller profile of about 6 per cent thickness.[1] The dotted line corresponds to the values of

$$1/\lambda = 1/\sqrt{1 - M_0^2}.$$

It is seen that the rough theory underestimates slightly the influence of compressibility for $M_0 < 0.6$, while at higher M_0 the character of the curve changes thoroughly.   The same situation prevails, in a still more outspoken form, in other experiments.

A more detailed discussion of the compressible fluid flow is beyond the scope of this book.   It should be mentioned once more that the enormous difficulties of such a theory lie in the fact that the equations of motion for the compressible flow are no longer linear, so that the

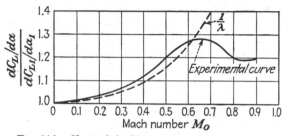

FIG. 214.—Slope of the lift curve vs. Mach number.

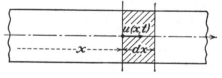

FIG. 215.

method of superposing solutions, which was used in Chap. VIII for the irrotational incompressible flow, is inapplicable.   Only very few results, at a great expense of mathematical work, have been reached so far.

It remains to explain why a is called the *velocity of sound*.   Consider a narrow cylindrical pipe filled with air engaged in axial oscillatory motion about a state of equilibrium.   Letting the x-axis coincide with the axis of the pipe, denote the velocity at the point x and at the time t by $u(x,t)$ (see Fig. 215).   According to Eq. (16), Chap. III, the acceleration equals $du/dt = \partial u/\partial t + u\, \partial u/\partial x$, while, according to the argument preceding Eq. (3), Chap. II, the resultant pressure force per unit of volume is $-\partial p/\partial x$.   The equation of motion therefore reads

$$\rho\left(\frac{\partial u}{\partial t} + u\frac{\partial u}{\partial x}\right) = -\frac{\partial p}{\partial x} \tag{17}$$

[1] Quoted from Th. v. Kármán, Compressibility Effects in Aerodynamics, *Jour. Aeronaut. Sci.*, **8**, 345 (1941).

The condition of continuity requires that the time rate of decrease of the mass contained in the portion between the fixed control sections x and $x + dx$ equals the surplus of the rate at which mass leaves this portion over the rate at which mass enters it. Now per unit of cross-sectional area this surplus equals $[\partial(\rho u)/\partial x]\, dx$, while the time rate of decrease of mass is $-(\partial\rho/\partial t)\, dx$ times cross section. The condition of continuity thus takes the form

$$\frac{\partial(\rho u)}{\partial x} + \frac{\partial\rho}{\partial t} = 0 \quad \text{or} \quad \rho\frac{\partial u}{\partial x} + u\frac{\partial\rho}{\partial x} + \frac{\partial\rho}{\partial t} = 0 \qquad (18)$$

If the velocity u is considered as small, terms such as $\partial\rho/\partial x$ are small, too. The terms $u\,\partial u/\partial x$ and $u\,\partial\rho/\partial x$ then are of a higher order and may be neglected. Replacing $\partial p/\partial x$ by $a^2\,\partial\rho/\partial x$ [see (7)], we write (17) and (18) as

$$\begin{aligned}
\frac{\partial u}{\partial t} &= -\frac{a^2}{\rho}\frac{\partial\rho}{\partial x} = -a^2\frac{\partial\log\rho}{\partial x} \\
\frac{\partial u}{\partial x} &= -\frac{1}{\rho}\frac{\partial\rho}{\partial t} = -\frac{\partial\log\rho}{\partial t}
\end{aligned} \qquad (19)$$

Here a^2 may be considered as a constant equaling the value of $dp/d\rho$ for the state of equilibrium about which the air is oscillating. In fact, the variation of $dp/d\rho$ due to the small changes of pressure and density occurring in the motion under consideration leads to terms that are small of higher order. Differentiating the first equation (18) with respect to x, the second with respect to t, and subtracting, we obtain

$$\frac{\partial^2\log\rho}{\partial t^2} - a^2\frac{\partial^2\log\rho}{\partial x^2} = 0$$

This differential equation is satisfied when $\log\rho$ is any function of one variable $z = x - at$. In fact, if $\log\rho = f(z)$ it follows that

$$\frac{\partial\log\rho}{\partial t} = \frac{df}{dz}\frac{dz}{dt} = -a\frac{df}{dz}, \quad \frac{\partial^2\log\rho}{\partial t^2} = -a\frac{\partial}{\partial t}\frac{df}{dz} = a^2\frac{d^2f}{dz^2}$$

and, in the same way, the second derivative with respect to x equals d^2f/dz^2. Now, if ρ depends on z only, p and u also will have the same values at such points of the x-t-plane for which $x - at$ is the same. But $z = x - at$ is the abscissa with respect to a reference system that moves along at the velocity a in the x-direction. This shows that, whatever phenomenon (ρ, p, u-distribution) prevails in the pipe, it will be steady with respect to a reference system which progresses at the rate a. Sound is such an oscillation phenomenon, and thus it is proved that, in a one-dimensional continuum, sound is propagated at the rate $a = \sqrt{dp/d\rho}$.

By a more general analogous computation it can be shown that the same is true for the propagation in any direction of a three-dimensional space. Under adiabatic conditions, with $p/\rho^{1.4} = C$, since $p/\rho = RT$ we have

$$a^2 = \frac{dp}{d\rho} = C \frac{d(\rho^{1.4})}{d\rho} = 1.4C \frac{\rho^{1.4}}{\rho} = 1.4 \frac{p}{\rho} = 1.4RT$$

This gives, with $R = 53.3$ ft./°F., the sound velocity $a = 1116$ ft./sec. for $T = 518°$ (sea level) and $a = 995$ ft./sec. for $T = 411°$ (30,000 ft. altitude). The values of the sound velocity in standard atmosphere are given in the last column of Table 1 (page 10).

Problem 3. Compare the aerodynamic characteristics of thin, slightly cambered airfoils at $V = 180$ ft./sec. and $V' = 670$ ft./sec.

Problem 4. Indicate the manner in which the dimensional analysis of Chap. IV must be modified when compressibility is taken into account.

Problem 5. Determine the velocity of sound in standard atmosphere at 10,000 and 20,000 ft. altitude.

*Problem 6. An airfoil moves with a speed of 550 ft./sec. at a given angle of attack, at sea level and at 10,000 ft. altitude. Estimate the ratio of the lift coefficients and the ratio of the lift values.

Part Three

PROPELLER AND ENGINE

CHAPTER XI
THE PROPELLER

1. Basic Concepts. A propeller blade may be considered as a strongly twisted wing. The cross sections of the blade are essentially of the same shape as those of a wing, with well-rounded leading end and sharp trailing end. Propeller sections, however, particularly those near the hub, have usually a greater relative thickness than wing sections, and profiles with a straight lower contour are preferred. Aside from such minor details, the difference between the geometry of a propeller blade and that of a wing consists mainly in the fact that the orientation of the profiles of a propeller changes considerably as one proceeds from the hub toward the blade tip. The angle β that the chords of the blade sections form with a plane perpendicular to the axis of the propeller is much greater for the sections near the hub than for those near the blade tip (Fig. 216). As will be seen later, this large twist of the blade is necessary in order to ensure that each blade section operates at a favorable angle of attack. Structural considerations are responsible for increasing the thickness of the profiles toward the hub.

In action, a propeller performs a twofold motion: it participates in the forward motion of the airplane (velocity V) and rotates about its axis. Under normal conditions the direction of this axis of rotation may be considered as coinciding with the direction of flight. Any blade section, then, has the velocity component V in the direction of the propeller axis and a rotational velocity component u parallel to the plane of the cross section and perpendicular to the propeller axis. In the cross sections shown in Fig. 216, the velocity components V and u appear as vertical and horizontal, respectively. The component V obviously has the same value for all cross sections, while the component u is proportional to the distance r from the propeller axis. The rotational speed of the propeller can be given by either the number n of revolutions per second or the angular velocity $\omega = 2\pi n$. Thus,

$$u = r\omega = 2\pi r n \tag{1}$$

285

The angle γ between the plane of rotation and the resultant velocity is determined by

$$\tan \gamma = \frac{V}{u} = \frac{V}{r\omega} = \frac{V}{2\pi rn} \tag{2}$$

This angle is thus seen to decrease with increasing r. By β we denote the angle between the plane of rotation and some well-defined reference line of the blade section (the so-called "chord"). If the blade element between the radii r and $r + dr$ is considered as a wing element of the spanwise extension dr, the angle of attack for this blade element will be $\alpha = \beta - \gamma$. If the reference line coincides with the zero lift direction of the profile, its angle with the plane of rotation will be called β' and then $\alpha' = \beta' - \gamma$ is the angle of incidence. For a given propeller the distribution of the values of β or β' along the radius of the blade is known. The angle of attack for any blade element is then determined when one value of γ, for example, the value at the blade tip, is known.

If d denotes the diameter of the propeller, the angle at the blade tip is given by

$$\tan \gamma_1 = \frac{2V}{\omega d} = \frac{1}{\pi} \frac{V}{nd}$$

and the values of γ, α, and α' at the distance r from the axis are found from

$$\tan \gamma = \frac{V}{2\pi rn} = \frac{d}{2\pi r} \frac{V}{nd} = \frac{d}{2r} \tan \gamma_1 \tag{3}$$

and

$$\alpha = \beta - \gamma = \beta - \text{arc tan} \left(\frac{d}{2\pi r} \frac{V}{nd} \right),$$

$$\alpha' = \beta' - \gamma = \beta' - \text{arc tan} \left(\frac{d}{2\pi r} \frac{V}{nd} \right) \tag{3'}$$

The dimensionless quantity $V/nd = \pi \tan \gamma_1$ is called the *advance ratio* of the propeller. If two propellers of the diameters d and d' are geometrically similar, corresponding blade sections (*i.e.*, blade sections at such distances r and r' from the axes that $r'/r = d'/d$) have the same shape and the same orientation (determined by β), while their chord lengths are proportional to the propeller diameters. The angles of attack of corresponding blade sections will then have the same value if and only if the advance ratio

$$J = \frac{V}{nd} \tag{4}$$

is the same for both propellers.

The concept of the advance ratio makes it possible to apply the dimensional analysis of Sec. IV.2 to propellers. Consider a series of geometrically similar propellers operating each at the same advance ratio J. According to the second equation (3), corresponding blade sections then possess the same angle of attack. Moreover, the magnitude of the resultant velocity at corresponding blade sections is proportional to nd. In fact, the translational velocity V equals the product of J and nd, and the rotational velocity u at the distance r from the propeller axis equals the product of $2\pi r/d$ and nd. The propeller diameter d may be taken as the reference length occurring in the definition of the Reynolds number and d^2 as the reference area occurring in formula (5), Sec. IV.2. The laws of similarity require, then, that any component of the force that the air exerts on the propeller operating at a given advance ratio J can be written as the product of the reference area d^2, the dynamic pressure corresponding to the velocity nd, and a coefficient that depends, for a given shape, only on the Reynolds number.

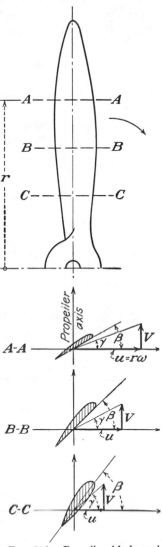

The *propeller thrust* T, that is, the force in the direction of the propeller axis, is of primary importance. According to what has just been explained the thrust can be written as

$$T = \rho(nd)^2 d^2 C_T = \rho n^2 d^4 C_T \qquad (5)$$

where the factor $\frac{1}{2}$ occurring in the definition of the dynamic pressure has been absorbed in the *thrust coefficient* C_T. For propellers of a given shape operating at a given advance ratio J the coefficient C_T depends only on the Reynolds number. That is, *the propeller thrust T equals the product of $\rho n^2 d^4$ and the thrust coefficient, which depends on the shape of the propeller, on the advance ratio J, and on the Reynolds number.* The advance ratio J

Fig. 216.—Propeller blade and sections.

here plays a role similar to that of the angle of attack α in airfoil theory. In fact, according to (3), the angles of attack of all blade sections are uniquely determined by J when the shape of the propeller is given.

Another important quantity is the resulting moment, with respect to the propeller axis, of the forces that the air exerts on the propeller. This moment is known as the *propeller torque* and will be denoted by Q. According to the laws of similarity, the formula for Q has an analogous structure to that for T but contains an additional factor which has the dimension of a length (page 80). If the propeller diameter d is taken as this factor, the torque appears as

$$Q = \rho n^2 d^5 C_Q \tag{6}$$

where the *torque coefficient* C_Q depends on J, Re, and the shape of the propeller.

It is in most instances preferable to consider the product ωQ rather than Q. This product represents the rate at which work is done by the torque Q; it is called the *propeller power* and will be denoted by P. Since $\omega = 2\pi n$,

$$P = 2\pi n Q = \rho n^3 d^5 2\pi C_Q = \rho n^3 d^5 C_P \tag{7}$$

where the *power coefficient* C_P has been introduced for the product $2\pi C_Q$. According to (7), *the propeller power equals the product of* $\rho n^3 d^5$ *and the power coefficient* C_P, *which depends on the shape of the propeller, the advance ratio, and the Reynolds number.*

It is easily seen that the forces which the air exerts on the propeller are statically equivalent to the thrust T and the couple of moment Q acting in a plane perpendicular to the propeller axis. In fact, consider an element dS of the surface of a two-blade propeller (Fig. 217a). The force exerted on dS can be resolved into the components dF_1 and dF_2, the first being parallel and the second perpendicular to the propeller axis. This is true irrespective of whether the force exerted on dS is normal to this surface element or not. The attitude of a propeller blade with respect to the axial and rotational motion does not change if the blade is rotated about the axis. The force acting on the element simply participates in the rotation. Now, in the case of a two-blade propeller the second blade coincides with the first when this is rotated through 180°. Therefore, the force exerted on the element dS' of the second blade that corresponds to dS has the components dF_1' equal to dF_1 and dF_2' equal and opposite to dF_2. The forces dF_1 and dF_1' have the resultant $2\,dF_1$ acting along the propeller axis. The line joining the points of application of the forces dF_2 and dF_2' is perpendicular to the propeller axis. These two forces are therefore equivalent to a couple acting in a plane perpendicular to the axis. Summing up the contributions from all pairs of symmetrically situated surface elements, the forces exerted on the propeller are seen to be equivalent to a resultant force acting along the propeller axis and a resultant couple acting in a plane perpendicular to this axis. A similar

line of reasoning can be applied to a propeller with three or more blades
if it is supposed that the blades are identical in shape and make equal
angles with one another (see Fig. 217b).   Here, the force components
dF_2, dF_2', dF_2'' parallel to the plane of rotation and acting on corresponding
blade elements sum up to an equilateral triangle, their vector sum is
zero, and their action results in a moment in the plane of rotation.

Another notation that is useful in the discussion of propeller problems
is the *propeller efficiency*.   The power that must be transmitted to the
propeller in order to obtain the desired angular velocity was seen to be

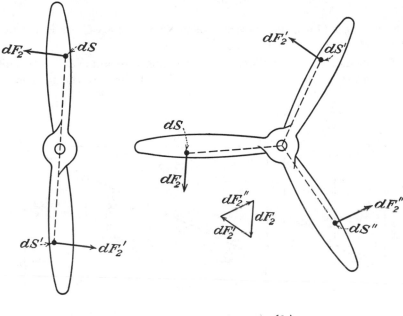

(a)   (b)

FIG. 217.—Symmetry of forces upon the blades.

$P = \omega Q$.   On the other hand, the product of thrust T and the velocity
of flight V measures the power that is available for the propulsion of the
airplane.   The ratio of power output to power input

$$\eta = \frac{TV}{P} = \frac{T}{Q}\frac{V}{\omega} = \frac{1}{2\pi}\frac{T}{Q}\frac{V}{n} \tag{8}$$

is called the efficiency of the propeller.   According to (5) and (7),

$$\eta = \frac{\rho n^2 d^4 C_T V}{\rho n^3 d^5 C_P} = \frac{C_T}{C_P}\frac{V}{nd} = \frac{C_T}{C_P}J \tag{9}$$

The propeller efficiency equals the product of the advance ratio and the ratio of the coefficients of thrust and power. The efficiency η here plays a role similar to that of the ratio D/L or the gliding angle ϵ in airfoil theory.

Problem 1. If a propeller is considered as an airfoil of area A (= blade area) moving at the velocity of the blade element that has the distance $3d/8$ from the axis, how should one define the thrust, torque, and power coefficients?

Problem 2. If blade setting and angle of attack at the point $r = 3d/8$ are called β_1 and α_1, study the relation between advance ratio J and α_1 for constant values of $\beta_1 = 15, 30,$ and $45°$.

Problem 3. Assume that the blade setting of a propeller is determined by

$$\tan \beta = \frac{p}{2\pi r}$$

with constant p. Compute the average value of α for a given J, neglecting terms of higher order in α.

2. Geometry of Propellers. In its shape the propeller, often called the airscrew, is closely related to helical surfaces. If the motion of a

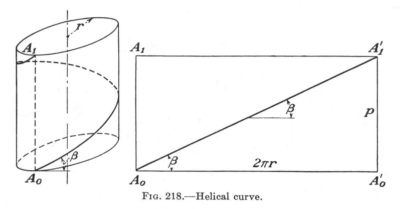

Fig. 218.—Helical curve.

rigid body can be considered as composed of a rotation about a certain axis and a simultaneous proportional translation in the direction of this axis, the body is said to perform a *helical motion.* During one complete revolution, all points of the body progress by the same amount p in the direction of the axis of rotation. The length p is called the *pitch* of the helical motion. Consider a point A of the body that has the distance r from the axis of rotation. The path of A is a *helix* of the pitch p, traced on the circular cylinder of the radius r whose axis coincides with the axis of rotation. Let A_0 and A_1 be two consecutive points on this helix lying on the same generatrix of the cylinder (Fig. 218). If the cylindrical surface between the cross sections through A_0 and A_1 is cut open along the generatrix A_0A_1 and developed on a plane, the helix is transformed into the diagonal of a rectangle of the basis $2\pi r$ and the height p. The slope β of the helix against any cross-sectional plane of the cylinder is thus seen to be connected with the pitch p and the radius r by

$$p = 2\pi r \tan \beta \qquad (10)$$

Since p has the same value for the helical paths of all points of a rigid body performing a helical motion, the slope of the various paths is proportional to $1/r$.

Any rigid segment of a straight or curved line performing a helical motion generates a *helical surface*. The simplest case is that of the helical surface generated by a straight line that intersects the axis of rotation at right angles. Consider a narrow strip of this helical surface, lying between two neighboring positions of the generating line (Fig. 219). The points $A, B, C, \ldots$ on this line describe helices that, within the narrow strip under consideration, may be regarded as short straight segments AA' BB', CC', $\ldots$ forming the respective angles β with a plane perpendicular to the axis of rotation. This strip of a helical surface could be used as a sort of skeleton for a propeller blade, the lines AA', BB', CC', $\ldots$ carrying the chords of the blade sections. If a propeller blade of infinitesimal thickness designed in this manner

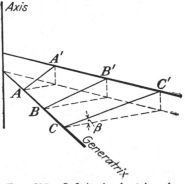

Fig. 219.—Infinitesimal strip of a helical surface.

operates at the advance ratio J, the angle of attack of the blade section at the distance r from the propeller axis is given by the first equation (3'), where, according to (10), $\beta = $ arc tan $p/2\pi r$. Thus,

$$\alpha = \text{arc tan} \frac{p}{2\pi r} - \text{arc tan} \frac{Jd}{2\pi r} \qquad (11)$$

If, in particular, the advance ratio is such that $Jd = p$ or $J = p/d$, the angle of attack becomes zero for all blade elements. In this case the points $A, B, C, \ldots$ move in the directions AA', BB', CC', $\ldots$, respectively; the helical strip used as the skeleton of the propeller blade moves along the generating helical surface in very much the same way as a plane strip could move along its own plane. Equation (11) shows that for any value of the advance ratio different from p/d the angle α will necessarily vary with r.

Actually, propeller blades are not shaped exactly in the manner outlined. They are not exact helical surfaces with a constant pitch of all elements but rather are composed of helical elements with slightly varying pitch. The reason for this is as follows: It has been seen that for each airfoil section there exists an optimum angle of attack for which the ratio of lift to drag attains its maximum value.   The designer of a propeller may try to obtain for every blade section the corresponding optimum angle when the propeller operates at the *design advance ratio*, or at the advance ratio at which the maximum efficiency is desired. If the same profile is adopted for all blade sections, the optimum angle of attack has the same value for all sections.   Now, according to (11), a

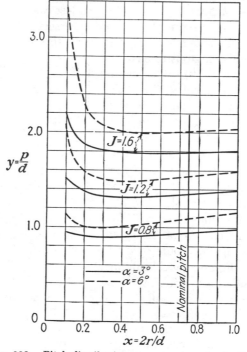

FIG. 220.—Pitch distribution at constant angle of attack.

constant value of α (except $\alpha = 0$) cannot be obtained with a constant pitch p.   In fact, since $\alpha = \beta - \gamma$ and $\tan \beta = p/2\pi r$, $\tan \gamma = Jd/2\pi r$,

$$\tan \beta = \tan (\alpha + \gamma) = \frac{\tan \alpha + \tan \gamma}{1 - \tan \alpha \tan \gamma}$$

or

$$p = 2\pi r \frac{\tan \alpha + Jd/2\pi r}{1 - (Jd/2\pi r) \tan \alpha} \tag{12}$$

In order to obtain a constant angle of attack α, the pitch p must vary with r according to (12).   If the dimensionless quantities $y = p/d$ and $x = 2r/d$ are introduced, Eq. (12) becomes

$$y = \pi x \frac{\tan \alpha + J/\pi x}{1 - J \tan \alpha/\pi x} = J \frac{1 + \pi x \tan \alpha/J}{1 - J \tan \alpha/\pi x} \tag{13}$$

Usually, the values of β and p at $r = 3d/8$ are given as the *nominal values* of blade setting and pitch for the entire propeller. These β- and d-values are connected by $p/d = \frac{3}{4}\pi \tan \beta$.

In Fig. 220 the p/d-values according to (13) are plotted against x for $\alpha = 3°$ and $6°$ and various J-values. The nominal p/d appears on the vertical at $x = 0.75$.

Figure 221 gives the comparison between the theoretical curve and the pitch distribution of an actual propeller. For $\alpha = 4.5°$ and the three J-values 0.35, 0.45, and 0.55, Eq. (13) leads to the three (nearly straight) lines seen in the graph. The curved solid line represents the pitch

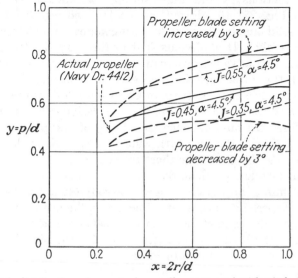

Fig. 221.—Pitch distribution of an actual propeller compared with pitch distribution at constant angle of attack.

distribution of a propeller tested by the NACA[1] with nominal $p/d = 0.65$, nominal $\beta = 15.4°$. There are several reasons why the actual pitch distribution differs from that corresponding to (13). Contrary to what has been assumed in the foregoing, the blade sections generally are not formed according to the same profile. It has already been mentioned that for structural reasons the relative thickness of the blade sections increases considerably toward the hub. The optimum angle of attack thus varies with r. Furthermore, a certain compromise must be worked out between the aim to obtain the highest possible efficiency for the design advance ratio and the desire to avoid too sharp a drop of the efficiency for other values of the advance ratio.

[1] *NACA Tech. Rept.* 340 (1930).

The propeller whose pitch distribution is shown in Fig. 221 would best fit the condition of a constant $\alpha = 4.5°$ at an advance ratio $J = 0.45$, that is, for example, $V = 180$ ft./sec., $d = 10$ ft., $n = 40/\text{sec.}$ It would be still useful within a range of J of about 0.40 to 0.50. But, if a much wider range of J has to be covered, it proves impossible to arrive at a compromise that is satisfactory. Therefore, in modern airplanes, *controllable-pitch propellers* are used more and more widely. The blade setting of such a propeller can be changed during the flight by rotating each blade about an axis that intersects the propeller axis at right angles. The effect of such a rotation consists in changing the β-values of all blade sections by the angle through which the blade has been rotated. In Fig. 221 the dotted curved lines correspond to the same blade as the solid one, but to a blade setting increased by 3° and decreased by 3°, respectively. It is obvious that in this way the theoretical conditions for varying advance ratio can be much better satisfied.

Although the pitch distribution may be considered as the most important geometric characteristic of a propeller design, the questions of *profile* and *plan-form*, predominant in wing design, play a certain role, too. Structural reasons influence the choice of the profiles to a much greater extent than is the case with wing profiles. Near the hub, rather thick profiles must be used since there the centrifugal forces as well as the bending moments are particularly high. Fortunately, the region near the axis has minor importance for the aerodynamic forces, as will be seen later. Experience also shows that the influence of the plan-form on the aerodynamic characteristics of a propeller is rather small. If the chord length of the blade sections were constant, the plan-form, which is defined as the contour of the projection of the blade on a plane perpendicular to the propeller axis, would slightly widen from the axis toward the tip since the sections near the tip are set at smaller angles than those near the hub. Generally, the chords are made to decrease rapidly over the outermost quarter of the blade radius in order to avoid too high centrifugal stresses and bending moments. Thus the typical plan-form is obtained, which is widest at about $r = 3d/8$ and narrows toward the hub as well as toward the blade tip.

A sufficiently detailed representation of the shape of a propeller blade requires a considerable amount of drafting. Figure 222 shows the example of a metal blade of 10 ft. 5 in. diameter. Besides the projections on a plane normal to the propeller axis (plan-form) and on a plane parallel to the axis as well as to the main direction of the blade, 10 blade sections are shown. Near the hub the developed plan-form is also given. This diagram shows the variation of the true chord length with r, whereas the plan-form gives the projections of the chord lengths on a plane perpendicular to the axis. The drawings must be supplemented by tables

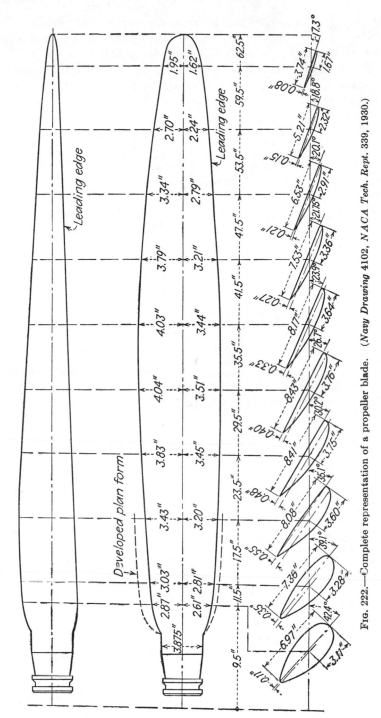

Fig. 222.—Complete representation of a propeller blade.   (*Navy Drawing 4102, NACA Tech. Rept. 339, 1930.*)

giving offsets for the upper and lower boundaries of the profiles. Figure 223 shows the distributions of chord length c and maximum thickness t of the blade sections, as well as the pitch distribution.

For wooden propeller blades, which are built up from boards of about 1 in. thickness, the *longitudinal sections* of the blade, *i.e.*, the intersections of the blade surface with planes normal to the axis, must also be drawn.

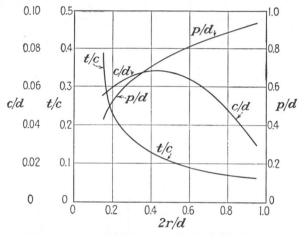

FIG. 223.—Pitch, thickness, and chord distribution.

Figure 224 shows an example. In the design of these longitudinal sections, attention should be paid to the careful fairing of all transitions since the flow is extremely sensitive to any lack of continuity in the curvature of the blade surface.

Problem 4. A variable-pitch propeller is designed in such a way that at the advance ratio $J = 0.8$ the angle of attack has the constant value $\alpha = 4°$. If at the advance ratios 0.6 and 1.0, respectively, the propeller blade is set so as to give $\alpha = 4°$ at the point $r = d/4$, what are the highest and lowest values of α within the range $r = 0.1d$ to $r = 0.5d$?

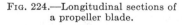

FIG. 224.—Longitudinal sections of a propeller blade.

3. Propeller Characteristics. It has been stated in the foregoing that the coefficients of thrust and power (or torque) of a given propeller depend on the advance ratio J and the Reynolds number Re. Since the influence of Re is of minor importance, the curves showing C_T *and* C_P (or C_Q) as functions of J are called the *propeller characteristics*. The points of these curves correspond to experimental results obtained with a fixed value of Re.

According to (5) and (7) the two coefficients are defined by

$$C_T = \frac{T}{\rho n^2 d^4}, \qquad C_P = \frac{P}{\rho n^3 d^5} \qquad (14)$$

Let us first consider a propeller with a moderate nominal blade setting (up to about 15°) so that the angles of incidence involved are essentially below stalling.   As a first result we may state the following: *With increasing advance ratio, the coefficients C_T and C_P both decrease.* This is seen in Fig. 225, which gives the characteristic curves of the propeller of Fig. 222.[1]   According to (3) the angle of incidence α' of all blade elements of a given propeller decreases with increasing J.   Since the aerodynamic forces on an airfoil increase with α' (the lift linearly, the drag according to a quadratic law), the decrease in C_T and C_P with increasing J is to be expected.

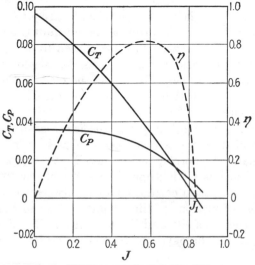

Fig. 225.—Characteristic curves of the propeller of Fig. 222 at a nominal blade setting of 15.5°.

As is seen from Fig. 225 the line representing the *thrust coefficient* is only slightly curved and, as a first approximation, may be considered as a straight line that intersects the axis at the abscissa J_1.   In a certain way the value J_1, where $C_T = 0$, may be compared to the zero lift angle of an airfoil.   Assume that the propeller blade is a narrow strip of a helical surface of thickness zero and of constant pitch p_1.   It has already been seen that, if such a helical strip moves at the advance ratio $J_1 = p_1/d$, the angle of incidence is zero over the entire blade, *i.e.*, each straight blade section progresses in its own direction through the surrounding air, very much like an indefinitely thin plate moving in its own plane.   In this case, if no skin friction interferes, T as well as Q, and consequently P, would vanish at the advance ratio p_1/d and would change signs when this value of J is surpassed.

[1] *NACA Tech. Rept.* 339 (1930).

Since the assumptions of zero friction, zero thickness, and constant pitch are not fulfilled for actual propeller blades, some modifications take place. The values of J that correspond to $C_T = 0$ and to $C_P = 0$, respectively, will not coincide. Skin friction at least will absorb a torque moment even where no thrust exists. For positive values of C_T and C_P the efficiency η is defined by (9). Since necessarily $\eta < 1$, it follows that $JC_T < C_P$. Now, for advance ratios exceeding the value J_1 for zero thrust the definition of η loses its meaning, but up to $J = J_1$ the relation $JC_T < C_P$ holds. Thus it is seen that the value J_2 of the advance ratio at which C_P becomes zero must be greater than J_1 (see Fig. 225).

The product $p_1 = J_1 d$ is often called the *effective pitch* of the propeller because for an indefinitely thin propeller of constant pitch p_1 and diameter d the advance ratio for zero thrust was seen to be $J_1 = p_1/d$. The effective pitch may be considered as an aerodynamically defined average of the various pitch values occurring along the blade. For conventional designs the effective pitch does not differ too much from the nominal pitch, *i.e.*, the pitch at $r = 3d/8$.

The curve representing the *power coefficient* as a function of the advance ratio is of rather parabolic character as seen in Fig. 225. The vertex of the parabola corresponds generally to comparatively small values of J. Insofar as C_T is approximately linear and C_P a quadratic function of J, there exists a certain analogy between the curves showing these coefficients as functions of the advance ratio and the curves representing the coefficients of lift and drag of an airfoil as functions of the angle of attack. This analogy is by no means accidental, as will become clear later.

The dotted curve in Fig. 225 represents the *efficiency* η defined by (9) as

$$\eta = \frac{TV}{P} = \frac{C_T}{C_P} J$$

It is obvious that $\eta = 0$ for $J = 0$ as well as for $J = J_1$. Somewhere between these two values of the advance ratio the efficiency η must have a maximum, which, of course, is smaller than 1. As a rule, this maximum lies nearer to $J = J_1$ than to $J = 0$. If J_m denotes that value of the advance ratio for which the efficiency assumes the maximum value η_m, the difference $J_1 - J_m$ usually is not greater than about 25 per cent of J_1. In general, values of the advance ratio occurring in flight will be rather close to J_m. The difference $J_1 - J$ [or the ratio $(J_1 - J)/J_1$ expressed in per cent] is known as the *slip* of the propeller. If the advance ratio of the propeller is conceived as the counterpart of the angle of attack of an airfoil, the slip $J_1 - J$ corresponds to the angle of incidence $\alpha' = \alpha - \alpha_0$.

For zero slip, thrust and efficiency are both zero.   For not too large slip values, the thrust and, in a less pronounced way, the efficiency are proportional to the slip.

The higher the speed for which a propeller is designed, the greater the angles β at which the blades are set.   Now, with increasing values of β, the pitch p of all elements and, consequently, the effective pitch p_1 and the advance ratio for zero thrust $J_1 = p_1/d$ will increase, too. Figure 226 shows the characteristics of such a high-speed propeller.   The C_T-curve differs in shape from that of Fig. 225: the almost linear increase of C_T with decreasing J toward the left from $J = J_1$ ends at a certain point; from there on toward $J = 0$ the thrust coefficient remains practically constant.   This behavior can easily be understood from the analogy between the C_T vs. J-curve of a propeller and the C_L vs. α-curve of an airfoil.   For small values of J the angle defined by (3) is nearly

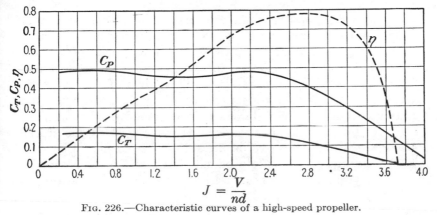

$$J = \frac{V}{nd}$$

FIG. 226.—Characteristic curves of a high-speed propeller.

proportional to J.   For large values of β and small values of J the angle of attack $\alpha = \beta - \gamma$ will therefore *exceed the stalling angle*.   When this occurs, the lift coefficient of the blade section will not increase with a further increase of α, that is, with a further decrease of J.   Since C_T depends essentially on C_L, this explains the fact that the C_T vs. J-curve runs more or less horizontal to the left of a certain J-value.   The C_P vs. J-curve is likewise affected by the transition to "beyond stalling" conditions.   The shape of both curves to the left of the stalling J varies strongly; it is very sensitive to changes of profile and plan-form, and depending on the blade setting.   No general properties have been established, but most experiments agree in showing both C_T and C_P first slightly decreasing, then again increasing toward the left (see next section).   This indefiniteness is of course connected with the strongly varying character of the C_L vs. α-curves of airfoil sections when the stalling point is exceeded, but it is also caused by the fact that the stalling conditions are not reached at the same time by all blade elements. For regular flight conditions only the region below stalling need be considered for propellers as well as for wings.

One important difference between propeller and airfoil coefficients must not be overlooked. Wing and airfoil coefficients refer to the actual area of the wing and can therefore be expected to change only slightly with the plan-form (aspect ratio). But in the definitions of thrust and power coefficients, C_T and C_P, the square of the diameter, d^2, was used as reference area rather than the blade area. It follows that both C_T and C_P, at least within a restricted range, will be proportional to the ratio A/d^2 (propeller-blade area to diameter squared). In fact, this holds true if with the usual blade forms a two-blade propeller is replaced by a propeller with three blades of the same shape. Experiments show that in this case the ratio C_T/C_P or the efficiency $C_T J/C_P$ drops not more than about 2 per cent, while C_T and C_P approximately multiply by $\frac{3}{2}$. This does not mean, however, that one can increase ad libitum the amount of thrust and power simply by using more blades or higher blade width c (chord length of the profiles) since the mutual interference of the blades becomes more and more marked in both cases. Average values of A/d^2 for two- and three-blade propellers will be given in the next section. More than four blades have almost never been used in practical flight.

Various secondary circumstances influence the propeller characteristics to a certain extent. For example, in modern planes the propeller tip speed reaches or even exceeds the velocity of sound. Under such conditions compressibility effects become more and more pronounced, and the similarity laws that underlie the definition of the propeller coefficients are no longer valid. Experiments on full-scale metal propellers of low pitch under standard pressure have shown that the efficiency η remains practically unaffected up to tip speeds of 1000 ft./sec. Above this limit the efficiency relative to that at lower speeds falls off at a rate of about 10 per cent per 100 ft./sec. increase. Apart from the loss of efficiency, tip speeds approaching and exceeding the velocity of sound have to be avoided because of the inconvenient noise such propellers produce.

A very serious problem, which can only be mentioned here, arises from the interference of the fuselage on which the propeller is mounted. If the shapes of both the propeller and the fuselage are kept constant, the extent of mutual interference will depend on the relative size of the two bodies. It can be expected that, mounted on the same fuselage, propellers of greater diameter will be less affected. If experiments of this kind are carried out in the wind tunnel, care must be taken to establish an appropriate notion of combined efficiency η_c. The propeller thrust T measured as the tension in the crankshaft at the section AA (**Fig. 227**) will differ from the resultant force T_r acting on the propeller-fuselage combination and measured, say, at BB. The difference $T - T_r$ is the actual drag exerted on the fuselage under the combined influence of the air flow with an air speed V and of the rotating propeller. If the first component D_p, the parasite drag of the fuselage, is known from ordinary drag tests, the increment D' can be found from the equation

$$T - T_r = D_p + D' \qquad \text{or} \qquad T - D' = T_r + D_p \qquad (15)$$

Each of the quantities $T - D'$ and $T_r + D_p$ can be·considered as the combined thrust value, and therefore the efficiency η_c can be defined as

$$\eta_c = \frac{T - D'}{P} V = \frac{T_r + D_p}{P} V \qquad (16)$$

Tests on a regular open cockpit fuselage with a maximum (rectangular) cross section of about 35 to 45 in. and propeller diameter of 8 to 10.5

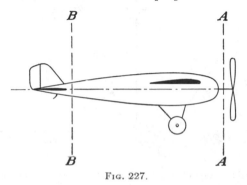

B   A

FIG. 227.

ft. showed that the variations in the propeller coefficients were quite small. The combined efficiency increased by about 1 per cent for a 10 per cent increase in diameter.

Problem 5. A propeller of diameter d that develops the thrust T when operating at the advance ratio J and at N r.p.m. is to be replaced by a pair of equal propellers of the same shape, operating at the advance ratio J and producing together the thrust T. The velocity V being the same in both cases, determine the diameter d' and the rotational speed N' of these two propellers. Prove that the total power required by the two propellers equals the original propeller power.

Problem 6. Define thrust and power coefficients C'_T and C'_P referring to the actual propeller blade A and to the actual velocity at the distance $r = (\lambda/2\pi)d$. The solidity, i.e., the ratio A to $d^2\pi/4$, may be given as $8\mu/\pi$. Instead of J, use the value of $\tan \gamma$ at the indicated r as the variable parameter. Give the relations between these new characteristics and the C_T, C_P, and η vs. J-curves.

Problem 7. Assume that the C_T vs. J-curve is a straight line and the C_P vs. J-curve a parabola with its vertex at $J = 0$. Discuss the corresponding η-curves, and find the limits for the slope of the straight line and the parameter of the parabola, in particular under the assumption that $J_1 = 0.8$ and $J_2/J_1 = 1.06$ (J_1, J_2 abscissas of the intersections with the J-axis).

Problem 8. How does the η-curve change, if the lines considered in the preceding problem start only at $J = J_{st}$ and both C_T and C_P are constant between $J = 0$ and $J = J_{st}$?

4. Quantitative Analysis. In the case of airfoil characteristics it was possible to give a fairly complete quantitative description for the various curves, C_L vs. α, C_D vs. α, etc. Three parameters alone are involved as long as conditions below stalling only are considered—the slope k of the C_L vs. α-line, the zero lift drag a, and the coefficient $1/b$ of the quadratic term in the expression for C_D (see Chap. VII). All these parameters vary only in a narrow range; if the aspect ratio R is given, we can even consider k and b as entirely known, the first being about 2π divided by $1 + 2/R$, the second equal to about πR. Now, in the problem of propeller characteristics there is a much greater variety. Even if we assume that the same profile is used for all blade sections, the

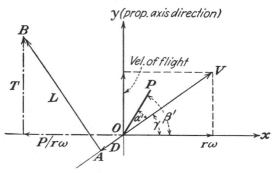

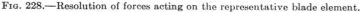

Fig. 228.—Resolution of forces acting on the representative blade element.

values of blade width c and of blade setting β, both varying along the blade according to some formula, add to the diversity. However, a rough survey of all possible forms covering the essential features of the curves can be obtained in the following way.

We assume, for the purpose of a first approximation, that the entire propeller acts as an airfoil the total area of which is concentrated at a certain distance r from the axis, which is set at the definite angle β against the plane of rotation, and which participates in the motion combined of an axial translation with velocity V and a rotation with angular speed ω. That is, all blades are replaced by *one blade element* at the distance r to which is attributed the blade area of the propeller or rather a fictitious propeller area A. This approach may be called the *method of a representative blade element*. It is of course not a method of precise performance computation for a propeller but a simple way to obtain general information about the shape of the characteristic curves.

In Fig. 228 the short solid line OP represents the propeller section or, better, its zero lift direction, forming the angle β' with the direction

Ox of the rotational velocity $r\omega$. The resultant velocity in direction OV, under the angle γ, has the magnitude $V/\sin\gamma = r\omega/\cos\gamma$ [see Fig. 216 and Eq. (2)]. The airfoil element is subject to a drag in the direction OA of magnitude $C_D(\rho/2)A(r\omega/\cos\gamma)^2$ and to a lift in the direction AB the magnitude of which is $C_L(\rho/2)A(r\omega/\cos\gamma)^2$. The components of these forces in the direction Oy, which represents the direction of the forward speed V, are

$$-C_D \frac{\rho}{2} A \left(\frac{r\omega}{\cos\gamma}\right)^2 \sin\gamma \quad \text{and} \quad C_L \frac{\rho}{2} A \left(\frac{r\omega}{\cos\gamma}\right)^2 \cos\gamma \quad (17),$$

while the components in direction xO, that is, opposite to the rotational speed $r\omega$, equal

$$C_D \frac{\rho}{2} A \left(\frac{r\omega}{\cos\gamma}\right)^2 \cos\gamma \quad \text{and} \quad C_L \frac{\rho}{2} A \left(\frac{r\omega}{\cos\gamma}\right)^2 \sin\gamma \quad (17')$$

The two forces (17) furnish the propeller thrust T:

$$T = C_T \rho n^2 d^4 = (C_L \cos\gamma - C_D \sin\gamma) \frac{\rho}{2} A \left(\frac{r\omega}{\cos\gamma}\right)^2 \quad (18)$$

The forces (17') when multiplied by $r\omega$ give the propeller power P:

$$P = C_P \rho n^3 d^5 = (C_L \sin\gamma + C_D \cos\gamma) \frac{\rho}{2} A \left(\frac{r\omega}{\cos\gamma}\right)^2 r\omega \quad (18')$$

From relations (18) and (18') we can find expressions for C_T and C_P, using $\omega = 2\pi n$ and the abbreviations

$$\lambda = \frac{2\pi r}{d}, \qquad \mu = \frac{A}{2d^2} \quad (19)$$

as follows:

$$C_T = \frac{C_L \cos\gamma - C_D \sin\gamma}{\cos^2\gamma} \lambda^2 \mu, \qquad C_P = \frac{C_L \sin\gamma + C_D \cos\gamma}{\cos^2\gamma} \lambda^3 \mu \quad (20)$$

Here γ equals $\beta' - \alpha'$ where β' is the blade setting and α' the variable angle of incidence (see Fig. 228). As C_D and C_L are known functions of α', Eq. (20) determine $C_T/\lambda^2\mu$ and $C_P/\lambda^3\mu$ as functions of α'. Adding the relation that follows from Eq. (2),

$$J = \frac{V}{nd} = \frac{r\omega \tan\gamma}{nd} = \frac{2\pi r}{d} \tan\gamma = \lambda \tan(\beta' - \alpha') \quad (21)$$

we have J/λ also as a function of α'. Thus we are in a position to compute and to plot the C_T vs. J- and C_P vs. J-curves, except for the scale factors $\lambda^2\mu$ for C_T, $\lambda^3\mu$ for C_P, and λ for J. The efficiency curve η vs. J can be found from

$$\eta = \frac{C_T J}{C_P} = \frac{C_L \cos\gamma - C_D \sin\gamma}{C_L \sin\gamma + C_D \cos\gamma} \frac{\lambda^2\mu}{\lambda^3\mu} \lambda \tan\gamma = \frac{C_L - C_D \tan\gamma}{C_L + C_D \cot\gamma} \quad (22)$$

in connection with (21). The η-formula includes no scale factor.

How the computation can be carried out is shown in Fig. 229, and the results are given in Fig. 230. In Fig. 229 the polar diagram of the

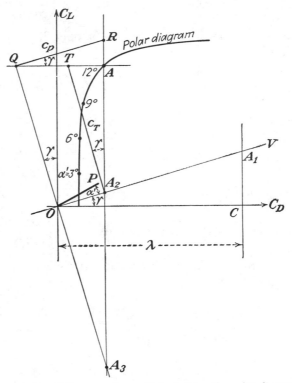

Fig. 229.—Computation of the propeller coefficients from the polar diagram of the representative blade element.

airfoil profile is plotted with equal scales for C_D and C_L and with the points marked corresponding to $\alpha' = 0, 3, 6°$, etc. A vertical straight line in the distance $\overline{OC} = \lambda$ is added. To any point A of the polar curve corresponds a definite velocity direction OV, which forms the respective angle α' with OP and which intersects the vertical line through C in A_1 and the vertical line through A in A_2. Then $\overline{CA_1} = \lambda \tan\gamma = J$ and $\overline{A_2 A} = C_L - C_D \tan\gamma$. If OA_3 is drawn perpendicular to OV and A_3 lies on the vertical line through A, we have $\overline{AA_3} = C_L + C_D \cot\gamma$

and thus $\eta = \overline{AA}_2/\overline{AA}_3$. To obtain $C_T/\lambda^2\mu$ and $C_P/\lambda^3\mu$, one has to draw $A_2T \perp OV$, giving $\overline{A_2T} = C_T/\lambda^2\mu$ and $QR \parallel OV$, supplying $\overline{QR} = C_P/\lambda^3\mu$.

This approach supplies fairly good information about the character of the C_T, C_P, and η vs. J-curves of a propeller, as long as below-stalling conditions prevail over the main part of the blade, i.e., for not too large

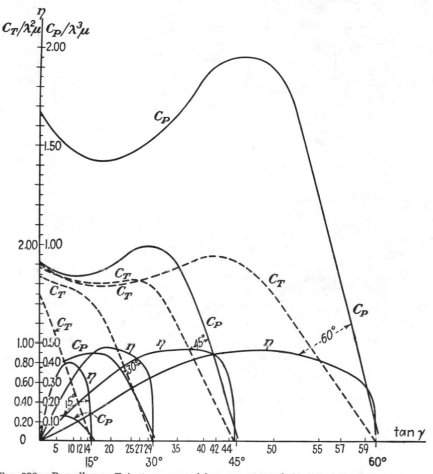

FIG. 230.—Propeller coefficients computed from an assumed characteristic of the representative section (Eqs. 20 to 22).

values of the slip $J_1 - J$. The curves in Fig. 230 are plotted under the assumptions $C_L = 0.1\alpha'$, $C_D = 0.02 + 0.0002\alpha'^2$ (α' in degrees) up to about $\alpha' = 13°$. This corresponds to an average airfoil of an aspect ratio of about 15. The three families of curves show in the indicated region all the characteristic properties seen in Fig. 231, which gives the experimental results on a three-blade propeller with Clark Y-section and 10 ft. diameter.[1] As to the angles indicated in the two figures, it must be noted that β in Fig. 231 is the nominal blade setting, *i.e.*, the angle of the profile chord at $r = 3d/8$, while β' in Fig. 230 is the angle of the zero lift direction at the representative section. Since this section

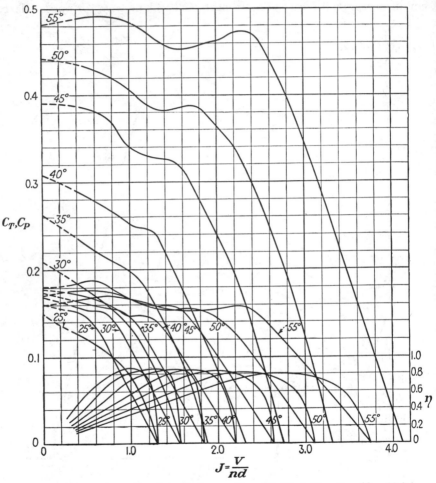

FIG. 231.—Coefficients of a three-blade propeller with Clark Y-section ($d = 10$ ft.).

is in general a little closer to the axis, a difference $\beta' - \beta$ up to about 5° may occur under average conditions. The main result is that, *below stalling, the values of C_T and C_P, and therefore of η, as functions of J depend essentially on three parameters, the "solidity" expressed in μ, the relative distance of the representative element expressed in λ, and the blade setting β'*

[1] *NACA Tech. Rept.* 658 (1939).

at the representative element; the parameters λ *and* μ *enter the equations only as scale factors.*

Comparison with various test results show that, for modern propellers, λ is close to 2.2, about in the range 2.1 to 2.3, and that μ varies between about 0.012 and 0.016 for a two-blade propeller and 0.018 to 0.024 in the case of three blades. With $\mu = 0.014$ the fictitious propeller area A follows, from (19), as $0.028d^2$, that is, considerably smaller than the actual blade area. This can easily be understood since the blade region near the axis has a comparatively small aerodynamic effect.

The situation is much more complicated where incidences beyond the stalling value are involved. The sharp drop of the C_L vs. α-line at the stalling point is due, as was seen in Sec. X.1, to the fact that the flow separates on the upper side of the profile. Now, in the case of a propeller the stalling condition will not be reached simultaneously for all blade elements. As J decreases, a dead-air region may develop at one section and then gradually spread over the whole blade. Within this transition the lift coefficient will still increase for one part of the blade, while it decreases for the remainder farther outside toward the blade tips. Since the more distant elements act at greater speed, it can be expected that the resultant C_L-value will rather increase up to a certain limit. In order to keep in accordance with the experimental C_T-curves the left-hand parts of the curves in Fig. 230 are plotted under the assumption that C_L goes up very slowly to about 1.7, while the C_D-values follow a parabolic law $C_D = 0.0005\alpha^2 - 0.002\alpha - 0.017$, according to experiments on plates at higher angles of incidence. It must be admitted that our knowledge about the behavior of a high-pitch propeller at small values of the advance ratio is very incomplete, but the practical interest in this behavior, far removed from the usual working conditions, is not great.

Returning now to the region below stalling, the fact that all C_T- and C_P-curves are nearly straight lines in this region suggests the following simplification of Eqs. (20) and (21), which will be useful for the purpose of airplane performance computation. In this range one can obviously consider α' as a small angle, thus replacing sin α' by α' and cos α' by 1. Moreover, the zero lift drag coefficient a may be regarded as a small quantity of the same order. If we write $C_L = k\alpha'$ and then disregard all magnitudes of higher order, the three expressions (20) and (21) become linear in α':

$$J = \lambda \left(\tan \beta' - \frac{\alpha'}{\cos^2 \beta'} \right),$$

$$C_T = \lambda^2 \mu \frac{k\alpha' - a \tan \beta'}{\cos \beta'}, \qquad C_P = \lambda^3 \mu \frac{k\alpha' \tan \beta' + a}{\cos \beta'} \qquad (23)$$

Here α' can be eliminated, and the following *linear equations for* C_T *and* C_P result:

$$C_T = \lambda\mu k \cos \beta'(J_1 - J) \text{ with } J_1 = \lambda \tan \beta' \left(1 - \frac{a}{k \cos^2 \beta'}\right)$$
$$C_P = \lambda^2\mu k \sin \beta'(J_2 - J) \text{ with } J_2 = \lambda \tan \beta' \left(1 + \frac{a}{k \sin^2 \beta'}\right) \tag{24}$$

It is seen that $J_2 > J_1$, as it must be. For not too small β', Eqs. (24) give a fairly good estimate for the most important parts of the characteristics. Note that the slopes $C_T/(J_1 - J)$ and $C_P/(J_2 - J)$ are independent of a. We can derive from (24) an expression for the efficiency as function of J:

$$\eta = \frac{C_T J}{C_P} = \frac{\cot \beta'}{\lambda} \frac{J(J_1 - J)}{J_2 - J} \tag{25}$$

This gives the general type of the correct curve with $\eta = 0$ at $J = 0$ and $J = J_1$ and one maximum in between, the value of which is smaller than 1.

For performance computation and other purposes, also, the *polar diagram of the propeller* will prove useful. We define it as a line the coordinates of which are

$$x = \frac{C_T}{J^2}, \qquad y = \frac{C_P}{J^2} \tag{26}$$

Such coefficients as C_T/J^2 and C_P/J^2, which refer the propeller forces to V^2 rather than to $(nd)^2$, are used in propeller theory (see Sec. XII.3) and in certain practical problems (see Sec. XVI.1, etc.). If the method of a representative blade element is applied, it follows, from (20) and (21), that

$$x = \mu \frac{C_L \cos \gamma - C_D \sin \gamma}{\sin^2 \gamma}, \qquad y = \lambda\mu \frac{C_L \sin \gamma + C_D \cos \gamma}{\sin^2 \gamma} \tag{27}$$

Below stalling, with small α' and a and not too small β', and if all terms of higher order are disregarded, this amounts to

$$x = \frac{\mu}{\sin \beta'} (k\alpha' \cot \beta - a), \qquad y = \frac{\lambda\mu}{\sin \beta'} (k\alpha' + a \cot \beta') \tag{28}$$

and hence, by elimination of α',

$$y = \lambda \tan \beta' \left(x + \frac{\mu a}{\sin^3 \beta'}\right) \tag{29}$$

To check this equation we have plotted in Fig. 232 the C_P/J^2-values against C_T/J^2 corresponding to the experimental results[1] on the propeller

[1] *NACA Tech. Rept.* 339 (1930), p. 15.

of Fig. 222 with a nominal blade setting of 15.5°. A fairly well approximating straight line, like that in the figure, has a slope of $\lambda \tan \beta' = 0.64$ and an intersection point on the y-axis at the distance

$$\lambda \mu a \frac{\tan \beta'}{\sin^3 \beta'} = 0.01$$

This agrees with the assumptions $\lambda = 2.1$, $\mu = 0.015$, $a = 0.026$, $\beta' = 17°$, which seem to fit the data of the actual propeller.

The computations of this section are not meant to supply a theory of the propeller. They rather draw the conclusions from the geometry of a blade element only and want to serve a double purpose. First, they explain the general form of the characteristic curves as found by experiments. Second, they will furnish a basis for simple assumptions

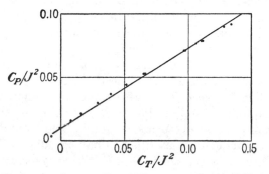

Fiu. 232.—Polar diagram for the propeller of Fig. 222 at a nominal blade setting of 15.5°.

about propeller performance to be used in the performance computation of the airplane.

Problem 9. If a two-blade propeller of 10 ft. diameter whose solidity is given by $A/2d^2$ lying between 0.013 and 0.015 rotates at $n = 25$ r.p.s. and the blade setting is chosen such that the propeller operates each time at 20 per cent slip, that is,

$$(J_1 - J)/J_1 = 0.20,$$

what thrust and power values can be expected at the velocities $V = 150$, 200, 250, and 300 ft./sec.?

Problem 10. Compute and plot the curves for $C_T/\lambda^2\mu$ and $C_P/\lambda^3\mu$ vs. J/λ under the assumption that up to $\alpha' = 0.25$ the lift and drag coefficients are given as

$$C_L = 5.3\alpha', \qquad C_D = 0.022 + 0.03C_L^2.$$

Blade settings of the representative element $\beta' = 12$ and 24°.

Problem 11. Derive the polar diagram from the results of Prob. 10, for $\lambda = 2.2$ and $\mu = 0.18$, and compare it with the straight line obtained by the simplified formula (29).

5. Propeller Sets and Variable-pitch Propeller. Propeller Charts.

For a propeller of a given shape, the curves C_T vs. J and C_P vs. J contain all the information that is required in performance computations. It has been seen that these curves depend on the shape of the propeller but are independent of its size as measured by the diameter. Among the parameters determining the shape of a propeller, the blade setting has the most marked influence on the characteristics: with increasing blade setting the advance ratio for maximum efficiency is increased, i.e., the practically important parts of the characteristics are shifted toward the right. Next comes the solidity ratio (μ), which constitutes a measure for the mean value of mc/d, where m is the number of blades and c the width (chord) of the blade: the coefficients of thrust and power increase nearly proportional to the solidity ratio, at least within the usual range of this ratio. The distribution of the values of β and of c/d along the blade (which find their expression in the parameter λ) and the shape of the blade profiles are of minor importance. In the preceding sections the influence of the blade setting as well as that of the solidity ratio has already been discussed to a certain extent.

A more systematic experimental study of the effect of changing the blade setting is desirable since this question dominates the choice of a suitable propeller for specified performance requirements. The simplest way of exploring the situation is to determine experimentally the characteristics of a so-called "propeller set." The individual propellers of such a set are frequently described as having different blade settings but as being "otherwise the same shape." What is meant by this somewhat vague expression is that, at any given value of r/d, the shape of the blade section and the value of c/d are the same for all members of the set, while the blade settings β have different values. The general type of the variation of β along the blade would have to be more or less the same for all members of a set.

An equivalent for a propeller set is a single propeller all of whose blades can be rotated simultaneously through the same angle about their longitudinal axes, i.e., axes running along the blades and perpendicular to the propeller axis. In this case the settings of all blade sections are changed by the same amount, and the members of the set are characterized by the position of the blades, defined, for example, by the nominal blade setting β_n at $r = 3d/8$. Today such propellers are widely used. If the blade setting can be modified only on the ground, the propeller is called an *adjustable-pitch propeller*. If the pilot can modify the blade setting in flight, for example, when changing from level flight to climb, the propeller is called a *controllable-pitch propeller*. Both cases are included in the expression "variable-pitch propeller." A diagram representing the characteristics for a propeller set or for one variable-pitch propeller at various blade settings is called a *propeller chart.*

Figure 233 shows the simplest form of a propeller chart—the two families of curves C_T vs. J and C_P vs. J for an adjustable-pitch propeller (Hamilton-Standard 1C1-0). The propeller has three blades and a

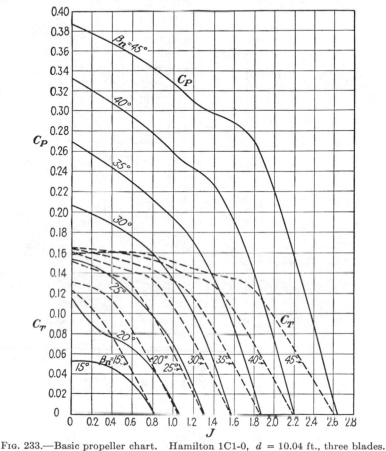

FIG. 233.—Basic propeller chart. Hamilton 1C1-0, d = 10.04 ft., three blades.

diameter of 10.04 ft.; with the exception of the vicinity of the hub the blade sections have the Clark Y-profile. Figure 234 shows the distribution of the chord length c and the thickness t of the blade sections and the pitch distributions corresponding to various values of the nominal blade setting β_n.*

* *NACA Tech. Rept.* 594 (1937). It should be noticed that the left-hand parts of the curves for higher pitch are hardly covered by the experiments, which mainly extend over the regions of incidences below stalling. No use is made of these parts in the following computations.

For practical purposes it is often preferable to use charts for some combinations of the quantities C_T, C_P, J instead of the basic C_T and C_P vs. J-curves. The efficiency $\eta = C_T J / C_P$ is such a combination. Another useful quantity is the *power-speed coefficient* C_S defined as

$$C_S = \frac{J}{\sqrt[5]{C_P}} = \frac{V}{nd}\sqrt[5]{\frac{\rho n^3 d^5}{P}} = V\sqrt[5]{\frac{\rho}{Pn^2}} \tag{30}$$

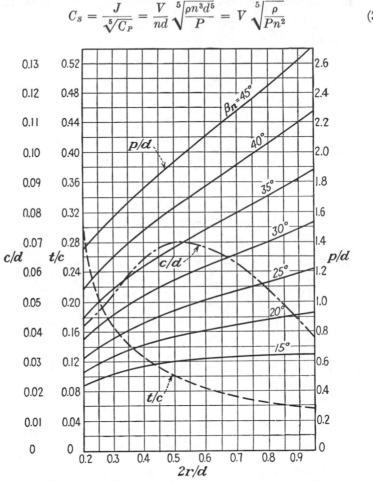

Fig. 234.—Geometric characteristics for the propeller of Fig. 233.

This coefficient has the advantage that for each ρ its value is known from the airplane design data (speed V, brake horsepower P, and number n of revolutions per second) before the choice of the propeller diameter has been made. If the basic propeller characteristics are known, the curve C_S vs. J can immediately be derived from the curve C_P vs. J. For a given C_S it then furnishes the corresponding advance ratio $J = V/nd$ from which the propeller diameter can be determined as $d = V/nJ$.

Combined with the curve η vs. J, the C_s graph (that is C_s and η vs. J) gives the thrust T of the propeller as $T = \eta P/V$.   Figure 235 shows the curves C_s and η vs. J corresponding to the same adjustable-pitch propeller as the curves of Fig. 233.   Note that a considerable portion of the

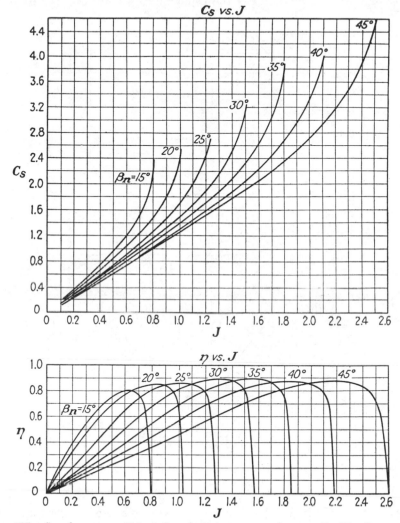

Fig. 235.—Speed-power coefficient C_s and efficiency η vs. advance ratio J for the propeller of Fig. 233.

curve C_s vs. J is practically a straight line passing through the origin. This is to be expected, since C_s is proportional to J and only to the fifth root of $1/C_P$.   Since C_P decreases with increasing J, the curve C_s vs. J deviates upward from the straight line through the origin.   The advance ratio for which this deviation becomes marked increases with the nominal pitch of the propeller.

For certain purposes another way of representing propeller characteristics is useful. Two relations between three variables may be represented graphically either in the usual way by two curves or by one

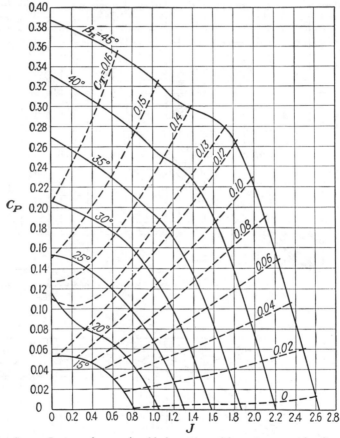

Fig. 236.—C_P vs. J curves for varying blade setting with contour map for C_T. Propeller of Fig. 233.

curve that relates in its coordinates two variables and carries a scale giving the corresponding values of the third variable. The polar diagram of an airfoil is of this kind, the curve showing C_L vs. C_D and the scale on the curve giving the values of the angle of attack α. In the case of a propeller, C_P may be plotted vs. J, and this curve may be graduated according to C_T. If a set of such graduated curves is plotted corresponding to the variation of a fourth parameter, the points carrying the same value of the third variable can be joined and then form another family of curves (*contour map* for the third variable).

A chart of this kind is shown in Fig. 236. Here the points on the various curves C_P vs. J that correspond to the same value of C_T are joined by (dotted) curves. In the figure these curves are drawn for the values $C_T = 0.02, 0.04, 0.06$, etc. The chart thus consists of two families of curves plotted in a coordinate system with J as abscissa and C_P as

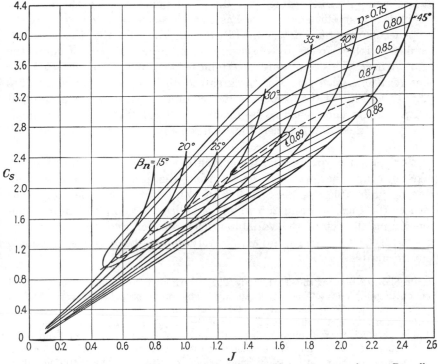

FIG. 237.—C_s vs. J curves for varying blade setting with contour map for η. Propeller of Fig. 233.

ordinate. The curves of the first family correspond each to a certain blade setting and show C_P vs. J. Along the curves of the second family, the blade angle varies, and C_T has a constant value.

Figure 237 shows another example of a propeller chart of the same type. Here two families of curves are drawn in a coordinate system with J as abscissa and C_S as ordinate. The curves of the first family correspond again each to a different blade setting and show C_S as function of J, while along each curve of the second family η has a constant round value.

In addition to the curves of these two families, the *best performance curve* (dotted line in Fig. 237) can be drawn, joining the points with horizontal tangent on the curves $\eta = \text{const.}$ A chart of this kind may be drawn either for a set of individual propellers or for a controllable-pitch propeller. In both cases the various curves C_S vs. J correspond to certain round values of the nominal pitch.

The chart (Fig. 237) can be used in the following way: For the given speed of flight V, the brake horsepower P of the engine, the engine speed n, and the density ρ at the altitude under consideration, the value of C_S is first computed according to (30). The point A with the ordinate C_S on the best efficiency curve has as abscissa $J = V/nd$, so that the propeller diameter d is determined by $d = V/nJ$. Finally, the necessary nominal pitch is found by observing on or near which C_S-J-curve the point A lies. This procedure determines that propeller of the set under consideration which possesses the greatest efficiency for the given values P, n, V, and ρ. If the propeller pitch cannot be controlled during flight and a definite choice has to be made for all conditions, a compromise must be worked out between the requirements of level flight, take-off, climbing, etc.

For special purposes, *e.g.*, for computing the maximum range or the take-off run, other ways of representing the propeller characteristics may be preferable (see Secs. XVI.1 and 2). But it may be noted that each of the Figs. 233 and 235 to 237 supplies the complete information concerning the set or the variable-pitch propeller under consideration. If one of these charts is given, the others can be derived by simple transformations.

Problem 12. Using the chart of Fig. 237, determine the optimum diameter d and the optimum blade setting β_n of an adjustable-pitch propeller if $P = 550$ hp., $n = 24$ r.p.s., and $V = 200$ m.p.h. at 5000 ft. altitude.

Problem 13. Determine the efficiency of the propeller of the preceding problem. How is this efficiency influenced by (*a*) an increase of 5° in β_n; (*b*) a drop of 25 hp. in P?

Problem 14. Determine the efficiency of the same propeller for $V = 175$ m.p.h. and 225 m.p.h. How do these efficiency values compare with the optimum values obtainable by choosing the optimum diameters corresponding to the given velocities?

CHAPTER XII
OUTLINE OF PROPELLER THEORY

1. Blade-element Theory. It was stated at the beginning of Chap. XI that a propeller blade may be considered as a strongly twisted wing. The adoption of this point of view leads to a manner of computing the forces acting on a propeller blade from experimental airfoil data that is known as the *blade-element theory*.

Figure 238 shows a cross section of the blade in the same way as it was considered in Fig. 216 and later in Fig. 228. The force which the air exerts on the blade element of the spanwise extension dr is the resultant of two component forces, lift and drag, perpendicular and

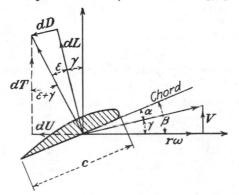

Fig. 238.—Resolution of forces acting on a propeller section.

parallel, respectively, to the velocity with which the element moves through the air. This velocity appears in Fig. 238 as a vector. with the horizontal component $r\omega = 2\pi rn$ and the vertical component V. The square of the resulting velocity equals

$$V^2 + r^2\omega^2 = V^2 + (2\pi nr)^2 = n^2d^2\left[J^2 + \left(\frac{2\pi r}{d}\right)^2\right] \tag{1}$$

where J denotes the advance ratio introduced in Eq. (4), Chap. XI. The angle between the velocity vector and the horizontal axis is

$$\gamma = \text{arc tan } \frac{V}{r\omega} = \text{arc tan } \frac{J}{2\pi r/d} \tag{2}$$

If β denotes again the angle between the plane of rotation and some well-defined reference line of the blade section (the so-called "chord"), the angle of attack is $\alpha = \beta - \gamma$. The force exerted on the blade

317

element under consideration could then be computed if the aerodynamic characteristics of the blade profile, *i.e.*, the coefficients C_L and C_D as functions of α, were known. Since the characteristics of an airfoil depend on the aspect ratio, the question arises which aspect ratio should be assumed in the actual case. One may think of defining the aspect ratio by means of the blade area and the blade length in the same way in which the aspect ratio of a wing is defined by area and span. However, all sections of a wing move with the same velocity, while for a propeller the sections near the blade tip move with a much higher velocity than those near the hub. Because of the corresponding difference in the flow patterns, there is no reason to believe that the formulas representing the influence of the aspect ratio on the aerodynamic characteristics of an airfoil (see Sec. VII.2) would be of much value in connection with a propeller blade. In fact, the difficulty in appreciating the influence of the aspect ratio, or, in other terms, the mutual interference of the various parts of one propeller, furnishes a serious objection against the blade-element theory. In the next two sections a theory will be developed that partly fills this gap and that finally leads to an improved method of computation (Sec. 4 of this chapter) taking into account the mutual interference of the blade elements. At present, approximate values of the airfoil coefficients corresponding to some large aspect ratio may be used for computing the forces acting on the blade. The influence of the Reynolds number may be neglected in this approximation. (A similar problem arises when the experimental results obtained with a rectangular airfoil without twist are to be applied to a twisted wing of arbitrary plan-form. Here Prandtl's wing theory gives the solution. It will be seen that the momentum theory discussed in Sec. 4 of this chapter leads to implications of a similar kind in the case of a propeller.)

In the following it is assumed that, for all blade sections, C_L and C_D are known functions of the angle α. The aerodynamic forces can then be computed in the same way as it was seen in Sec. XI.4 for the "representative" blade element. Since β is given by the known shape of the blade and γ can be computed from (2), the angle of attack $\alpha = \beta - \gamma$ can be determined for any blade section. According to the lift and drag formulas of Sec. VI.1 the blade element between the radii r and $r + dr$ experiences the following infinitesimal lift and drag forces:

$$dL = C_L \frac{\rho}{2} n^2 d^2 \left[J^2 + \left(\frac{2\pi r}{d} \right)^2 \right] c \, dr$$

$$dD = C_D \frac{\rho}{2} n^2 d^2 \left[J^2 + \left(\frac{2\pi r}{d} \right)^2 \right] c \, dr$$

(3)

where c denotes the chord length of the blade section under consideration and the dynamic pressure has been computed with the velocity value from (1). As is seen from Fig. 238, the force dL, which, by definition, is perpendicular to the velocity vector, makes the angle γ with the direction of the propeller axis; the force dD is parallel to the resulting velocity and perpendicular to dL. It will be useful to decompose the resultant of these two forces into the components dT and dU parallel and perpendicular to the propeller axis. These components are given by (see Fig. 238)

$$dT = dL \cos \gamma - dD \sin \gamma, \qquad dU = dL \sin \gamma + dD \cos \gamma \qquad (4)$$

The component dT has the direction of V, that is, the direction of flight, and thus represents the contribution of the blade element to the propeller thrust. The total propeller thrust T is obtained by summing the contributions dT from all elements of the m blades of the propeller. Substituting from (3) in (4) and taking the integral over one propeller blade, we have

$$T = m \frac{\rho}{2} n^2 d^2 \int_0^{d/2} \left[J^2 + \left(\frac{2\pi r}{d} \right)^2 \right] (C_L \cos \gamma - C_D \sin \gamma) c \, dr \qquad (5)$$

Here C_L and C_D are supposed to be known functions of $\alpha = \beta - \gamma$, which, in turn, is a known function of r by virtue of (2) and the known propeller shape (distribution of β). The chord length c is also a known function of r. Thus the integral in (5) can be evaluated. Using the dimensionless ratio r/d as the variable of integration, we rewrite (5) as

$$T = m \frac{\rho}{2} n^2 d^4 \int \left[J^2 + \left(2\pi \frac{r}{d} \right)^2 \right] (C_L \cos \gamma - C_D \sin \gamma) \frac{c}{d} d \left(\frac{r}{d} \right) \qquad (6)$$

Compared with Eq. (5), Sec. XI.1, this furnishes the thrust coefficient

$$C_T = \frac{m}{2} \int \left[J^2 + \left(2\pi \frac{r}{d} \right)^2 \right] (C_L \cos \gamma - C_D \sin \gamma) \frac{c}{d} d \left(\frac{r}{d} \right) \qquad (7)$$

This equation shows that C_T depends exclusively on the advance ratio J, once the propeller shape (*i.e.*, the blade profiles, the distributions of β and c/d along the blade, and the number of blades m) is given and the influence of Reynolds number on C_L and C_D is disregarded.

A certain difficulty arises from the fact that in the region extending from $r = 0$ (propeller axis) to a certain r_1 (root of the blade) the cross sections of the propeller normal to the longitudinal axis of the blade do not have the shape of an airfoil section at all. For this region no values of α, β, C_L, etc., can be assigned. The integral of (6) extended from r_1/d to $\frac{1}{2}$ will furnish the thrust of the blades proper. The core of the propeller body, consisting of the hub and the transitions between

hub and blades, experiences an axial air resistance that must be sub-
tracted from the thrust furnished by the blades. (This has nothing to
do with the so-called "effective thrust," which is defined as the difference
between T and the increase in drag that the slip stream of the propeller
produces on the airplane body.)

The drag of the propeller core may be written as

$$D_1 = \frac{\rho}{2} V^2 A_1 C_{D_1} \tag{8}$$

where A_1 is the frontal area of the propeller core and C_{D_1} a drag coefficient
that will not differ much from 1 (see Sec. V.2). The corrected value of
C_T becomes, therefore,

$$C_T = \frac{m}{2} \int_{r_1/d}^{\frac{1}{2}} \left[J^2 + \left(2\pi \frac{r}{d} \right)^2 \right] (C_L \cos \gamma - C_D \sin \gamma) \frac{c}{d} d\left(\frac{r}{d}\right) - \tfrac{1}{2} J^2 \frac{A_1}{d^2} \tag{9}$$

where C_{D_1} has been assumed equal to 1. In most cases the additional
term is insignificant. For example, assume A_1 to be a rectangle of 18 in.
length and 6 in. width (that is, $r_1 = 9$ in.), and let the propeller diameter
equal $d = 10$ ft. At $J = 0.7$ the second term on the right side of (9) is
then about 0.002.

The other component, dU, of the force acting on the blade element
of Fig. 238 has the moment $r\,dU$ with respect to the propeller axis. It
contributes to the propeller torque Q and to the power P, which can be
written as

$$Q = m \int r\,dU, \qquad P = \omega Q = \omega m \int r\,dU = 2\pi n m \int r\,dU$$

Proceeding in the same way as above, we find that

$$P = 2\pi n m \frac{\rho}{2} n^2 d^2 \int_0^{d/2} \left[J^2 + \left(2\pi \frac{r}{d} \right)^2 \right] (C_L \sin \gamma + C_D \cos \gamma) rc\,dr \tag{10}$$

Here the same difficulty as before arises because of the propeller
core. However, since the elements of the core move with a comparatively
small velocity and the arm r of the drag encountered by such elements is
small, the core will not furnish any considerable contribution to the
propeller torque. Accordingly, the region from $r = 0$ to $r = r_1$ may
simply be omitted from the integral of Eq. (10). Introducing r/d
again as the variable of integration and taking account of the definition
of the power coefficient [Eq. (7), Sec. XI.1], we obtain

$$C_P = \pi m \int_{r_1/d}^{\frac{1}{2}} \left[J^2 + \left(2\pi \frac{r}{d} \right)^2 \right] (C_L \sin \gamma + C_D \cos \gamma) \frac{r}{d} \frac{c}{d} d\left(\frac{r}{d}\right) \tag{11}$$

Here again the coefficient C_P appears as a function of the advance ratio only, once the propeller shape is known.

Equation (7) shows the prevailing influence of the lift coefficient on C_T. In fact, except for the elements near the hub, the angle γ is usually small so that $\cos \gamma$ is considerably greater than $\sin \gamma$. In (7) the lift coefficient C_L is multiplied by $\cos \gamma$, while the drag coefficient C_D, which is much smaller than C_L, is multiplied by the comparatively small $\sin \gamma$. Thus C_T depends largely on the lift coefficient of the blade sections. No similar statement can be made concerning the power coefficient C_P, since in (11) the larger coefficient C_L is multiplied by the smaller factor $\sin \gamma$ and the smaller coefficient by the greater factor $\cos \gamma$.

For the purposes of actual computation it is preferable to replace the factor $J^2 + (2\pi r/d)^2$ by $J^2(1 + \cot^2 \gamma) = J^2/\sin^2 \gamma$ and to write (9) and (11) in the form

$$C_T = \frac{J^2}{2d^2} \left(m \int_{r_1}^{d/2} \frac{C_L \cot \gamma - C_D}{\sin \gamma} c \, dr - A_1 \right)$$

$$C_P = \pi m \frac{J^2}{d^3} \int_{r_1}^{d/2} \frac{C_L + C_D \cot \gamma}{\sin \gamma} c \, r \, dr \tag{12}$$

As an example, consider the adjustable-pitch propeller shown in Fig. 222. This figure contains 10 blade sections corresponding to $r = 9.5$, 11.5, 19.5, 23.5, . . . , 59.5 in. The propeller was tested in a wind tunnel at various values of the advance ratio J. In the experiment the blade was set so that $\beta = 15.5°$ at $r = 47.5$ in. Since the figure gives $\beta = 20.1°$ for this value of r, all β-values of the figure must be diminished by 4.6°. If the effective portion of the blade is assumed to begin with $r = 8.5$ in., the profile given for $r = 9.5$ in. may be considered as the mean profile of the blade portion between $r = 8.5$ in. and $r = 10.5$ in., the profile for $r = 11.5$ in. as the mean profile of the portion between $r = 10.5$ in. and $r = 14.5$ in., etc. While these first two portions of the blade have the lengths of 2 in. and 4 in., respectively, the following portions are 6 in. long, with the exception of the last one, whose effective length may be estimated as 5 in. because of the rounding off of the blade tip. The computation shown in Table 8 is carried out for the advance ratio $J = 0.6$. The values of $\cot \gamma$ are computed according to $\cot \gamma = 2\pi r/Jd$. From this the values of $\cot \gamma$ and $1/\sin \gamma$ are determined. The angles of attack are computed according to $\alpha = \beta - \gamma$.

The aerodynamic characteristics of the various blade profiles are not given in the *NACA Report* 339, from which the other data for this example are taken. For the purposes of the present illustrative computation it has been assumed that $C_L = k\alpha'$ with $k = 0.1$ for α' in degrees

Table 8.—Propeller Computation (Blade-element Theory)

No.	(1) r in.	(2) cot γ	(3) γ°	(4) 1/sin γ	(5) β°	(6) α°	(7) α'°	(8) C_L	(9) C_D	(10) $C_L \cot \gamma - C_D$	(11) $C_L + C_D \cot \gamma$	(12) c in.	(13) c Δr in.²	(14) A in.²	(15) B in.³
1	9.5	0.795	51.5	1.28	37.8	−13.7	−10.7	−1.07	.04	−0.89	−1.04	6.97	13.94	−15.9	−176
2	11.5	0.965	46.0	1.39	34.5	−11.5	−8.5	−0.85	.04	−0.86	−0.81	7.36	29.44	−35.2	−381
3	17.5	1.47	34.2	1.78	30.5	−3.7	0.7	−0.07	.03	−0.13	−0.03	8.08	48.48	−11.2	−45
4	23.5	1.97	26.9	2.21	25.6	−1.3	+1.7	+0.17	.03	+0.31	+0.23	8.41	50.46	+34.6	+604
5	29.5	2.47	22.0	2.67	22.1	−0.1	2.9	0.29	.02	0.70	0.34	8.43	50.58	94.6	1355
6	35.5	2.97	18.6	3.13	19.3	+0.7	3.7	0.37	.02	1.08	0.43	8.17	49.02	166	2340
7	41.5	3.47	16.1	3.60	17.15	1.05	4.05	0.41	.02	1.40	0.48	7.53	45.18	228	3240
8	47.5	3.98	14.1	4.10	15.5	1.4	4.4	0.44	.02	1.73	0.52	6.53	39.18	278	3970
9	53.5	4.58	12.3	4.69	14.2	1.9	4.9	0.49	.02	2.17	0.58	5.21	31.26	318	4550
10	59.5	4.98	11.3	5.10	12.7	1.4	4.4	0.44	.02	2.16	0.54	3.74	18.70	206	3070
														1325	19129
														−62	−602
														1263	18527

and that $\alpha' = \alpha + 3°$. The first assumption corresponds almost to an infinite aspect ratio [see Eq. (12a), Sec. VII.2]; the second represents a resonable estimate of the value of α_0 and has been found to yield results in good agreement with the measured thrust and power coefficients. Since C_D is of minor importance, it has been assumed to have the constant value 0.02 over the larger part of the blade, which represents an average value of the profile drag of profiles of the kind used for the propeller blade. The assumption of a constant C_D, too, corresponds to an infinite aspect ratio. This, however, should not imply that C_D is supposed to be so small at the high angles of attack near the hub. But these blade elements with their small r contribute very little to the resultant forces.

The columns giving $C_L \cot \gamma - C_D$ and $C_L + C_D \cot \gamma$ are computed according to these assumptions. Upon taking into account the chord lengths c and the lengths Δr of the blade portions, the expressions

$$A = \left(\frac{1}{\sin \gamma}\right)(C_L \cot \gamma - C_D)c \, \Delta r$$

and

$$B = \left(\frac{1}{\sin \gamma}\right)(C_L + C_D \cot \gamma)cr \, \Delta r$$

are computed as shown in columns (14) and (15) of Table 8. Summing up of the entries in these two columns furnishes approximate values of the integrals of Eqs. (12): 1,263 in.² for the first and 18,527 in.³ for the second. With $J = 0.6$ and $d = 125$ in. and an estimated $A_1 = 90$ in.², this leads to

$$C_T = \frac{0.6^2}{2 \times 125^2}(2 \times 1{,}263 - 90) = 0.0281$$

$$C_P = 2\pi \frac{0.6^2}{125^3} 18{,}527 = 0.0214$$

in sufficiently good accordance with the experimental results shown in Fig. 225. The corresponding value of the efficiency is

$$\eta = J \frac{C_T}{C_P} = 0.79$$

a little less than the experimental value of almost 0.81.

A careful study of this and similar examples will reveal how the blade-element theory works and what modifications must be made in order to fulfill various conditions. A few remarks about the computation of the propeller efficiency from the standpoint of the blade-element theory may be added.

In Sec. XI.1 the propeller efficiency was introduced as $\eta = VT/P$. The blade-element theory suggests an analogous notion, the *element efficiency*

$$\eta_{el} = \frac{V\,dT}{dP} \tag{13}$$

where dT and dP are the contributions that the blade portion between r and $r + dr$ furnishes to thrust and power, respectively. In Sec. VII.1

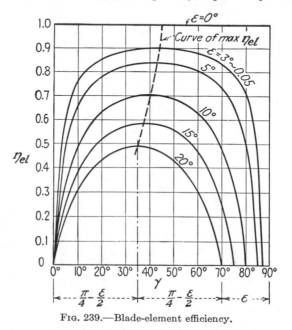

FIG. 239.—Blade-element efficiency.

the gliding angle ϵ of a profile has been defined by $\tan \epsilon = C_D/C_L$. The use of this concept permits Eqs. (4) to be written in the form (see Fig. 238)

$$
\begin{aligned}
dT &= dL(\cos \gamma - \tan \epsilon \sin \gamma) = dL\,\frac{\cos (\gamma + \epsilon)}{\cos \epsilon} \\
dU &= dL(\sin \gamma + \tan \epsilon \cos \gamma) = dL\,\frac{\sin (\gamma + \epsilon)}{\cos \epsilon}
\end{aligned}
\tag{4'}
$$

Now, $dP = r\omega\,dU$ and $\tan \gamma = V/r\omega$. Therefore,

$$\eta_{el} = \frac{V\,dT}{r\omega\,dU} = \frac{V}{r\omega}\cot (\gamma + \epsilon) = \frac{\tan \gamma}{\tan (\gamma + \epsilon)} \tag{14}$$

Some curves η_{el} vs. γ for constant values of ϵ are shown in Fig. 239. For $\epsilon = 0$, we have $\eta_{el} = 1$ independently of γ. It can easily be proved that the curves are symmetrical with respect to the line $\gamma = \pi/4 - \epsilon/2$.

For $\gamma = 0$ and $\gamma = \pi/2 - \epsilon$, we have $\eta_{el} = 0$. For $\gamma = \pi/4 - \epsilon/2$, the element efficiency η_{el} assumes its maximum value

$$(\eta_{el})_{\max} = \frac{\tan\left(\dfrac{\pi}{4} - \dfrac{\epsilon}{2}\right)}{\tan\left(\dfrac{\pi}{4} + \dfrac{\epsilon}{2}\right)} = \frac{\left(1 - \tan\dfrac{\epsilon}{2}\right)^2}{\left(1 + \tan\dfrac{\epsilon}{2}\right)^2} \sim \frac{1 - \epsilon}{1 + \epsilon} \sim 1 - 2\epsilon \qquad (15)$$

the last two expressions being approximate values valid for small ϵ only.

The η_{el}-curve for any $\epsilon < \epsilon_1$ lies above the η_{el}-curve for ϵ_1 in the entire range of γ from 0 to $\pi/2$. On the other hand, it follows from (4') that dU and hence dP are positive as long as $\epsilon < \pi/2$. Since the propeller efficiency equals

$$\eta = \frac{TV}{P} = \frac{\int V \, dT}{\int dP} = \frac{\int \eta_{el} \, dP}{\int dP}$$

this efficiency cannot exceed the greatest value that η_{el} assumes along the blade, as long as all dP are positive. Since for $\epsilon = 0$ the element efficiency is $\eta_{el} = 1$, the propeller efficiency, too, would be equal to 1 if ϵ were equal to zero all along the blade. That is, within the framework of the blade-element theory, the propeller efficiency would equal 1 if $C_D = 0$ for all blade sections. On the other hand, when ϵ has a small positive value for all blade sections, it is seen from (15) that the greatest value of η_{el} cannot exceed the value of $1 - 2\epsilon_{\max}$, where $\epsilon_{\max}$ denotes the greatest ϵ occurring along the blade. Therefore, an upper limit for the propeller efficiency would be $\eta < 0.92$ for $\epsilon \leq 0.04$ and $\eta < 0.90$ for $\epsilon \leq 0.05$. But even if ϵ has the constant value 0.05 over the entire blade length, η can never reach the value 0.90 since γ cannot assume the corresponding value $\pi/4 - \epsilon/2$ for more than one blade section. The values of γ start from about 90° at the blade root and drop to rather small amounts near the blade tip, so that an average ordinate rather than the maximum of the respective curve should be taken.

In the case of the propeller considered in the foregoing, ϵ is far from being constant and small. In fact, for $\alpha' = 0$, we have $\epsilon = 90°$, and for negative values of α' the gliding angle exceeds 90°. For such values of ϵ the curve η_{el} vs. γ has a much more complicated shape (see Prob. 2 below). According to (13) dU and hence dP then are negative, which means that the blade element under consideration works as a wind motor rather than as a propeller.

Summing up, we may state that *the blade-element theory furnishes the values of C_T and C_P to within ± 10 per cent if suitable airfoil data are used.* As has been already explained, it seems impossible to reach a higher accuracy because of the uncertainty regarding the proper values of the aspect ratio. Moreover, the blade-element theory does not give a satisfactory account of how the energy losses (appearing in the difference $1 - \eta$) are to be subdivided into friction losses and losses determined by the geometry of the problem; the value of ϵ taken from airfoil experiments includes both kinds of losses. Finally, the theory is open to the objection that it furnishes C_T and C_P proportional to the blade width c and to the number m of blades. It would therefore seem that, by increasing the blade area, thrust and power of a propeller could be increased in the same ratio, a conclusion that, obviously, is unacceptable. The approach discussed in the following sections will provide certain answers to these questions.

Problem 1. How would the assumption $\alpha' = \alpha + 2°$ (instead of $\alpha' = \alpha + 3°$) influence the result of the sample computation carried out in the foregoing?

Problem 2. Plot and discuss the curves η_{el} vs. γ corresponding to $\epsilon = 90, 105,$ and $120°$. Discuss also the meaning of η for negative γ.

Problem 3. Compute thrust and power for the propeller shown in Fig. 222 at the advance ratios $J = 0.5$ and 0.7. Take the blade setting as in the text, $\beta = 15.5°$ at $r = 47.5$ in.

Problem 4. Compute thrust and power for the same propeller at $J = 0.8$ under the assumption that all blade angles are increased by a constant so as to make the nominal blade setting $30°$.

2. Momentum Theory, Basic Relations. A satisfactory propeller theory should enable the designer to compute the characteristic curves (C_T vs. J and C_P vs. J) for a propeller of a given shape. The blade-element theory solves this problem by assuming rather arbitrarily that each blade element operates independently of all the other elements as if it were an element of an airfoil of infinite (or large) span engaged in a uniform rectilinear motion. It is obvious that this theory cannot give more than a rough approximation. Moreover, the blade-element theory does not furnish any indication concerning the limits that are set to the propeller efficiency independently of the propeller shape, while general experience leaves no doubt that such limits exist. In the case of a wing of finite span one finds a minimum value of the drag (the induced drag) depending on the aspect ratio but independent of the shape of the profile and of the viscosity of the fluid. Similarly, in the motion of a propeller there is a certain loss of energy determined by the diameter and by the advance ratio only. In fact, it will be seen that the greater part of the energy loss in the operation of a propeller would occur even if the propeller were to operate in a perfect fluid.

The *momentum theory*, which will be discussed in this and the following sections, assumes the air to be a perfect fluid. It gives a propeller efficiency that, under all conditions, is less than 1. On the other hand, the momentum theory does not attempt to solve the complete propeller problem stated above and must be supplemented, for practical purposes, by certain concepts of the blade-element theory. What the momentum theory achieves is to furnish *one relation between the three variables* C_T, C_P, *and* J, a relation that is supposed to be valid for any propeller independently of the number of blades, their profiles, plan-form, etc. No account whatever is taken of the shape of the propeller body. The theory replaces this body by an ideal mechanism, the so-called "actuator disk." The action that this disk is supposed to exert on the fluid represents the main features of the action exerted by the propeller. The momentum theory was inaugurated by W. J. M. Rankine in 1865.

Figure 240 serves to explain the idealization made in replacing the propeller by an actuator disk. The figure shows a longitudinal section of the stream of air in front and in the rear of the propeller. All cross sections are supposed to be of circular form. In particular, the cross section a-a where the propeller is located has the diameter d (= propeller

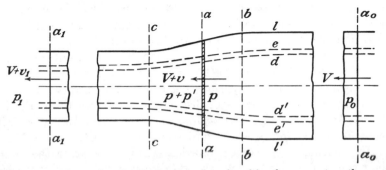

Fig. 240.—Illustrating the idealizations introduced by the momentum theory.

diameter), and the disk of area $S = \pi d^2/4$ is supposed to represent the propeller. In flight the propeller would move toward the right with the velocity V. In Fig. 240 the inverse motion is shown where the propeller is at rest and air, coming from the right, streams toward it with the initial velocity V. The cross section a_0-a_0 is taken at a great distance upstream of the actuator disk, mathematically speaking at infinity, and the section a_1-a_1 at a great distance downstream. It is assumed that all particles of air which have passed or will pass through the actuator disk form a body of revolution whose longitudinal section is bounded by the lines l and l' of the figure. Since the velocity of the air is increased by the propeller, this body of revolution must have a smaller diameter downstream than upstream.

In the case of an actual propeller rotating with the angular velocity $\omega = 2\pi n$ the flow is not strictly steady but is quasi-steady (Sec. III.4), even periodic, with mn periods per second where m is the number of blades. However, the idealized flow pattern studied in the momentum theory is supposed to be strictly steady, at least outside an infinitesimal region surrounding the actuator disk. It has been seen in Sec. III.5

that in a strictly steady continuous flow no exchange of power between the perfect fluid and a rigid body immersed in it is possible. To account for the energy exchange occurring in the case of a propeller, the momentum theory has to introduce another assumption: The regions upstream and downstream of the propeller are supposed to be *separated by an infinitesimal region in which sudden changes of pressure and velocity occur.* That is, the flow between the boundaries shown in Fig. 240 is assumed to be continuous except in the immediate neighborhood of the actuator disk, where the following discontinuities are admitted:

1. The pressure has different values on the two sides of this disk, p on the upstream side and $p + p'$ downstream. The integral $\int p' \, dS$ extended over the disk area represents the propeller thrust acting toward the right.

2. The fluid particles passing through the disk region can here abruptly change their velocity component u tangential to the circular paths described by the points of the rotating propeller.

The assumption of a steady (not periodic) motion outside the disk region will obviously be the better justified the larger the number of blades, since the number of periods per second increases with m. This is why the theory of the actuator disk is often referred to as the theory of a *propeller with an infinite number of blades.* Note that even within the concept of an infinite m the flow inside the zone swept by the propeller blades cannot converge toward a steady flow since the pressure on the two sides of each blade preserves different values. Otherwise, no torque moment would exist.

In the cross section a-a, on the downstream side of the actuator disk, the velocity components parallel to the propeller axis, in the radial, and in the tangential directions will be denoted by $V + v$, w, and u, respectively. On the upstream side only $V + v$ and w exist, while the tangential component u vanishes as will be proved later. The subscripts 0 and 1 will be used to denote the values that all variables assume at the sections a_0-a_0 and a_1-a_1, respectively. At the first of these two sections the pressure equals the atmospheric pressure p_0. Furthermore, one has to assume $v_0 = 0$, $w_0 = 0$, $u_0 = 0$, since at a large distance upstream of the actuator disk the velocity has the direction of the propeller axis and the magnitude V.

The flow is supposed to have rotational symmetry in the sense that the variables v, w, u, p are functions of the abscissa x, measured in the direction of the propeller axis, and of the radial distance r from this axis, but are independent of the position of the point under consideration on the circle of radius r. Accordingly, any streamline will coincide with certain other streamlines when rotated about the propeller axis. Any such group of streamlines covers a surface of revolution which in Fig. 240

is represented by pairs of curves like d, d' or e, e'. Because of the rotational symmetry of the flow pattern, a ring of the radius r and the width dr can be taken as the typical element of the disk area S. The area of this ring equals $dS = 2\pi r\, dr$.

The following theorems can be applied to the flow under consideration: Bernoulli's theorem; the momentum theorem for the direction of the propeller axis; the moment of momentum theorem with respect to this axis; the energy theorem for quasi-steady flow.

1. Consider first the *moment of momentum theorem* for the direction of the propeller axis. Let the control surface consist of the surfaces of revolution generated by the neighboring streamlines d and e and of the corresponding annular elements in the plane a_0-a_0 and in some other cross-sectional plane b-b upstream of the actuator disk. The normals of this control surface either intersect the propeller axis or are parallel to it. As the fluid is supposed to be perfect, the thrusts exerted by the surrounding fluid on the mass within this control surface act in the direction of the normals, and their moment with respect to the propeller axis will thus be zero. If dQ denotes the flux of mass across the annular element, the moment of momentum theorem states that the change of $ru\, dQ$ between the entrance and exit sections equals the sum of all moments of thrust, *i.e.*, equals zero in the present case. As $u_0 = 0$, it follows that *the tangential velocity component u is zero everywhere upstream of the actuator disk.*

A similar result is obtained when two cross sections downstream of the actuator disk are considered instead of two upstream sections. The difference between $r_c u_c\, dQ$ at some downstream section c-c and $r_1 u_1\, dQ$ at the section a_1-a_1 must vanish. That is, *downstream of the actuator disk the moment of the velocity vector with respect to the propeller axis has a constant value for each particle.*

$$r_c u_c = r_1 u_1 = ru \tag{16}$$

If one of the cross sections of the control surface is taken upstream and the other downstream of the actuator disk, the two surfaces of revolution together with the two annular elements in the planes b-b and c-c do not constitute a complete control surface. The part of the propeller surface lying between two radii r and $r + dr$ must be added. If dM is the moment of the thrusts that the fluid exerts on this part of the rotating rigid body, the moment of momentum theorem states that the increase of moment of momentum between the entrance and exit sections equals dM.[*] As $u = 0$ upstream of the disk, this increase equals $r_c u_c\, dQ$ or $ru\, dQ$, where dQ is given by

$$dQ = \rho(V + v)\, dS = 2\pi\rho(V + v)r\, dr. \tag{17}$$

[*] Since the letter Q is used in connection with the flux, the propeller torque will be denoted by M in this section.

Thus the third and last conclusion drawn from the moment of momentum theorem reads

$$dM = \rho(V + v)ru\, dS \qquad M = \rho\!\int(V + v)ru\, dS \qquad (18)$$

2. Next consider *Bernoulli's theorem* applied first to a streamline between the cross-sectional planes a_0-a_0 and a-a (immediately upstream of the actuator) and then to the prolongation of this streamline between a-a (immediately downstream of the actuator) and a_1-a_1. With the notations already introduced, we have, for the first part,

$$\frac{\rho}{2} V^2 + p_0 = \frac{\rho}{2}[(V + v)^2 + w^2] + p \qquad (19)$$

and, for the downstream part,

$$\frac{\rho}{2}[(V + v)^2 + w^2 + u^2] + p + p' = \frac{\rho}{2}[(V + v_1)^2 + w_1^2 + u_1^2] + p_1 \qquad (20)$$

Bernoulli's equation cannot be applied to the total length of the streamline between the cross-sectional planes a_0-a_0 and a_1-a_1 because of the unsteady character of the flow in the immediate vicinity of the actuator disk. (As regards the moment of momentum theorem the situation is different since this theorem applies to quasi-steady flows, too.)

The distance between the cross-sectional planes a_0-a_0 and a-a as well as that between a-a and a_1-a_1 is supposed to be very large. If, at the same time, the differences between the cross-sectional areas S_0, S_1, and S are supposed to be small, the equation of continuity shows that the velocities v and v_1 are small, too. Therefore, the slope of the streamlines against the propeller axis, which is proportional to the differences $r_0 - r$, $r - r_1$, etc., and inversely proportional to the distances between a_0-a_0, a-a, a_1-a_1 is small of the second order. Accordingly, the radial velocity components w which are proportional to this slope are small of the second order, too. The squares w^2 and w_1^2 can therefore be neglected in (19) and (20), even if terms of the second order like v^2 are retained.

The tangential velocity components u and u_1 will be supposed to be of the same order as v and v_1. Then, according to (16), the difference

$$u^2 - u_1^2 = u^2\left(1 - \frac{r^2}{r_1^2}\right) = u^2(r_1 - r)\frac{r_1 + r}{r_1^2}$$

is small of the third order and can be neglected in (20).

Equations (19) and (20) thus reduce to

$$\frac{\rho}{2} V^2 + p_0 = \frac{\rho}{2}(V + v)^2 + p \qquad (19')$$

and

$$\frac{\rho}{2}(V + v)^2 + p + p' = \frac{\rho}{2}(V + v_1)^2 + p_1 \tag{20'}$$

Adding and solving for p', we find

$$p' = \rho v_1 \left(V + \frac{v_1}{2}\right) + p_1 - p_0 \tag{21}$$

Since the thrust dT exerted on the element dS of the actuator disk equals $p'\, dS$, the result supplied by the Bernoulli equations is

$$dT = \rho v_1 \left(V + \frac{v_1}{2}\right) dS + (p_1 - p_0)\, dS$$

$$T = \rho \int v_1 \left(V + \frac{v_1}{2}\right) dS + \int (p_1 - p_0)\, dS \tag{22}$$

The significance of the second integral will be discussed below.

3. Next consider the *momentum theorem for the direction of the propeller axis*, applied to the two control surfaces shown in Fig. 241, *i.e.*, to

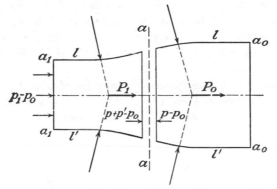

Fig. 241.—Control surfaces for the application of the momentum equation.

the total bulk of the fluid upstream and downstream of the disk zone. In the case of the moment of momentum equation the fluid pressures acting on control surfaces of this type did not furnish any contribution, since their moment with respect to the propeller axis is zero. Now, however, these thrusts must be taken into account. The resultant of these thrusts equals the resultant produced by the overpressures $p - p_0$, since a uniform pressure p_0 acting on all elements of a closed surface has the resultant zero. Let P_0 and P_1 denote the resultants of the overpressures acting on the surfaces of revolution l, l' upstream and downstream and on the cross sections a_0-a_0 and a_1-a_1, respectively.

Because of the rotational symmetry these resultants have the direction of the propeller axis. For the two control surfaces the resultant thrusts, considered as positive if acting in the direction of flight, then equal $\int (p - p_0) \, dS + P_0$ and $-\int (p + p' - p_0) \, dS + P_1$, respectively. The flux of momentum across the upstream control surface equals

$$-\int (V + v) \, dQ + \int V \, dQ,$$

where the first integral corresponds to the section a-a and the second to the section a_0-a_0 [see the discussion preceding Eq. (4), of Sec. III.2]. With $dQ = \rho(V + v) \, dS$, the momentum equation applied to the upstream control surface furnishes

$$\int (p - p_0) \, dS + P_0 = -\rho \int v(V + v) \, dS \qquad (23)$$

Similarly, for the downstream control surface,

$$-\int (p + p' - p_0) \, dS + P_1 = -\rho \int (v_1 - v)(V + v) \, dS \qquad (24)$$

By adding these equations a new expression for $T = \int p' \, dS$ is obtained:

$$T = \int p' \, dS = \rho \int (V + v) v_1 \, dS + P_0 + P_1 \qquad (25)$$

(It might be asked why the momentum theorem is not applied to a fluid mass including the zone of the actuator disk. Since there is no sudden increase of the axial velocity component at the disk, the momentum theorem applied to the region between two planes immediately upstream and downstream of the disk would only yield $dT = p' \, dS$, a relation that has already been used.)

Information concerning P_0 and P_1 may be obtained by solving (23) and (24) with respect to these quantities and introducing $p - p_0$ from (19') and $(p + p')$ from (20'). Thus, after some rearranging of terms, one finds

$$P_0 = -\frac{\rho}{2} \int v^2 \, dS, \qquad P_1 = \frac{\rho}{2} \int (v - v_1)^2 \, dS + \int (p_1 - p_0) \, dS \qquad (26)$$

These relations show that the overpressure along l, l' cannot simply be assumed to vanish. This assumption, made in some presentations of the momentum theory, would give $P_0 = 0$ in contradiction to the first equation (26), which shows that P_0 is negative. On the other hand, some assumptions concerning P_0 and P_1 and the value of $p_1 - p_0$ have to be made within the framework of the present theory, which does not attempt to give a complete solution, *i.e.*, an evaluation of the distribution of velocity and pressure based on the differential equations of hydrodynamics. It seems plausible that at a large distance downstream of the propeller the pressure in the wake will approach the outside pressure p_0. At least the average (integral) value of $p_1 - p_0$ may be supposed to vanish. If one accepts this, both quantities, P_0 and P_1, become small of the second order and of opposite sign. It may then appear justified to neglect the sum $P_0 + P_1$ in (25). As a matter of fact, the usual momentum theory is based *on the assumptions that the integral of* $(p_1 - p_0) dS$ *and the sum* $P_0 + P_1$ *can be regarded as zero.*

Comparing now the thrust values as given by (22) and (25), we have

$$\int v_1 \left(V + \frac{v_1}{2} \right) dS = \int v_1 (V + v) \, dS \quad \text{or} \quad \int v_1 v \, dS = \frac{1}{2} \int v_1^2 \, dS \quad (27)$$

If v and v_1 can be assumed to be constant over the cross section, the last equation would furnish

$$v = \tfrac{1}{2} v_1 \tag{28}$$

Another way of arriving at this result is to apply the momentum equation to the fluid within some interior stream surface d, d' (Fig. 240) and to assume that for any such surface the expressions corresponding to $(p_1 - p_0) dS$ and to $P_0 + P_1$ are zero. The relation $dT = \rho(V + v)v_1 \, dS$, corresponding to (25), would then hold true for any annular element dS, and comparison with the first equation (22) would furnish (28). We thus have the following important result: If it is assumed either that the additional velocities v and v_1 are uniformly distributed over the circular cross section or that there is no resultant mutual pressure interference between the annular layers of the fluid, one can conclude that *one half of the total increase v_1 in axial velocity produced by the propeller occurs upstream of the actuator, the other half downstream.*

4. Finally apply the energy equation, in the form derived in Eq. (27), Sec. III.5, to the total bulk of fluid between the cross-sectional planes a_0-a_0 and a_1-a_1. This flow, as a whole, can be considered as quasi-steady (periodic). The increase in total head between the sections a_0-a_0 and a_1-a_1 equals

$$\frac{p_1 - p_0}{\gamma} + \frac{(V + v_1)^2 + w_1^2 + u_1^2}{2g} - \frac{V^2}{2g}$$

$$= \frac{p_1 - p_0}{\gamma} + \frac{v_1(2V + v_1) + w_1^2 + u_1^2}{2g} \tag{29'}$$

where w_1^2 is small of the fourth order and may be neglected, while u_1^2 may be replaced by u^2 since the difference between these two quantities is of the third order. The pressure difference contributes to the energy equation a term of the form $(1/\rho)\int(p_1 - p_0) \, dQ = \int(p_1 - p_0)(V + v) \, dS$. If v is assumed to be constant over the cross section, this integral must vanish according to the hypothesis already introduced. In the general case, $\int(p_1 - p_0) \, dQ = 0$ would constitute an independent assumption. The gravity of the fluid being neglected, the power imparted to the fluid by the propeller may then be written as

$$P = \rho \int (V + v) \left[v_1 \left(V + \frac{v_1}{2} \right) + \frac{u^2}{2} \right] dS \tag{29}$$

Equations (18), (22), (27), and (29) are the basic relations of the momentum theory. They relate the velocities v, v_1, u (all of them functions of r) to the thrust T, the torque M, and the power P of the propeller.

Problem 5. Show that the equations of the momentum theory as given in the foregoing include the assumption that the flow upstream of the actuator disk is irrotational but do not imply this assumption for the wake.

*Problem 6. Show, from the form of the streamlines, that the pressure cannot possibly be constant along the border l, l' of the upstream flow.

3. Momentum Theory, Conclusions. The basic equations of the momentum theory established in the preceding section are not sufficient or intended to determine the unknown functions v, v_1, u of the radius r. To determine these unknowns, it would be necessary to set up the differential equations governing the motion of the fluid and to solve them on the basis of a given propeller shape. However, Eqs. (18), (22), (27), and (29) yield relations between T, M, and P if appropriate assumptions concerning $v(r)$, $v_1(r)$, and $u(r)$ are introduced.

The simplest set of plausible assumptions is the following: (1) The axial velocity components v and v_1 are assumed to be constant over the respective cross sections. (2) The tangential velocity components u immediately downstream of the actuator disk are assumed to be proportional to the distance r from the propeller axis: $u = r\omega'$. These assumptions obviously amount to replacing the variables v, v_1 and u/r by certain average values. The first assumption leads to $v = v_1/2$ by virtue of (27), as already explained. According to the second assumption one can compute

$$\int u^2\, dS = 2\pi \int_0^{d/2} u^2 r\, dr = 2\pi\omega'^2 \int_0^{d/2} r^3\, dr = 2\pi\omega'^2 \frac{1}{4}\left(\frac{d}{2}\right)^4 = \frac{\omega'^2 d^2}{8}\, S$$

$$\int ru\, dS = 2\pi \int_0^{d/2} r^2 u\, dr = 2\pi\omega' \int_0^{d/2} r^3\, dr = \frac{\omega' d^2}{8}\, S \tag{30}$$

Let us now collect the equations needed in the further development of the momentum theory: (I) Eq. (22), where the integration can be performed since v_1 is assumed to be constant with respect to r; (II) Eq. (18), transformed by introducing $v = v_1/2$, the second relation (30), and $P = M\omega$; (III) Eq. (29), transformed by introducing $v = v_1/2$ and the first relation (30). Thus,

$$T = \rho S\left(V + \frac{v_1}{2}\right)v_1 \tag{I}$$

$$P = \rho S \left(V + \frac{v_1}{2} \right) \omega \omega' \frac{d^2}{8} \tag{II}$$

$$P = \rho S \left(V + \frac{v_1}{2} \right) \left[\left(V + \frac{v_1}{2} \right) v_1 + \omega'^2 \frac{d^2}{16} \right] \tag{III}$$

Elimination of v_1 and ω' between these equations furnishes a relation between T, P, V, ω, and d that embodies the main result of the momentum theory. For given values of V, ω, and d a relation between T and P will thus be obtained.

Let us first consider a simple transformation of (III). In (III) the first term on the right-hand side is a multiple of the right-hand side of (I), while the second term is a multiple of the right-hand side of (II); thus,

$$P = T \left(V + \frac{v_1}{2} \right) + P \frac{\omega'}{2\omega} \tag{IIIa}$$

From this equation the value of TV/P can be computed. In the preceding discussion this expression was called the efficiency of the propeller. However, since in the present argument all frictional losses are disregarded, it seems preferable to introduce a special term for the value of TV/P furnished by the momentum theory. In the following this value is called the *induced efficiency* and denoted by η_i. Making use of (IIIa), we find the following first expression for the induced efficiency:

$$\eta_i = \frac{TV}{P} = \frac{1 - \omega'/2\omega}{1 + v_1/2V} \tag{31}$$

Another expression for η_i can be obtained by multiplying (I) by V and dividing by (II). With $J = V/nd$ and $\omega = 2\pi n$, the result can be written as

$$\eta_i = \frac{TV}{P} = \frac{8Vv_1}{\omega \omega' d^2} = \frac{4Jv_1}{\pi \omega' d} = \frac{2}{\pi^2} J^2 \frac{v_1/V}{\omega'/\omega} \tag{32}$$

The Equations (31) and (32) show the importance of the ratios v_1/V and ω'/ω. Note that v_1 is the axial and ω' the angular velocity that the actuator disk imparts to the fluid.

Equations (I) to (III) may be given a dimensionless form. Instead of the former thrust and power coefficients new dimensionless parameters will be used, the *thrust loading* τ and the *power loading* σ. These parameters are defined by

$$\tau = \frac{2T}{\rho V^2 S}, \qquad \sigma = \frac{2P}{\rho V^3 S} \tag{33}$$

(It is seen that these definitions follow the general pattern for dimensionless force and power coefficients more closely than the definitions of

C_T and C_P. However, τ and σ lose their meaning for $V = 0$, that is, in the case of a propeller running while the airplane is at rest on the ground.) The parameters C_T and C_P are connected with τ and σ by

$$
\begin{aligned}
C_T &= \frac{T}{\rho n^2 d^4} = \frac{TJ^2}{\rho V^2 d^2} = \frac{\pi}{8} J^2 \tau \\
C_P &= \frac{P}{\rho n^3 d^5} = \frac{PJ^3}{\rho V^3 d^2} = \frac{\pi}{8} J^3 \sigma
\end{aligned}
\tag{34}
$$

Note that the ratio τ/σ equals the induced efficiency

$$
\eta_i = \frac{VT}{P} = \frac{JC_T}{C_P} = \frac{\tau}{\sigma}
\tag{34'}
$$

Introducing the dimensionless parameters τ and σ, Eqs. (I) to (III) can be written as

$$
\tau = 2\left(1 + \frac{v_1}{2V}\right)\frac{v_1}{V}
\tag{I'}
$$

$$
\sigma = 2\left(1 + \frac{v_1}{2V}\right)\frac{\pi^2}{2J^2}\frac{\omega'}{\omega}
\tag{II'}
$$

$$
\sigma = 2\left(1 + \frac{v_1}{2V}\right)\left[\left(1 + \frac{v_1}{2V}\right)\frac{v_1}{V} + \frac{\pi^2}{4J^2}\frac{\omega'^2}{\omega^2}\right] = \tau\left(1 + \frac{v_1}{2V}\right) + \sigma\frac{\omega'}{2\omega}
\tag{III'}
$$

Now we can carry out the elimination of v_1/V and ω'/ω in order to find the announced relation between τ, σ, and J, which represent, in a dimensionless form, the quantities T, P, V, d, and ω. First solve (I') for v_1/V.

$$
\left(\frac{v_1}{V}\right)^2 + 2\frac{v_1}{V} = \tau \qquad \text{or} \qquad \frac{v_1}{V} = \sqrt{\tau + 1} - 1
$$

where only the positive root of the quadratic equation for v_1/V may be retained. Then introduce this value of v_2/V into (II'), and solve with respect to ω'/ω.

$$
\frac{\omega'}{\omega} = \frac{J^2}{\pi^2}\frac{\sigma}{1 + v_1/2V} = \frac{2J^2\sigma}{\pi^2(\sqrt{\tau + 1} + 1)} = \frac{2J^2}{\pi^2}\frac{\sigma}{\tau}(\sqrt{\tau + 1} - 1)
$$

since $(\sqrt{\tau + 1} + 1)(\sqrt{\tau + 1} - 1) = \tau$. Finally, divide both sides of the second form of (III') by τ, and introduce the values for v_1/V and ω'/ω:

$$
\frac{\sigma}{\tau} = \frac{1}{\eta_i} = \tfrac{1}{2}(\sqrt{\tau + 1} + 1) + \frac{J^2}{\pi^2}\frac{1}{\eta_i^2}(\sqrt{\tau + 1} - 1)
\tag{35}
$$

This is the desired relation between τ, η_i, and J. As τ and σ are equivalent to C_T and C_P, this result may be stated in the following form: *The momentum theory leads to a definite relation between the coefficients of thrust* and power, or between one of these coefficients and the induced efficiency, for each value of the advance ratio. This relation is independent of the propeller shape.

In general, the last term in (35) is small as compared with the preceding term. Thus, as a first approximation,

$$\eta_i = \frac{2}{\sqrt{\tau + 1} + 1} \tag{35a}$$

A second approximation may be obtained by introducing (35a) on the right-hand side of (35). Thus

$$\frac{1}{\eta_i} = \tfrac{1}{2}(\sqrt{\tau + 1} + 1) + \frac{J^2}{4\pi^2}(\sqrt{\tau + 1} + 1)^2(\sqrt{\tau + 1} - 1)$$

$$= \tfrac{1}{2}(\sqrt{\tau + 1} + 1)\left(1 + \frac{J^2\tau}{2\pi^2}\right)$$

Since $J^2\tau/2\pi^2$ will be small as compared with 1, this relation is equivalent to

$$\eta_i = \frac{2}{\sqrt{\tau + 1} + 1}\left(1 - \frac{J^2\tau}{2\pi^2}\right) \tag{35b}$$

Relation (35a) follows from (I) and (III) if the rotational velocity u is completely neglected from the beginning; it has been derived in this way by W. J. M. Rankine. Relation (35b) has been established by Th. Bienen and Th. v. Kármán. Figure 242, which corresponds to (35) rather than to (35b), shows some curves η_i = const. in a diagram using J^2 as abscissa and τ as ordinate. The dotted hyperbolas are curves of constant C_T. Since in the momentum theory frictional losses are disregarded entirely, relation (35) and Fig. 242 must be interpreted, not as giving efficiency values in accordance with actual experiments, but rather as giving the highest efficiency that, eventually, could be approached under particularly favorable conditions. The result already stated can also be expressed in this form: *The momentum theory gives for each value of thrust loading and advance ratio an upper limit for the efficiency that cannot be exceeded whatever the shape of the propeller.* For example, it is seen from Fig. 242 that at an advance ratio $J = 0.4$ a propeller with the thrust loading $\tau = 1.18$ ($C_T = 0.074$) cannot possibly have an efficiency exceeding 80 per cent. If, for the same advance ratio, the thrust loading is raised to $\tau = 4.0$ ($C_T = 0.251$), the greatest possible efficiency is lowered to about 60 per cent. In a general way, the diagram shows that higher efficiencies are bound to small load factors.

Experimental evidence agrees with the conclusions of (35). Figures 243 and 244 compare the results obtained from Eq. (35) with the experimental results obtained in the case of the propeller whose characteristics

were given in Fig. 225.   In Fig. 243 the curve marked η represents the
actual efficiency computed from the experimental characteristics according to $\eta = JC_T/C_P$; the curve marked η_i corresponds to Eq. (35), the
thrust loading $\tau = 8C_T/\pi J^2$ being computed, for each advance ratio,
from the observed values of C_T.   It is seen that η_i is everywhere greater

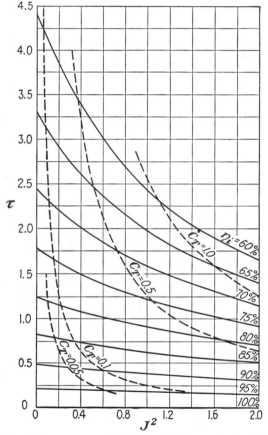

Fig. 242.—Thrust loading vs. square of advance ratio for constant efficiency according to
Eq. (35).

than η.   In Fig. 244 the observed values of C_P are confronted with the
values of JC_T/η_i that represent according to the present theory the values
of C_P if no frictional losses occurred.   As is to be expected, JC_T/η_i is
everywhere smaller than C_P, the difference constituting a measure for the
frictional losses.

Problem 7.   Compute the induced efficiency as function of J under the assumption that the C_T vs. J-curve is a straight line $C_T = a(J_1 - J)$.   Compare the values

of η_i at $J = 0.75J_1$ with the η-values estimated according to the method of the representative blade element (Sec. XI.4).

Problem 8. Plot and discuss the curves η_i vs. τ for constant J. Also, draw the lines $C_P = $ const. in this diagram.

Problem 9. If Eq. (35) is solved for η_i, one finds two values. Explain why one must be discarded, and give the explicit formula for the other.

Problem 10. Setting $\eta_i = \sigma/\tau$, Eq. (35) becomes a relation between σ and τ for each constant J. Plot these curves σ vs. τ and discuss their relation to the polar diagram of the propeller (Sec. XI.4).

Problem 11. Develop a formula for η_i as function of τ for small values of τ.

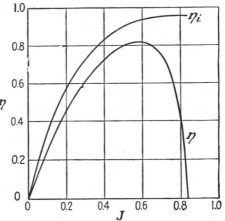

FIG. 243.—Induced efficiency computed from observed C_T-values, compared with the actual efficiency.

4. Modified Momentum Theory. Various attempts have been made to improve on the momentum theory by taking into account the finite number of blades, the variation of v, v_1, and ω' over the cross section, certain kinds of friction losses, etc. The following trait is common to all these investigations: The integral form of Eqs. (I) to (III) is replaced by a certain differential form. That is, *it is stated that for each annular disk element relations for dT and dP exist which are analogous to Eqs. (I) to (III).* This implies the introduction of certain new hypotheses.

As regards (I), the correct differential form was given in the first equation (22). If it is now assumed that p_1 equals p_0 on all streamlines, the simple formula (I'') as given below is obtained. As to (II), the correct expression for $dM = dP/\omega$ was derived in (18); with the exception of the replacement of v by $v_1/2$, this

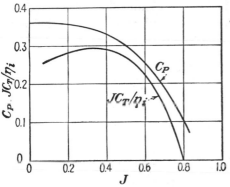

FIG. 244.—Power coefficient as computed from induced efficiency, compared with the actual power coefficient.

agrees with the differential form of (II) to be used now. The equality $v = v_1/2$ could be based, as already mentioned, on the assumption that v and v_1 are constant over the entire cross section. However, this would not serve the purpose of the generalization that is now intended. On the other hand, for variable v and v_1, it has already been seen that the relation $v = v_1/2$ can be justified by assuming that there is no resultant pressure interference between adjacent st eam layers.

The following two assumptions will now be made: (1) On each streamline the final pressure p_1 equals the pressure value p_0 of the undisturbed air. (2) The resultant thrust between two adjacent stream layers of rotational symmetry is zero. Under these conditions, Eq. (18) can be written in the form (II″) below. As to Eq. (III) there is no difficulty in applying the energy theorem for quasi-steady motion, as derived in Sec. III.5, to each annular element of the disk and the corresponding stream layer separately. Once the assumption $p_1 = p_0$ has been made, the pressure term in (29′) drops out and (29) can be written down for the power element dP by omitting the integral sign.

In this highly hypothetical differential form the momentum theory of the propeller is based on the following three equations, where $dS = 2r\pi\, dr$ refers to an annular element of the disk area:

$$dT = \rho\left(V + \frac{v_1}{2}\right)v_1\, dS \tag{I″}$$

$$dP = \rho\left(V + \frac{v_1}{2}\right)r^2\omega\omega'\, dS \tag{II″}$$

$$dP = \rho\left(V + \frac{v_1}{2}\right)\left[\left(V + \frac{v_1}{2}\right)v_1 + \frac{r^2\omega'^2}{2}\right]dS = \left(V + \frac{v_1}{2}\right)dT + \frac{\omega'}{2\omega}\, dP \tag{III″}$$

Here v_1 and ω' may still depend on r. Corresponding to (31) and (32) one now obtains two formulas for the *induced efficiency of the disk element*, one from (III″) and one dividing (I″) by (II″).

$$\eta_{i\,el} = \frac{V\, dT}{dP} = \frac{1 - \omega'/2\omega}{1 + v_1/2V} = \frac{Vv_1}{r^2\omega\omega'} \tag{36}$$

The latter equation defines a relation between the two ratios $v_1/2V$ and $\omega'/2\omega$. Upon introducing $J = V/nd$, this relation can be written as

$$\frac{\omega'}{2\omega}\left(1 - \frac{\omega'}{2\omega}\right) = \frac{J^2}{\pi^2}\frac{v_1}{2V}\left(1 + \frac{v_1}{2V}\right)\left(\frac{d}{2r}\right)^2 \tag{36'}$$

Thus, the differential form of the momentum theory restricts the range of admissible functions $v_1(r)$ and $\omega'(r)$. For example, it is not possible to have both v_1 and ω' constant over the disk. Moreover, if v_1 has any finite value at $r = 0$, the ratio ω'/ω would become infinite. It is very doubtful whether these results are legitimate. One can try to find suitable forms for the functions $v_1(r)$ and $\omega'(r)$ that satisfy (36'); one can even ask for the "best" distributions, *i.e.*, for those leading to the highest resultant efficiency η_i. But all such computations are open to the objection that they are based on the rather artificial assumption of the complete absence of interference between the stream layers. In the following, use will be made of (36) only for the purpose of deriving a modified set of equations, replacing (I'') and (II''), which takes into account friction forces in a certain approximate way.

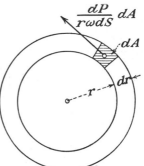

The last equality in (36) admits of a simple geometric interpretation. If the ring of the radius r and the width dr transmits the power dP, it experiences on its area dS a moment $dM = dP/\omega$. Since the flow pattern is supposed to possess rotational symmetry, the moment will be uniformly distributed over the ring area. This means that at each point of the ring a tangential force of the magnitude $dP/r\omega\,dS$ per unit of area is acting (Fig. 245). On the other hand, the same annular element of the disk experiences an axial force of the

Fig. 245.—Tangential force component supplying the torque moment dM.

magnitude dT/dS per unit of area acting normal to the ring plane which is seen in Fig. 245. Denote by γ' the angle that the resultant of these two forces makes with the propeller axis. Then, using (36),

$$\cot \gamma' = \frac{dT}{dP/r\omega} = \frac{r\omega}{V}\,\eta_{i\,el} = \frac{r(\omega - \omega'/2)}{V + v_1/2} = \frac{v_1}{r\omega'} \tag{37}$$

The last two expressions follow from (36).

Figure 246 shows how the direction γ' can be found by plotting a right triangle ABC with the sides $\overline{AB} = r(\omega - \omega'/2)$ and $\overline{BC} = V + v_1/2$. The force that the air exerts on the blade element is therefore seen to be perpendicular to a velocity vector that, in the partly inverse flow considered in the present analysis, has the components $V + v_1/2$ and $r(\omega - \omega'/2)$ parallel and perpendicular, respectively, to the propeller axis. In the actual flow, the relative velocity of the air with respect to the propeller in the two cross sections a_0-a_0 upstream and a_1-a_1 downstream, has the axial components V and $V + v_1$ and the tangential components $r\omega$ and $r(\omega - \omega')$, respectively. If the force that a perfect fluid exerts on a rigid body is considered as a "lift," i.e., as acting perpendicular to an effective velocity, Eq. (37) and its geometrical interpretation show that, in the present case, this *effective velocity equals the arithmetic mean of the relative velocities upstream and downstream of the propeller.* The analogy between this effective velocity and that appearing in the theory of the wing of finite span will be discussed later.

In Fig. 246 the points E and D are determined by $\overline{EC} = V$, $\overline{DE} = r\omega$, and DE perpendicular to EC, so that the angle CDE equals γ as defined by Eq. (2), Sec. XI.1.   The angle ACD then is $\gamma' - \gamma$, and (37) shows that AD is perpendicular to AC.   Indeed, in the triangle ADF we have $\overline{AF} = v_1/2$, $\overline{DF} = r\omega'/2$ and thus $\cot DAF = v_1/r\omega'$, that is, $\angle DAF = \gamma'$.

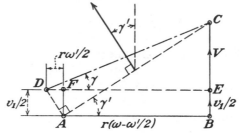

Fig. 246.—Directions of air reaction and effective relative velocity according to Eq. (36).

If the resultant force were perpendicular to the velocity vector resulting from V and $r\omega$, instead of from $V + v_1/2$ and $r(\omega - \omega'/2)$, the element efficiency would equal 1, since, in this case,

$$\frac{T}{P/r\omega} = \frac{r\omega}{V} \qquad \text{that is} \qquad VT = P \qquad \text{and} \qquad \frac{TV}{P} = 1$$

The angular deviation $\gamma' - \gamma$ thus accounts for the loss of energy which

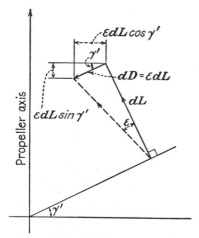

Fig. 247.—Additional drag force parallel to the effective relative velocity.

is admitted in the momentum theory, although this theory so far neglects all frictional losses.   This suggests a way to supplement the theory by an additional assumption designed to take account of the influence of friction.

It seems reasonable to assume that the force perpendicular to the mean relative velocity corresponds to a lift dL experienced by the blade element and *that the addition of a drag dD parallel to this velocity would correspond to the actual presence of friction forces.*   The drag dD will change the resulting force as indicated in Fig. 247.   The line of action is rotated through an angle ϵ and the magnitude becomes $dL/\cos \epsilon$, where $\tan \epsilon = dD/dL$.   The figure shows that, because of the drag dD, the thrust component of the force is diminished by $dL \sin \gamma' \tan \epsilon$, the tangential component increased by $dL \cos \gamma' \tan \epsilon$.

But $dL \sin \gamma' = dP/r\omega$ and $dL \cos \gamma' = dT$, where dP and dT are the uncorrected values. The new values consequently are given by

$$dT' = dT - \epsilon \frac{dP}{r\omega}, \qquad dP' = dP + \epsilon r\omega \, dT \qquad (38)$$

where ϵ is supposed to be small so that $\tan \epsilon$ could be replaced by ϵ.

Introducing the values of dT and dP given by (I'') and (II'') and assuming v_1, ω', and ϵ to be constant over the disk radius, we find

$$T' = \rho \left(V + \frac{v_1}{2} \right) \int (v_1 - \epsilon\omega'r) \, dS = \rho S \left(V + \frac{v_1}{2} \right) \left(v_1 - \frac{\epsilon\omega'd}{3} \right)$$

$$P' = \rho \left(V + \frac{v_1}{2} \right) \int r\omega(r\omega' + \epsilon v_1) \, dS = \rho S\omega \left(V + \frac{v_1}{2} \right) \left(\frac{\omega'd^2}{8} + \frac{\epsilon v_1 d}{3} \right) \qquad (38a)$$

The assumptions $v_1 = $ const. and $\omega' = $ const. are now allowed since with the introduction of friction forces the energy equation (III'') and thus the first expression for $\eta_{i \, el}$ in (36) lose their validity. The corrected propeller efficiency accordingly equals

$$\eta = \frac{T'V}{P'} = \frac{V}{\omega d} \frac{v_1 - \epsilon\omega'd/3}{\omega'd/8 + \epsilon v_1/3} = \frac{8Vv_1}{\omega\omega'd^2} \frac{1 - \epsilon\omega'd/3v_1}{1 + 8\epsilon v_1/3\omega'd} \qquad (39)$$

Comparing with (32), we find that the first factor on the right-hand side equals η_i and that the ratio $v_1/\omega'd$ can be replaced by $\pi\eta_i/4J$. Thus, taking account of the fact that ϵ is small, we may write the expression for η as

$$\eta = \eta_i \frac{1 - 4J\epsilon/3\pi\eta_i}{1 + 2\pi\eta_i\epsilon/3J} \sim \eta_i \left[1 - \epsilon \left(\frac{4}{3\pi} \frac{J}{\eta_i} + \frac{2\pi}{3} \frac{\eta_i}{J} \right) \right] \qquad (39a)$$

This corrected value of the efficiency is denoted simply by η, since it contains friction losses and thus is comparable to the efficiency of an actual propeller. The values of η as given by (39) agree to a certain extent with the experimental evidence. In the case of the numerical example of Sec. 1 of this chapter the observed value of C_T at the advance ratio $J = 0.6$ equals $C_T = 0.028$. According to (34) the corresponding value of τ equals $\tau = 8C_T/\pi J^2 = 0.198$, and (35b) gives $\eta_i = 0.95$. Equation (39a) then furnishes $\eta = 0.80$ for $\epsilon = 0.05$ and $\eta = 0.83$ for $\epsilon = 0.04$, while the observed value was 0.81. It appears justified to assume that the average value of ϵ for the profiles of the propeller under consideration lies close to 0.05, although values of ϵ beyond 90° occur near the hub. [It would seem correct to identify the observed value of C_T with that corresponding to T' rather than to T. However, the first equation (38a) can be written as

$$T' = T \left(1 - \epsilon \frac{\omega'd}{3v_1} \right) = T \left(1 - \epsilon \frac{4J}{3\pi\eta_i} \right) \qquad (40)$$

This shows that in the case of the present example the difference between T and T' is only about one per cent.]

Figures 248 and 249 show the curves η = const. for $\epsilon = 0.04$ and $\epsilon = 0.05$ in the coordinate system used in Fig. 242. This diagram can be used for determining the propeller diameter so as to obtain the best efficiency under given conditions. If a propeller is to be selected for an airplane, the required thrust T, the velocity of flight V, and the number of revolutions per second n may be considered as given. From these data alone neither τ nor J but only the ratio τ/J^2 can be computed.

$$\frac{\tau}{J^2} = \frac{2T}{\rho S V^2} \frac{n^2 d^2}{V^2} = \frac{8}{\pi} \frac{Tn^2}{\rho V^4} \tag{41}$$

In Figs. 248 and 249 this ratio determines a straight line through the origin. It is seen that for the point of contact of this line with a certain

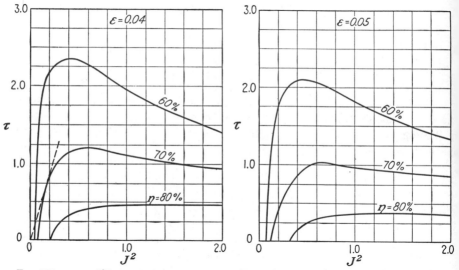

FIG. 248.— τ vs. J^2 for constant η at $\epsilon = 0.04$.   FIG. 249.— τ vs. J^2 for constant η at $\epsilon = 0.05$.

curve η = const. the efficiency η reaches its greatest value. The corresponding propeller diameter can be computed from either the abscissa or the ordinate of this point since

$$d = \frac{V}{nJ} = \sqrt{\frac{8}{\pi} \frac{T}{\rho V^2 \tau}} \tag{42}$$

If this value of d is admissible from the structural point of view, it represents the best possible choice of the diameter. Since the diameter is the only parameter characterizing the propeller shape that appears in the momentum theory, no further indications regarding the best possible propeller can be expected.

As already mentioned, other refinements of the momentum theory aim at taking account of the variation of v, v_1, and ω' over the propeller radius. Note, for example, that with constant v_1 and ω' the element efficiency $\eta_{i\,el}$ is not constant over the radius. It may be asked what relation must be assumed between τ and η in order to obtain a constant element efficiency. It is doubtful whether the considerable work required by this kind of computation is really worth while. It should be kept in mind that Eqs. (I'') to (III'') contain the assumption of the mutual independence of the annular disk elements or the corresponding stream tubes. The approximate character of this assumption sets a limit to the accuracy of the results that may be obtained from any theory based on relations (I'') to (III'').

Problem 12. Compute the efficiency of a propeller according to (39a) if its diameter is 10.5 ft., the engine speed 20 r.p.s., the air speed $V = 200$ ft./sec., and the thrust $T = 820$ lb.

Problem 13. Discuss the relation between η and J that follows from (39a) if the C_T vs. J-curve is assumed as $C_T = $ const. $(J_1 - J)$. Compute, in particular, the efficiency at $J = 0.75J_1$.

*Problem 14. Find the relation between η_i and τ that replaces (35) in the case where the relations (I'') and (III'') are accepted and $\eta_{i\,el}$ is assumed to be constant over the blade section.

*Problem 15. Find the relation between η, C_T, and J that results from the following assumptions: Equation (I) of the momentum theory is accepted. Instead of (II) and (III) it is assumed that the power P is spent to produce an increase of kinetic energy in the wake that corresponds to the axial velocity increase v_1 and a rotational velocity $\omega' = \lambda_n$. Discuss suitable values for λ.

5. The Two Theories Combined. A combination of both the blade-element and the momentum theories can be based on the following idea: It has been assumed in the first approach that each blade element between two radii r and $r + dr$ experiences a lift and a drag force, dL and dD, acting normal and parallel to the vector of relative velocity of the air against the element. This velocity was assumed to consist of the axial component V and the rotational component $r\omega$ and to make the angle γ with the plane of rotation so that thrust and power, dT and dP, can be computed according to (4) with $dP = r\omega\,dU$. Now, in the differential form of the momentum theory, it was seen that if no friction is admitted the thrust on the propeller element acts perpendicular to a velocity vector composed of an axial velocity $V + v_1/2$ and a rotational velocity $r(\omega - \omega'/2)$, forming an angle γ' with the plane of rotation. Moreover, the quantities v_1 and ω' were connected with dT and dP by relations (I'') and (II''). Confronting these two approaches, one may venture the following working hypothesis:

Assume, first, that dT and dP can be computed according to (4) from such lift and drag forces as are produced by a velocity with the components

$V + v_1/2$ and $r(\omega - \omega'/2)$; *second, that relations* (I'') *and* (II'') *between* dT, dP, *and* v_1, ω' *still hold.*

Using the abbreviations for the *inflow factors*

$$a = \frac{v_1}{2V}, \qquad b = \frac{\omega'}{2\omega} \tag{43}$$

and setting $dS = 2\pi r \, dr$, Eqs. (I'') and (II'') take the form

$$dT = 4\pi\rho V^2(1 + a)ar \, dr, \qquad dP = 4\pi\rho\omega^2 V(1 + a)br^3 \, dr \tag{44}$$

On the other hand, Eqs. (4) combined with (3), if γ is replaced by γ' and V_r written for the resultant velocity, become (Fig. 250)

$$dT = m\frac{\rho}{2} V_r^2(C_L \cos \gamma' - C_D \sin \gamma')c \, dr$$

$$dP = r\omega \, dU = m\frac{\rho}{2} \omega V_r^2(C_L \sin \gamma' + C_D \cos \gamma')cr \, dr \tag{45}$$

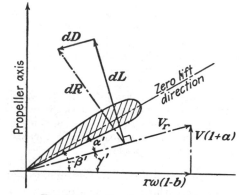

Fig. 250.—Inflow factors a and b.

Here, the factor m (number of blades) was added. The coefficients C_L and C_D are considered as given functions of the incidence, and for this value the effective incidence $\alpha'' = \beta' - \gamma'$ has to be substituted.

Introducing the solidity of the propeller area at the radius r,

$$s = \frac{mc}{2\pi r} \tag{46}$$

and using the obvious relations

$$\frac{V}{V_r} = \frac{\sin \gamma'}{1 + a}, \qquad \frac{r\omega}{V_r} = \frac{\cos \gamma'}{1 - b} \tag{47}$$

the comparison of (44) and (45) leads to

$$\frac{a}{1 + a} = \frac{s}{4 \sin^2 \gamma'} (C_L \cos \gamma' - C_D \sin \gamma')$$

$$\frac{b}{1 - b} = \frac{s}{4 \sin \gamma' \cos \gamma'} (C_L \sin \gamma' + C_D \cos \gamma') \tag{48}$$

If the shape of the propeller is given, *i.e.*, the value of s for each r and the angle β' which the zero lift direction of each profile makes with the plane of rotation, we have in (48) two equations for the three unknowns a, b, and γ'. A third equation is supplied by the relation between γ' and the advance ratio,

$$\tan \gamma' = \frac{V}{r\omega}\frac{1+a}{1-b} = J\frac{d}{2\pi r}\frac{1+a}{1-b}, \qquad J = \frac{2\pi r}{d}\frac{1-b}{1+a}\tan \gamma' \quad (49)$$

Now the computation of thrust and power for a propeller of given shape at a given advance ratio can be carried out in the following way:

For some tentative values of γ', the effective incidence $\alpha'' = \beta' - \gamma'$, the two inflow factors a and b from (48), and the right-hand side in the second equation (49) are computed. The values of the latter expression are plotted vs. γ', and from this graph the γ'-value corresponding to the given J is found. The actual values of a and b are thus determined. Then, if thrust and power coefficients for the element are introduced by

$$dC_T = \frac{dT}{\rho n^2 d^4}, \qquad dC_P = \frac{dP}{\rho n^3 d^5} \quad (50)$$

and x is written for $2r/d$, Equations (44) supply

$$\frac{dC_T}{dx} = \pi J^2 a(1+a)x, \qquad \frac{dC_P}{dx} = \pi^3 Jb(1+a)x^3 \quad (51)$$

The curves showing the right hand sides of (51) as functions of x are known as the *thrust and power grading curves*. If the effective blade begins at $r = r_1$, $2r_1/d = x_1$, one has

$$C_T = \int_{x_1}^{1} \frac{dC_T}{dx}\,dx, \qquad C_P = \int_{x_1}^{1} \frac{dC_P}{dx}\,dx \quad (52)$$

The coefficients C_T and C_P are thus found as the areas between the grading curves and the axis of abscissas, within the range of abscissas from x_1 to 1.

As an example the same propeller may be taken of which the design was described in Fig. 222 and for which the coefficient C_T and C_P at the advance ratio $J = 0.6$ have been computed in Sec. 1 of the present chapter. The same assumptions for the aerodynamic characteristics of the profile will be used. Table 9 shows the computation

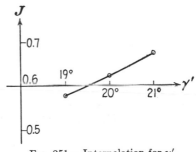

Fig. 251.—Interpolation for γ'.

of dC_T and dC_P for the blade section No. 6, $r = 35.5$ in., $c = 8.17$ in., $\beta = 19.3°$. For $J = 0.6$ the angle γ for this section is $18.6°$ as given in

TABLE 9.—COMPUTATION FOR A PROPELLER ELEMENT
Blade section No. 6

$r = 35.5$ in. $\beta = 19.3°$ $c = 8.17$ in.

$x = 2r/d = 0.568$ $\beta' = 22.3°$

$\pi x = 1.785$ $s = \dfrac{2 \times 8.17}{2\pi \times 35.5} = 0.0732$

γ'°	$\cot \gamma'$	α''°	C_L	C_D	$C_L \times \cot \gamma' - C_D$	$C_L + C_D \times \cot \gamma'$	$\dfrac{a}{1+a}$ Eq. (48)	$\dfrac{b}{1-b}$ Eq. (48)	a	b	J Eq. (49)
19	2.90	3.3	.33	.02	.937	.388	.0526	.00752	.0555	.00746	.579
20	2.75	2.3	.23	.02	.613	.285	.0328	.00555	.0339	.00552	.624
21	2.60	1.3	.13	.02	.318	.182	.0162	.00356	.0165	.00354	.674
19.5	2.82	2.8	.28	.02	.770	.336	.0422	.00653	.0440	.00649	.600

column (3) of Table 8. Accordingly, 19, 20, and 21° are taken as tentative values for γ'. The values of J obtained in the last column are plotted against γ' in Fig. 251, and $\gamma' = 19.5°$ is found to correspond to the given

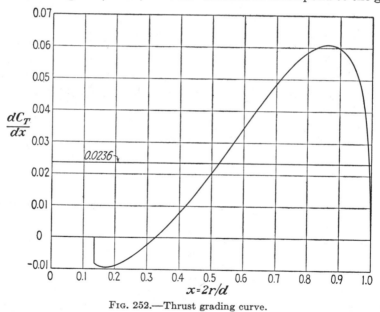

FIG. 252.—Thrust grading curve.

$J = 0.6$. The last line, corresponding to $\gamma' = 19.5°$, is then added to the table. With the values $a = 0.0440$, $b = 0.00649$, Eqs. (51) furnish

$$\frac{dC_T}{dx} = 0.0295, \qquad \frac{dC_P}{dx} = 0.0231$$

The results of these computations, found in the same way for all blade sections, are plotted against x in Figs. 252 and 253. As previously, the effective blade is assumed to begin with $r_1 = 8.5$ in., that is, with $x_1 = 0.136$. Evaluation of the two areas between $x = 0.136$ and $x = 0$ gives, finally,

$$C_T = 0.0236, \qquad C_P = 0.0191$$

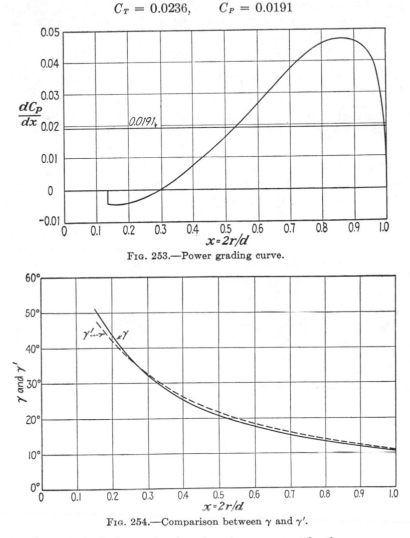

FIG. 253.—Power grading curve.

FIG. 254.—Comparison between γ and γ'.

If the drag of the hub portion is taken into account in the same way as in Sec. 1 of this chapter, the thrust coefficient will be diminished to $C_T = 0.0226$.

Figure 254 gives the comparison between the angles γ used in the primitive blade-element theory and the angles γ' used in the present refined theory. It is seen that, for the major part of the blade, γ'

exceeds γ by almost 1°. The effective angle of incidence α'' is therefore smaller than the α' used in the first approach. This explains why the new value of C_T is somewhat smaller than that found by the primitive blade-element theory.

An actual test of the accuracy of the predictions of the two theories would require that aerodynamic characteristics be used which are based on experiments for each single profile occurring in the blade.

Problem 16. Compute power and thrust for the propeller given in Fig. 222 at the advance ratios J = 0.5 and 0.7 using the method of the preceding section, and compare the results with those of Prob. 3.

Problem 17. Make the same computation for J = 0.8 under the assumption that all blade angles are increased by a constant so as to have the nominal blade setting equal to 30°.

6. Additional Remarks. A complete solution of the propeller problem would require finding the velocity and pressure distribution in the whole space as caused by the helicoidal motion of the propeller. Such an investigation should be based, as it was in the case of the wing theory (Chaps. VIII and IX), on the principles of the mechanics of a perfect fluid (hydraulic hypothesis, Sec. IV.3). Not much has been done in this direction; but Joukowski, later Prandtl, and in a more detailed way S. Goldstein extended the ideas that proved efficient in the *three-dimensional wing theory* to find some interesting results for the case of the propeller.

According to the classical potential theory the disturbance caused by the motion of a rigid body of regular shape in a perfect fluid at rest can be considered as induced by a vortex sheet coinciding with the surface of the body. If the body is finite, it can be seen that the thrusts exerted by the fluid on it can result in a couple only but do not lead to a resultant force (D'Alembert's paradox, Chap. IX). Then, in the case of a wing with a sharp trailing edge, it has been discovered (Lanchester, Prandtl), that the perfect fluid motion must show a certain kind of discontinuity: the vortex sheet that induces the disturbance velocities must be assumed to extend into the free stream and tend toward infinity so that its effect is the same as that of an infinite body whose surface consists of the actual wing surface plus the two (coinciding) sides of the trailing vortex sheet (see Sec. IX.3).

In the case of a moving propeller the vortex lines on the body run essentially on both sides of the blades in a radial direction. At the blade tip they must turn backward and form in the slip stream a kind of helicoidal line (Fig. 255). If a great number of blades is assumed, the whole cylinder of diameter d (or some surface of revolution with the same axis) behind the propeller will be covered by a vortex sheet. This vortex sheet or the corresponding vorticity can then be analyzed into one sheet consisting of vortex lines parallel to the axis of the propeller and another sheet consisting of circles in the cross-sectional planes. For reasons of symmetry the first type of vortex will induce circumferential velocities and the second type velocities essentially parallel to the axis. The same is true if, in closer analogy to the wing theory, vortex lines are assumed to spring from all points of the trailing edge of the blade rather than from the tip only. Now the basic hypothesis of Prandtl's wing theory comes into play. It will be assumed that the flow around each profile is such as if it were the cross section of an airfoil of infinite span subject to an outside flow whose velocity is composed of the actual velocities of the element, V and $r\omega$, plus the induced components $v_1/2$ in the axial and $-r\omega'/2$ in the tangential direction.

Thus it is seen that this theory, if carried out completely, should supply the distribution over the blade of the v_1- and ω'-values, which were introduced in the preceding sections. On the other hand, the relations

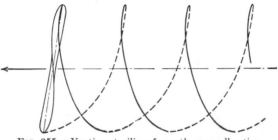

FIG. 255.—Vortices trailing from the propeller tips.

between v_1, ω' and the thrust and power elements dT, dP, as derived from the momentum theorems and Bernoulli's equation, must remain valid as far as the simplifying assumptions used in their derivation are justified. It follows, therefore, that *the outcome of the whole vortex conception cannot go much beyond what was presented in Sec. 5 of this chapter as the combined blade-element and momentum theory.* The use of airfoil characteristics for infinite span in that computation may be regarded as established by the present point of view. Besides, the expressions "induced" velocities for v_1 and $r\omega'$ and "induced" efficiency for the efficiency as computed on the basis of these additional velocities (Sec. 4 of this chapter), now appear legitimate. Note that the distributions of v_1 and ω' over the blade length are completely determined by the shape of the propeller if the method described in Sec. 5 of this chapter is accepted.

Another question which may be briefly discussed here concerns the mutual interference between the successive blades of one propeller, the so-called "multiple interference." A part of this effect can be studied, in first approximation, as a problem in two-dimensional wing theory. Consider in the x-y-plane, instead of a single airfoil section, a so-called "cascade" of airfoils, *i.e.*, an endless sequence of equal and equally spaced airfoils (Fig. 256). Such a cascade corresponds roughly to the successive blade elements of a propeller if the spacing s is made to equal $2\pi r/m$ where m is the number of blades. If the cascade is subject to a uniform airflow with the velocity V making an angle α' with the zero lift direction of the profile, there must be a lift force L' normal to $\overrightarrow{V}$, acting on each single airfoil per unit of span. The lift coefficient

$$C_L = \frac{2L'}{\rho V^2 c}$$

will then depend on the shape of the profile, the angle β' between the profile and the axis of the cascade, the ratio c/s, and the incidence α'. The mathematical problem of finding C_L was solved by Joukowski under the assumption that the profile reduces to a straight line of length c

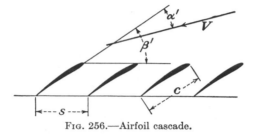

FIG. 256.—Airfoil cascade.

under the angle β'. In this case, as was seen in Sec. VIII.6, the lift coefficient for a single profile and small α' is $2\pi\alpha'$. The exact solution given by Joukowski for a cascade with $s = 2\pi r/m$ is

$$C_L = 2\pi \sin \alpha' \frac{4r}{mc} \frac{\tanh x}{\sin \beta'} \tag{53}$$

where x has to be computed by eliminating p and q from the following three equations

$$\sinh x = \sin \beta' \tan q, \qquad \sin q = \frac{\sin p}{\cos \beta'}, \qquad x \sin \beta' + p \cos \beta' = \frac{mc}{4r} \tag{54}$$

The result of this computation is represented in Fig. 257, which gives the values of $C_L/2\pi \sin \alpha'$ for various blade settings β' and solidity ratios $mc/2\pi r$. In view of these results and the actual inaccuracy of all assumptions used in conventional propeller computations, it appears to be legitimate to neglect the multiplane interference in all practical cases of two- three-, or four-blade propellers. Another approach to the interference problem, based on concepts of the three-dimensional wing theory and supplying an effect of opposite sign, confirms this conclusion.

This section may be concluded with some remarks on the *propeller moving at advance ratio* 0, that is, the propeller rotating in a bulk of air at rest. In the case of a vertical axis this is the problem of the lifting airscrew, which plays the decisive role in the design of a helicopter. There is no difficulty in applying the primitive blade-element theory (Sec. 1 of this chapter) to the case $V = 0$, $J = 0$ if the airfoil characteristics are known for the total range of incidences from zero up to the maximum β'-value. As to the momentum theory, one of its basic hypotheses fails, *viz.*, the assumption that v_1/V is a small quantity. Nevertheless, it is usual, and apparently not quite unjustified, to apply its principal results, simply setting $V = 0$ in the equations. If we do

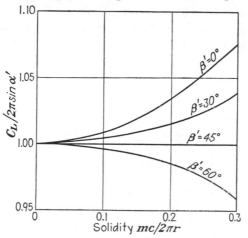

Fig. 257.—Multiplane interference. (*From H. Glauert in W. F. Durand, "Aerodynamic Theory," Vol. 4, p.* 230.)

so and call T_0, P_0 the values of thrust and power at $V = 0$, Eqs. (I) to (III) now read

$$T_0 = \rho S \frac{v_1^2}{2} \tag{I_0}$$

$$P_0 = \rho S \frac{v_1}{2} \omega\omega' \frac{d^2}{8} \tag{II_0}$$

$$P_0 = \rho S \frac{v_1}{2} \left(\frac{v_1^2}{2} + \omega'^2 \frac{d^2}{16} \right) \tag{III_0}$$

Solving the first equation for v_1 and the second for ω', we obtain

$$v_1 = \sqrt{\frac{2T_0}{\rho S}}, \qquad \omega' = \frac{16 P_0}{\rho S v_1 \omega d^2} = \frac{16 P_0}{\omega d^2} \sqrt{\frac{1}{2\rho S T_0}} \tag{55}$$

and, introducing this in the third,

$$P_0 = T_0 \sqrt{\frac{T_0}{2\rho S}} \left(1 + \frac{P_0^2}{T_0^2} \frac{8}{\omega^2 d^2} \right) \tag{56}$$

The dimensionless parameters η, σ, τ used in Sec. 3 of this chapter cannot be applied here since η equals zero and σ, τ become infinite. The thrust and power coefficients, however, C_T and C_P, remain useful; besides, an *induced figure of merit* is conveniently introduced as

$$\mu_i = \frac{T_0}{P_0}\sqrt{\frac{T_0}{2\rho S}} = \sqrt{\frac{2}{\pi}}\frac{C_T^{3/2}}{C_P} \tag{57}$$

The dimensionless form of (56) is thus found to be

$$\frac{1}{\mu_i} = 1 + \frac{2}{\pi^2}\frac{C_P^2}{C_T^2} = 1 + \frac{4}{\pi^3}\frac{C_T}{\mu_i^2} \tag{58}$$

which replaces Eq. (35). The second term in (58) is small as compared with the first. Thus μ_i is in first approximation equal to 1, and in second approximation one has

$$\mu_i = 1 - \frac{4C_T}{\pi^3} \tag{58'}$$

In Fig. 258 the induced figure of merit is plotted vs. C_T according to (58). (There exists another branch of this curve, which is omitted since it corresponds to values of ω' greater than ω.)

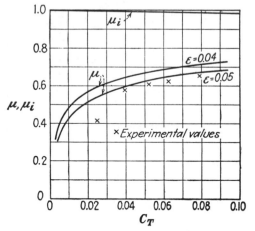

Fig. 258.—Figure of merit, experimental and theoretical values.

The figure of merit for small C_T-values changes entirely if friction losses are taken into account according to the modified momentum theory discussed in Sec. 4 of this chapter. If V is set zero in Eqs. (38a) and a definite figure of merit introduced by

$$\mu = \frac{T'}{P'}\sqrt{\frac{T'}{2\rho S}} \tag{59}$$

one finds immediately that

$$\mu = \mu_i \frac{(1 - \epsilon\omega'd/3v_1)^{3/2}}{1 + 8\epsilon v_1/3\omega'd} \tag{60}$$

On the other hand, from (55) and (57), the ratio $\omega'd/2v_1$ can be computed,

$$\frac{\omega'd}{2v_1} = \frac{1}{\mu_i}\sqrt{\frac{8C_T}{\pi^3}} = \frac{t}{\mu_i} \tag{61}$$

and this gives for μ, from (60),

$$\mu = \mu_i \frac{(1 - \frac{2}{3}\epsilon t/\mu_i)^{3/2}}{1 + \frac{4}{3}\epsilon\mu_i/t} \sim \frac{1 - \epsilon t/\mu_i}{1 + \frac{4}{3}\epsilon\mu_i/t} \tag{60a}$$

The curved lines in Fig. 258 show μ vs. C_T for $\epsilon = 0.04$ and $\epsilon = 0.05$. The five points marked by crosses correspond to experimental results found by H. Glauert.[1] A complete computation of a helicopter airscrew, as required for design purposes, would have to follow the lines of the method given in Sec. 5 of this chapter.

Problem 18. A propeller of 12 ft. diameter supplies a thrust of 2400 lb. when rotating at $n = 20$/sec. without forward speed. Compute the required power (*a*) neglecting all friction losses; (*b*) assuming $\epsilon = 0.05$. Give also the figure of merit in both cases.

Problem 19. What thrust would be supplied by the propeller of the preceding problem if at the same speed the power dropped to 70 per cent of its first value?

*****Problem 20.** Give an estimate of how the thrust coefficient C_T of a propeller would change because of the two-dimensional blade interference as given in Eqs. (53) and (54). Assume a three-blade propeller with a representative blade element of $\beta' = 22°$, $\alpha' = 5°$, and $c/r = \frac{1}{5}$.

---

[1] *Rept. and Mem.* 1132, Aeronautical Research Committee, London, 1928.

CHAPTER XIII

THE AIRPLANE ENGINE

1. The Engine at Sea Level. The airplane engine converts the thermal energy produced by burning of fuel into mechanical work. This work is done by rotating the propeller against the resisting aerodynamic torque moment. The propeller is driven either directly on the engine shaft or indirectly by means of a reduction gear. At present the only engine type employed in aircraft is the internal-combustion reciprocating engine.[1] Gas turbines, which seem promising in many respects, are not yet beyond the experimental stage. Most widely used is the four-stroke-cycle gasoline engine with carburetor and spark ignition. Diesel engines burning heavier oil, without carburetor and without spark ignition, operating at a four- or two-stroke cycle, have also been used successfully.

Gasoline is the most volatile part of the petroleum distillate, improved by certain chemical processes. The usual gasoline is a mixture of several components, hexane (C_6H_{14}) being the prevailing one. Hexane is the lightest part of the petroleum distillate; its specific gravity with respect to water is 0.685. The various commercial brands of gasoline have slightly higher specific weights, ranging up to 0.70. In the process of the distillation of petroleum, hexane is followed by kerosene; the heavier components that come next include Diesel oil of a specific weight of 0.84 to 0.88.

Heat energy is measured in British thermal units (B.t.u.), 1 B.t.u. being defined as the heat energy necessary to raise the temperature of 1 lb. of water by 1°F. One B.t.u. is equivalent to 778 ft.-lb. mechanical energy. The complete combustion of 1 lb. petroleum supplies 19,000 to 20,000 B.t.u., which corresponds to 14,800,000 to 15,560,000 ft.-lb. The exact values for hexane are 19,250 B.t.u. or 14,980,000 ft.-lb. The combustion energy H of 1 lb. of a fuel has the dimension energy per pound and can be measured either in B.t.u. per pound or in foot pounds per pound, that is, in feet. For gasoline, $H = 19,300$ B.t.u./lb. or 15,000,000 ft. may be taken as an average value.

For the complete combustion of a given fuel a definite amount of oxygen is required. The theoretical value for 1 lb. hexane is 3.53 lb. oxygen. Since, under standard conditions, 1 lb. air contains 0.232 lb. oxygen

[1] Jet propulsion, which makes the use of a propeller unnecessary, is not discussed here. Likewise, the fact that the engine power is often used partly for other than propulsion purposes is disregarded.

(Sec. I.5), the amount of air required for the complete combustion of 1 lb. hexane would be $3.53/0.232 = 15.2$ lb. air. Experience shows that a somewhat higher ratio x of weight of air to weight of gasoline gives better results. For modern engines, $x = 18$ may be taken as an average value.

The quantities H and x determine the *size of an engine* designed to supply a given power. Let n be the number of revolutions and n' the

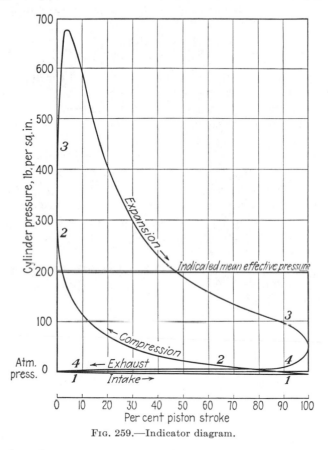

FIG. 259.—Indicator diagram.

number of cycles per second (for the two-stroke-cycle, $n' = n$; for the four-stroke-cycle, $n' = \frac{1}{2}n$). In each cycle a certain volume v, the *displacement of the pistons*, is filled by a mixture of fuel and air. If γ denotes the specific weight of the outside air, $\gamma v/x$ is approximately the weight of the fuel entering the cylinders in each cycle. However, for the following reasons a correction factor must be introduced: The resistances in the intake valves, the carburetor, etc., cause a difference of pressure between the air outside and inside the cylinders. Moreover, the temperature in the cylinders differs from that of the outside air. Both these facts make the effective value of the specific weight for the air that fills the cylinders a little smaller than γ. Finally, the air entering the cylinders necessarily contains fluid particles (about 2 per cent by volume). All these circumstances are taken into account by the introduction of the *volumetric efficiency* η_v, so that the weight of fuel in each cycle has to be written $\eta_v \gamma v / x$. For modern engines, η_v can be taken as about 80 per cent. The thermal energy produced per second thus equals

$$P_t = \eta_v \frac{H\gamma}{x} n'v \tag{1}$$

With the average values of H, x, and η_v given and the standard value of $\gamma = 0.0765$ lb./ft.^{3}, this energy equals $65.6n'v$ B.t.u./sec. $= 51,030n'v$ ft.-lb./sec., where v must be expressed in cubic feet.

The work done by the piston is shown in the so-called "indicator diagram," in which the pressure in the cylinder is plotted against the position of the piston. Figure 259 shows the indicator diagram of a normal four-stroke gasoline engine. The four branches (1 to 4) of the curve correspond to the four strokes—intake, compression, expansion (firing), exhaust. The ratio of the gas volumes at the beginning and the end of the compression stroke is called the *compression ratio*. The area enclosed by the two loops of the curve (difference of the upper and lower loops) gives the work done in one cycle per unit of piston area. If the length of the stroke is denoted by s, the inner diameter of the cylinder by d, and the number of cylinders by m, the displacement v introduced above equals

$$v = ms \frac{\pi d^2}{4} \tag{2}$$

If p_i is the so-called "indicated mean effective pressure," *i.e.*, if the area of the indicator diagram equals $p_i s$, the work per cylinder and cycle is $\pi d^2 p_i s / 4$. The total work per second equals therefore

$$P_i = mn' \frac{\pi}{4} d^2 p_i s = p_i n'v \tag{3}$$

This is called the *indicated power* since it is computed from the information contained in the indicator diagram.

The number of cylinders used for one engine is determined by the limitations imposed on both stroke and diameter for various reasons. The stroke must be kept within certain limits since, for a given engine speed, the accelerations and inertia forces of the reciprocating parts increase proportionally to the stroke. The diameter must be not too large in order to ensure that the heat produced at the piston head is carried off to the walls, which are under the influence of the cooling system.   A power of about 100 hp. is at present considered as the practical limit for one cylinder.

The power P_{br} delivered at the crankshaft (brake power) is smaller than P_i because of various friction losses at the piston, connecting rod, crankpin, etc.   The ratio

$$\eta_m = \frac{P_{br}}{P_i} \tag{4}$$

is called the *mechanical efficiency*.   The value $\eta_m = 0.85$ may be taken as an average.

The ratio of the brake power P_{br} to the total thermal energy produced in fuel burning is known as the *thermal efficiency*.   From Eqs. (1) to (4) the thermal efficiency P_{br}/P_t is found as

$$\eta_t = \frac{P_{br}}{P_t} = \eta_m \frac{P_i}{P_t} = \eta_m \frac{x}{\eta_v H \gamma} p_i \tag{5}$$

The highest value of mean indicated pressure p_i reached so far in gasoline engines of the usual type is about 200 lb./in.^{2} = 28,800 lb./ft.^{2} With the average values given above, Eq. (5) then furnishes a maximum thermal efficiency η_t of about 45 per cent.   With an average value of $p_i = 165$ lb./in.^{2} the thermal efficiency is reduced to 37 per cent.   The practical range of η_t, including carburetor losses, etc., extends from 25 to 35 per cent, in exceptional cases even more.

An important formula for computing the power from the dimensions of the engine is obtained by combining (3) and (4).   Introducing the *brake mean effective pressure* $p_e = \eta_m p_i$, we write

$$P_{br} = \eta_m P_i = \eta_m p_i n' v = p_e n' v \tag{6}$$

It is customary to express p_e in pounds per square inch, v in cubic inches, and P in horsepower.   Since 1 hp. = 550 ft.-lb./sec., Eq. (6) then takes the form

$$P_{br}^{HP} = \frac{p_e n' v}{6600} \quad (p_e \text{ in lb./in.}^2 \text{ and } v \text{ in in.}^3) \tag{6'}$$

If 2400 r.p.m. is taken as an average value of the engine speed, we have $n' = 20$ for the four-stroke cylinder.   With $p_e = 165$ lb./in.^{2} (a rather high value of the effective mean pressure), (6') gives $P_{br}^{HP} = v/2$ (v in cubic inches).   That is, the brake horsepower is numerically equal to one-half the displacement, expressed in cubic inches.   This simple formula gives a rough idea of the power obtainable from an engine of a given size.

The precise value of p_i depends on the details of the thermal process in the cylinder, the most important factor being the compression ratio.   By increasing this ratio from 5 to 7.5 the mean indicated pressure may be raised by as much as 35 per cent. The detonation tendency of the fuel sets a limit to the application of high compression ratios. The resistance against detonation is measured by the so-called "octane rating." An octane rating of 100 per cent means that, with respect to knocking, the fuel behaves exactly as a certain standard liquid, the iso-octane. If the knocking properties of a fuel are that of a mixture of p per cent iso-octane with $(100 - p)$ per cent heptane, the fuel has the octane rating p. A marked progress in aircraft engine performance at sea level can at present be expected only from an improvement along this line, *i.e.*, with respect to the antiknock properties of the fuel.

The power output of an engine depends on two operation variables, the speed n and the throttle setting, which controls the manifold pressure.

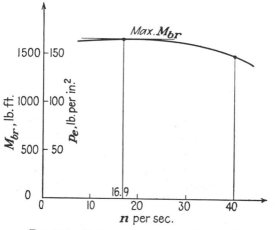

Fig. 260.—Brake moment vs. engine speed.

Recent engines sometimes also have controllable air-fuel mixture. We may leave this out of consideration and assume that no changes in mixture occur other than unavoidable ones, caused by the change of speed and the intake manifold pressure.

At first glance it may seem that the speed n (or the number n' of cycles per second) should not influence p_i. At least, the carburetor is designed with the aim of supplying the cylinder with a gas mixture of constant air to fuel ratio and constant pressure independent of the number of cycles per second and thus to secure an invariable indicator diagram for all speeds. However, a certain variation cannot be avoided, since valves and channels are designed for certain speeds and will work less efficiently under different conditions. Moreover, the mechanical efficiency will vary with the speed. The mean effective pressure p_e as a function of n thus will possess a maximum for a certain design speed and will decrease slightly for smaller as well as for higher speeds (Fig. 260).

According to (6) the torque moment M_{br} transmitted to the crankshaft equals

$$M_{br} = \frac{P_{br}}{\omega} = \frac{P_{br}}{2\pi n} = \frac{1}{2\pi} \frac{n'}{n} v p_e \tag{7}$$

where n'/n equals $\frac{1}{2}$ for the four-stroke cycle and 1 for the two-stroke cycle. If P_{br} is plotted against n (Fig. 261), the slope of the tangent from the origin to the curve P_{br} vs. n indicates the maximum torque, and the abscissa of the point of contact gives the corresponding speed. The design speed, or *rated speed* (for the throttle setting under consideration), usually exceeds the speed corresponding to M_{max}. If P_0, n_0 denote the rated values of P_{br} and n and if P_1 corresponds to $n_1 = 0.8n_0$, the ratio

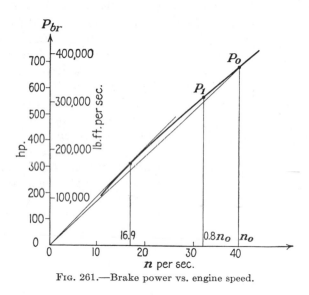

Fig. 261.—Brake power vs. engine speed.

P_1/P_0 is often called the *power-drop factor*. This factor usually ranges from 0.75 to 0.85. If the straight line joining the points P_0, n_0 and P_1, n_1 is considered to coincide practically with the curve P_{br} vs. n at the point P_0, n_0, the maximum torque will be reached for a speed smaller than n_0, equal to n_0, or greater than n_0, according to whether the power-drop factor is greater than, equal to, or smaller than 0.80, respectively. In most cases, within the comparatively narrow range of speeds used in aircraft operation the curve P_{br} vs. n can be considered rectilinear.

A complete picture of the operation conditions needed in connection with performance computations is given by the family of the curves P_{br} vs. n for all manifold pressures obtainable by throttling. Figure 262 shows these curves for a 600-hp. engine of modern design. In the charts supplied by engine manufacturers another type of representation is frequently preferred. For a constant value of the speed in r.p.m. the power P_{br} is plotted against the manifold pressure. Such curves are shown in Fig. 263; they refer to the same engine as the curves of Fig. 262.

A quantity that is of importance for some performance computations is the *fuel consumption per unit of power and time* (work). This quantity c has the dimension of weight/work or 1/length; it depends on all variables determining the mean effective pressure, above all on the compression ratio and on the ratio x that characterizes the mixture of air and fuel. Experience shows that from the point of view of fuel consumption the optimum operating conditions correspond to a comparatively leaner mixture than that which would be best from the point of view of power output. In the present context it will suffice to state that the specific fuel consumption expressed in pounds per hp-hour varies between 0.45 and 0.55 for modern aircraft engines of the carburetor and spark-ignition type. Under exceptionally good conditions even smaller values have been reached with modern high-powered engines using high-octane fuel. To this must be added a consumption of lubrication oil of about 0.02 to 0.05 lb./hp.-hr.

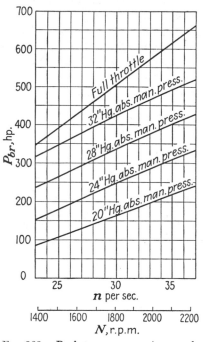

Fig. 262.—Brake power vs. engine speed for various manifold pressures.

For the purpose of more accurate computation of range and endurance the variability of the ratio c (that is, the fuel consumption per unit of power and time) with engine speed and engine output must be considered (see Sec. XVI.1).

Finally, the *engine weight per unit of power output* is of interest in aircraft design. This figure has dropped enormously since the early days of aviation. The ratio of engine weight to rated brake horsepower was about 6 to 8 lb./hp. in 1910 and about 2 in 1920 and now is actually, for modern air-cooled engines, below 1.2 lb./hp. Still more favorable values can be expected with the recently inaugurated use of fuels of higher octane rating.

Problem 1.   Compute the thermal efficiency of an airplane engine of 12 cylinders, 6 in. diameter, and 8 in. stroke, which supplies 700 hp. at 1250 r.p.m.   The combustion energy of the fuel is $H = 19,000$ B.t.u.; the air to fuel ratio 18:1.

Problem 2.   An engine has to be designed that supplies 1000 hp. at 2400 r.p.m. The ratio bore to stroke is limited to 4:5.   Give suitable number and dimensions of the cylinders, if the mean effective pressure is known to be 145 lb./in.²

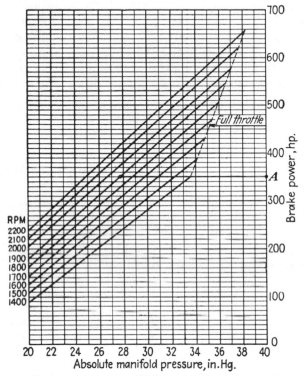

Fig. 263.—Brake power vs. manifold pressure for various engine speeds.

Problem 3.   Prove that the theoretical value of fuel consumption per horsepower-hour is

$$c = 13,750 \frac{\gamma}{x \mu_e} \eta_v$$

where γ is the specific weight of the air in pounds per cubic foot, p_e the mean effective pressure in pounds per square inch, x the air to fuel ratio, and η_v the volumetric efficiency.

Problem 4.   If the power P is considered a linear function of ω in the range from $n_1 = 0.8n_0$ to $n_2 = 1.25n_0$ and the power-drop factor is f, prove that

$$\frac{M_1}{M_0} = \frac{5f}{4}, \qquad \frac{M_2}{M_0} = \mathbf{2} - f$$

2. The Engine at Altitude. The power output of an internal-combustion engine depends on the atmospheric conditions under which the engine operates. From thermodynamic considerations the mean indicated pressure must be expected to vary in proportion to the quotient of pressure p and the temperature T of the mixture at the beginning of the compression stroke. If these quantities are assumed to have essentially the same values as in the surrounding air, the mean indicated pressure p_i, and hence for a given engine speed n the indicated power P_i, should be proportional to the quotient p/T of the surrounding air. By the equation of state this quotient, in turn, is proportional to the density of the surrounding air. If the ratio of P_i at altitude h to P_i at sea level at the same speed is called the *indicated power-altitude factor* and denoted by $\psi(h)$, it would then be

$$\psi(h) = \frac{P_i(h)}{P_i(0)} = \frac{p_i(h)}{p_i(0)} = \frac{\rho}{\rho_0} = \sigma \tag{8}$$

The value of the density ratio σ for any given altitude in the standard atmosphere can be found from Table I (page 10).

The experimental evidence suggests that ψ varies in proportion to σ^n, where n is slightly higher than 1. H. L. Stevens found P_i to be proportional to $p^{1.05}$. Since, in the standard atmosphere, the pressure p is proportional to $\sigma^{1/(1-\lambda R)} = \sigma^{1.235}$ [see Eq. (11), Sec. I.2], the indicated power-altitude factor is then seen to equal

$$\psi(h) = (\sigma^{1.235})^{1.05} = \sigma^{1.29} \tag{9}$$

Some American tests gave $\psi(h)$ proportional to pressure and inversely proportional to the square root of T. This is the same as proportional to the square root of p/ρ, according to the equation of state. Some investigators found ψ to be proportional to $p^{1-a}\rho^a$, with a ranging from $\frac{1}{3}$ to $\frac{1}{2}$. This gives for standard atmosphere with the above given relation between p and σ

$$\psi(h) = (\sigma^{1.235})^{1-a}\sigma^a = \sigma^{1.235-0.235a} = \begin{cases} \sigma^{1.117} & \text{for } a = \frac{1}{2} \\ \sigma^{1.157} & \text{for } a = \frac{1}{3} \end{cases} \tag{9'}$$

Figure 264 shows the curves representing the various expressions for $\psi(h)$. In view of the complexity of the phenomenon and the wide range of varying engine design it is not surprising that there is some discrepancy between the results found by the various investigators.

For performance computations the brake power P_{br} is more important than the indicated power P_i. If the mechanical losses are assumed to be independent of the atmospheric conditions, the formula of the *brake power-altitude factor* $\varphi(h)$, that is, the ratio of brake power at altitude h

to brake power at sea level at the same engine speed, can be seen to have
the form (see Prob. 6)

$$\varphi(h) = \frac{p_e(h)}{p_e(0)} = \frac{M_{br}(h)}{M_{br}(0)} = \frac{P_{br}(h)}{P_{br}(0)} = \frac{\psi(h) - C}{1 - C} \tag{10}$$

In order to agree with the experiments on certain types of ordinary
engine the C in this formula should be taken as follows:

$C = 0.15$ if ψ is computed according to $\psi = \sigma$

$C = 0.05$ to 0.08 if ψ is computed according to $\psi = \sigma^{1.117}$

$C = 0.02$ to 0.06 if ψ is computed according to $\psi = \sigma^{1.157}$

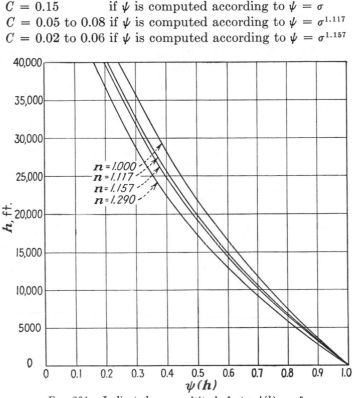

FIG. 264.—Indicated power-altitude factor $\psi(h) = \sigma^n$.

The formula

$$\varphi(h) = \frac{P_{br}(h)}{P_{br}(0)} = \frac{\sigma - 0.15}{1 - 0.15} = 1.176\sigma - 0.176 \tag{11}$$

may be considered as an acceptable approximation under average conditions. Figure 265 shows how the factor φ depends on the altitude h
in the standard atmosphere. The factor is seen to equal $\frac{1}{2}$ for an altitude
of about 20,000 ft. and 0.11 at 40,000 ft. Such a reduction of the power
output with altitude would make high-altitude flying practically impossible. In order to keep the power output at a reasonable level some

provisions have to be made for adapting the engine to the atmospheric conditions at high altitudes.   A number of different patterns have been developed during the last thirty years.   Today the most widely employed method is known as the *supercharging* of aircraft engines.

Supercharging is also a means of improving the power output of an engine at sea level.   The idea is to increase the charge weight per cycle by compressing the mixture before it reaches the cylinder.   In a normal engine the mixture is inducted into the cylinder by the suction of the piston, and the pressure within the cylinder during the suction stroke

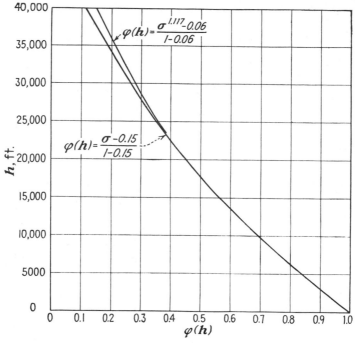

$$\varphi(h) = \frac{\sigma^{1.117} - 0.06}{1 - 0.06}$$

$$\varphi(h) = \frac{\sigma - 0.15}{1 - 0.15}$$

FIG. 265.—Brake power-altitude factor.

is a little below the atmospheric pressure.   Therefore, the weight of the charge in one cycle is a little smaller than the amount of air-fuel mixture that would fill the cylinders at atmospheric pressure, the small pressure difference being the main reason for the introduction of the volumetric efficiency in (1).   It is clear that much more favorable conditions will result if the mixture is forced into the cylinder by overpressure.   In this way, by applying small overpressure only, the volumetric efficiency can be brought up to 100 per cent.   With higher values of the induction pressure, produced in an appropriate type of compressor, even considerably higher charges can be inducted into the cylinder at each cycle. When used on the ground this method of increasing the power per cycle often proves superior to the use of a higher compression rate, since the value of the pressure at the end of the compression stroke can be kept at a lower level.

When used in altitude flying, supercharging makes the fuel intake practically independent of the atmospheric conditions, as long as the compressor delivers the air at a constant pressure to the cylinder and the power consumption by the compressor does not rise too high. Operated at sea level a compressor of the usual supercharger size requires

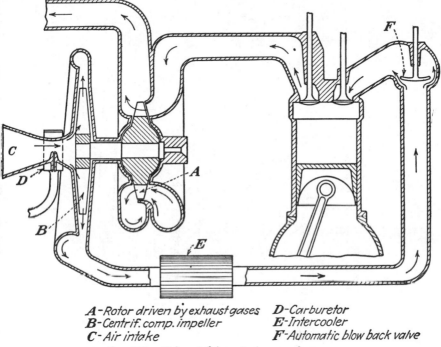

A-*Rotor driven by exhaust gases*   *D*-*Carburetor*
B-*Centrif. comp. impeller*   *E*-*Intercooler*
C-*Air intake*   *F*-*Automatic blow back valve*
Fig. 266.—Exhaust-driven turbosupercharger.

only 6 to 10 per cent of the total power output of the engine. An ideal solution of the supercharging problem is supplied by a turbocompressor driven by the energy of the exhaust gases of the engine. The higher the altitude, the greater the difference in pressure between the exhaust gases and the outside and the greater, consequently, the compressor effect. The scheme of a power plant with exhaust turbosupercharger, as used for the first time about twenty-five years ago and now developed to high perfection, is shown in Fig. 266. Using this arrangement one can keep the power output of the engine nearly constant up to a considerable altitude.

A useful device, often employed in modern aircraft to boost the intake pressure, takes advantage of the so-called "ram effect." It consists in exposing the intake opening to the pressure of the relative wind that is produced by the velocity of flight. With appropriate design the ram effect may appreciably increase the quantity of air going into the supercharger.

It is obvious that no compressor can deliver the air at a constant pressure completely independent of its intake pressure. The best superchargers of aircraft engines supply a constant or nearly constant

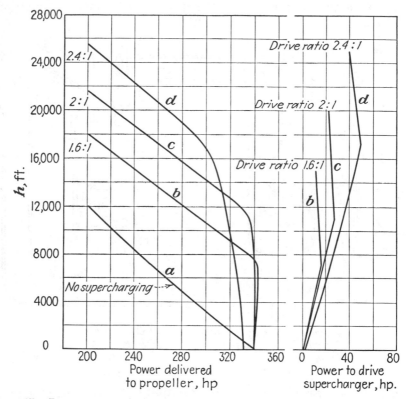

Fig. 267.—Power output of supercharged engine; power consumption of supercharger.

power output up to a definite altitude, which is called the *critical altitude* of the engine. From there on, the power output decreases with increasing altitude in very much the same way as it decreases for a normal engine from $h = 0$ on. Figure 267 shows the curves altitude vs. brake power for an engine operating at a constant speed, (*a*) without supercharger; (*b*), (*c*), (*d*) supercharged with increasing drive ratio, corresponding to critical altitudes of 7000, 11,500, and 17,000 ft. Note that the upper parts of these curves are practically parallel: the decrease in power output beyond the critical altitude occurs at about the same rate as for the nonsupercharged engine. If σ_1 denotes the density ratio at the critical altitude, the power output at higher altitudes ($\sigma < \sigma_1$) can be given approximately by a formula

$$\varphi(h) = 1 \text{ for } \sigma \geqq \sigma_1, \qquad \varphi(h) = \frac{P_{br}(h)}{P_{br}(0)} = \frac{\sigma - c}{\sigma_1 - c} \text{ for } \sigma \leqq \sigma_1 \qquad (12)$$

which is represented in Fig. 268 for $c = 0.15$ and σ_1 corresponding to $h = 10,000, 15,000, 20,000,$ and $25,000$ ft. The curves on the right-hand side of Fig. 267 give the power required for driving the compressor at

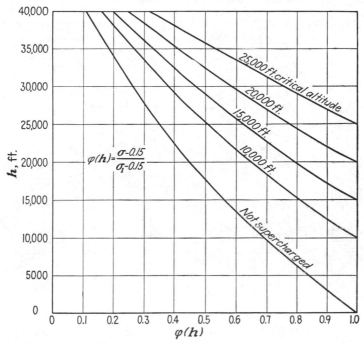

Fig. 268.—Brake power-altitude factor for an engine supercharged to various critical altitudes.

the various altitudes. The power output of the engine decreases slightly even before the critical altitude is reached. This is due to a series of minor factors, like the imperfect balance between the power available in the exhaust and the power required by the compressor.

Figure 269 shows the power-altitude chart of the Pratt and Whitney Wasp R-1340 (without supercharger). This chart supplements the sea-level charts of Fig. 263. In Fig. 269 the curves falling from left to right correspond to various engine speeds and to fully opened throttle. For a given speed, the corresponding curve represents brake power P_{br} (ordinate) as a function of the altitude h in the standard atmosphere (abscissa). The curves falling from right to left correspond to various values of the manifold pressure (expressed in inches mercury). The chart thus gives the brake power for full throttle at a given altitude and speed and indicates at the same time the corresponding manifold pressure (point B). In order to obtain the power corresponding to another throttle setting, the following procedure is recommended: From the sea-level diagram Fig. 263 determine the power corresponding to the given speed and to the manifold pressure as already found. On the ordinate axis of Fig. 269 mark the point A corresponding to this power value. The straight line connecting this point to the point B already found is assumed to repre-

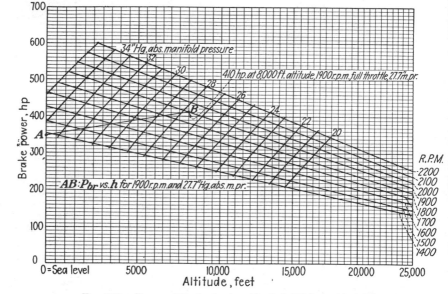

Fig. 269.—Power-altitude chart. (*Pratt & Whitney Aircraft.*)

sent the power P for all intermediate altitudes h at constant speed and constant manifold pressure.[1]

Problem 5. Certain experiments show that the mean indicated pressure p_i in an engine changes proportional to the 1.15 power of the atmospheric pressure if the temperature is constant and proportional to the -0.5 power of the absolute temperature in the atmosphere when tested at constant atmospheric pressure. What follows for the power-altitude factor $\psi(h)$ in standard atmosphere if it is assumed that the effects of p and T are independent?

Problem 6. Prove that if the mechanical losses in an engine are independent of the altitude the relation

$$\varphi(h) = \frac{\psi(h) - C}{1 - C}$$

must hold between the two altitude factors. Explain the significance of C.

[1] Wasp, Jr. B, Wasp Hl and Hornet E Series Engines, "Operator's Handbook" 3d ed., p. 72, Pratt & Whitney Aircraft, East Hartford, Conn.

Problem 7. What critical altitude must be used for a supercharged engine if it is required to supply 70 per cent of its sea-level power at 40,000 ft. altitude?

Problem 8. Plot the curves showing $\sqrt{\sigma}\,P_{br}$ vs. n for a supercharged engine, critical altitude 17,500 ft. Assume that M_{br} vs. n is parabolic with its maximum at $n_0 = 40$ per second and dropping 5 per cent when n increases to 45 per second.

3. Engine Vibrations. In performance computation and in all similar problems the torque moment delivered at the crankshaft of an

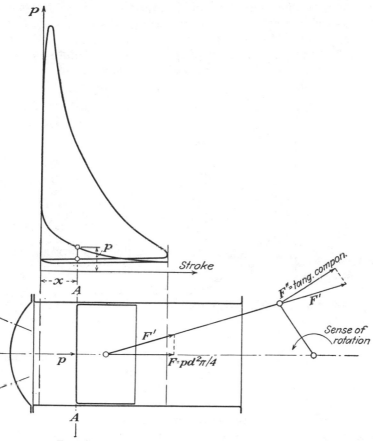

Fig. 270.—Power transfer from piston to crankshaft.

engine under invariable conditions (constant speed, constant air density, etc.) is considered as constant in time. It is actually not exactly constant, however, but is subject to small periodic changes within each revolution of the engine. These changes cause torsional oscillations, which may be of great importance in connection with the elastic properties of the crankshaft and other parts of the plane. Control of engine vibrations is one of the vital problems in the design of aircraft power plants.

The periodic variations of the torque moment M are essentially due to two facts. (1) According to the indicator diagram (see Fig. 259), the thrust exerted on the piston inside the cylinder changes considerably during each cycle. (2) The transfer of power from the piston to the crankpin is different in each position of the piston. In Fig. 270 the skeleton of the transfer mechanism for a single cylinder is shown. In the position AA, at the distance x from the left end of the stroke, with the crank rotating counterclockwise, the gas pressure on the working side of the piston corresponds to the ordinate p, at the abscissa x, of either the compression or the exhaust branch of the diagram. It is seen from the figure that in this position of the piston the torque moment acting on the crankshaft is negative (clockwise), while its mean value, the average over a full cycle, must necessarily be positive. When the piston

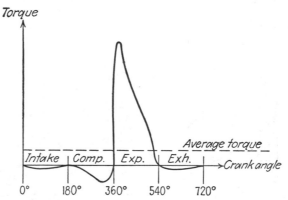

FIG. 271.—Torque moment vs. crank angle for a single cylinder of a four-stroke engine.

reaches one of its extreme positions, at the beginning or end of each stroke, the torque is zero.

During one full cycle the torque hits the zero mark four times, goes through negative values during the intake, the compression, and the exhaust strokes, and is positive only in the expansion, or firing, stroke (Fig. 271). The relation between the instantaneous values of gas pressure p or piston thrust $pd^2\pi/4$ and torque moment M is determined by the elementary rules of statics and need not be explained here. Suffice it to say that, using a well-known procedure devised by J. v. Radinger (1892), one can also take into account the inertia forces connected with the reciprocating masses and thus find the precise magnitude of the torque M at any position ϑ of the crank arm, at least in the case of uniform rotation.

Such large variations of torque as those seen in Fig. 271 would be most detrimental to the purpose that an aircraft engine has to serve. In fact, no internal-combustion engine consists of one cylinder only.

If several pistons work on the same crankshaft, care is taken to space the firing strokes and thus shift the cycles in such a way that a much more uniform torque diagram results. For example, in the case of six cylinders the cycles will start at distances of 120° crank angle, and the superposition of six curves equal to that of Fig. 271, each shifted by 120°, gives the picture shown in Fig. 272. Here the torque oscillates between two positive values only, and the period of the oscillations is reduced to one-third of one revolution. In the case of m cylinders the frequency of oscillations is $m/2$ times the frequency of revolutions. With in-line engines the firing order is achieved by appropriately spacing the directions of the crank arms. With radial engines where all cylinders of one row operate on the same crankpin, an odd number of cylinders is used in each row, and then the angle between adjacent cylinders is made equal to half the firing interval.

The effect of the torque on the motion of the engine can be judged by the following: Let M be the periodically changing moment given as a function of the crank angle ϑ, and M_0 its mean value over a period,

$$M_0 = \frac{m}{4\pi} \int_0^{4\pi/m} M(\vartheta)\, d\vartheta \tag{13}$$

The force opposing the engine moment is the propeller torque Q, which may show a slight periodic change due to the interference with the airplane body but can practically be considered as a constant. Its value then must equal M_0, and $M - M_0$ is the resulting moment producing angular acceleration of the rotating mass. If J denotes the moment of gyration of the crankshaft and all masses connected with it, the equation of motion reads

$$J \frac{d^2\vartheta}{dt^2} = M - M_0 \tag{14}$$

If we multiply the right-hand side by $d\vartheta$ and the left-hand side by $(d\vartheta/dt)\, dt$, we have to the left the derivative of $\frac{1}{2}(d\vartheta/dt)^2$, and integration supplies

$$\int_{\vartheta_1}^{\vartheta_2} (M - M_0)\, d\vartheta = \frac{1}{2}J \left\{ \left(\frac{d\vartheta}{dt}\right)_2^2 - \left(\frac{d\vartheta}{dt}\right)_1^2 \right\} = \frac{1}{2}J(\dot\vartheta_2^2 - \dot\vartheta_1^2) \tag{15}$$

In Fig. 273 the curve $M - M_0$ vs. ϑ corresponding to the example given in Fig. 272 is supplemented by the integral curve $\int(M - M_0)\, d\vartheta$. Equation (15) expresses the fact that the difference between the greatest and the smallest ordinates of this integral curve, multiplied by $2/J$, equals the maximum difference of the (squared) angular velocities which

occur during each period.   To keep this velocity oscillation on a low
level there exists one classical method—to increase J by using a flywheel.
As this is unpractical in the case of an airplane engine because of the
considerable weight of an efficient flywheel, other methods have been
sought.   One of the most successful devices currently used in modern

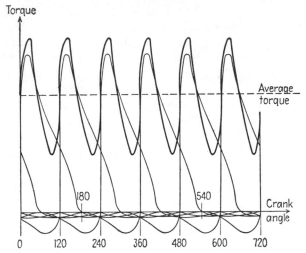

Fig. 272.—Resultant torque vs. crank angle for a six-cylinder four-stroke engine.

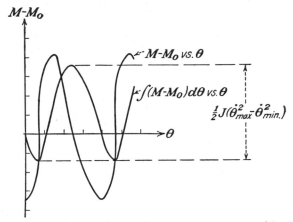

Fig. 273.—Torque variation relative to average torque, and integral.

aircraft is the so-called "tuned pendulum," first presented by E. S. Taylor
in 1936.

The tuned pendulum in its simplest form consists of a comparatively
small mass m_1 (B in Fig. 274), which is allowed to move along a circular
path B_1B_2 in a slit incised into a disk rigidly connected with the rotating
crankshaft.   The center A of the circular arc B_1B_2 has a distance r from

the center O of the rotation.   The radius of the arc may be called l.
The scheme of this motion is shown in Fig. 275, where $\overline{OA} = r$, $\overline{AB} = l$,
and ϑ is the angle OA with a fixed direction and φ the angle between OA
and AB.   The reaction of m_1 (which
also may be considered a mathematical pendulum suspended in A)
on the disk consists of a force S
in the direction of AB, if friction in
the slit is neglected.   The moment
of this force with respect to the
center O is $rS \sin \varphi$, so that Eq. (14)
has to be replaced by

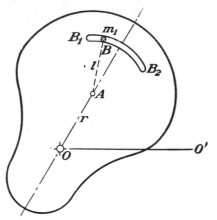

$$J \frac{d^2\vartheta}{dt^2} = M - M_0 + rS \sin \varphi \qquad (16)$$

To find S we have to apply to the
mass m_1 Newton's equation, stating
that the product of mass times the
acceleration vector equals the sum of

Fɪɢ. 274.—Pendulum damper reduced to
its simplest form.

all forces acting on the mass.   The
forces are S in the direction BA and gravity m_1g in the vertical direction.
The acceleration of the end point B of the vector OB, which is the sum
of the vectors OA and AB, can be computed in the following way: First,

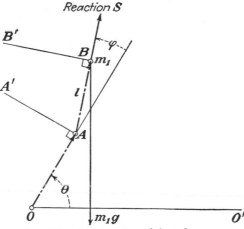

the vector OA of length r
rotates with the angular ve-
locity $d\vartheta/dt = \dot{\vartheta}$.   The acceleration of its end point A
therefore consists of a tan-
gential component $a_1 = r\ddot{\vartheta}$ in
the direction AA' normal to
OA and of a radial compo-
nent $a_2 = r\dot{\vartheta}^2$ in the direction
AO.   In the same way the
vector AB of length l rotating with the angular velocity
$d(\vartheta + \varphi)/dt = \dot{\vartheta} + \dot{\varphi}$ adds
an acceleration $a_3 = l(\ddot{\vartheta} + \ddot{\varphi})$
in the direction BB' normal
to AB and an acceleration

Fɪɢ. 275.—Scheme of pendulum damper.

$a_4 = l(\dot{\vartheta} + \dot{\varphi})^2$ in the direction BA.   The resulting acceleration of the
endpoint B has thus the components
In direction BB':

$$a_1 \cos \varphi + a_2 \sin \varphi + a_3 = r\ddot{\vartheta} \cos \varphi + r\dot{\vartheta}^2 \sin \varphi + l(\ddot{\vartheta} + \ddot{\varphi})$$

In direction BA:

$$-a_1 \sin \varphi + a_2 \cos \varphi + a_4 = -r\ddot{\vartheta} \sin \varphi + r\dot{\vartheta}^2 \cos \varphi + l(\ddot{\vartheta} + \ddot{\varphi})^2$$

If the reference line OO' is considered as horizontal, the angle between the gravity and the direction BB' is $(\vartheta + \varphi)$. Accordingly, the two equations of motion for the mass m_1 are

$$m_1[r\ddot{\vartheta} \cos \varphi + r\dot{\vartheta}^2 \sin \varphi + l(\ddot{\vartheta} + \ddot{\varphi})] = -m_1 g \cos (\vartheta + \varphi)$$
$$m_1[-r\ddot{\vartheta} \sin \varphi + r\dot{\vartheta}^2 \cos \varphi + l(\ddot{\vartheta} + \ddot{\varphi})^2] = S + m_1 g \sin (\vartheta + \varphi) \quad (17)$$

These equations must be used to express φ and S in terms of ϑ; then (16) will supply the equation for ϑ.

The following assumptions may be introduced in order to simplify the mathematical treatment: (1) We neglect the influence of gravity since its amount is small as compared with the inertia forces. (2) We consider φ and its derivatives as small, thus writing φ for $\sin \varphi$ and 1 for $\cos \varphi$. (3) We assume that ϑ differs only slightly from ωt where $\omega = 2\pi n$ is the mean angular speed of the shaft. Thus $\dot{\vartheta}$ is nearly ω and $\dot{\vartheta}^2 \varphi$ has to be replaced by $\omega^2 \varphi$ if terms of higher order are discarded. The first equation (17) then takes the form

$$r\ddot{\vartheta} + r\omega^2 \varphi + l(\ddot{\vartheta} + \ddot{\varphi}) = 0 \quad (18)$$

while the second supplies $m_1(r + l)\omega^2 = S$, all other terms being of higher order. Combined with (16) this gives

$$J\ddot{\vartheta} - m_1 r(r + l)\omega^2 \varphi = M - M_0 \quad (19)$$

Equations (18) and (19) can be integrated if the torque disturbance $M - M_0$ is resolved into harmonic components, i.e., into terms of the form $A_\nu \sin \nu\vartheta \sim A_\nu \sin \nu\omega t$, with appropriate values for ν. The main term will correspond to $\nu = m/2$ according to what was said in the foregoing. Upon introducing one single term of this kind on the right-hand side of (19) the solution can be written as

$$\varphi = \varphi_0 \sin \nu\omega t, \qquad \vartheta - \omega t = \vartheta_0 \sin \nu\omega t \quad (20)$$

where the amplitudes φ_0 and ϑ_0 follow from

$$-\nu^2 \omega^2 r\vartheta_0 + r\omega^2 \varphi_0 - \nu^2 \omega^2 l(\vartheta_0 + \varphi_0) = 0$$
$$-\nu^2 \omega^2 J\vartheta_0 - m_1 r(r + l)\omega^2 \varphi_0 = A_\nu \quad (21)$$

Equations (21) are derived by introducing (20) in (18) and (19) and dividing all terms by $\sin \nu\omega t$. The first equation (21) yields

$$\varphi_0 = \frac{\vartheta_0 \nu^2 (r + l)}{(r - \nu^2 l)},$$

and this combined with the second gives the solution for the amplitude ϑ_0 of the torsional vibration:

$$\vartheta_0 = -\frac{A_\nu}{\nu^2\omega^2}\left[J + \frac{m_1(r + l)^2}{1 - \nu^2 l/r}\right]^{-1} \tag{22}$$

If no pendulum is used, i.e., for $m_1 = 0$, this reduces to $\vartheta_0 = -A_\nu/\nu^2\omega^2 J$, which, of course, could also be derived immediately from (16) or from (19) with $m_1 = S = 0$. The influence of the mass m_1 is thus seen to consist of an *apparent increase of the moment of gyration J*. The increase is proportional to m_1 and inversely proportional to $1 - \nu^2 l/r$, and becomes infinite if the ratio r/l is made to equal ν^2. In this case, $\vartheta_0 = 0$; that is, no torsional oscillation results from a torque disturbance proportional to $\sin \nu\omega t$. Thus we see that *the pendulum can be "tuned" in such a way as to make completely inefficient the torque disturbance of one certain frequency, while the effect of all disturbance components with lower frequencies is weakened by an apparent increase of J, which is proportional to the mass m_1 of the pendulum.*

The actual construction of the tuned pendulum must be somewhat modified in order to achieve a possibly frictionless relative motion of the mass m_1 in a pathway of very small radius of curvature, etc. But the mechanical conditions remain essentially the same as those developed in the preceding argument.

There exists another source of engine vibrations, also. If the effective length of the crankshaft, i.e., in the case of a radial engine the distance between the propeller (or the reduction gear) and the middle of the crankpin, is considerable, the *elasticity of the shaft* must be taken into account. The angle of rotation for the propeller, δ, is then different from the ϑ for the crankshaft and the torque moment conferred on the propeller proportional to $\vartheta - \delta$. Let K denote the torque that would produce a twist of 1 rad. and J_p the moment of gyration of the propeller. The equation of motion for the propeller is then

$$J_p\ddot{\delta} = -K(\delta - \vartheta) - M_0 \tag{23}$$

while Eq. (19) for the crank becomes

$$J_c\ddot{\vartheta} = K(\delta - \vartheta) + M + m_1 r(r + l)\omega^2\varphi \tag{24}$$

Here J_c is the moment of gyration of the crank jaws, counterweights, etc. The mass of the shaft itself is disregarded. As a third equation, for the damper, we have (18) unchanged. For a single harmonic term $A_\nu \sin \nu\omega t$ in the expression for the torque deviation $M - M_0$ the solution of (18), (23), and (24) has the form

$$\delta - \omega t = \delta_0 \sin \nu\omega t - \frac{M_0}{K}, \quad \vartheta - \omega t = \vartheta_0 \sin \nu\omega t, \quad \varphi = \varphi_0 \sin \nu\omega t \tag{25}$$

The amplitudes δ_0, ϑ_0, φ_0 of the forced vibrations are determined by three linear equations, which we obtain by introducing (25) in the three differential equations and dividing by sin $\nu\omega t$.

$$(K - \nu^2\omega^2 J_p)\delta_0 - K\vartheta_0 = 0$$
$$-K\delta_0 + (K - \nu^2\omega^2 J_c)\vartheta_0 - m_1 r(r + l)\omega^2\varphi_0 = A_\nu \qquad (26)$$
$$-(r + l)\nu^2\omega^2\vartheta_0 + \omega^2(r - \nu^2 l)\varphi_0 = 0$$

We multiply the first equation by $K/(K - \nu^2\omega^2 J_p)$ and the last by $m_1(r + l)/(1 - \nu^2 l/r)$ and sum up. This gives

$$\left[-\frac{K^2}{K - \nu^2\omega^2 J_p} + (K - \nu^2\omega^2 J_c) - \frac{m_1(r + l)^2}{1 - \nu^2 l/r} \nu^2\omega^2 \right] \vartheta_0 = A_\nu$$

and after rearranging, with $J = J_c + J_p$,

$$\vartheta_0 = -\frac{A_\nu}{\nu^2\omega^2} \left[J + \frac{m_1(r + l)^2}{1 - \nu^2 l/r} + J_p \frac{1}{K/\nu^2\omega^2 J_p - 1} \right]^{-1} \qquad (27)$$

Here the first two terms in the bracket are the same as in (22). The last term, which vanishes for a rigid shaft ($K = \infty$), is in general negative, except for very small engine speed ω. The main result is that *a pendulum tuned for a certain frequency ν, that is, with $r/l = \nu^2$, makes the corresponding shaft vibration inefficient, independently of the torsional elasticity* (since the second term in the bracket becomes infinite).

For a brief numerical discussion of torsional vibrations we choose a nine-cylinder single row radial engine of about 1000 rated horsepower at 1800 r.p.m. Adequate values for the moments of gyration would be $J_p = 175$ (metal propeller) and $J_c = 15$ lb.-in./sec.^{2}, while K may be estimated as 5,000,000 lb.-in. The principal frequency of the torque fluctuation is given by $\nu = \frac{9}{2}$. Then, if no damper is used ($m_1 = 0$), Eq. (27) can be rearranged in the form

$$\vartheta_0 = -\frac{A_\nu}{4\pi^2\nu^2 n^2 J_c} \frac{n^2 - K/4\pi^2\nu^2 J_p}{n^2 - KJ/4\pi^2\nu^2 J_p J_c} \qquad (28)$$

In our case this gives

$$\frac{\vartheta_0}{A_\nu} = -0.0000836 \frac{n^2 - 35.7}{n^2(n^2 - 452)}$$

These values are shown in Fig. 276 against engine speed n up to $n = 40$. It is seen that at $n = \sqrt{452} = 21.3$ sec.$^{-1}$ the ordinate becomes infinite. This is the *resonant point* where the natural frequency of torsional vibrations coincides with νn. The average torque moment at $n = 30$, $P_{br} = 1000$ hp., is $12 \times 550,000/60\pi \sim 35,000$ lb.-in. If we assume that the amplitude of the torque oscillation reaches 30 per cent of the mean value, we have

$$A_\nu = 10,500 \text{ lb.-in.}, \qquad \vartheta_0 = -0.0019 \quad \text{for } n = 30 \text{ sec.}^{-1}$$

The static twist angle would be $35,000/K = 0.007$. Since the first Eq. (26) shows that δ_0 is small as compared with ϑ_0, the additional stress in the crankshaft the upper limit of which is proportional to $|\vartheta_0 - \delta_0|$ comes near to one-third the stress as computed for static conditions. (The negative sign of ϑ_0 indicates that the moment and twist oscillations have a phase difference of one-half period.)

If a pendulum damper, tuned to $r/l = (\frac{9}{2})^2$, is applied, the amplitudes δ_0, ϑ_0 should theoretically be zero for all engine speeds. The theory,

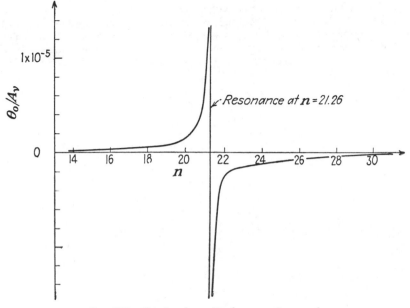

Fig. 276.—Torsional amplitude vs. engine speed.

however, disregards several minor influences, such as friction and finite values of all deviations. Thus, we can expect only that the amplitudes will be kept at a low level and the danger of the resonant point removed. A quantitative conclusion can be drawn from Eq. (27) if we ask for the influence of this damper on a disturbance of a different frequency. In most cases a first order disturbance, $\nu = 1$, will arise from the master rod construction used in most radial engines. If a pendulum of 10 lb. weight is used with $r = 8$ in. and consequently $l = 0.395$ in., the second term in the bracket of (27) will be $10 \times 8.395^2/0.951 \times 32.17 = 23$. The resonant engine speed according to (28) lies now at a $4\frac{1}{2}$ times higher value, i.e., at $n = 4.5 \times 21.3 = 95.8$ sec.$^{-1}$, and the last term in the bracket has the negative value -845. Thus it is seen that there is no danger of resonance in the case of a first order disturbance and that the

damper will have a slight unfavorable influence decreasing the amount
845 − 190 = 655 by 23, that is, by 3.5 per cent.

*Problem 9. Compute the difference between maximum and minimum speed for
an engine of m cylinders under the assumption that the torque contribution from one
cylinder is given by

$$M = 0 \qquad\qquad 0 < \vartheta < \pi \text{ intake}$$
$$M = M_1 \sin \vartheta \qquad \pi < \vartheta < 2\pi \text{ compression}$$
$$M = M_2 \sin \vartheta \qquad 2\pi < \vartheta < 3\pi \text{ expansion}$$
$$M = 0 \qquad\qquad 3\pi < \vartheta < 4\pi \text{ exhaust}$$

Discuss the influence of the number of cylinders and of the ratio $(M_2 + M_1)/M_0$.

*Problem 10. Compute the amplitude of the main torsional vibration due to the
elasticity of the crankshaft under the assumptions of the preceding problem. The
engine speed is $n = 21$ per sec., the number of cylinders $m = 9$, the nominal engine
power 700 hp., the moments of gyration $J_p = 150$, $J_c = 20$ lb. in. sec.^{2}, and the stiffness factor $K = 5,000,000$ lb. in.

Part Four

AIRPLANE PERFORMANCE

CHAPTER XIV

THE GENERAL PERFORMANCE PROBLEM

1. Introduction. The performance of an airplane is the result of the cooperation of three vital factors, the wing, the propeller, and the engine. Two passive forces have to be overcome, the weight and the parasite resistance of the airplane body. It is the characteristic difference between the airplane and any other kind of vehicle (cars, boats, airships) that power must be supplied to balance the weight. This implies that the character of the performance problem in aviation is entirely different from that in the other cases.

Only steady, straight flight, with all forces in equilibrium, is considered in the performance computation.

Speaking first of *level flight*, *i.e.*, flight in a straight line perpendicular to the direction of gravity, the weight (W) is balanced by the lift force (L), which is due to the motion of the wing. As we know, the wing implies a certain drag (D) invariably connected with the lift. The wing drag combined with the parasite drag of the airplane gives the total drag (D_{to}). To overcome this force, which acts parallel and opposite to the direction of flight, a propeller is used which supplies a thrust (T) in the direction of flight. Now, the propeller thrust, as we learned in Part Three, is invariably connected with the torque moment (Q) of the propeller, a moment acting about the propeller axis in the sense opposite to that of rotation. We have to balance this moment, and we do this by using an engine. The brake moment (M_{br}) of the engine acts in the sense of the rotation. Therefore, three equations must be fulfilled, if uniform level flight is to be maintained. These equations are

$$W = L, \qquad D_{to} = T, \qquad Q = M_{br} \tag{1}$$

A fourth equation, which states the equilibrium of the pitching moments on the airplane, will be discussed later (Chap. XVII). It has no immediate relation to the performance problem.

Out of the six quantities in (1), four, *i.e.*, all except the first and the last, are aerodynamic forces which we consider (neglecting the influence

381

of the Reynolds number) as proportional to the density of the air and to the square of the velocity. The first and the last, W and M_{br}, are physical forces. As the drag is connected with the lift and the propeller thrust with the propeller torque, the three equations link weight and engine moment. Even in the ideal case of no parasite drag and no friction (wing drag reduced to induced drag only) a certain engine power would be necessary to carry the weight, while, in the case of all land vehicles, boats, and airships in horizontal uniform motion, power is required only for overcoming the resistance forces due to friction, *i.e.*, solid-body friction, and water and air resistance, respectively.

A complete symmetry in the performance theory would prevail if the torque moment M_{br} supplied by the engine were a constant, as is the weight W, instead of being dependent on the air density ρ and the engine speed n. Since the four inner variables L, D_{to}, T, Q are all proportional to ρ and to a velocity square, V^2 and n^2, respectively, the system (1) would then admit a solution with constant values of ρV^2 and ρn^2, whatever ρ is. This means that *with constant M_{br} we could fly at any altitude,* but that flight speed and engine speed would both increase proportionally to the square root of $1/\sigma$ (σ = density ratio ρ/ρ_0). Note that in this idealization the presence of air friction is not excluded. In reality, however, a limit is set to the ascension to indefinite heights, with ever increasing speeds, even if the engine could be kept working invariably at any level, since it is impossible to have an engine supply a torque independent of the engine speed. This is due to the friction between the moving parts of the engine, including throttling losses in the channels through which the fuel, the fuel-air mixture, and the exhaust pass. In the present state of development the friction, particularly that between the piston rings and cylinder walls, proves the decisive factor in limiting further progress in airplane performance.

In our general performance computation, we shall consider M_{br} as a function of both density ρ and engine speed n, as laid down in the preceding chapter. In many restricted problems, however, the assumption of a constant M_{br} or an M_{br} depending on ρ only will be used as a sufficient approximation.

It is usual, but of course unessential, to multiply the second and third of Eqs. (1) by the factors V and $\omega = 2\pi n$, respectively, so as to have equations between power quantities rather than forces or moments,

$$W = L, \qquad VD_{to} = VT, \qquad \omega Q = \omega M_{br} \text{ or } P = P_{br} \qquad (1')$$

where P is written for ωQ in accordance with Chap. XI and P_{br} is the brake power of the engine. The product VD_{to} is called the *required power* and the product VT the *available power*. The three level-flight equations can then be reduced to a single one by the following consideration.

For a given ρ, the two aerodynamic forces L and D_{to} can both be expressed as functions of the two variables V and α, where α is the angle of attack (or angle of incidence) of the wing. It is supposed, of course, that the wing area S and the parasite area S_p of the plane are known as well as the characteristics of the wing. If we use the equation $W = L$ to eliminate α, we find D_{to} and thus also VD_{to} as a function of V alone. On the other hand, for a given ρ, the quantities T and Q are functions of n and J, where J is the advance ratio of the propeller. Here, of course, the propeller diameter and the characteristics of the propeller must be known. Instead of J and n we may introduce J and V, since $V = Jnd$ (d = propeller diameter). For a given ρ, the engine moment M_{br} or the engine power P_{br} are known functions of n, or, what is the same, of J and V. Thus we can use the last of the three equations to eliminate J and to express T and VT as function of V alone. Then the only remaining equation to be fulfilled in level flight is

$$VD_{to} = VT \qquad \text{or} \qquad P_{re} = P_{av} \qquad (2)$$

where both sides of the equations, for a given ρ, are functions of V and of the airplane and propeller parameters. The curves showing P_{re} and P_{av}, respectively, as functions of V are known as the *power curves* of the plane. This method can be summarized as follows:

If the first equation (1) is used to eliminate α in the expression for D_{to} and the last to eliminate J in the expression for T, the level flight is determined by the single equality (2), power required equal to power available, where both P_{re} and P_{av} are functions of V, of ρ, of the parameters S, S_p, and d, depending on the characteristic curves of the wing, the propeller, and the engine.

It will be shown in the next section how the functions P_{re} and P_{av} can be computed from the given airplane, propeller, and engine data, supplying an equation for V that gives the exact answer to the performance problem for level flight. This will also include the answer to the question of *ceiling altitude*, which can be defined as the altitude beyond which no level flight is possible. Meanwhile, a remark may be added on how the two functions $P_{re}(V)$ and $P_{av}(V)$ can be used, at least to a certain degree of approximation, in the case of a *slightly inclined flying path*, climbing or flat descent.

Figure 277a shows the vector diagram of the four forces W, L, D_{to}, and T in equilibrium at level flight. If the velocity direction is no longer perpendicular to the direction of gravity but is slightly inclined at the angle ϑ as shown in Fig. 277b, the propeller thrust can still be supposed to act (nearly) in the direction of V. Lift L and drag D_{to} are defined as the components of the aerodynamic forces normal and parallel to V. If the motion is to be uniform, the four forces must be in equilibrium and thus combine to the trapezoid shown in Fig. 277b. It follows from this diagram that

$$T - D_{to} = W \sin \vartheta \tag{3}$$

If we multiply both sides by V and consider that $V \sin \vartheta$ is the vertical component of the velocity or the *climbing rate*, which can be written as

$$V \sin \vartheta = \frac{dh}{dt} \tag{4}$$

we find

$$P_{av} - P_{re} = W \frac{dh}{dt} \tag{5}$$

As to the first term (P_{av}), there is not much objection to using the same function of V as in the level-flight computation. Only the small

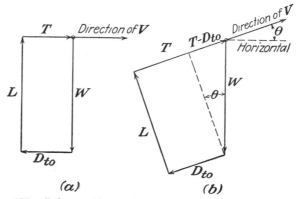

Fɪɢ. 277.—Balance of forces, (a) in level flight, (b) in a flat climb.

angle between propeller axis and velocity V, which is due to the change in angle of attack, is neglected. But in the expression for P_{re} in (2) the relation $W = L$ was used, which now should be replaced by

$$L = W \cos \vartheta \tag{6}$$

as seen in Fig. 277b. However, if ϑ is small, $\cos \vartheta$ differs from 1 only by a quantity of second order. Thus a small error only will be incurred if we use in (5) for P_{re} the function of V found in the way already described for the case of level flight. Summing up, we have the following statement:

For slightly inclined uniform flight, Eq. (5), or

$$\frac{dh}{dt} = \frac{P_{av} - P_{re}}{W} \tag{5'}$$

can be used for determining the (positive or negative) climbing rate, with the expressions for P_{av} and P_{re} as introduced in the case of level flight.

This practically solves the problem for uniform climbing since usually small ϑ only are involved in this case. In particular, one can answer in this way the third fundamental performance question asking for the *maximum climbing rate* at each altitude (which leads to another definition of ceiling altitude, *viz.*, the height at which the maximum climbing rate = 0). As to the conditions of downward flight, the results have to be corrected, in general, with respect to Eq. (6), unless the (negative) values of ϑ are sufficiently small. The case of gliding and diving at a considerable angle will be discussed in Sec. XVI.3.

2. Power-required and Power-available Curves. According to what was said in the preceding section, the power required, that is, $P_{re} = VD_{to}$, can be computed as a function of V if the following data are known: the total weight W, the wing area S, and the two wing characteristics showing C_L and C_D (lift and drag coefficients) as functions of incidence; then the parasite area S_p of the plane; and finally the air density ρ.

In performance computation, the density ρ is considered as invariably linked to the level altitude h, according to the ρ, p, T-distribution in the standard atmosphere (Sec. I.3). Consequently, ρ or the density ratio $\sigma = \rho/\rho_0$, where $\rho_0 = 0.002378$ slug/ft.³, is regarded as indicating the altitude at which the horizontal (or slightly inclined) flight takes place.

The parasite area S_p refers to the complete airplane except the wing. We disregard the fact that the shape of the airplane body does not exactly satisfy the symmetry conditions stated in Sec. V.1, which led to the definition of parasite drag. That is, we assume that, in spite of its deviating from the supposed double symmetry (existence of a symmetry axis coinciding with the V-direction), the total aerodynamic action exerted on the airplane in uniform flight reduces to a single force parallel and opposite to V. How its magnitude can be computed or estimated for a given airplane design has been discussed in Sec. V.5. The slight changes of this force due to the variation of the angle of attack are disregarded. (An additional pitching moment due to the deviation between the fuselage axis and the direction of flight will be discussed in Sec. XVII.3.) If the magnitude of the parasite drag is written as

$$D_p = \frac{\rho}{2} V^2 S_p \tag{7}$$

according to Eq. (3), Chap. 5, the factor S_p is called the *parasite area of the airplane.* Numerical data about actual values of S_p are given in Sec. XV.4.

The wing area S is the area to which the lift and drag coefficients refer. The forces

$$L = \frac{\rho}{2} V^2 S C_L, \qquad D = \frac{\rho}{2} V^2 S C_D \tag{8}$$

normal and parallel to the velocity direction may be taken from experiments with a model of the actual wing or computed on the basis of wing theory or—in most cases—may be found by combining both methods.   We suppose here that the two coefficients C_L and C_D are given as functions

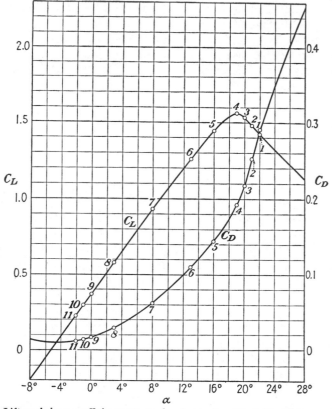

FIG. 278.—Lift and drag coefficients vs. angle of attack.   Clark Y airfoil, aspect ratio 6.

of the angle of attack α or the angle of incidence α' by a diagram, Fig. 278. It will be found, however, that a little less is needed.   The polar diagram alone, without indication of the α-values (*i.e.*, the correspondence between simultaneous C_L- and C_D-values alone) is sufficient for the present purpose.

   The two formulas to be used in deriving the power-required curve P_{re} vs. V are

$$L = \frac{\rho}{2} V^2 S C_L(\alpha) \qquad \text{and} \qquad D_{to} = \frac{\rho}{2} V^2 [S C_D(\alpha) + S_p] \tag{9}$$

The procedure of finding successive points of the power-required curve runs as follows:

Step 1: Assume an α-value, take from the C_L vs. α-curve the value of C_L, and compute from $W = L$:

$$V = \sqrt{\frac{2W}{\rho S C_L}} \qquad (a)$$

Step 2: Take from the C_D vs. α-curve the C_D-value for the same α, and compute from $P_{re} = V D_{to}$:

$$P_{re} = \frac{\rho}{2} V^3 [S C_D(\alpha) + S_p] \qquad (b)$$

using, of course, the V-value found in step 1. Thus one pair of cor-

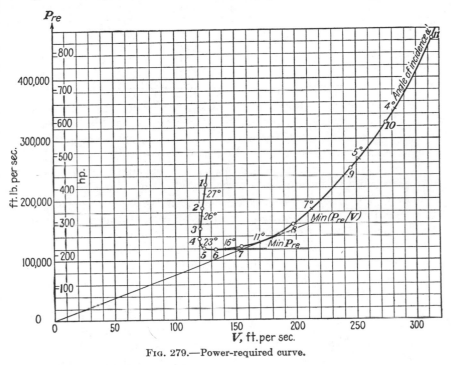

Fig. 279.—Power-required curve.

responding coordinates V, P_{re} is determined. Continuing in this way with α-values chosen to cover the whole range given in the wing character-istics, one can find the P_{re} vs. V-curve as shown in Fig. 279. The procedure is carried out here for the typical characteristics of Fig. 278 using the values $\rho = \rho_0 = 0.0024$, $W = 8000$ lb., $S = 300$ ft.², $S_p = 9$ ft.² For some purposes it may be useful to add an α- or α'-scale to the points of the V-axis. The angles run opposite to V. The numerical results are shown in Table 10.

The first three columns include the given figures. The fourth is computed from Eq. (a) and the fifth from Eq. (b). The curve is shown in Fig. 278.

TABLE 10.—POWER-REQUIRED CURVE

Point No.	α, deg.	C_L	C_D	V, ft./sec.	P_{re}, ft.-lb./sec.
1	22	1.42	0.290	125	225,500
2	21	1.48	0.252	122.5	186,500
3	20	1.53	0.216	120.5	154,500
4	19	1.56	0.192	119.5	135,500
5	16	1.44	0.144	124	119,500
6	13	1.26	0.110	133	118,000
7	8	0.93	0.062	154.5	122,000
8	3	0.58	0.029	196	159,500
9	0	0.37	0.017	245	252,000
10	−1	0.30	0.014	274	326,000
11	−2	0.23	0.012	313	464,000

The following three properties of the P_{re} vs. V-curve may be emphasized. (1) Since C_L has a maximum value (stalling value), it follows from Eq. (a) that V has a minimum. That is, no solution of the performance equation exists (no level flight is possible) at velocities below a certain minimum. This minimum velocity is called the *stalling speed* of the plane, and it depends, as we see, on wing and plane only, not on propeller and engine. If we assume the C_L vs. α-curve beyond the stalling point as a horizontal line, the corresponding section of the power-required curve is a vertical straight line, since V remains constant. If C_L drops with higher increase of α (as is the case in Fig. 278), the P_{re} vs. V-curve is convex toward the left, as seen in our example. (2) If we turn to smaller angles of attack, down to zero lift angle, the velocity V increases indefinitely as is seen from Eq. (a). At the same time, P_{re} must also go to ∞, since the expression in the brackets is positive and has a positive lower bound. The right-hand branch of the P_{re}-curve behaves for large V like a cubic parabola $y = $ const. $\times x^3$. (3) Since the left branch of the curve (small velocities, high incidence) drops with increasing V while the right branch (larger velocities, smaller angles) increases, there must be a point at which the required power is a minimum. Besides, there must be a point (to the right side of the former) at which a tangent to the curve drawn from the origin contacts the curve. The latter point marks the minimum of the quotient P_{re}/V or of D_{to}.

It can easily be seen how instead of the C_L and C_D vs. α-curves some other representative of the wing characteristics could be used for deriving the power-required curve, e.g., the C_L vs. α- and ϵ vs. α-curves. In particular, the use of the polar diagram will be discussed in the next section.

We turn now to the *power available* $P_{av} = VT$, which depends on the two parameters ρ and d, air density and propeller diameter, on the propeller characteristics, and on the engine chart, which shows P_{br} as function of n at the given ρ (Fig. 280). For propeller characteristics we may use any combination of two curves discussed in Chap. XI, *e.g.*,

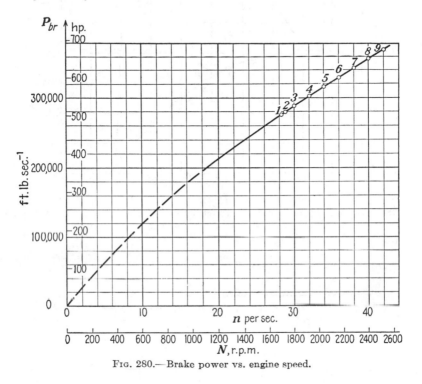

FIG. 280.—Brake power vs. engine speed.

the power coefficient C_p and the efficiency η, both plotted vs. advance ratio J (Fig. 281). The two equations to be used then are

$$P = \rho n^3 d^5 C_p(J) \quad \text{and} \quad P_{av} = P\eta(J) \tag{10}$$

The successive points of the P_{av} vs. V-curve can now be found in three steps:

Step 1: Assume a value for propeller speed n, take from the engine chart the value of P_{br} (for this n if no reduction gear is used; otherwise, for the respective n'), and compute C_P from the condition $P = P_{br}$:

$$C_P = \frac{P_{br}}{\rho n^3 d^5} \tag{c}$$

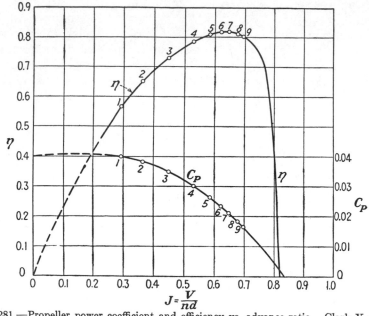

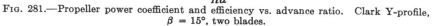

$$J = \frac{V}{nd}$$

Fig. 281.—Propeller power coefficient and efficiency vs. advance ratio.   Clark Y-profile, $\beta = 15°$, two blades.

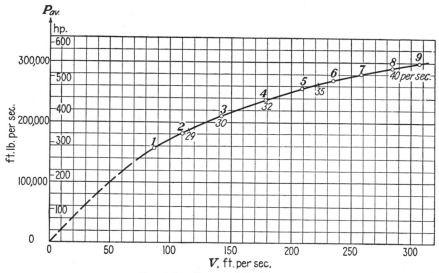

Fig. 282.—Power-available curve.

Step 2: Take from the C_P vs. J-curve of the propeller the value of J that corresponds to this C_P, and compute

$$V = Jnd \qquad (d)$$

Step 3: Take from the η vs. J-curve the efficiency value η for the same J as found in the preceding step, and compute the available power,

$$P_{av} = \eta P_{br} \qquad (e)$$

using, of course, the P_{br} secured in step 1.

Thus, in (d) and (e) a pair of corresponding coordinates V, P_{av} is found. Continuing in this way and varying the choice of n within the total range of engine speeds, we obtain the whole P_{av} vs. V-curve. The procedure is carried out in Fig. 282 for $d = 10.5$ ft., a typical propeller diagram (Fig. 281, with $\eta_{max} = 0.82$ and $C_P = 0.20$ at $J = 0.65$), and a normal engine supplying 650 b.hp. at 2400 r.p.m. of the propeller (according to Fig. 280). For some purposes the n-values may be ascribed to the abscissas V. They run the same way as V. The numerical results are included in Table 11.

TABLE 11.—POWER-AVAILABLE CURVE

Point No.	n, sec.$^{-1}$	P_{br}, ft.-lb./sec.	C_P	J	η	V, ft./sec.	P_{av}, ft.-lb./sec.
1	28.3	276,000	0.0398	0.295	0.565	87.5	156,000
2	28.8	279,000	0.0382	0.365	0.650	110.5	181,500
3	30	288,000	0.0349	0.405	0.730	142	210,500
4	32	302,500	0.0301	0.531	0.783	178	237,000
5	34	316,000	0.0262	0.585	0.808	209	255,500
6	36	330,000	0.0230	0.622	0.817	235	269,500
7	38	344,000	0.0210	0.648	0.817	258.5	281,000
8	40	357,500	0.0182	0.677	0.811	284.5	290,000
9	42	371,000	0.0164	0.696	0.803	307	298,000

It can easily be seen how instead of the C_P and η vs. J-curves any other equivalent representation, for example, C_S and η vs. J (C_S = speed power coefficient) could be used. But, at variance to the P_{re} case, in general, a simplified procedure using the relation between C_P and C_T or C_P and η only, without the indication of the J-values, is not possible. This is feasible, however, as will be seen below, under an assumption that is approximately fulfilled in almost all cases.

As to the general character of the power-available curve the following can be stated: (1) If the engine chart is prolonged down to $n = 0$, the corresponding $P_{br} = 0$; then, according to Eq. (d), $V = 0$ and, according to Eq. (e), $P_{av} = 0$. This means that the prolongation of the P_{av} vs. V-curve passes through the origin. (2) On the other hand, P_{br} has a maximum and goes to zero if C_P decreases. It follows that P_{av} must decrease for larger V. This part of the curve is not shown in Fig. 282. Generally speaking, P_{av} vs. V will be, within the actual range, upward convex as seen in our example. (A section upward concave at the beginning can occur.)

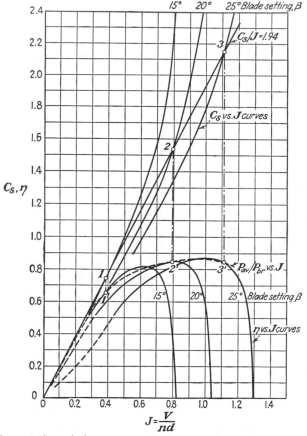

Fig. 283.—Computation of the power-available curve for a constant-speed propeller.

The P_{av}-curve thus described holds for the case of an invariable propeller. If the *propeller pitch is controllable* by the pilot, no definite P_{av}-value is determined as long as no definite rule of pitch control is adopted. Only one prescription is generally used today: to *control the pitch in such a way as to keep the engine (and propeller) speed constant.* In this case the P_{av}-curve can be found in the following manner: Consider the propeller chart given in one of the forms discussed in Sec. XI.4. Assume, for example, that the efficiency η and the speed power coefficient C_S are both plotted vs. the advance ratio J for each blade setting (Fig.

283). To the constant engine speed n there corresponds now at any altitude or ρ-value a definite brake power $P_{br} = P$. As C_S and J are defined by

$$C_S = V \sqrt[5]{\frac{\rho}{Pn^2}}, \qquad J = V \frac{1}{nd}$$

we know the value of

$$\frac{C_S}{J} = d \sqrt[5]{\frac{\rho n^3}{P}} \qquad (f)$$

This equation represents a straight line of given slope through the origin

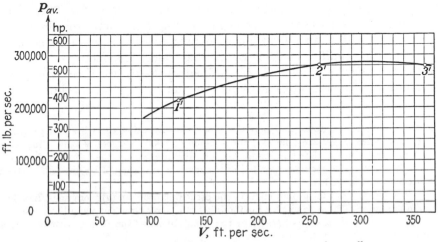

Fig. 284.—Power-available curve for a constant-speed propeller.

in the C_S vs. J-diagram. The P_{av}-curve can now be found by the following two steps:

Step 1: Assume a value for the instantaneous blade setting β, and mark on the corresponding C_S vs. J-curve the point of intersection with the straight line (f). Its abscissa multiplied by the constant nd gives V.

Step 2: Take from the η vs. J-curve, for the assumed β, the value of η at the abscissa of the marked point, and compute $P_{av} = \eta P$.

This procedure is carried out in Figs. 283 and 284 for an engine of $P_{br} = 600$ hp. at $n = 31$ and a propeller diameter $d = 10.5$ ft. With $\rho = 0.0024$ we have, from Eq. (f),

$$\frac{C_S}{J} = 10.5 \sqrt[5]{\frac{0.0024 \times 31^3}{600 \times 550}} = 1.94$$

The straight line through the origin, passing through the point $J = 1.0$, $C_S = 1.94$, is drawn in Fig. 283. The points of intersection with the C_S vs. J-curves for these blade settings are marked 1, 2, 3. Each of these points is projected on the respective η-curve, and the projections are marked 1', 2', 3'. The following values have been found:

TABLE 12.—POWER AVAILABLE FOR CONSTANT-SPEED PROPELLER

Point No.	β	J	V, ft./sec.	η	P_{av}, ft.-lb./sec.
1	15	0.385	125	0.655	214,500
2	20	0.795	259	0.850	280,500
3	25	1.110	361	0.845	279,000

The corresponding values of V and P_{av} are shown in Fig. 284. The curve has the same general type as in the case of constant propeller pitch, increasing with V and slightly convex upward within the range of practical flight conditions. Except for the scales (the factor nd for the abscissas and the factor P for the ordinates) the P_{av} vs. V-curve appears immediately in the propeller chart as the locus of the points 1', 2', 3', ...

Problem 1. Compute and plot the power curves, using the wing and propeller characteristics (Figs. 278 and 281) for a plane of 6000 lb. gross weight, wing area 180 ft.², parasite area 6 ft.², flying at 8,000 ft. altitude. Take the propeller diameter $d = 10$ ft. and the engine power for any engine speed n equal to one-half of that indicated in Fig. 280.

Problem 2. Let the gross weight in the preceding problem (a) increase to 7000 lb.; (b) decrease to 5000 lb., and give the modified power-required curves in both cases.

Problem 3. If the aspect ratio of the wing changes from 6 to 8, how does this influence the power-required curve?

***Problem 4.** Explain the procedure of finding the power-available curve in the case that two equal propellers are used instead of one.

Problem 5. Find the power-available curve for a controllable-pitch propeller with the characteristics of Fig. 283, working at 1200 r.p.m. with an engine supplying 450 hp.

3. Dimensionless Performance Analysis. For many purposes a slightly modified procedure, which may be called a method of *dimensionless performance analysis*, proves useful. It will be briefly discussed in this section.

As to power required, P_{re}, we introduce a basic velocity value V_1,

$$V_1 = \sqrt{\frac{2W}{\rho S}} \tag{11}$$

which is, as can be seen immediately from Eq. (a), the flight velocity at lift coefficient $C_L = 1$. Then we set

$$V = xV_1 \quad \text{and} \quad P_{re} = yWV_1 \tag{12}$$

where obviously $x = V/V_1$ and $y = P_{re}/WV_1$ are dimensionless quantities.   The y vs. x-curve will be the dimensionless representative of the power-required curve.   In the left part of Fig. 285 the polar diagram of the wing is shown under the following assumptions: Both scales, for C_D and for C_L, are equal; the C_D are plotted toward the left; the curve is shifted to the left by the amount S_p/S where S_p is the parasite area of

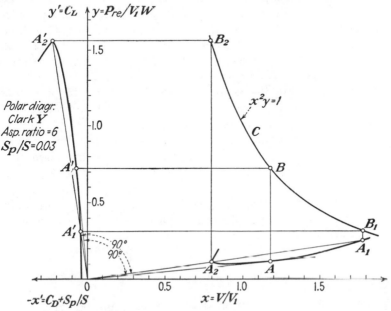

FIG. 285.—Power-required dimensionless analysis.

the airplane (wing excluded).   The coordinates of any point on this curve will be called x', y'.   Thus,

$$-x' = C_D + \frac{S_p}{S}, \qquad y' = C_L \tag{13}$$

Now, the equilibrium conditions $W = L$ and $P_{ro} = V(D + S_p\rho V^2/2)$ furnish, if V_1 is introduced,

$$W = \frac{\rho}{2}\left(\frac{V}{V_1}\right)^2 V_1^2 S C_L = \frac{\rho}{2}x^2\frac{2W}{\rho S} S C_L = x^2 W y'$$

$$P_{re} = VD_{to} = \frac{\rho}{2}\left(\frac{V}{V_1}\right)^3 V_1^3 S\left(C_D + \frac{S_p}{S}\right) = \frac{\rho}{2}x^3\frac{2W}{\rho S}V_1 S\left(C_D + \frac{S_p}{S}\right)$$
$$= -x^3 W V_1 x' \tag{14}$$

The first equation reduces to $x^2y' = 1$, while the second, with $P_{re}/WV_1 = y$, leads to $y = -x^3x'$.   These two relations can be written as

$$x^2y' = 1 \qquad \text{and} \qquad \frac{y}{x} = -x^2x' = -\frac{x'}{y'} \tag{15}$$

They determine the transformation of the polar diagram y' vs. x' into the dimensionless power-required curve y vs. x (see Fig. 285). The second equation (15) expresses the fact that the radius vector OA from the origin to the point x, y is perpendicular to the radius vector OA' which leads to the corresponding point of the polar curve. The first equation can be interpreted in the following way: We plot, once for all, on our sheet the curve $x^2y = 1$ (C in Fig. 285), which has the x- and y-axes as asymptotes. Then we draw $A'B$ parallel to x (B on the curve C) and through B a straight line parallel to y; this vertical intersects the line perpendicular to OA' at the point A on the power-required curve. In this manner, the entire power-required curve can be deduced, point by point, from the polar curve of the wing. In particular, the stalling abscissa and the point of minimum P_{re}/V-value can be found immediately, as seen in the figure.

Essentially the same procedure can be followed for deriving the dimensionless power-available curve from the polar diagram of the propeller if the assumption is made that *the torque moment of the engine, M_{br}, is constant within the speed range under consideration.* As a matter of fact, almost all performance computations are carried out under this assumption. If we plot again the propeller polar curve toward the left so as to have

$$-x_1' = \frac{C_T}{J^2}, \qquad y_1' = \frac{C_P}{J^2} \tag{16}$$

(for the definition of the propeller polar diagram, see Sec. XI.4), the equilibrium conditions

$$M_{br} = Q = \rho \frac{C_P}{2\pi} n^2 d^5 = \rho \frac{1}{2\pi} \frac{C_P}{J^2} V^2 d^3$$
$$P_{av} = TV = \rho C_T n^2 d^4 V = \rho \frac{C_T}{J^2} V^3 d^2 \tag{17}$$

can be transformed by using a new velocity unit

$$V_2 = \sqrt{\frac{2\pi M_{br}}{\rho d^3}} \tag{18}$$

and the dimensionless variables

$$x_1 = \frac{V}{V_2}, \qquad y_1 = \frac{P_{av}}{2\pi M_{br}} \frac{d}{V_2} \tag{19}$$

Here, V_2 is obviously the flight speed corresponding to the value $C_P/J^2 = 1$, and y_1 is $1/\pi$ times the ratio blade tip speed to V_2, if the propeller speed is computed for $\eta = 1, P_{av} = P$. Introducing x_1, y_1, and x_1', y_1' in (17) we find

$$M_{br} = \rho \frac{1}{2\pi} y_1' \left(\frac{V}{V_2}\right)^2 \frac{2\pi M_{br}}{\rho d^3} d^3 = x_1^2 M_{br} y_1'$$

$$P_{av} = -\rho x_1' \left(\frac{V}{V_2}\right)^3 \frac{2\pi M_{br}}{\rho d^3} V_2 d^2 = -x_1^3 2\pi M_{br} \frac{V_2}{d} x_1'$$

(20)

This gives the same relations between x_1, y_1 and x_1', y_1' as those found in Eq. (15):

$$x_1^2 y_1' = 1 \quad \text{and} \quad \frac{y_1}{x_1} = -x_1^2 x_1' = -\frac{x_1'}{y_1'} \tag{21}$$

Thus, the dimensionless power-available curve could be derived from the propeller polar diagram in the same way as was shown in Fig. 285 for the power-required curve.

This procedure, however, would have a twofold disadvantage. First, while the value $C_L = 1$ falls into the usual range of lift coefficients (about 0 to 1.5), the values of C_P/J^2 are much smaller than 1. This would imply that far-distant points of the curve $x^2 y = 1$ have to be used. On the other hand, we wish to have finally both P_{re} and P_{av} plotted in the same coordinate system, *i.e.*, in the same scales, while the indicated method would use abscissas V/V_1 in the first case and V/V_2 in the second, etc. Both inconveniences are avoided if we introduce, instead of the original propeller polar diagram, a distorted one, plotting

$$-x' = \frac{C_T}{J^2} \frac{2d^2}{S}, \qquad y' = \frac{C_P}{J^2} \frac{W d^3}{\pi M_{br} S} \tag{22}$$

rather than x_1' and y_1' as defined in (16) (see Fig. 286). If, then, the power available is represented in the coordinate system x, y, which was used for P_{re},

$$x = \frac{V}{V_1}, \qquad y = \frac{P_{av}}{W V_1} \tag{23}$$

the relations between x, y and x', y' are again the same as in (15). In fact, introducing (22) and (23) in (17), we have

$$M_{br} = \rho \frac{1}{2\pi} \frac{C_P}{J^2} V_1^2 \left(\frac{V}{V_1}\right)^2 d^3 = \rho \frac{1}{2\pi} \frac{C_P}{J^2} \frac{2W}{\rho S} x^2 d^3 = x^2 M_{br} y'$$

$$P_{av} = \rho \frac{C_T}{J^2} V_1^2 \left(\frac{V}{V_1}\right)^3 V_1 d^2 = \rho \frac{C_T}{J^2} \frac{2W}{\rho S} x^3 V_1 d^2 = -x' x^3 W V_1$$

(24)

whence (15) follows immediately. In Fig. 286 the transformation, according to the procedure already explained for Fig. 285, is carried out under the assumption $2d^2/S = 0.735$, $Wd/\pi M_{br} = 18.3$. This corresponds to an engine that furnishes 600 hp. at $n = 36$, to a propeller diameter $d = 10.5$ ft., and to the values already given of W and S.

Summarizing, we can state the following: *The polar diagram of the plane and the polar diagram of the propeller, the latter plotted with modified scales, give, when subject to the same simple transformation* (15), *the power-required and power-available curves in dimensionless form.* For the power available the method is restricted to the assumption that the engine supplies a sensibly constant brake moment within the speed range under consideration.

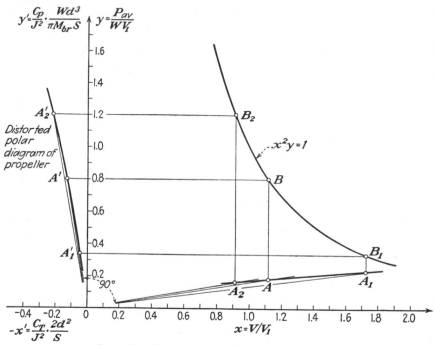

Fig. 286.—Power-available dimensionless analysis.

Problem 6. Plot the polar diagram for the profile represented in Fig. 278, but with the aspect ratio 8 instead of 6, and deduce the dimensionless power-required curve.

Problem 7. Develop the distorted polar diagram and the dimensionless power-available curve for the propeller represented in Fig. 281 and the data of Prob. 1, with an average $M_{br} = 760$ ft.-lb.

*Problem 8. Develop the theory of the dimensionless power diagram for the case that more than one propeller is used. Assume (a) two equal engines with equal propellers; (b) two pairs of engines and propellers, each pair consisting of equal pieces.

4. Discussion of Sea-level Flight. By putting together the two curves of power required and power available (Fig. 287), important conclusions can be reached. It was seen in Sec. 1 of this chapter that by using the functions $P_{re}(V)$ and $P_{av}(V)$ the three equilibrium conditions for level flight reduce to one: $P_{av} = P_{re}$. Therefore, horizontal flight at a constant velocity is possible if and only if the ordinates of the two curves are equal. This is the case at two different velocity values where the P_{re}- and P_{av}-curves intersect. We call these the high speed and the low speed of level flight. It will be seen that the high speed only is used in normal flight. The respective abscissas are marked in Fig. 287 as V'_l (left) and V_l (right). As we have learned in the preceding section how the two power curves can be derived from airplane, propeller, and engine data, the first performance problem, which asks for the

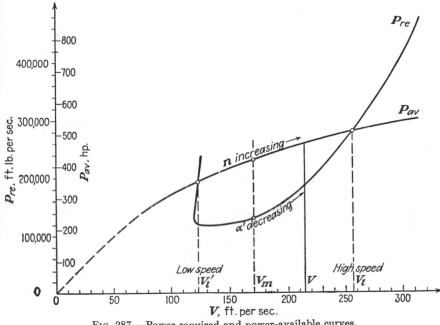

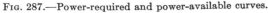

FIG. 287.—Power-required and power-available curves.

magnitude of level-flight velocity, is solved, except for the ambiguity of the solution. In our example, as given in the previous Figs. 279 and 282 (fixed-pitch propeller) and now combined in Fig. 287, the two velocity values are $V'_l = 122.5$, $V_l = 257$ ft./sec.

We find the complete answer, which eliminates the ambiguity, by considering first the slightly modified diagram, Fig. 288. Here, the left end of the power-required curve is a little changed so as to give an inter-section A' within the region where P_{re} decreases with increasing V. In this case, the total range of possible V-values, from stalling speed upward, can easily be divided into three distinct parts, from V_{st} to V'_l, from V'_l to V_l, and from V_l upward. In the first and in the third of these intervals, P_{re} is greater than P_{av}, while in the second P_{av} prevails. According to Eqs. (5) and (5′) the inner interval is a region of climbing (positive ϑ), and the two outer intervals are regions of descent (negative ϑ). To each V there corresponds a certain angle of attack α, and it is known that α decreases when V increases.

Now, suppose a pilot is flying at the *high speed of level flight* and wants to change to a slight positive climbing. What he can actually do is to operate the elevator in such a sense as to direct the nose of the airplane upward. In the first moment, before the velocity is changed, the air impinges upon the wing at a larger angle of incidence and produces an

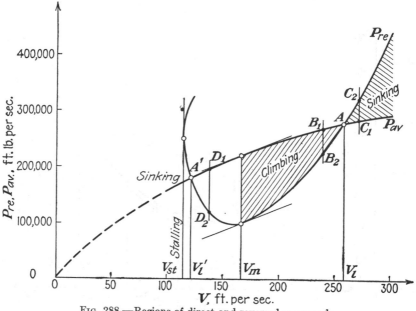

Fig. 288.—Regions of direct and reversed commands.

increased lift and an increased drag. The excess in lift gives the plane an upward acceleration, while the increased drag slows down the forward speed. Thus the transition is found to a new state of steady flight with an upward component of velocity, with smaller V and higher angle of incidence, corresponding to points B_1, B_2 of the power curves to the left of A. As seen in Fig. 289a and b the angle through which the plane is rotated is the sum of ϑ and the increase in incidence. Exactly the opposite occurs if the pilot chooses to descend and turns the nose of the airplane down. The new state of steady flight then corresponds to points C_1, C_2 of the power curves at an abscissa to the right of A. Figure 289c shows how the angles change.

This way of using the elevator does not lead to the desired effect in case the pilot is originally flying at the *low speed of level flight*. In fact, if he intends to climb and turns the airplane nose up, he will lose speed, as before, but there is no state of steady climbing at a velocity lower than V_i'.   In the region $V < V_i'$ the required power exceeds the available power and only descent is possible.   In order to change from level flight to a climb the pilot would have to push the elevator stick forward to gain speed and thus create the conditions for a climbing flight corresponding to points D_1, D_2 of the power curves to the right of A'.   This phenomenon is known as the *reverse of commands*.   As to the level-flight speed, we have reached the following conclusion:

There is only one level-flight speed for a definite airplane with definite power plant under normally working commands.   It corresponds to the

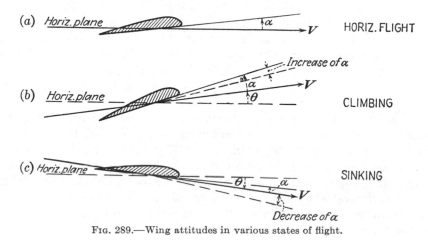

Fig. 289.—Wing attitudes in various states of flight.

right-hand intersection of the two power curves and is called the high speed of level flight.

The situation of reversed commands subsists for all flying speeds to the left of a certain value V_m between V' and V.   Let us suppose a pilot is flying upward at the speed V indicated in Fig. 289, *i.e.*, with a positive climbing rate

$$V \sin \vartheta = \frac{dh}{dt} = \frac{P_{av} - P_{re}}{W} \tag{25}$$

and he wants to increase this rate.   Then, the normal use of commands will consist in pulling the stick to lift the nose of the airplane.   This will lead to the desired effect if a smaller velocity and a higher angle of incidence are connected with a higher climbing rate, *i.e.*, with a greater distance between the two power curves.   Thus it is seen that *all points to the left of the abscissa of maximum distance between the power curves are points of reversed commands*.   Only the shadowed area in Fig. 288 can be used for normal flight.   If we return to Fig. 287, the conditions are the same as for Fig. 288 at least in the V-range from V_l down to V or close to V'_l. One cannot reach the low speed of level flight from the region of normal flight without passing through stages of reversed commands. The slight change in the left end of the P_{re}-curve cannot have any decisive influence.

What is said here about the use of commands is all we can conclude from considering the equilibrium conditions as developed in the present chapter. Any transition, however, from one state of motion to another is a nonuniform motion connected with accelerations. From this point of view the problem will be discussed later in Chap. XVIII.

The limit between normal and reversed commands is given by the abscissa V_m, which obviously is the velocity of greatest climbing rate. In the graph, it can be found as the abscissa where the two power curves have parallel tangents. In our example (Fig. 287) we have $V_m = 170$ ft./sec. and the corresponding $P_{av} = 230,000$ ft.-lb./sec., $P_{re} = 130,000$ ft.-lb./sec., and thus, with $W = 8000$ lb., $(dh/dt)_{max} = 12.5$ ft./sec. It is seen that V_m is larger than the abscissa of the lowest point of the P_{re}-curve.

The maximum climbing rate must not be confounded with the maximum climbing angle or steepest climbing path. To find the latter we had better use the curves representing $T = P_{av}/V$ and $D_{to} = P_{re}/V$, which are preferable to the power curves in dealing with various special questions. Figure 290 shows the curves for our example (Fig. 287). Their ordinates, which are the quotients of the coordinates of the power curves, can easily be found by a well-known graphical procedure. According to (3),

$$\sin \vartheta = \frac{T - D_{to}}{W} \tag{26}$$

Therefore, the points of maximum distance here give the maximum ϑ and the abscissa $V'_m = $ flying speed for steepest climbing. In our example, $V'_m = 146$ ft./sec., $\vartheta_{max} = 4.6°$. In general, V'_m will be smaller than V_m.

Climbing or sinking rates can be computed in this way from the two power curves as long as the ϑ involved is not too large. Up to $\vartheta = \pm 15°$, the errors are negligible. As pointed out in Sec. 1 of this chapter, the errors are due to the fact that in plotting the P_{re} vs. V-curve, L was equated to W instead of to $W \cos \vartheta$. This affects the abscissas as well as the ordinates of the P_{re}-curve. For $\vartheta = 30°$, the correct P_{re}-curve would correspond to a weight value reduced by 13.5 per cent. If we plot the modified curve (dotted line in Fig. 291), which gives P_{re} for the gross weight $0.865W$, we can use the diagram—by means of interpolation—up to angles of 30°.

The P_{av}-curve in each case is drawn for a definite engine, with a definite throttle setting. If, for example, a curve is plotted for full

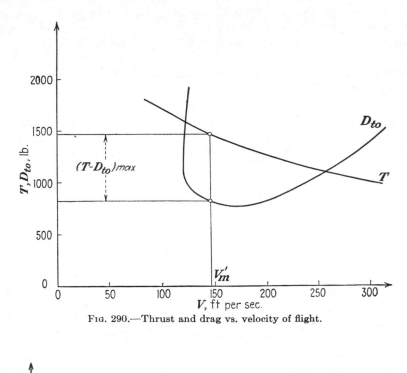

Fig. 290.—Thrust and drag vs. velocity of flight.

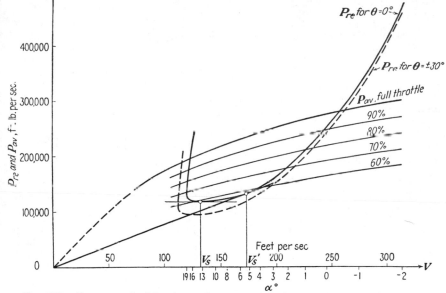

Fig. 291.—Power required for $\vartheta = 30°$; power available for various throttle settings.

throttle, the curves for *part-throttle* opening will run below the first one. Using an 80, 60, or 40 per cent curve, etc., would give all information required in practical computations. Some of these curves are shown in Fig. 291.

The case of *zero power* (engine shut off) is of particular interest. In this case, the axis of abscissas replaces the P_{av}-curve, and the corresponding flying modes are called *gliding*. It is obvious that here the difference $P_{av} - P_{re}$ is always negative; thus, with negative ϑ, sinking only is possible. With the restriction to not too large ϑ-values, the ordinates of the P_{re}-curve are immediately a representation of the sinking

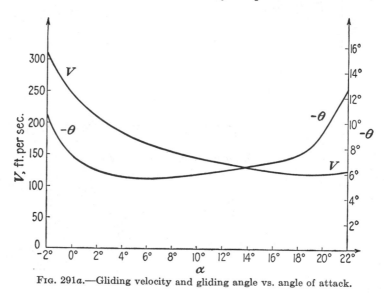

Fig. 291a.—Gliding velocity and gliding angle vs. angle of attack.

rate. The *slowest sinking* is given by the abscissa V_S, where P_{re} has its minimum. The *flattest gliding* occurs at V_S', where a tangent from the origin contacts the curve. In our example, $V_S = 131$, $V_S' = 172$ ft./sec., and the corresponding values of $-dh/dt$ and of ϑ are 14.6 and 16.4 ft./sec., 6.5 and 5.5° (Fig. 291). A correction of these values with respect to $\cos \vartheta < 1$ is hardly necessary.

The angles of incidence of the various states of motion can be read off in our diagram if there is added to the V-scale on the horizontal axis the α- (or α'-) scale. The latter scale can be determined by the computations that lead to the P_{re} vs. V-curve according to the procedure described in Sec. 2 of this chapter. Gliding angle ϑ and velocity V vs. angle of attack α are shown in Fig. 291a in the range up to about $-\vartheta = 15°$. The conditions of gliding under higher angles have to be computed with due consideration of the modified equilibrium equation $L = W \cos \vartheta$.

This will be discussed later (Sec. XVI.3) in connection with landing problems.

For a further discussion of the performance problem we have to study the *modifications of the power curves due to changes in the parameters*. The influence of ρ, that is, of altitude, will be the subject matter of a particular study in the next section. Here, we shall discuss briefly the influence of a change in weight and in wing area (or wing loading) on the P_{re}-curve and the influence of a change of propeller diameter, engine moment, and propeller pitch on P_{av}.

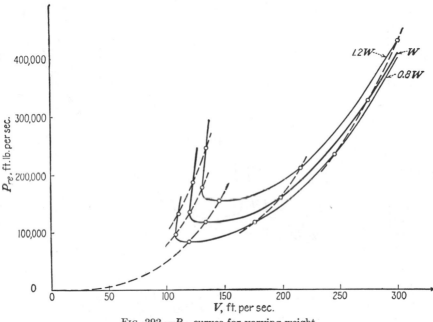

FIG. 292.—P_{re}-curves for varying weight.

Let, first, the *weight W vary*. Equations (a) and (b) show that, if a point of the P_{re}-curve is identified by the α-value, its abscissa changes in proportion to $\sqrt{W}$ and its ordinate proportional to $\sqrt{W^3}$. Accordingly, with changing W all points of the curve move along cubic parabolas, *i.e.*, along curves the ordinates of which are proportional to the cube of the abscissas (Fig. 292). The figure gives the P_{re}-curves for $0.80W$ and $1.20W$ corresponding to the former example. Owing to the increase of the ordinates with increasing W, the climbing rates always go down. Thus the maximum load an airplane can carry is determined by the requirement of a certain climbing rate for the start. The influence of a change in weight on level-flight speed is much smaller, particularly for fast airplanes. In the latter case, the second term in Eq. (b), namely,

$V^3 S_p$, prevails in the range of high speed, and then it is seen that the curve $P_{re} \sim \rho S_p V^3 / 2$ is approximately independent of W (the W influences the first term only, through C_D). In general, a slight decrease in level speed will be the consequence of increased W.

If the weight is left constant and the *wing area increased*, i.e., the wing loading diminished, Eq. (a) shows that a point with given α moves toward the left, its abscissa being proportional to $1/\sqrt{S}$. In Eq. (b) the first term is proportional to $1/\sqrt{S}$ also, but the second to $S^{-3/2}$; thus the ordinate changes in the same sense, but at a higher rate than the abscissa. Figure 293 shows the P_{re}-curves for a wing loading changed by

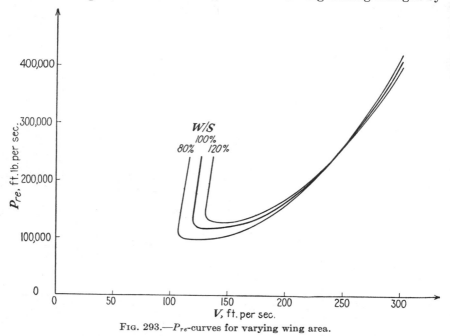

Fig. 293.—P_{re}-curves for varying wing area.

± 20 per cent as compared with that of our current example. The influence on level-flight speed is small (as can be explained as for the case of changing weight). The maximum climbing rate and still more the range of admissible V (range between maximum climbing and high-speed level flight) are, in general, reduced with increase of the wing loading W/S.

Turning now to changes in the available power, we first assume a *fixed-pitch propeller*. The effect on P_{av} of a change in engine power (throttling effect) can be estimated in the case of an invariable-pitch propeller in the same way as the influence of weight on the required power. We use the assumption—approximately correct within the speed range of interest—that the brake moment M_{br} of the engine is independent of the engine speed.   Then we obtain from Eqs. (c), (d), and (e), with $P_{br} = 2\pi n M_{br}$,

$$n = \sqrt{\frac{2\pi M_{br}}{\rho C_P d^5}}, \quad V = J\sqrt{\frac{2\pi M_{br}}{\rho C_P d^3}}, \quad P_{av} = \eta P_{br} = \eta 2\pi M_{br}\sqrt{\frac{2\pi M_{br}}{\rho C_P d^5}} \quad (27)$$

Here C_P and η are functions of the advance ratio J.   It follows that the forward speed V is proportional to $\sqrt{M_{br}}$ for any fixed J and P_{av} is proportional to M_{br} times n, that is, to $\sqrt{M_{br}^3}$.   Thus, each point of the P_{av} vs. V-curve moves along a cubic parabola.   In Fig. 294 the power curves for the moment changed by ± 20 per cent are plotted.   It is

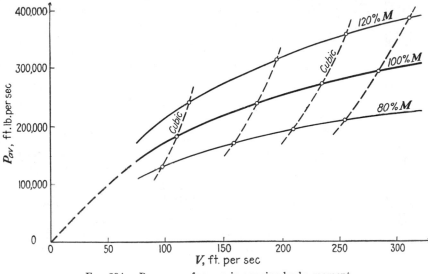

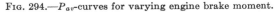

Fig. 294.—P_{av}-curves for varying engine brake moment.

immediately seen that level-flight speed as well as climbing rates increase with M_{br}.

A change of the propeller diameter d, with the brake moment M_{br} unchanged, has a similar effect on P_{av} as had the change of wing area on P_{re}.   We see from the foregoing equation that the abscissa V of a point identified by its J-value changes proportional to $d^{-3/2}$.   On the other hand, according to Eq. (e), with constant moment M_{br}, the P_{av}-value is proportional to n, that is, to $d^{-5/2}$.   Thus both coordinates decrease with increasing d, the ordinates at a higher rate than the abscissas. Figure 295 shows the power curves of our example for a 9 and 12 ft. propeller diameter as compared with the first assumption of $10\frac{1}{2}$ ft.   It is seen that, with the smaller propeller, level-flight speed and climbing rate are higher; but it should be noted that since M_{br} is assumed to be unchanged the engine power is greater.

Finally, the influence of changing the propeller pitch on the P_{av}-curve should be discussed. We defer the discussion in detail to the next chapter, where analytical methods of performance computation will be studied. But a glance at Fig. 286, which gives the dimensionless power curve derived from the propeller polar diagram, reveals the fact that increasing pitch essentially shifts the P_{av} vs. V-curve toward the right. It seems obvious that, all other conditions being equal, higher pitch propellers correspond to higher flight speed values.

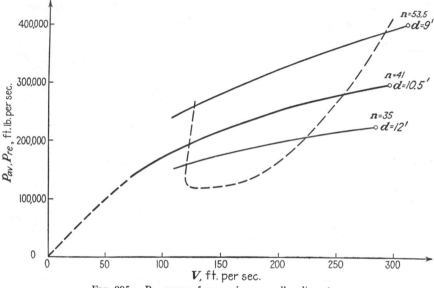

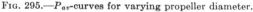

FIG. 295.—P_{av}-curves for varying propeller diameter.

The case of a *variable-pitch propeller* with the engine speed kept constant is governed by Eq. (*f*) and the definitions of the advance ratio and the efficiency, which can be written as

$$\frac{C_S}{J} = d\sqrt[5]{\frac{\rho n^2}{2\pi M_{br}}}, \qquad V = Jnd, \qquad P_{av} = \eta 2\pi n M_{br} \qquad (28)$$

It has already been seen how the P_{av}-curve can be derived from a propeller chart, including the C_S vs. J- and η vs. J-curves for various blade settings. If one of the constants ρ, d, n, M_{br} is changed, the whole procedure as already described must be repeated in order to find the modified P_{av} vs. V-curve. Only if several parameters change in such a way that $n^2 d^5/M_{br}$ is kept constant may the same straight line through the origin be used, and then the modified P_{av}-curve can be found by stretching the abscissas of the original curve proportional to nd and the ordinates proportional to nM_{br}. Figure 296 shows the η vs. J-curves for three values of the quantity

$\rho n^2 d^5 / M_{br}$.   This family of curves represents, except for the scales, all P_{av}-curves that can occur with propellers characterized by the same propeller chart.

Problem 9. In Prob. 1 find the high and low speeds of level flight and the maximum climbing rate. Compute the angle of attack and the advance ratio of the propeller for each of these three states of flight. In the state of high-speed level flight, what are the ratio of wing drag to total drag and the ratio of induced drag to total parasite drag?

Problem 10. Make the same computations for a gross weight of 5000 lb. and 7000 lb., respectively, as in Prob. 2.

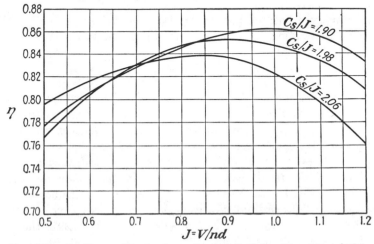

Fig. 296.—P_{av}/P_{br} vs. J for various values of C_s/J (see Figs. 283 and 284).

*Problem 11. Try to find the maximum weight that the airplane of Prob. 1 can carry if a climbing rate of 3 ft./sec. is required.

Problem 12. How are the answers to Probs. 9 and 11 affected if the aspect ratio of the wing is changed from 6 to 8?

Problem 13. In Prob. 1 assume that the engine power for all n is (a) reduced to 75 per cent; (b) increased to 125 per cent of its original value; and compute the change in level-flight speed and maximum climbing rate.

Problem 14. In Prob. 1 discuss numerically the influence of a change in propeller diameter by ± 1 ft.

Problem 15. If in the example in the text two variable-pitch propellers are used, both at $n = 31$, find the propeller diameter that gives for $V = 300$ ft./sec. the same available power as the single propeller.

5. Altitude Flight. The most important modification of the two power curves is that which takes place when the air density ρ is changed. The discussion of flight at higher altitudes is based on this modification. The questions to be answered here are, first, the same as in the case of sea-level flight: At what speed can we fly in a horizontal path, and what is the highest climbing rate? In addition, the problem of *ceiling* *altitude* arises: What is the maximum altitude at which flying with a specified airplane is possible?

The influence of air density ρ upon the *power-required* curve can easily be described using the Eqs. (a) and (b). We multiply each of these equations by $\sqrt{\sigma} = \sqrt{\rho/\rho_0}$. Then they read

$$V\sqrt{\sigma} = \sqrt{\frac{2W}{\rho_0 SC_L}}, \qquad P_{re}\sqrt{\sigma} = \frac{\rho_0}{2}(V\sqrt{\sigma})^3(SC_D + S_p) \qquad (29)$$

It is usual to call the two products

$$V\sqrt{\sigma} = V_i \quad \text{and} \quad P_{re}\sqrt{\sigma} = P_{rei} \qquad (30)$$

the *indicated speed* and the *indicated power required*, respectively. Upon introducing these quantities the equations take the form

$$V_i = \sqrt{\frac{2W}{\rho_0 SC_L}}, \qquad P_{rei} = \frac{\rho_0}{2}V_i^3(SC_D + S_p) \qquad (30')$$

The two expressions to the right are now identical with the expressions given in Eqs. (a) and (b) for V and P_{re} in the case of sea-level flight, *i.e.*, in the case $\rho = \rho_0$. Thus the following conclusion has been reached: *The curve representing P_{re} as function of V for sea-level conditions coincides with the plot of P_{rei} vs. V_i, indicated power required vs. indicated velocity, for any altitude.* This result can also be put in the following form: In order to obtain the P_{re} vs. V-curve for the altitude with density ρ, both the abscissas and the ordinates of each point of the sea-level curve must be multiplied by the factor $\sqrt{1/\sigma} = \sqrt{\rho_0/\rho}$. In Fig. 297, the power-required curves are shown for the altitudes 10,000, 20,000, 30,000, 40,000, and 50,000 ft., corresponding to the σ-values 0.74 to 0.15.

Let us assume for a moment that the power available would be independent of ρ. In this case, it is seen in Fig. 297 that the high speed of level flight (right point of intersection of the two power curves) first increases with increasing altitude and later becomes smaller. The distances between the two power curves, and thus the maximum climbing rate, decrease. There is a definite P_{re}-line, which has only one point in common with the P_{av}-curve (dotted line, in our case corresponding to σ about 0.144, that is, altitude of 51,200 ft.). On this line the high speed and the low speed of level flight coincide and no positive difference $P_{av} - P_{re}$ exists. At the altitude that corresponds to this P_{re}-curve, no climbing is possible: it is the *ceiling altitude* h_c.

The curves A in Fig. 298 show the level-flight velocities (high speed and low speed) as well as the maximum climbing speed w_m as abscissas to the ordinates h, as deduced from Fig. 297. The ceiling value h_c is characterized by two properties, $w_m = 0$ and the coinciding of high and low speeds. Although this diagram is based on the assumption of an

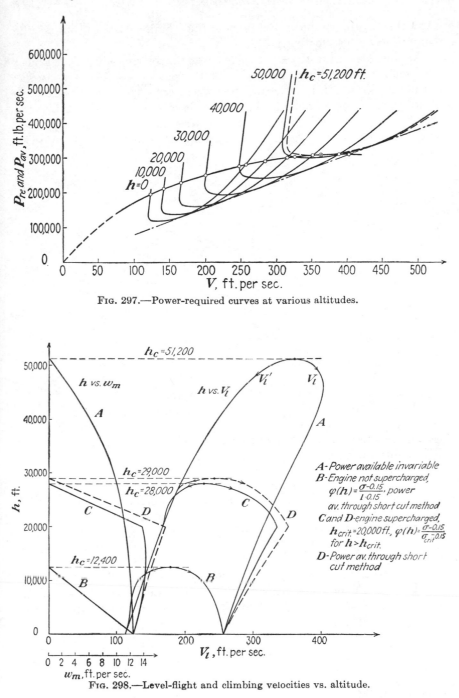

Fig. 297.—Power-required curves at various altitudes.

Fig. 298.—Level-flight and climbing velocities vs. altitude.

invariable available power, it shows the characteristic traits of the diagram for actual conditions.

An analysis of the *power-available* curve for varying altitude (varying ρ) is much more difficult than that of the power required, for, as was learned in Chap. XIII, the reaction of an internal-combustion engine to a change of air pressure and temperature is rather complicated. Let us first take up the case of a nonsupercharged engine combined with a fixed-pitch propeller. As a first, approximate *working hypothesis* we

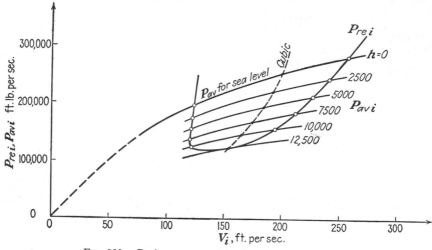

Fig. 299.—Performance at altitude h, short-cut method.

assume that for not too large altitudes h the brake-power altitude factor $\varphi(h)$ as introduced in Eqs. (10) and (11), Chap. XIII,

$$\varphi(h) = \frac{M_{br}(h)}{M_{br}(0)} = \frac{\psi(h) - C}{1 - C} = \frac{\sigma - 0.15}{0.85} \tag{31}$$

can be applied to each of the (slightly varying) moment values, which correspond to the various points of the power-available curve for sea level. If we then consider Eqs. (c), (d), and (e), which define the P_{av} vs. V-curve, either in their original form or rearranged in (27), we see that for a given value of J the velocity V is proportional to $\sqrt{M_{br}/\rho}$ and P_{av} proportional to $M_{br}\sqrt{M_{br}/\rho}$. Thus, if all values for sea level are denoted by the subscript zero,

$$\frac{V}{V_0} = \sqrt{\frac{M_{br}}{M_0}\frac{\rho_0}{\rho}}, \qquad \frac{P_{av}}{P_0} = \frac{M_{br}}{M_0}\sqrt{\frac{M_{br}}{M_0}\frac{\rho_0}{\rho}} \tag{32}$$

Introducing the ratios σ and φ and passing to the indicated values of velocity and power, $V_i = V\sqrt{\sigma}$, $P_{av\,i} = P_{av}\sqrt{\sigma}$, we find that

$$\frac{V_i}{V_0} = \sqrt{\varphi(h)}, \qquad \frac{P_{av\,i}}{P_0} = \sqrt{\varphi^3(h)} \qquad (32')$$

This means that each point of the P_{av} vs. V-curve plotted for sea-level conditions has to be moved on a cubic parabola through a distance determined by $\varphi(h)$. By using this short-cut method the performance computation for any altitude can be carried out in the following comparatively simple way (Fig. 299):

We start with the two power curves for sea-level flight and note that abscissas and ordinates will from now on mean $V\sqrt{\sigma}$ and $P\sqrt{\sigma}$ instead of V and P. The P_{re}-curve remains entirely unchanged, as we have already learned. To find the P_{av}-curve for any altitude h we compute the value $\varphi(h)$ and then multiply, for each point of the sea-level curve, the abscissa by the square root of $\varphi(h)$ and the ordinate by the third power of this root. This is carried out in Fig. 299 for the altitudes $h = 2500$, 5000, 7500, 10,000, and 12,500 ft. To find the level-flight speeds (low and high) at each altitude, we have to take the points of intersection of the respective P_{av}-curve with the invariable P_{re}-curve and divide the abscissas by $\sqrt{\sigma}$. Likewise, the maximum climbing rate is found by dividing the greatest vertical distance between the two curves by $W\sqrt{\sigma}$. The results of this analysis for the numerical example taken up in the preceding section are given in the following table. For the relation between σ and h the standard-atmosphere (Table I, page 10) is used.

	$h =$	0	2500	5000	7500	10,000 ft.
	$V_l =$	257	251	245	238	226 ft./sec.
	$V_l' =$	122.5	128	130	134	141 ft./sec.
$w_m = \left(\dfrac{dh}{dt}\right)_{max}$	$=$	12.5	10.1	7.4	4.9	2.4 ft./sec.

These values are shown in the curves B of Fig. 298. It will be noticed that the h vs. w_{max}-curve is almost exactly linear. The ceiling altitude is found by interpolation, or from the curves B, to be 12,400 ft.

The same method can be used in the case of a supercharged engine where $\varphi(h)$ is to be replaced by unity up to the critical altitude and then by the expression (12), Chap. XIII,

$$\varphi(h) = \frac{M_{br}(h)}{M_{br}(0)} = \frac{\sigma - 0.15}{\sigma_1 - 0.15} \quad \text{for } \sigma < \sigma_1,\ h > h_1 \qquad (33)$$

where σ_1 is the density at the critical altitude. The results of this computation are shown in the two curves D, for level-flight velocity and for maximum climbing rate, in Fig. 298. However, the working hypothesis above introduced, which led to what was called the short-cut method, does not supply a sufficiently good approximation when higher altitudes are involved.

In fact, in order to derive the P_{av}-curves for altitudes $h > 0$ directly from the P_{av}-curve for sea level we had to assume that in multiplying the value $M_{br}(0)$ by $\varphi(h)$ the value $M_{br}(h)$ is obtained that is valid at the same advance ratio J. This is not the exact meaning of the power-altitude factor φ. As was stated in Chap. XIII, the product $M_{br}(0)\varphi(h)$ gives the brake moment at the altitude h for the same engine speed n rather than for the same advance ratio. As long as it can be assumed that $M_{br}(0)$ varies only slightly with n and $\varphi(h)$ is not too far from 1, the difference may be insignificant. At any rate, there is one and only one

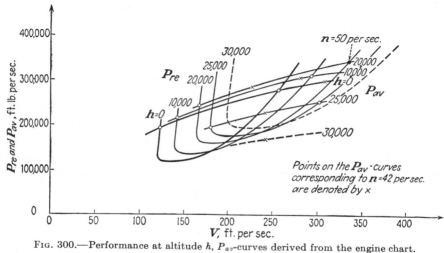

Fig. 300.—Performance at altitude h, P_{av}-curves derived from the engine chart.

way to find the correct solution to our problem, and that is to repeat the construction of the P_{av}-curve for each altitude, using for each h the $M_{br}(h)$ vs. n- or $P_{br}(h)$ vs. n-curve of the engine in the same way as it was done for $h = 0$. In Fig. 300 these P_{av}-curves are shown, together with the P_{re}-curves, the latter following the similarity law (similarity factor $\sqrt{\sigma}$) as already stated. The P_{av}-curves in Fig. 300 are derived, in the way described in Sec. 2 of this chapter, from an engine chart giving the curves $P_{br}(h) = \varphi(h)P_{br}(0)$ as functions of n. The results as to level-flight speeds and maximum climbing rate are shown in the curves C of Fig. 298. It is seen here that the high speed of level flight, the climbing rate, and the ceiling altitude all have slightly smaller values than those obtained through the short-cut method. The approximately linear character of the h vs. w_m-plot (beyond the critical altitude) appears as a common characteristic in each approach.

In discussing the results of performance computations at high altitudes it must be kept in mind that the successive points on a power-available curve correspond to increasing engine speeds.   On the curves in Fig. 300 the points are marked for which n equals 42.   It is seen that the high speeds of level flight from about $h = 10,000$ ft. on are obtainable only if the engine is allowed to exceed that limit.   It is different for the maximum climbing rate, which, in all cases, is reached at smaller forward speeds and smaller engine speeds.

From a knowledge of the maximum climbing rate w_m at each altitude we may gain an approximate picture of how a *continued climbing flight*, starting at sea level or higher, proceeds.   The climb will be represented

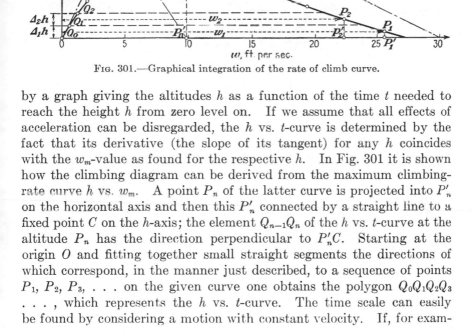

Fig. 301.—Graphical integration of the rate of climb curve.

by a graph giving the altitudes h as a function of the time t needed to reach the height h from zero level on.   If we assume that all effects of acceleration can be disregarded, the h vs. t-curve is determined by the fact that its derivative (the slope of its tangent) for any h coincides with the w_m-value as found for the respective h.   In Fig. 301 it is shown how the climbing diagram can be derived from the maximum climbing-rate curve h vs. w_m.   A point P_n of the latter curve is projected into P'_n on the horizontal axis and then this P'_n connected by a straight line to a fixed point C on the h-axis; the element $Q_{n-1}Q_n$ of the h vs. t-curve at the altitude P_n has the direction perpendicular to $P'_n C$.   Starting at the origin O and fitting together small straight segments the directions of which correspond, in the manner just described, to a sequence of points P_1, P_2, P_3, . . . on the given curve one obtains the polygon $Q_0 Q_1 Q_2 Q_3$ . . . , which represents the h vs. t-curve.   The time scale can easily be found by considering a motion with constant velocity.   If, for example, the point C is connected to the point on the horizontal axis marked 30 ft./sec., the perpendicular through O to this straight line cuts the horizontal through C at a distance $\overline{CC}_1$ that corresponds to the time $\overline{OC}:30$. In our case, since $\overline{OC}$ represents the altitude 30,000, the quotient is $30,000:30 = 1000$ sec. In this way the points marked 1,000, 2,000, . . . on the upper horizontal in Fig. 301 are found.

In Fig. 302 the two climb curves for our example, both with and without use of a supercharger, are shown. They can be found in the way

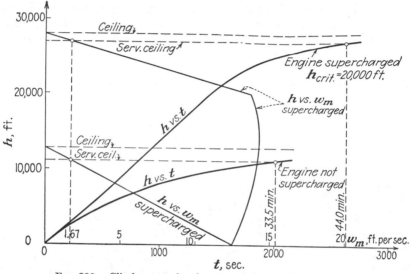

FIG. 302.—Climb curves for the numerical example of the text.

just described or by means of any other graphical or numerical method of integration.[1] The graphs for w_m are taken, respectively, from the curves B and C of Fig. 298.

It can be seen by simple reasoning that the ceiling altitude cannot be reached within a finite time. In fact, it follows from

$$w_m = \frac{dh}{dt}, \qquad dt = \frac{dh}{w_m}, \qquad t = \int \frac{dh}{w_m} \tag{34}$$

that t becomes infinite as w_m approaches zero under a finite slope of the h vs. w_m-curve. This makes it unpractical to use the real ceiling altitude as a criterion for the climbing ability of a plane. It is a widely used convention to introduce the *service ceiling*, defined as the altitude at which the maximum climbing rate is reduced to 100 ft./min. = 1.67 ft./sec., and to give the time needed for reaching this altitude as a criterion of climbing ability. The service ceiling can be read off from

[1] For a rapid method of integration see Sec. XVI.5.

the h vs. w_m-curve before the integration is carried out.   In our two cases the service ceiling is seen to be 11,100 and 26,950 ft., respectively,

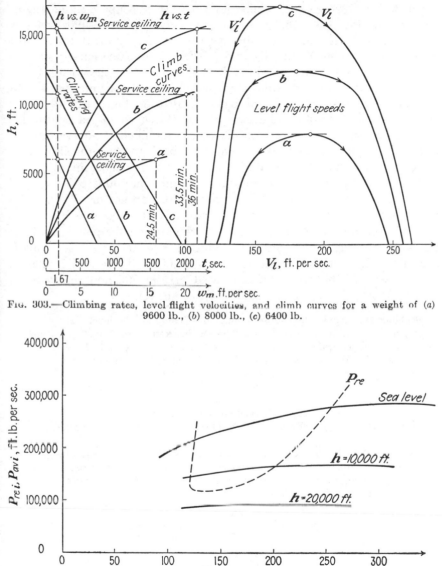

Fig. 303.—Climbing rates, level flight velocities, and climb curves for a weight of (a) 9600 lb., (b) 8000 lb., (c) 6400 lb.

Fig. 304.—Controllable-pitch propeller; indicated power available for various altitudes.

and the time needed to reach it, as found by the integration process, is 33.5 min. in the first and 44 min. in the second case.

It is to be expected that high-altitude flight is strongly influenced by a *change of weight*. The answer to this problem is supplied by combining the power-required curves for overweight and underweight, as developed in Fig. 292, with the power-available curves for various altitudes as given in Fig. 299 or 300. The results for the nonsupercharged engine are shown in Fig. 303. One group of curves gives the level-flight velocity V_l, the second the maximum climbing rate w_m for all altitudes and the gross weight values $0.80W$, W, and $1.20W$. The three curves starting at the origin are the integrated climb curves h vs. t for these three cases.

In the case of a variable-pitch propeller with engine speed kept constant, the answer to the problem of altitude flight can be found only by the repeated construction of the P_{av}-curve for each altitude. It is not possible in this case to derive the modified P_{av}-curve directly from that for $\sigma = 1$; instead, one must go back to the propeller chart. This chart supplies the family of η vs. J-curves for constant $\rho n^2 d^5/M_{br}$ as shown in Fig. 296. Now, for each altitude the value of ρ/M_{br} and consequently that of $\rho n^2 d^5/M_{br}$ is supposed to be known. We have to take the respective η vs. J-curve and multiply the abscissas by $nd \sqrt{\sigma}$ and the ordinates by $2\pi n M_{br}(h) \sqrt{\sigma}$ in order to find the power-available curve in the diagram of indicated values $V \sqrt{\sigma}$ and $P_{av} \sqrt{\sigma}$. This is shown in Fig. 304 for the altitudes $h = 10{,}000$, $20{,}000$, and $30{,}000$ ft. on the basis of the propeller chart used in Fig. 296 and the value of $\varphi(h) = M_{br}(h)/M_{br}(0)$ given in Eq. (10), Chap. XIII.

Problem 16. Apply the short-cut method to compute the level-flight velocities and maximum climbing rates at all altitudes for the example of Prob. 1. Assume the engine not supercharged and the brake power decreasing with $\varphi = 1.15\sigma - 0.15$.

Problem 17. How do the results of Prob. 16 change if the engine is supercharged, the critical altitude being $10{,}000$, $20{,}000$, or $30{,}000$ ft.?

Problem 18. Apply the correct method to solve Prob. 16, using the same formula for φ, and compare the results with those of the short-cut method. Make the same comparison in the cases of supercharged engines, Prob. 17.

Problem 19. Plot the climb curve h vs. t for one of the examples for which w_m vs. h has been found.

***Problem 20.** In the case of a nonsupercharged engine the maximum climbing rate w_m can be assumed to drop linearly from the ground value w_0 to zero at the ceiling altitude h_c. Give, under this assumption, the expressions for the service altitude h_s and the time t_s needed to reach it.

CHAPTER XV

ANALYTICAL METHODS OF PERFORMANCE COMPUTATION

1. Analytic Expressions for the Power Curves. In Sec. XIV.2 graphical methods were developed for deriving the power-required curve from the wing characteristics and the power-available curve from propeller and engine charts. These methods are equivalent to certain transformation formulas that were given in Sec. XIV.3, under the restriction that the brake moment of the engine can be considered independent of the engine speed. If this restriction is maintained, the engine chart (for full throttle) is replaced by giving the constant brake moment M_0 at sea level and the function $\varphi(h)$, which was introduced in Eq. (10), Chap. XIII, as the power-altitude factor:

$$M_{br} = \varphi(h)M_0 \tag{1}$$

In order to obtain now a full analytical representation of the power curves one has to introduce analytic expressions for both the wing and the propeller characteristics. As only approximations can be used to this end, it is obvious that *the analytical method of performance computation is less accurate, in principle, than the graphical one.* Its advantage is to supply the results in the form of functions of the data and thus to allow a much easier discussion of the various influences

Let us first take up the representation of the *power-required* curve. In Eqs. (3) and (5), Chap. VII, empirical expressions were given for lift and drag coefficients of an airfoil below the stalling angle:

$$C_L = k\alpha', \qquad C_D = a + \frac{k^2}{b}\alpha'^2 \tag{2}$$

By eliminating α' (the angle of incidence) the equation of the polar diagram is found in the form

$$C_D = a + \frac{C_L^2}{b} \tag{3}\cdot$$

This represents, as we know, a parabola with its vertex on the C_D-axis at the distance a from the origin. The constant a is the drag coefficient for zero lift. The three-dimensional wing theory suggests that in the case of a single wing (monoplane) with elliptic lift distribution the product $\pi\!R$ (R = aspect ratio) shall be set for b. This was shown in Sec. IX.5 [see, in particular, Eq. (19) in that section]. In the performance computation a much more general use is made of Eq. (3).   It is assumed that the C_L vs. C_D-curve, that is, the polar diagram, can be sufficiently approximated, in the range below stalling, by a parabola of the form (3), even if the lifting system is a multiplane with any lift distribution.   The values a and b are then taken from the experimental draft, *i.e.*, by drawing the parabola that fits best the empirical curve (Fig. 305).   If R is an appropriately defined aspect ratio, the square root of the ratio $b/\pi R$ is often called *Munk's span factor*.   This adaptation factor includes both the

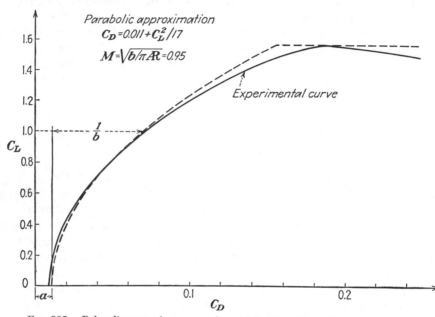

Fig. 305.—Polar diagram of a rectangular airfoil (Clark Y-profile, aspect ratio 6).

reduction from multiplane to monoplane and the influence of modified lift distribution, etc. (see Sec. IX.6).   Usual values of $b/\pi R$ for monoplanes are 0.90 to 0.95.

Beyond the stalling point the polar curve may be considered as a horizontal straight line.

For the purpose of performance computation we have to know, besides the wing characteristics, the values of weight W, of wing area S, and of parasite area S_p.   Using the notations introduced in Eq. (13), Chap. XIV,

$$-x' = C_D + \frac{S_p}{S}, \qquad y' = C_L$$

we have now

$$-x' = a + \frac{S_p}{S} + \frac{y'^2}{b} = a' + \frac{y'^2}{b} \tag{4}$$

where $a' = a + (S_p/S)$. This relation must be introduced in the formulas given in Chap. XIV for transforming the polar diagram x', y' into the power-required curve x, y. Using the first two equations (15), Chap. XIV, we find

$$y' = \frac{1}{x^2}, \qquad y = -x^3 x' = x^3 \left(a' + \frac{y'^2}{b}\right) = x^3 \left(a' + \frac{1}{bx^4}\right) \tag{5}$$

The last expression for y, which can be written as

$$y = a' x^3 + \frac{1}{bx} \tag{A}$$

represents, in dimensionless form, the power-required curve that will be used throughout the present chapter. In fact, according to the definitions in Eq. (12), Chap. XIV, x stands for V and y for P_{re}.

$$x = \frac{V}{V_1}, \qquad y = \frac{P_{re}}{WV_1} \qquad \text{with } V_1 = \sqrt{\frac{2W}{\rho S}} \tag{6}$$

Going back to the original variables V and P_{re}, we can write Eq. (A) as

$$P_{re} = W \left(\frac{a'}{V_1^2} V^3 + \frac{V_1^2}{b} \frac{1}{V}\right) \tag{A'}$$

The two coefficients V_1^2/a' and V_1^2/b, which have both the dimension of a velocity squared, are proportional to the wing loading W/S and inversely proportional to the density ρ or the density ratio σ. Thus W. B. Oswald, who was the first to apply in a systematic way the parabolic expression (3) in the performance calculation, introduced two loading factors, or *loadings*, λ_p and λ_s by the following definitions:

$$\frac{V_1^2}{a'} = \frac{2}{\rho} \frac{W}{a'S} = \frac{1}{\sigma} \frac{2}{\rho_0} \frac{W}{a'S} = \frac{\lambda_p}{\sigma}, \qquad \frac{V_1^2}{b} = \frac{2}{\rho} \frac{W}{bS} = \frac{1}{\sigma} \frac{2}{\rho_0} \frac{W}{bS} = \frac{\lambda_s}{\sigma} \tag{7}$$

The product $a'S$ equals, according to the definition of a', the sum $S'_p = aS + S_p$ and can therefore be considered as a combined parasite area S'_p of the airplane, including the parasite drag of the wing as well as of all other parts, like fuselage, tail, and landing gear. Except for the factor $2/\rho_0$ which is an absolute constant,

$$\lambda_p = \frac{2}{\rho_0} \frac{W}{a'S} = \frac{2}{\rho_0} \frac{W}{S'_p} \tag{7'}$$

is the weight per unit of parasite area and may therefore be called the *parasite loading*. On the other hand, the theoretical value of bS in the case of a monoplane is $\pi \mathcal{R} S = \pi B^2$, where B denotes the wing span. In general, one may introduce an effective span B_e (equal to actual span times Munk factor) by setting $bS = \pi B_e^2$. Then

$$\lambda_s = \frac{2}{\rho_0} \frac{W}{bS} = \frac{2}{\rho_0} \frac{W}{\pi B_e^2} \tag{7''}$$

represents, except for the absolute constant $2/\rho_0$, the weight per unit of a circular area of the radius B_e. This is why the expression *span loading* is in use for λ_s.

Introducing (7) we find a third form of Eq. (A) or (A'),

$$\frac{P_{re}}{W} = \sigma \frac{V^3}{\lambda_p} + \frac{1}{\sigma} \frac{\lambda_s}{V} \tag{A''}$$

Whether we use Eq. (A), (A'), or (A''), the shape of the power-required curve can easily be described. The first term on the right-hand

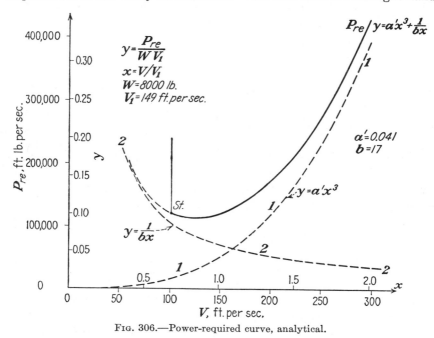

Fig. 306.—Power-required curve, analytical.

side gives a cubic parabola (1 in Fig. 306) starting at the origin with horizontal tangent and passing toward infinity. The second term supplies a hyperbola (2 in Fig. 306) starting at infinity and dropping to zero with increasing abscissas. The solid line in Fig. 306 corresponds, to the right of the point St, to the sum of the two terms 1 and 2. Beyond the stalling point where a constant C_L and thus constant V are assumed, the curve is replaced by a vertical straight line. The whole shape of the curve is then seen to be in satisfactory agreement with what was found in the preceding chapter by graphical methods.

For the dependence on altitude we find, multiplying by $\sqrt{\sigma}$,

$$\frac{P_{re}\sqrt{\sigma}}{W} = \frac{(V\sqrt{\sigma})^3}{\lambda_p} + \frac{\lambda_s}{V\sqrt{\sigma}} \qquad \text{or} \qquad \frac{P_{re\,i}}{W} = \frac{V_i^3}{\lambda_p} + \frac{\lambda_s}{V_i} \qquad (A''')$$

which confirms that the relation between the indicated values

$$P_{re\,i} = P_{re}\sqrt{\sigma} \qquad \text{and} \qquad V_i = V\sqrt{\sigma}$$

is the same as that between P_{re} and V in the case $\sigma = 1$.

We turn now to an analogous analysis of the *power-available* curve in the case of a fixed-pitch propeller. As a starting point, we take the formulas developed in Sec. XI.4 for the thrust and power coefficient C_T and C_P of a propeller under the assumption of a representative blade element at a distance r from the axis. Denoting the blade area by S_P (instead of A) and keeping all the other notations of Chap. XI unchanged, we write

$$\frac{C_T}{J^2} = \frac{1}{2}\frac{S_P}{d^2}\frac{C_L \cos \gamma - C_D \sin \gamma}{\sin^2 \gamma}, \qquad \frac{C_P}{J^2} = \frac{1}{2}\frac{S_P}{d^2}\frac{2\pi r}{d}\frac{C_L \sin \gamma + C_D \cos \gamma}{\sin^2 \gamma} \qquad (8)$$

In these equations we introduce $\gamma = \beta - \alpha = \beta' - \alpha'$ and consider the angle of incidence α' as small so as to have $\sin \alpha' = \alpha'$, $\cos \alpha' = 1$. (It should be remembered that β' is the blade setting angle with respect to the zero lift direction of the blade profile.) Consequently, we have to take $C_L = k\alpha'$ (below stalling), and C_D which, in general, would be of the form (3) has to be identified with the first constant term since C_L^2 is of second order in α' and, moreover, the denominator b, which is proportional to the aspect ratio, has a rather large value. The constant term, *i.e.*, the zero lift drag coefficient of the propeller profile, may be called a_P and considered as a quantity of the order α'. Thus, upon using

$$\begin{aligned}
C_L &= k\alpha', & \sin \gamma &= \sin \beta'(1 - \alpha' \cot \beta') \\
C_D &= a_P, & \cos \gamma &= \cos \beta'(1 + \alpha' \tan \beta')
\end{aligned} \qquad (9)$$

the two expressions (8) become, if terms of second order in α' are neglected,

$$\frac{C_T}{J^2} = \frac{S_P}{2d^2}\frac{k\alpha' \cos \beta' - a_P \sin \beta'}{\sin^2 \beta'}, \qquad \frac{C_P}{J^2} = \frac{S_P}{2d^2}\frac{2\pi r}{d}\frac{k\alpha' \sin \beta' + a_P \cos \beta'}{\sin^2 \beta'} \qquad (10)$$

The two ratios C_T/J^2 and C_P/J^2 were used in Sec. XIV.3 in the definition (22) of the dimensionless quantities x' and y'. Accordingly, we now have

$$x' = -\frac{S_P}{S}\frac{k\alpha' \cos \beta' - a_P \sin \beta'}{\sin^2 \beta'}, \qquad y' = \frac{S_P}{S}\frac{Wr}{M_{br}}\frac{k\alpha' \sin \beta' + a_P \cos \beta'}{\sin^2 \beta'} \qquad (11)$$

In order to find the equation of the distorted polar diagram of the propeller as used in Sec. XIV.3, we have to eliminate α' from these two equations. Multiplying the first by Wr/M_{br} and sin β', the second by cos β', and adding both, we find

$$\frac{Wr}{M_{br}} x' \sin \beta' + y' \cos \beta' = \frac{S_P}{S} \frac{Wr}{M_{br}} \frac{a_P}{\sin^2 \beta'} \qquad (12)$$

This linear equation or the straight line that it represents takes the place of the parabola (4) in the case of the power-required analysis. If

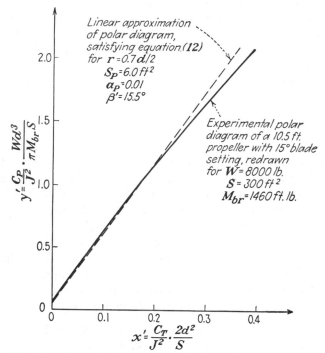

Fig. 307.—Propeller polar diagram for analytic performance computations.

the polar curve of the propeller is given, the best-fitting straight line would supply appropriate values for the coefficients of Eq. (12) (see Fig. 307).

The last step consists in transforming, according to the transformation formulas (15), Chap. XIV, the variables x', y' into x and y, which are the dimensionless representatives of V and P_{av}. Writing $1/x^2$ for y' and $-y/x^3$ for x', as stated in Eq. (15), Chap. XIV, and multiplying both sides of (12) by x^3, we find that

$$-\frac{Wr}{M_{br}} y \sin \beta' + x \cos \beta' = \frac{S_P}{S} \frac{Wr}{M_{br}} \frac{a_P}{\sin^2 \beta'} x^3$$

and, finally, solving for y,

$$y = \frac{M_{br}}{Wr} \cot \beta' \, x - \frac{a_P S_P}{S} \frac{1}{\sin^3 \beta'} x^3 \qquad (B)$$

This is *in dimensionless form the equation of the power-available curve* for an engine of constant brake moment with a fixed-pitch propeller. If we go back to the original variables $V = xV_1$ and $P_{av} = yWV_1$, the

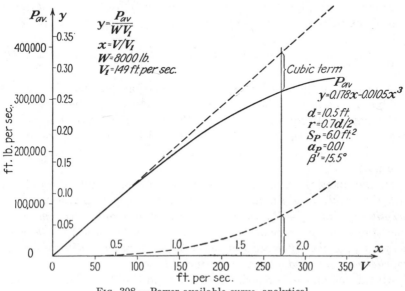

Fig. 308.—Power-available curve, analytical.

equation reads

$$P_{av} = \frac{M_{br}}{r \tan \beta'} V - \frac{a_P S_P}{S \sin^3 \beta'} \frac{W}{V_1^2} V^3 \qquad (B')$$

It is seen from (B) and (B') that the power-available curve can be found by subtracting from the ordinates of a straight line through the origin the ordinates of a cubic parabola (Fig. 308). If we take into account the value of $V_1^2 = 2W/\rho S$ and introduce in analogy to (7) two new loadings

$$\lambda_P = \frac{2}{\rho_0} \frac{W}{a_P S_P} \sin^3 \beta', \qquad \lambda_0 = \frac{Wr}{M_0} \tan \beta' \qquad (13)$$

Eq. (B') takes the form, according to (1),

$$\frac{P_{av}}{W} = \frac{\varphi(h)}{\lambda_0} V - \sigma \frac{V^3}{\lambda_P} \qquad (B'')$$

The *propeller parasite loading* λ_P is in obvious analogy to the λ_p as defined in (7'). Except for the factor $\sin^3 \beta'$, the parasite area of the airplane, S_p', is replaced by $a_P S_P$, which can be considered as the parasite area of the propeller. The *propeller power loading* λ_0 is dimensionless and connected with the power loading, which will be introduced later. As seen from Eq. (12), λ_0 is the slope of the distorted propeller polar curve, taken at sea level, or of the straight line approximating this curve. On the other hand, the value of $1/x'$ at $y' = 0$ gives, when multiplied by $2W/S\rho_0$, the magnitude λ_P.

Finally, multiplying both sides of Eq. (B'') by $\sqrt{\sigma}$, we obtain the form of the power-available expression corresponding to Eq. (A''').

$$\frac{P_{av}\sqrt{\sigma}}{W} = \frac{\varphi(h)}{\lambda_0} V\sqrt{\sigma} - \frac{(V\sqrt{\sigma})^3}{\lambda_P} \quad \text{or} \quad \frac{P_{av\,i}}{W} = \frac{\varphi(h)}{\lambda_0} V_i - \frac{V_i^3}{\lambda_P} \quad (B''')$$

Here, of course, the conclusion that the relation between the indicated variables $P_{av\,i}$ and V_i is the same for all altitudes cannot be drawn, owing to the power-altitude factor $\varphi(h)$. We learn, however, that only the linear term on the right-hand side is affected by the change of altitude, *i.e.*, only the slope of the straight line (Fig. 308) which determines the main part of P_{av}.

Summarizing the results of this analysis we may state the following: *With the usual simplifications for wing and propeller characteristics, the analytic expression for P_{re} appears as the sum of a cubic term (main term, depending on the parasite area) and a term proportional to $1/V$; the expression for P_{av}, as the difference of a linear term (main term, determined by the propeller moment) and a term proportional to V^3.*

All this, of course, is restricted to the condition that the wing and the propeller are working below stalling.

Problem 1. Derive Eq. (A'') directly from the equilibrium conditions, using the assumptions (2) about lift and drag coefficients, without referring to the transformation formulas of Sec. XIV.3.

Problem 2. Compute the expression for P_{re}/W as a function of V in the case of a monoplane with 32 lb./ft.² wing loading, aspect ratio 7.4, and wing profile according to Fig. 116 in Chap. VII. Assume that the frontal area of the fuselage (with the parasite drag coefficient 0.11) is one-fifth the wing area and that it contributes 70 per cent of the total parasite area of the plane.

Problem 3. If instead of λ_p and λ_s, so-called "engineering" forms of the loading,

$$l_p = \frac{W}{S_p'}, \quad l_s = \frac{W}{B_e^2}$$

are used, determine their dimensions and show that the numerical relations, in foot-pounds per second, are

$$\lambda_p = 841.0\, l_p, \quad \lambda_s = 267.7\, l_s$$

Problem 4. Compute the power available curve for an airplane of 8000 lb. gross weight with an engine supplying 600 hp. at 2000 r.p.m. (at sea level). Assume that the propeller is equivalent to an airfoil element of 3.6 ft.² area, with Clark Y-profile, working at a distance $r = 3.4$ ft. from the axis and set at such an angle that at the forward speed $V = 250$ ft./sec. the slip is about 20 per cent.

Problem 5. Give the analytic expressions for total drag and propeller thrust following from Eqs. (A'') and (B''). Try to find a simple direct derivation for the expression of T.

2. Gliding. Level Flight with Given Power. In this section, questions will be discussed that can be answered without knowing the shape of the power-available curve. This is the case, particularly, for gliding flight with the engine shut off.

It was learned in Chap. XIV that with $P_{av} = 0$ the sinking rate of an airplane equals the quotient P_{re}/W. Thus, upon using Eq. (A''), the sinking rate at the altitude h is determined by

$$w_s(h) = \sigma \frac{V^3}{\lambda_p} + \frac{1}{\sigma} \frac{\lambda_s}{V} \tag{14}$$

where σ is considered a known function of h. The smallest sinking rate at given h is found by differentiation with respect to V:

$$3\sigma \frac{V^2}{\lambda_p} - \frac{1}{\sigma} \frac{\lambda_s}{V^2} = 0, \qquad V = \sqrt[4]{\frac{\lambda_s \lambda_p}{3\sigma^2}} \tag{15}$$

If this V is introduced in Eq. (14), we obtain

$$[w_s(h)]_{min} = \frac{4\sigma}{\lambda_p} \left(\frac{\lambda_s \lambda_p}{3\sigma^2}\right)^{\frac{3}{4}} = 1.75 \frac{1}{\sqrt{\sigma}} \lambda_p^{-\frac{1}{4}} \lambda_s^{\frac{3}{4}} \tag{16}$$

On the other hand, the gliding angle ϑ can be computed from

$$w_s = V \sin \vartheta,$$

thus:

$$\vartheta \sim \sin \vartheta = \sigma \frac{V^2}{\lambda_p} + \frac{1}{\sigma} \frac{\lambda_s}{V^2} \tag{17}$$

The slope of flattest gliding follows by differentiation,

$$2\sigma \frac{V}{\lambda_p} - \frac{2}{\sigma} \frac{\lambda_s}{V^3} = 0, \qquad V = \sqrt[4]{\frac{\lambda_s \lambda_p}{\sigma^2}} \tag{18}$$

and, combining (18) with (17),

$$\vartheta_{min} \sim (\sin \vartheta)_{min} = \frac{2\sigma}{\lambda_p} \sqrt{\frac{\lambda_s \lambda_p}{\sigma^2}} = 2\sqrt{\frac{\lambda_s}{\lambda_p}} \tag{19}$$

Thus it is seen that *the minimum sinking rate increases with altitude, proportional to* $1/\sqrt{\sigma}$, *while the angle of flattest gliding remains unchanged;* both quantities depend on the parasite and span loading factors λ_p and λ_s. Besides, it follows from (15) and (18) that the forward speed at the flattest gliding is $\sqrt[4]{3} = 1.32$ times as great as the forward speed at the lowest sinking rate.

Another important problem that can be solved independently of the shape of the P_{av}-curve consists in finding the *level-flight speed* in the case in which the actual value of power available in level flight is given. Assume that the brake power P_0 of the engine and the efficiency η_0 of the propeller, both for the conditions of sea-level flight, are known in advance. Then, for this particular mode of flight we have

$$P_{av} = P_{re} = \eta_0 P_0 \tag{20}$$

and this substituted in Eq. (A'') gives, with $\sigma = 1$,

$$\frac{V^3}{\lambda_p} + \frac{\lambda_s}{V} = \frac{\eta_0 P_0}{W} \tag{21}$$

If this equation is solved for V, the speed V_l of sea-level flight is obtained. It is usual to introduce here a third loading factor, the *power loading*

$$\lambda_t = \frac{W}{\eta_0 P_0} \tag{22}$$

Evidently, $1/\lambda_t$ has the dimension of a velocity. Numerical values for λ_t will be discussed in Sec. 4 of this chapter. Upon writing V_l for V in (21) this equation becomes

$$\frac{V_l^3}{\lambda_p} + \frac{\lambda_s}{V_l} = \frac{1}{\lambda_t} \quad \text{or} \quad V_l\left(1 - \frac{\lambda_t}{\lambda_p} V_l^3\right) = \lambda_s \lambda_t \tag{23}$$

To solve Eq. (23) we may introduce a dimensionless substitute z for the unknown V_l by setting

$$z = V_l \sqrt[3]{\frac{\lambda_t}{\lambda_p}} \quad \text{or} \quad V_l = z \sqrt[3]{\frac{\lambda_p}{\lambda_t}} \tag{24}$$

This transforms (23) into

$$z(1 - z^3) = \lambda_s \lambda_t \sqrt[3]{\frac{\lambda_t}{\lambda_p}} = \lambda_p^{-1/3} \lambda_s \lambda_t^{4/3} \tag{25}$$

The expression on the right-hand side is known as the *fundamental performance parameter* Λ:

$$\Lambda = \lambda_p^{-1/3} \lambda_s \lambda_t^{4/3} \tag{26}$$

It is seen that Λ is a dimensionless quantity. The numerical discussion that follows in Sec. 4 will show that Λ has in all practical cases small values, not exceeding 0.20. Equation (25), which has to be solved for z, takes now the simple form

$$z - z^4 = \Lambda \tag{25'}$$

The function $z - z^4$ is plotted against z in Fig. 309. It vanishes at $z = 0$ and $z = 1$ and in between has positive values, the maximum of which is $\frac{3}{4} \times 4^{-\frac{1}{3}} = 0.472$.   On each horizontal line with an ordinate smaller than 0.472 there are two points of intersection with the curve.   For Λ-values smaller than 0.20, one point is found close to $z = 0$ and another point near to $z = 1$.   Only the latter, which corresponds to the high speed of level flight, is of interest since our expression for P_{re} loses its validity for small velocities, as was mentioned in the preceding section.   The abscissa z of the right-hand point can be found in the following way:

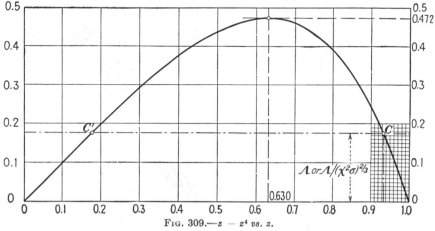

Fig. 309.—$z - z^4$ vs. z.

As z is near to 1 and Λ a small quantity, we may write $z = 1 + c_1\Lambda$ where c_1 is an undetermined constant.   Introducing this in (25′) and neglecting higher powers of Λ, we find since $(1 + c_1\Lambda)^4 = 1 + 4c_1\Lambda + 6c_1^2\Lambda^2 + \ldots$,

$$(1 + c_1\Lambda) - (1 + 4c_1\Lambda) = \Lambda, \qquad \text{that is, } c_1 = -\tfrac{1}{3}$$

Thus $1 - \Lambda/3$ is a first approximation for z.   To find a better one, we now assume $z = 1 - \Lambda/3 + c_2\Lambda^2$, introduce this in (25′), and neglect all terms of higher than second order.   This gives

$$\left(1 - \frac{\Lambda}{3} + c_2\Lambda^2\right)\left(1 - \frac{4\Lambda}{3} + \frac{6\Lambda^2}{9} + 4c_2\Lambda^2\right) = \Lambda, \quad \text{that is, } c_2 = -\tfrac{2}{9}$$

In this way, we may continue and compute another term, etc.   The result is

$$z = 1 - \frac{\Lambda}{3} - \frac{2\Lambda^2}{9} - \frac{20\Lambda^3}{81} - \cdots \tag{27′}$$

The fourth term cannot exceed 0.002, and in most cases even the third term is negligible.

Returning to (24) we find the *speed value* V_l *for sea-level flight*

$$V_l = \sqrt[3]{\frac{\lambda_p}{\lambda_t}}\left(1 - \frac{\Lambda}{3} - \frac{2\Lambda^2}{9} - \cdots\right) \tag{27}$$

expressed in terms of the three loading factors λ_p, λ_s, λ_t *or of two of them and* Λ.   Cutting off the power series after the second term would mean that the curve $z - z^4$ is replaced by its tangent at $z = 1$. A glance at Fig. 309 shows that this, in general, is a sufficiently good approximation. Omitting all terms, except the first, would make V_l completely independent of the span loading λ_s. This would mean that the induced wing drag is completely disregarded as being small compared with the

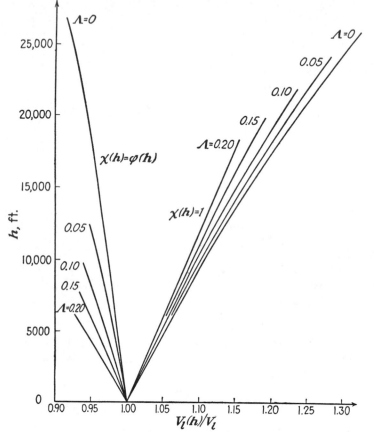

Fig. 310.—Altitude h vs. speed ratio $V_l(h)/V_l$, computed from the first three terms of the development.

total parasite drag. In the case of high-performance airplanes (very small Λ) this is justified.

The same method can be used, to a certain extent, for finding the level-flight speed $V_l(h)$ at an altitude h, if we assume that the power available at this altitude in level flight is known. Let us suppose that the corresponding ratio

$$\frac{P_{av}(h)}{P_{av}(0)} = \frac{\eta(h)P_{br}(h)}{\eta_0 P_0} = \chi(h) \tag{28}$$

which we may call the "power-available altitude factor for level flight," is given for all values of h. Instead of Eq. (21) we have to apply Eq. (A'''), which gives

$$\frac{V_i^3}{\lambda_p} + \frac{\lambda_s}{V_i} = \frac{\chi\sqrt{\sigma}}{\lambda_t} \tag{29}$$

This is equivalent to (23) if V_l is replaced by $V_i = V_l(h) \sqrt{\sigma}$ and λ_t by $\lambda_t/\chi\sqrt{\sigma}$. Therefore, (27) supplies

$$V_l(h) \sqrt{\sigma} = \sqrt[3]{\frac{\lambda_p}{\lambda_t}} \chi(h) \sqrt{\sigma} \left(1 - \frac{\Lambda}{3} \chi^{-4/3}\sigma^{-2/3} - \frac{2\Lambda^2}{9} \chi^{-8/3}\sigma^{-4/3} - \cdots \right)$$

or, divided by $\sqrt{\sigma}$,

$$V_l(h) = \sqrt[3]{\frac{\lambda_p}{\lambda_t}} \left(\chi^{2/3}\sigma^{-1/3} - \frac{\Lambda}{3} \chi^{-1}\sigma^{-1} - \frac{2\Lambda^2}{9} \chi^{-7/3}\sigma^{-5/3} - \cdots \right) \tag{30}$$

Figure 310 shows the graph of h vs. $V_l(h)/V_l$ for five Λ-values from 0 to 0.20, first, under the assumption that $\chi = 1$, that is, that the power available in level flight is the same for all altitudes. This may be the case with a supercharged engine and a variable-pitch propeller working at constant engine speed. In addition, in Fig. 310 the curves corresponding to (30) are plotted under the hypothesis that $\chi(h)$ can be identified with the power-altitude factor $\varphi(h)$ as defined in Eq. (10), Chap. XIII. It is seen that $V_l(h)$ increases with h in the first case and decreases in the second. Under intermediate conditions, with supercharged engines, it will be found that $V_l(h)$ first increases with h and later drops.

Equation (30), however, can be used for moderate altitudes only, since for large h the argument of the power series, $\Lambda\chi^{-4/3}\sigma^{-2/3}$, which is supposed to be a small quantity, increases more and more. Then the convergence of the series becomes unsatisfactory. However, using Table 13, which gives the solution of the equation

$$z - z^4 = Z$$

for all values of Z up to 0.4725, we can immediately find the answer for all altitudes. In fact, the method just described is equivalent to ootting

$$z = V_l(h) \sqrt[3]{\frac{\sigma\lambda_t}{\chi\lambda_p}}, \qquad V_l(h) = z \sqrt[3]{\frac{\chi\lambda_p}{\sigma\lambda_t}} \tag{30a}$$

and introducing this into (29). With $V_i = V_l(h) \sqrt{\sigma}$ we obtain

$$z - z^4 = (\sigma\chi^2)^{-2/3}\Lambda = Z \tag{30b}$$

Hence for given σ, χ, and Λ the quantity Z is known and z can be taken from Table 13 and $V_l(h)$ from (30a).

The correct curves h vs. $V_l(h)$ have a characteristic property that is not immediately reproduced by Eq. (30) and not seen in the graphs (Fig. 310). There cannot be any V_l-value beyond a certain $h = h_c$ which is the *ceiling altitude*. (Mathematically speaking, the power series becomes divergent for $h = h_c$.)

It is not difficult to find the ceiling altitude once the power factor $\chi(h)$ is known. Equation (29) shows that no solution for V_i exists if the

TABLE 13.—SOLUTIONS OF $z - z^4 = Z$

$Z = 0.01$	$z = 0.997$	$Z = 0.26$	$z = 0.891$
2	93	27	86
3	90	28	80
4	86	29	74
5	83	30	68
0.06	0.979	0.31	0.862
7	75	32	55
8	72	33	49
9	68	34	42
10	64	35	34
0.11	0.960	0.36	0.826
12	56	37	18
13	52	38	10
14	48	39	00
15	44	40	791
0.16	0.940	0.41	0.780
17	35	42	68
18	31	43	55
19	26	44	40
20	22	45	23
0.21	0.917	0.46	0.700
22	12	47	662
23	07	4725	630
24	02		
25	897		

right-hand side is smaller than the minimum of the expression to the left, which is nothing else than $[w_s(h)]_{min}$ for $\sigma = 1$. This value was found in (16) to equal $1.75\,\lambda_p^{-\frac14}\lambda_s^{\frac34}$. Thus the ceiling altitude is determined by

$$\chi\,\frac{\sqrt{\sigma}}{\lambda_t} = 1.75\lambda_p^{-\frac14}\lambda_s^{\frac34}$$

If we denote the ceiling values of σ and h by σ_c and h_c and reintroduce the fundamental parameter Λ, the last equation becomes

$$\chi(h_c)\,\sqrt{\sigma_c} = 1.75\Lambda^{\frac34} = 1.75\lambda_p^{-\frac14}\lambda_s^{\frac34}\lambda_t \tag{31}$$

The relation between h_c and Λ is shown in Fig. 311. The curve a corresponds to the ideal assumption $\chi = 1$, assumed as valid up to the critical altitude of a supercharged engine with constant-speed propeller. For $\Lambda = 0.1$, the ceiling is about 21,000 ft. for an engine without supercharger and about 37,000 ft. for an engine with critical altitude of 30,000 ft. As a result, we may state that *the ceiling altitude h_c depends exclusively on the fundamental performance parameter and the available power-altitude*

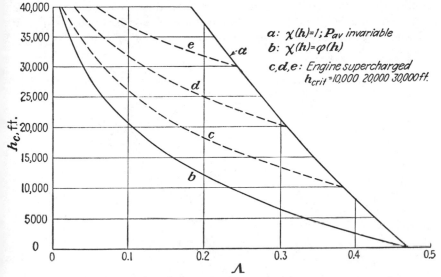

a: $\chi(h)=1$; P_{av} invariable
b: $\chi(h)=\varphi(h)$
c,d,e: Engine supercharged
$h_{crit}=10,000\ 20,000\ 30,000\ ft.$

FIG. 311.—Ceiling altitude h_c vs. fundamental performance parameter Λ.

factor, according to (31). All the curves in Fig. 311 can easily be plotted by solving (31) for Λ,

$$\Lambda = \tfrac{3}{4}\left(\frac{\sigma\chi^2}{2}\right)^{\tfrac{2}{3}} \tag{31'}$$

with σ and χ corresponding to any value of h.

Let us now discuss the exact relation between the level-flight speed $V_l(h)$ and the altitude h, supplementing formula (30), which can be used for smaller h only. First, we take up the case $\chi = 1$, power available constant. Upon introducing the velocity ratio v,

$$v = \frac{V_l(h)}{V_l} = \frac{V_i(h)}{\sqrt{\sigma}\,V_l} \tag{32}$$

Eq. (29) becomes

$$\chi\frac{\sqrt{\sigma}}{\lambda_t} = \frac{v^3 V_l{}^3 \sqrt{\sigma^3}}{\lambda_p} + \frac{\lambda_s}{v V_l \sqrt{\sigma}} \quad \text{or} \quad \chi\sigma = v^3 V_l{}^3 \frac{\lambda_t}{\lambda_p}\sigma^2 + \frac{\lambda_s\lambda_t}{v V_l} \tag{33}$$

Here we introduce z from (24) and Λ from (26) and find, with $\chi = 1$,

$$\sigma = v^3\sigma^2 z^3 + \frac{\Lambda}{vz} \qquad \text{or} \qquad \sigma^2 - \frac{\sigma}{v^3 z^3} + \frac{\Lambda}{v^4 z^4} = 0 \qquad (33')$$

Solving this quadratic equation for σ,

$$\sigma = \frac{1 \pm \sqrt{1 - 4\Lambda v^2 z^2}}{2v^3 z^3} \qquad (34)$$

supplies the answer to our question. In fact, for a given Λ the z can be computed from (25'), and then (34) gives for each abscissa v the value of σ and therefore the corresponding h for standard atmosphere.

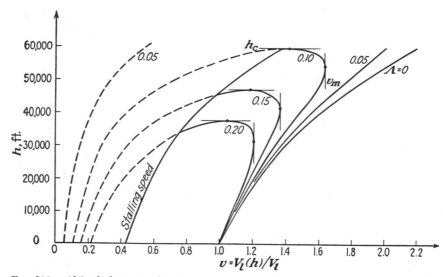

Fig. 312.—Altitude h vs. speed ratio v for various Λ-values and constant power available.

These h vs. v-curves are plotted in Fig. 312 for the same Λ-values as were used in Fig. 310. All curves except that for $\Lambda = 0$ have one point where v, and therefore $V_l(h)$, has a maximum and one point where h reaches its maximum h_c. The maximum v_m is obtained when the square root in (34) vanishes. This supplies the equations

$$1 = 4\Lambda v^2 z^2 \qquad \text{and} \qquad \sigma = \frac{1}{2v^3 z^3}$$

from which we draw

$$v_m = \frac{1}{2z\sqrt{\Lambda}} \qquad \text{and} \qquad \sigma_m = 4\Lambda^{3/2} \qquad (35)$$

To find the ceiling values v_c and σ_c, combine the first equation (33') and its derivative with respect to v:

$$vz\sigma = v^4 z^4 \sigma^2 + \Lambda, \qquad z\sigma = 4v^3 z^4 \sigma^2$$

These two equations solved for v and σ supply

$$v_c = \frac{\sqrt{3}}{4z\sqrt{\Lambda}} \quad \text{and} \quad \sigma_c = \frac{16}{3\sqrt{3}}\Lambda^{3/2} \tag{36}$$

The last relation coincides exactly with the earlier result (31) for $\chi = 1$.

The continuation of the h vs. v-curves to the left of the ceiling point corresponds evidently to the low speeds of level flight. Owing to stalling, these branches of the curves lose their validity at certain points. If C_L is assumed to remain constant beyond the stalling angle, the left end of each h vs. v-curve will have ordinates proportional to $1/\sqrt{\sigma}$. This is shown in Fig. 312 under the assumption that the stalling velocity at sea level is about $0.42V_l$. Note that the graph is based on the hypothe-

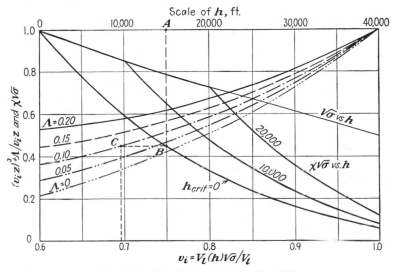

Fig. 313.—Graphical solution of Eq. (37).

sis that the available power is the same for all states of level flight under consideration, whether at the low or at the high speed.

In the general case of P_{av} changing with h, one may compute $V_l(h)$ by using Table 13 as described and from this find $v = V_l(h)/V_l$. Or the graph Fig. 313 can be applied to determine the level-flight velocity in the following way: If in deriving (33') from (33) the factor χ is not omitted, this equation reads

$$\chi\sigma = v^3\sigma^2z^3 + \frac{\Lambda}{vz} \quad \text{or} \quad \chi\sqrt{\sigma} = (v_iz)^3 + \frac{\Lambda}{v_iz} \tag{37}$$

where $v_i = v\sqrt{\sigma}$ is the indicated velocity ratio

$$v_i = \frac{V_l(h)\sqrt{\sigma}}{V_l} \tag{38}$$

For each Λ the value of z is known from (26); thus the right-hand side of (37) can be plotted against v_i. This is done in Fig. 313 for the five Λ-values already considered. Moreover, Fig. 313 includes the curve $\chi \sqrt{\sigma}$ vs. h, which must be supposed to be given. For any $h = \overline{OA}$, one has $\overline{AB} = \chi \sqrt{\sigma}$; and, with C on the horizontal through B and on the curve for the respective Λ, the abscissa of C supplies v_i and finally $v = v_i/\sqrt{\sigma}$.

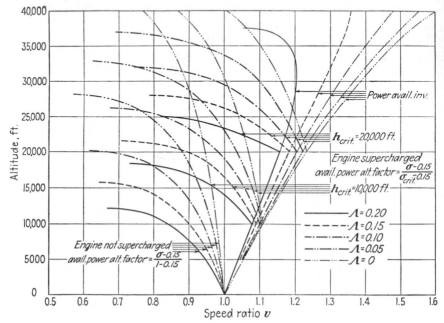

Fig. 314.—Altitude h vs. speed ratio v for various engine conditions and Λ-values; power available varying with altitude.

In Fig. 314 the results are given under the same assumption for $\chi(h)$ as that already used, for a nonsupercharged engine and for superchargers with the critical altitudes 10,000 and 20,000 ft. Only the high-speed velocities of level flight, up to the ceiling altitude, are plotted in the graph.

Problem 6. The parasite and span loadings of a plane are given: $\lambda_p = 700{,}000$, $\lambda_s = 800$ ft.^{2}/sec.^{2} Compute the minimum gliding rate, the minimum gliding angle, and the corresponding forward speeds at the altitudes 10,000, 20,000, and 30,000 ft. How does the slope of the gliding path at the minimum rate compare with the slope of the flattest gliding? How do the gliding rates compare in these two different motions?

*Problem 7. Compute, for given λ_p and λ_s, the time needed and the space covered in gliding from $h_1 = 20{,}000$ ft. to $h_2 = 10{,}000$ ft. under the condition (a) in slowest gliding; (b) of flattest gliding. Use data of Prob. 6.

Problem 8. Compute the level-flight velocity for an airplane with a power loading of 10 lb./b.hp., 84 per cent propeller efficiency, and the λ_p- and λ_s-values as in

Prob. 6, (a) at sea level; (b) at 15,000 and 30,000 ft. altitude under the assumption that the available power has dropped 25 per cent in either case.

Problem 9. If the loadings l_p and l_s as defined in Prob. 3 are combined with $l_t = \lambda_t$ in pounds per horsepower, show that the fundamental parameter in engineering units becomes

$$L = \frac{\Lambda}{0.006293}$$

*Problem 10. In what way must the available power change with altitude if the level-flight speed should be kept constant?

Problem 11. Compute the ceiling altitude under the assumption that the available power decreases linearly with altitude at a rate of 2.5 per cent per 1000 ft., starting with the critical altitude. Develop a chart corresponding to Fig. 311 under this assumption.

*Problem 12. Explain the geometrical relationship between the h vs. v-curve for $\chi = 1$ and the original P_{re} vs. V-curve. Show how the set of h vs. v-curves plotted for $\chi = 1$ can be used to determine the ratio $V_l(h)/V_l(0)$ for a given function $\chi(h)$.

3. The Ideal Airplane: Power Available Independent of Speed. If it is desired to compute the climbing rate of an airplane or to answer any

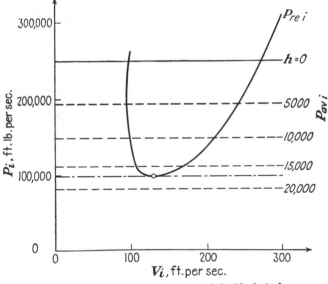

FIG. 315.—Indicated power curves of the ideal airplane.

question in connection with it, the shape of the power-available curve P_{av} vs. V must be known. It was seen in Chap. XIV that in the case of a fixed-pitch propeller (see, e.g., Fig. 282) this curve has a slope decreasing toward greater V and of moderate value in the range of usual velocities. A similar behavior has been found in the case of a variable-pitch propeller operated at constant engine speed (Fig. 296). Thus it seems worth while to try a first approximation by replacing the actual power-available curve by a horizontal straight line, *i.e.*, to assume that P_{av} is independent of V. Results reached under this hypothesis are often referred to as concerning an *ideal airplane*.

Note that the concept of the ideal airplane so far does not include any assumption about the relation between power and altitude. But since P_{av} is independent of V we may use the power factor $\chi(h)$ introduced in Eq. (28) in the form

$$P_{av}(h) = \chi(h)P_{av}(0) = \text{const. } \chi(h) \tag{39}$$

In Fig. 315 the indicated power curves for an ideal airplane are plotted.

According to Eq. (5′), Chap. XIV, the maximum climbing rate of any airplane is given by

$$w_m = \left(\frac{dh}{dt}\right)_{\text{max}} = \left(\frac{P_{av} - P_{re}}{W}\right)_{\text{max}} \tag{40}$$

Upon introducing P_{re} from Eq. (A'') and using (39) and the definition (22) of λ_t, the relation (40) supplies

$$w_m = \frac{P_{av}}{W} - \left(\frac{P_{re}}{W}\right)_{\text{min}} = \frac{\chi(h)}{\lambda_t} - \left(\sigma\frac{V^3}{\lambda_p} + \frac{\lambda_s}{\sigma V}\right)_{\text{min}} \tag{41}$$

The minimum of the expression in the parentheses was found in (16) to equal $1.75\lambda_p^{-\frac{1}{4}}\lambda_s^{\frac{3}{4}}/\sqrt{\sigma}$. Thus, according to (26),

$$w_m = \frac{\chi(h)}{\lambda_t} - \frac{1.75}{\sqrt{\sigma}}\lambda_p^{-\frac{1}{4}}\lambda_s^{\frac{3}{4}} = \frac{1}{\lambda_t}\left[\chi(h) - \frac{1.75}{\sqrt{\sigma}}\Lambda^{\frac{3}{4}}\right] \tag{41′}$$

If w_m is set zero, this equation confirms the previously found expression for the ceiling altitude [Eq. (31)].

If a climb is carried out with the maximum climbing rate at each level, the time needed to reach the altitude h from sea level is determined as the integral of dh/w_m, extended from 0 to h, as was seen in Eq. (34), Chap. XIV. In order to give this integral a simple form we may use for $\sqrt{\sigma}$ the approximation formula proposed in Eq. (14), Chap. I

$$\sqrt{\sigma} = \frac{c - h}{c} \quad \text{with } c = 81,000 \text{ ft.} \tag{42}$$

In the case of $\chi = 1$ (power available independent of altitude), (41) and (42) combined lead to the following relations for the ceiling altitude h_c:

$$0 = 1 - \frac{c}{c - h_c}1.75\Lambda^{\frac{3}{4}}, \quad h_c = c(1 - 1.75\Lambda^{\frac{3}{4}}), \quad 1.75\Lambda^{\frac{3}{4}} = \frac{c - h_c}{c} \tag{43}$$

The last expression introduced in (41) gives

$$w_m = \frac{1}{\lambda_t}\left(1 - \frac{c - h_c}{c - h}\right) = \frac{1}{\lambda_t}\frac{h_c - h}{c - h} \tag{44}$$

and thus for the time t needed for climbing

$$t = \int_0^h \frac{dh}{w_m} = \lambda_t \int_0^h \left[1 + \frac{c - h_c}{h_c - h} \right] dh$$

$$= \lambda_t \left[h - (c - h_c) \log \left(1 - \frac{h}{h_c} \right) \right] \quad (45)$$

The dotted lines in Fig. 316 show $\lambda_t w_m$ vs. h according to (44). In Fig. 317, climb curves corresponding to (45) with the abscissas t/λ_t are plotted for the five Λ-values already used.

The time t_{sc} needed to reach the *service ceiling* where

$$w_m = 1.67 \text{ ft./sec.} = w_{sc}$$

is found by computing from (44) the altitude h_{sc} of the service ceiling:

$$w_{sc} = \frac{1}{\lambda_t} \frac{h_c - h_{sc}}{c - h_{sc}}, \qquad h_{sc} = \frac{h_c - cw_{sc}\lambda_t}{1 - w_{sc}\lambda_t} = \frac{h_c - 135,000\lambda_t}{1 - 1.67\lambda_t} \quad (46)$$

This introduced in (45) leads to

$$t = \lambda_t \left[\frac{h_c - 135,000\lambda_t}{1 - 1.67\lambda_t} - (81,000 - h_c) \log \frac{\lambda_t}{h_c} \frac{135,000 - 1.67h_c}{1 - 1.67\lambda_t} \right] \quad (47)$$

Formula (45) for the climbing time can be used in the case of supercharged engines for h-values below the critical altitude. In this case, h_c is not the actual ceiling but an abbreviation for $c(1 - 1.75\Lambda^{3/4})$ according to the second relation (43). Beyond the critical altitude h_1 (in the case of a nonsupercharged engine $h_1 = 0$) the factor $\chi(h)$ is decreasing. We may assume, for convenience, a decrease of $\sqrt{\sigma}\,\chi(h)$ in the form of a hyperbolic curve,

$$\chi(h)\sqrt{\sigma} = k \frac{a - h}{b + h} \quad (48)$$

where the coefficients a, b, k can be adapted to the experimental evidence. If the values h_1, σ_1 correspond to the critical altitude, the relation

$$k \frac{a - h_1}{b + h_1} = \frac{c - h_1}{c} \quad (48')$$

will be satisfied. Figure 318 shows some curves according to (48) and (48') plotted for the parameter values ascribed in the graph. There is no doubt that sufficient accuracy can be reached by an appropriate choice of a, b, k. Then the ceiling altitude follows from $\chi\sqrt{\sigma} = 1.75\Lambda^{3/4}$,

$$h_c = \frac{ak - 1.75\Lambda^{3/4}b}{k + 1.75\Lambda^{3/4}} \quad (49)$$

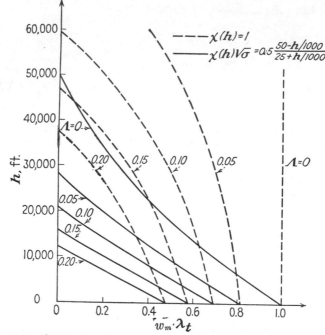

FIG. 316.—Altitude h vs. $\lambda_t \times$ maximum climbing rate for the ideal airplane.

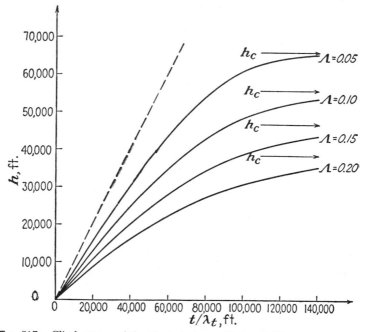

FIG. 317.—Climb curves of the ideal airplane with invariable power available.

and the climbing rate for $h > h_1$ can be written as

$$w_m = \frac{(a + b)ck}{\lambda_t(b + h_c)} \frac{h_c - h}{(c - h)(b + h)} \tag{50}$$

In Fig. 316 the climbing rates are given according to (50) for a non-supercharged engine, with $a = 50,000$, $b = 25,000$, $k = 0.5$. By integra-

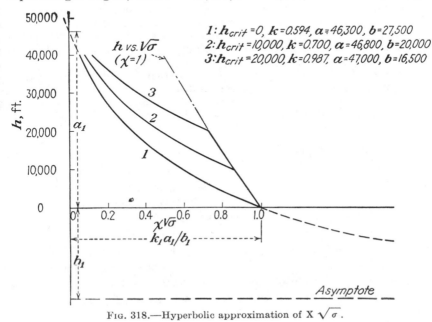

1: $h_{crit} = 0$, $k = 0.594$, $a = 46,300$, $b = 27,500$
2: $h_{crit} = 10,000$, $k = 0.700$, $a = 46,800$, $b = 20,000$
3: $h_{crit} = 20,000$, $k = 0.987$, $a = 47,000$, $b = 16,500$

FIG. 318.—Hyperbolic approximation of $X\sqrt{\sigma}$.

tion, the time needed for climbing from h_1 to any h is found to be

$$t_1 = \int_{h_1}^{h} \frac{dh}{w_m} = \frac{\lambda_t(b + h_c)}{ck(a + b)} \times$$

$$\left[(h - h_1)\left(b + \frac{h + h_1}{2} - c + h_c\right) - (b + h_c)(c - h_c) \log \frac{h_c - h}{h_c - h_1} \right] \tag{51}$$

which may be verified by differentiation. If, for $h_1 > 0$, this formula is combined with (45) applied to $h = h_1$, one obtains the total time needed to reach the altitude h from sea level. But, in this case, one must keep in mind that h_c in (45) stands for $c(1 - 1.75\Lambda^{3/4})$, not for the ceiling altitude. If 1.67 ft./sec. is substituted in (50) for w_m, one finds the service ceiling h_{sc}, and this introduced in (51) supplies the time required for reaching h_{sc}.

All these formulas are based on the assumption of an "ideal airplane," meaning that the available power is constant for all speeds of flight under consideration.

Problem 13. Compute the maximum climbing rate of an ideal airplane with $\lambda = 0.07$, $\lambda_t = 9$ lb./hp. in altitude steps of 5000 ft. up to the ceiling, under the assumption that χ is given as $(\sigma - 0.15)/0.85$. Find the service ceiling by interpolation.

Problem 14. Compute the service ceiling and the time needed to reach it for the airplane of Prob. 13 under the assumption that $\chi\sqrt{\sigma}$ follows Eq. (48). Take $a = 50,000$ ft., and determine k and b in such a way that the drop at sea level is 4.6 per cent per 1000 ft.

*Problem 15. Why is it not possible to make such an assumption about $\chi(h)$ that the ceiling altitude would be reached in finite time for all values of Λ?

Problem 16. Solve Prob. 14 for a supercharged engine with critical altitude 10,000 ft. Assume again $a = 50,000$ ft., and make the drop of $\chi\sqrt{\sigma}$ at the critical altitude equal to 4.6 per cent per 1000 ft.

*Problem 17. The graphs (Fig. 298, etc.) show that, beyond the critical altitude, w_m as function of h can be very well represented by a straight line. Compute the time needed to reach the service ceiling under the assumption $w_m = C(h_c - h)$. Study the power-altitude factor that corresponds to this assumption, and determine C by comparison with the previously given relations.

4. Numerical Data. Example.

To illustrate the results of the preceding sections we now discuss as an example a specified type of aircraft. Let us assume it is desired to design a transportation plane of 8000 lb. gross weight for more than moderate velocities, with an engine of about 580 rated horsepower at 2100 r.p.m.

The first thing to decide on is the wing area, or the *wing loading* W/S. The usual values of this ratio have been steadily increasing since the beginning of flight practice. Up to about 1915, values of 6 to 10 lb./sq.ft. were in use. Today the range is about 10 to 40, with preference for the interval 20 to 30, and with figures higher than 40 (and sometimes smaller than 10) for special purposes. In general, higher speeds are connected with higher wing loadings; on the other hand, the structural problems become more difficult and a more elaborate flying technique is required as W/S increases. In our example, in order to keep the velocities at not too low a level, we may choose a wing area of $S = 300$ sq. ft. corresponding to the wing loading $W/S = 26.7$ lb./sq. ft.

The next task is to estimate the parasite area S_p or the *parasite ratio* S_p/S, including the parasite drag of all parts of the airplane except the wing. This ratio has considerably diminished in the course of improving airplane design. While values as high as 0.1 and more occurred in earlier types, modern aircraft have a parasite ratio of 0.05 down to 0.02, the smallest values for extremely well streamlined bodies, liquid engine cooling, and retractable landing gear. It is often assumed that S_p/S increases, other things being equal, proportional to the wing loading W/S; this may be justified as long as the size of the airplane is supposed to be determined by its weight alone. In planning a new model one will take, for a first estimate, an S_p/S-value according to the experience previously obtained with similar types. Later, as the design work progresses, a better founded assumption can be made on the basis of summing up the drag contributions of the main component parts, as discussed in Chap. V.   Finally, a wind-tunnel model test, with careful consideration of scale effect (Reynolds number and surface texture), may supply a definite figure.   In our case, $S_p = 9$ sq. ft., *i.e.*, $S_p/S = 0.03$, would be a fair estimate under the conditions of perfect streamlining, no considerable trussing, and retractable landing gear.   In final computations of a more elaborate character, claiming higher accuracy, a change of parasite drag with the angle of attack may be taken into account also.

From S and S_p the *parasite loading* can be computed provided that the zero lift drag coefficient a for the wing profile is known.   We may assume $a = 0.01$ as a highly satisfactory value that can just be reached with very careful wing design.   (Note that the wind-tunnel test, carried out for the principal profile used in the wing design, is not decisive by itself.)   Now, from $S'_p = S_p + aS = 9 + 3 = 12$ sq. ft., according to Eq. (7') and $\rho_0 = 0.0024$,

$$\lambda_p = \frac{2}{\rho_0} \frac{W}{S'_p} = 555{,}600 \text{ ft.}^2/\text{sec.}^2$$

The geometrical aspect ratio B^2/S has always been and is today in the range of about 6 to 9, with preference for the middle values 7 to 8. In the case of a monoplane the effective span as used in the formula for the *span loading* may be supposed to be 0.92 times the geometric span. Thus, assuming $\mathcal{R} = 7.6$, Eq. (7'') supplies the span loading

$$\lambda_s = \frac{2}{\rho_0} \frac{W}{\pi B_e^2} = \frac{2}{\rho_0} \frac{W}{S\pi \times 0.92\mathcal{R}} = 1{,}011 \text{ ft.}^2/\text{sec.}^2$$

The two loadings λ_p and λ_s determine the power-required curve according to Eq. (A'') or (A''').   Its graph, shown in Fig. 319, is not necessary for the solution of the performance problem if the method developed in the present chapter is used.   The only thing we still need is the power loading as defined in (22).   As the engine power is given as 580 hp., the available power for normal flight at sea level will be known if an appropriate value for the propeller efficiency η_0 is assumed.   In accordance with our previous assumptions we choose $\eta_0 = 0.84$, a value that can be realized under favorable conditions.   In this case,

$$\lambda_t = \frac{W}{\eta_0 P_0} = \frac{8000}{0.84 \times 580 \times 550} = 0.0299 \text{ sec./ft.}$$

From the three loadings λ_p, λ_s, λ_t follows the *fundamental parameter* value

$$\Lambda = \lambda_s \lambda_t \sqrt[3]{\frac{\lambda_t}{\lambda_p}} = 0.114$$

according to the definition (26). The third root of λ_p/λ_t, which is the first approximation to V_t, equals 264.9 ft./sec.

Before proceeding with the computation of performance for our example, let us briefly discuss the *possible range* of the four parameters $\lambda_p, \lambda_s, \lambda_t,$ and Λ. If the wing loading W/S lies between 10 and 40 lb./sq. ft. and the parasite ratio S'_p/S between 0.03 and 0.07, the widest range for λ_p would be 120,000 to 1,100,000. Actually, $\lambda_p = 200,000$ to 900,000 may correspond to present-day conditions for conventional airplanes. As to λ_s, for monoplanes, with $\mathcal{R} = 6$ to 8 and the foregoing wing loadings, the widest limits would be 350 to 1700, and practical limits are about

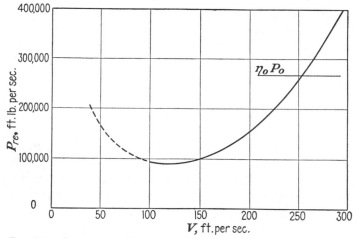

FIG. 319.—Power-required curve for the numerical example of the text.

400 to 1500. The power loading is usually expressed in pounds per brake horsepower; *i.e.*, the conventional figure refers to $550W/P_0$. Present-day limits of this magnitude are about 7 to 16, if extreme cases are excluded, corresponding (with $\eta_0 \sim 0.80$) to λ_t from 0.016 to 0.036. Combining these data we find the limits for $\lambda_s\lambda_t$ as 6.4 to 54; for the third root of λ_t/λ_p, 0.0024 to 0.0067; and for its reciprocal value, 150 to 420. The widest range for the fundamental performance parameter Λ would thus be 0.015 to 0.36. Note that a Λ-value greater than $\frac{3}{4} \times 4^{-1/3} = 0.472$ is impossible as Eq. (26) and Fig. 309 show. As a matter of fact, no values larger than 0.19 or smaller than about 0.02 occur with airplanes actually in use. The following limits are given by W. B. Oswald for various types of airplanes:

Pursuit planes	$\Lambda = 0.025$ to 0.070
Observation planes	0.045 to 0.090
Training planes	0.063 to 0.120
Bombardment and transport planes	0.063 to 0.126
Heavy boats	0.090 to 0.190

The value $\Lambda = 0.114$ found for our example of a transportation airplane fits into this scheme.

Using formula (27) we compute the sea-level flight velocity, thus:

$$V_l = 264.9(1 - 0.038 - 0.003 - \cdots) = 254.0 \text{ ft./sec.}$$

This holds for the air density $\rho_0 = 0.0024 \text{ slug/ft.}^3$ As far as the product of the engine output and the propeller efficiency can be considered as independent of the density ratio $\rho/\rho_0 = \sigma$, the level-flight speed at a different density value can be found from (30), with $\chi = 1$, as

$$264.9(\sigma^{-\frac{1}{3}} - 0.0380\sigma^{-1} - 0.0029\sigma^{-\frac{5}{3}} - 0.00037\sigma^{-\frac{7}{3}} \cdots)$$

This formula can be used for sea-level flight, considering the climatic change of ρ, as well as for altitude flight with the ρ-value corresponding to standard atmosphere conditions, or ρ taken from an immediate observation.

In the general case, the power-altitude factor $\chi(h)$ must be known if the level-flight speed for $h > 0$ is to be computed. Assuming that our present $\chi(h)$ can be identified with the former $\varphi(h)$ as given by formula (10), Chap. XIII, for a nonsupercharged engine, we find at the altitude $h = 15,000$ ft. with the standard $\sigma = 0.629$ the factor $\chi = 0.564$, $(\chi/\sigma)^{\frac{1}{3}} = 0.964$, $\Lambda\chi^{-\frac{1}{3}}\sigma^{-\frac{2}{3}} = 0.333$, and thus, from (30),

$$V_l(15,000) = 264.9 \times 0.964 \times (1 - 0.11 - 0.024) = 220 \text{ ft./sec.}$$

Using Eq. (12), Chap. XIII, with a critical altitude $h_1 = 10,000$, we find $\chi = 0.814$ and therefore

$$V_l(15,000) = 264.9 \times 1.089 \times (1 - 0.068 - 0.009) = 269 \text{ ft./sec.}$$

while the first solution, with $\chi = 1$, would give 291 ft./sec. for the same σ.

The *ceiling altitude* can be computed from Eq. (31) with

$$1.75\Lambda^{\frac{3}{4}} = 0.343$$

In the case $\chi = 1$ we find $\sigma_c = 0.1176$ corresponding to the altitude 55,400 ft. as seen from the σ vs. h-relation in Sec. I.3. This value, of course, is much too high. Under the alternative assumptions for $\chi(h)$ already introduced, the computation is less simple. But sufficiently precise values can be taken from the graph (Fig. 311). At the abscissa $\Lambda = 0.114$ we find $h_c = 19,200$ without supercharging (curve b) and $h_c = 24,800$ for a supercharger with critical altitude of 10,000 ft. (curve c). These results depend, of course, on the correctness of the formula used for the power-altitude factor.

The *maximum climbing rate* w_m at any altitude below ceiling can be found—under the hypothesis of an ideal airplane—from Eq. (41). For sea level we find $w_m = (1 - 1.75\Lambda^{\frac{3}{4}})/\lambda_t = 22 \text{ ft./sec.}$ At $h = 15,000$ ft.

the formula gives w_m = 19, 4.4, and 12.7 ft./sec. for the three different assumptions about the behavior of the engine. The graph (Fig. 320) shows the results concerning w_m and V_l for the case of a supercharged engine with the critical altitude 10,000 ft.

The *service ceiling* h_{sc}, in the case $\chi = 1$, is found by solving Eq. (44) for h with $\lambda_t = 0.03$ sec./ft., $w_m = 1.67$ ft./sec., and $h_c = 55,400$ ft.

$$h_{sc} = \frac{h_c - c\lambda_t w_m}{1 - \lambda_t w_m} = 54,000 \text{ ft.}$$

In the two cases of an engine with critical altitude $h_1 = 0$ and $h_1 =$

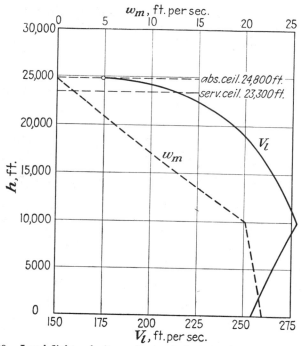

FIG. 320.—Level-flight velocity and rate of climb for the numerical example.

10,000 ft., respectively, the service ceiling can be computed by a trial-and-error method, *i.e.*, by evaluating w_m from (41) for various heights. With the power-altitude factor already used, $h_{sc} = 17,200$ and 23,300 ft. is obtained in the two cases.

The *time needed for climbing* up to an altitude h follows from (45). For an engine of constant power ($\chi = 1$) and $h = 10,000$ ft., $\lambda_t = 0.03$ we find

$$t = 0.03[10,000 - 25,600 \log (1 - 0.18)] = 453 \text{ sec.}$$

In the case of the supercharged engine with critical altitude $h_1 = 10,000$

the additional time t_1 for climbing from h_1 up to h_{sc} is given by Eq. (51). The parameters k, a, b may be taken as $k = 0.7$, $a = 46,800$, $b = 20,000$ ft. so as to give a close approximation for $\sqrt{\sigma}\,\chi(h)$ to the values following from Eq. (12), Chap. XIV.   Thus,

$$t_1 = 0.355 \left[(0.001h - 10)(0.0005h - 31.2) - 2520 \log \frac{24,800 - h}{14,800} \right]$$

This gives $t_1 = 1955$ sec. for $h = h_{sc} = 23,300$ ft.

Problem 18.   Carry out the computations indicated in this section for a fighter plane of the following specifications: gross weight $W = 13,000$ lb., brake power at sea level $P_0 = 2000$ hp., wing loading 54 lb./ft.², aspect ratio $R = 8$, total parasite area $S'_p = 3.9$ ft.²

Problem 19.   Carry out the same for a big transport plane with these specifications: gross weight $W = 90,000$ lb., brake power at sea level $P_0 = 9000$ hp., wing loading 38 lb./ft.², aspect ratio $R = 7.4$, total parasite area $S'_p = 58$ ft.²

Problem 20.   Carry out the same for a small airplane with the following specifications: $W = 4250$ lb., $P_0 = 450$ hp., $W/S = 14.7$ lb./ft.², $R = 7$, $S'_p = 9$ ft.²

Problem 21.   Show that if weight, power, span, and sea-level velocity of a fast airplane are known, its fundamental performance parameter can be estimated as

$$\Lambda \sim \frac{2W^2}{\pi \eta_0 \rho_0 B_e^2 P_0 V_l}$$

5. Small Variations.   Choice of Propeller.   A question that is often discussed in connection with analytical methods of performance computation is that of the *influence of small variations of the data* on the unknowns. This problem is solved by considering the derivatives of all equations involved.   We may explain the procedure for the sea-level flight speed V_l as the unknown and W, $\eta_0 P_0$, S'_p, and B_e^2 as the variable parameters.   Let us use for a moment the abbreviations

$$x = \frac{\lambda_p}{\lambda_t} = \frac{2}{\rho_0}\frac{\eta_0 P_0}{S'_p}, \qquad y = \lambda_s \lambda_t = \frac{2}{\rho_0}\frac{1}{\pi}\frac{W^2}{B_e^2(\eta_0 P_0)}$$

Then Eqs. (24), (25′), and (26) read

$$V_l = x^{1/3}z, \qquad z - z^4 = \Lambda, \qquad \Lambda = x^{-1/3}y$$

and their derivatives are

$$\frac{dV_l}{V_l} = \tfrac{1}{3}\frac{dx}{x} + \frac{dz}{z}, \qquad dz(1 - 4z^3) = d\Lambda, \qquad \frac{d\Lambda}{\Lambda} = -\tfrac{1}{3}\frac{dx}{x} + \frac{dy}{y}$$

Combining the last three expressions and considering that

$$4z^4 - z + \Lambda = 3z^4$$

we find that

$$\frac{dV_l}{V_l} = \frac{1}{z(4z^3 - 1)} \left(z^4 \frac{dx}{x} - \Lambda \frac{dy}{y} \right)$$

From the foregoing definitions of x and y, we conclude that

$$\frac{dx}{x} = \frac{d(\eta_0 P_0)}{\eta_0 P_0} - \frac{dS'_p}{S'_p}, \qquad \frac{dy}{y} = \frac{2\,dW}{W} - \frac{d(B_e^2)}{B_e^2} - \frac{d(\eta_0 P_0)}{\eta_0 P_0}$$

and thus, again using $z^4 + \Lambda = z$,

$$\frac{dV_l}{V_l} = \frac{1}{4z^3 - 1}\left[-\frac{2\Lambda}{z}\frac{dW}{W} + \frac{d(\eta_0 P_0)}{\eta_0 P_0} - z^3\frac{dS'_p}{S'_p} + \frac{\Lambda}{z}\frac{d(B_e^2)}{B_e^2} \right] \qquad (52)$$

Substituting here $z = 1 - \Lambda/3$ from (27) and neglecting all higher powers of Λ, we find that this reduces to

$$\frac{dV_l}{V_l} = \tfrac{1}{3}\left[-2\Lambda\frac{dW}{W} + (1 + \tfrac{4}{3}\Lambda)\frac{d(\eta_0 P_0)}{\eta_0 P_0} - (1 + \tfrac{1}{3}\Lambda)\frac{dS'_p}{S'_p} + \Lambda\frac{d(B_e^2)}{B_e^2} \right] \qquad (52')$$

The relations (52) and (52') are correct for infinitesimal variations. They can be applied, however, for small finite variations by substituting for dW/W, dV_l/V_l, etc., the respective changes in per cent. In our case, with $\Lambda = 0.114$, a 5 per cent increase of W and a simultaneous 5 per cent decrease of $\eta_0 P_0$ would give, according to (52'),

$$\tfrac{1}{3}[-2\Lambda \times 5 - (1 + \tfrac{4}{3}\Lambda)5] = -\tfrac{5}{3}(1 + \tfrac{10}{3}\Lambda) = -2.3 \text{ per cent}$$

as the change of V_l. It is noteworthy that the influence on V_l of the power available $\eta_0 P_0$ and of the parasite area S'_p is much stronger than the influence of weight and span, the latter two variations being multiplied by the small factor Λ. Formulas similar to (52) and (52') can also be set up for maximum climbing rate, ceiling altitude, etc.

Another problem closely connected with performance computation concerns the *selection of the propeller* for an airplane so as to fulfill given conditions. It was assumed, in the foregoing example, that a propeller working with an efficiency of 84 per cent can be found for the airplane under consideration. Whether this is correct or not can be found out by examining an appropriate propeller chart. Let us assume that parasite and span loadings λ_p and λ_s are given, and likewise the engine power and engine speed to be used. The difficulty then is that the propeller efficiency depends on the advance ratio J and thus on the speed of flight V_l, while, on the other hand, V_l cannot be computed before the efficiency is known.

A trial-and-error method that usually leads rapidly to the required result may be briefly described here. The type of propeller chart that is most convenient for the present purpose is a combined C_s vs. J- and η vs. J-diagram as shown in Fig. 235 or a chart that gives both η and J as functions of C_s. Figure 321 is such a chart, referring to a family of two-blade propellers with Clark Y-profiles, Navy Design 5868-9. Each pair of curves corresponds to one definite blade setting as denoted in the figure.   To use this chart, some reasonable value for η is first assumed, then $\lambda_t = W/\eta P_0$ and the approximate level-flight velocity $V_l = (\lambda_p/\lambda_t)^{1/3}$ are computed.   This gives $C_s = V_l(\rho/P_0 n^2)^{1/5}$.

The next step consists in selecting a suitable blade setting out of those represented in the chart.   Among the various η vs. C_s-curves in the chart there will be one that has its maximum at the abscissa that equals the just determined C_s-value.   If this one is chosen, one has the so-called "peak-efficiency propeller."   It is a general practice to prefer the use of the curve which envelops all η-curves and to pick out that

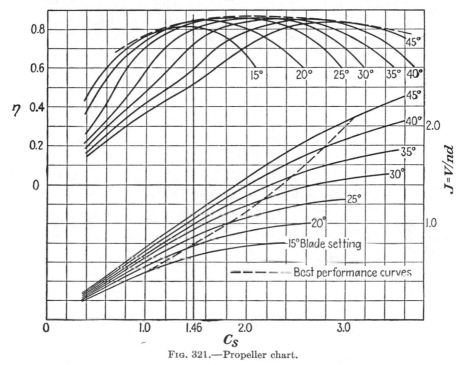

FIG. 321.—Propeller chart.

β-value the efficiency curve of which contacts the envelope at a point with the abscissa C_s.   This is obviously equivalent to choosing the greatest possible η for a given C_s.   In this case one speaks of the *best-performance propeller*.

Now, whichever way was followed, a possible η-value has been found, compatible with the previously computed C_s.   This value will be applied to give a corrected λ_t, a corresponding new value of V_l, and finally a modified C_s.   If this process is repeated, a C_s-value that does not sensibly change will soon be found and the corresponding η can be considered as definitive.   In our example, we may start with $\eta = 0.80$.   This gives $\lambda_t = 0.0314$, $V_l = 260$ ft./sec., and finally $C_s = 1.49$.   Figure 321 supplies as the maximum ordinate at the abscissa 1.49 the efficiency 0.84. This leads to $V_l = 254$, as has already been shown, and C_s is changed into 1.46. But no sensible change in the efficiency value follows from this shifting of C_s. Thus we are allowed to consider 0.84 as a possible efficiency compatible with the given conditions. The η-curve that contacts the envelope at the point with the coordinates 1.46 and 0.84 may be estimated to belong to $\beta = 21°$.

It is still necessary to determine the propeller diameter. This is what the J vs. C_s-curve is used for. The dotted line in the J vs. C_s-diagram connects those points which correspond to the points lying on the envelope of the η-curves. Thus it is easily seen (better than in the η-diagram) that $C_s = 1.46$ belongs to a blade setting of about 21°. The corresponding J is found to equal about 0.78, and this gives, with $V_l = 254$ and $n = 35$, the propeller diameter $d = 254/27.2 = 9.3$ ft. If the diameter found in such a procedure is not acceptable, a different propeller chart has to be used.

Problem 22. Compute the change in flight velocity at sea level for the airplane of Prob. 20 if the weight increases by 250 lb., the available power drops 8 per cent, and the parasite area is increased to 10 ft.^{2}

Problem 23. Extend formula (52') to the case of level flight at altitude h. In particular, give the percentage of change in V_l (h) due to a small change of h.

*Problem 24. Develop formulas analogous to (52') for the change in ceiling altitude, in maximum climbing rate, and in service ceiling.

Problem 25. Use Fig. 321 and the trial-and-error method described in this section to determine a propeller for the airplane of Prob. 20. Engine speed is 1200 r.p.m.

6. Power Available Varying with Speed. A better approximation to actual conditions than that supplied by the theory of an ideal airplane (with the power available independent of speed) can be obtained by using the expression for the available power as developed in Sec. 1 of the present chapter. Let us suppose that a definite propeller with invariable pitch has been selected for the propulsion of an airplane whose parasite and span loadings λ_p and λ_s are known. Then the propeller polar diagram, replaced by the best-fitting straight line, supplies the two coefficients of the linear equation (12) or, which is the same, the two loadings λ_P and λ_0 as defined in Eq. (13). Combining the expressions (A'') and (B'') for powers required and available, respectively, with the general performance statement (5'), Chap. XIV, we have

$$\frac{dh}{dt} = \frac{P_{av} - P_{re}}{W} = \frac{\varphi(h)}{\lambda_0} V - \sigma \frac{V^3}{\lambda_P} - \left(\sigma \frac{V^3}{\lambda_p} + \frac{1}{\sigma} \frac{\lambda_s}{V} \right) \tag{53}$$

This equation suggests that we should combine the two parasite loadings, λ_p for wing and airplane and λ_P for the propeller, to one *total parasite* loading λ_p' by the definition

$$\frac{1}{\lambda_p'} = \frac{1}{\lambda_p} + \frac{1}{\lambda_P} \tag{54}$$

By considering the definitions of λ_p in (7′) and λ_P in (13) it is seen that

$$\frac{1}{\lambda_p'} = \frac{\rho_0}{2} \frac{1}{W} \left(S_p' + \frac{a_P S_P}{\sin^3 \beta'} \right) = \frac{\rho_0}{2W} \left(aS + S_p + \frac{a_P S_P}{\sin^3 \beta'} \right) \tag{55}$$

The expression in the last parentheses can be called the "total parasite area" covering the parasite drag of the wing, all component parts of the airplane, and the propeller. The range of numerical values of λ_p' as well as of λ_0 will be discussed later. Now, Eq. (53) can be written as

$$\frac{dh}{dt} = \frac{\varphi(h)}{\lambda_0} V - \sigma \frac{V^3}{\lambda_p'} - \frac{1}{\sigma} \frac{\lambda_s}{V} \tag{53′}$$

Thus, *the general performance of an airplane with invariable-pitch propeller, under the assumption of the brake moment being independent of the engine speed, is completely determined by three loadings, viz., span loading, propeller power loading, total parasite loading, and by the power-altitude factor* $\varphi(h)$.

The first question that may be raised is that concerning the *level-flight speed* at sea level, V_l, or at altitude, $V_l(h)$. Equating the right-hand side of (53′) to zero we find, by a simple rearrangement,

$$V^4 \sigma^2 - \frac{\varphi \lambda_p'}{\lambda_0} V^2 \sigma + \lambda_p' \lambda_s = 0 \tag{56}$$

or, introducing the indicated speed $V \sqrt{\sigma} = V_i$,

$$V_i^4 - \frac{\varphi \lambda_p'}{\lambda_0} V_i^2 + \lambda_p' \lambda_s = 0 \tag{56′}$$

If $\varphi = 1$ is assumed (brake moment independent of altitude), the coefficients of the second equation are independent of σ and h. This means that V_i is a constant or that the level flight speed $V_l(h)$ is proportional to $1/\sqrt{\sigma}$. The result is in accordance with the general statement formulated in Sec. XIV.1 that under the assumption of constant brake moment an airplane can rise to any altitude with both engine and flying speed increasing proportional to $1/\sqrt{\sigma}$. The limitation of climbing is ascribed, in the present theory, to the decrease of the power-altitude factor $\varphi(h)$ only. There is no doubt that in reality the power-altitude relation is the essential factor in determining the ceiling, etc.

We introduce now a new fundamental parameter in Eq. (56),

$$\Lambda' = \frac{\lambda_0^2 \lambda_s}{\lambda_p'} \tag{57}$$

and write, in analogy to (24),

$$z' = V \sqrt{\frac{\sigma\lambda_0}{\varphi\lambda_p'}}, \qquad V = z' \sqrt{\frac{\varphi\lambda_p'}{\sigma\lambda_0}} \tag{58}$$

Then the relation for z' following from (56) reads

$$z'^4 - z'^2 + \frac{\Lambda'}{\varphi^2} = 0 \qquad \text{or} \qquad z'^2 - z'^4 = \frac{\Lambda'}{\varphi^2} \tag{59}$$

This biquadratic equation, which takes the place of (25') with $\varphi = 1$, can easily be solved. Keeping the positive sign of the root only (the

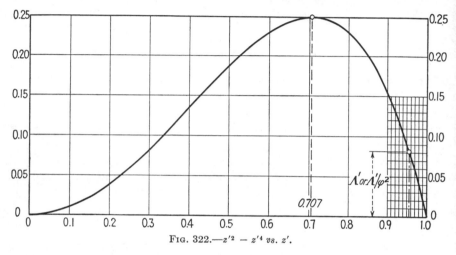

Fig. 322.—$z'^2 - z'^4$ vs. z'.

negative would correspond to the low speed at level flight) we have

$$z'^2 = \tfrac{1}{2}\left(1 + \sqrt{1 - \frac{4\Lambda'}{\varphi^2}}\right), \qquad V_l^2(h) = \frac{\varphi}{\sigma}\frac{\lambda_p'}{2\lambda_0}\left(1 + \sqrt{1 - \frac{4\Lambda'}{\varphi^2}}\right) \tag{60}$$

Another form of the solution, useful in the case of small Λ'/φ^2, can be found in the same way as was shown for Eq. (25'), by developing z' in a power series with respect to Λ'/φ^2. One may confirm by introducing the expression on both sides of (59) that

$$z' = 1 - \tfrac{1}{2}\frac{\Lambda'}{\varphi^2} - \tfrac{5}{8}\left(\frac{\Lambda'}{\varphi^2}\right)^2 - \tfrac{21}{16}\left(\frac{\Lambda'}{\varphi^2}\right)^3 \cdots \tag{61}$$

The result, which takes the place of (30), is accordingly

$$V_l(h) = \sqrt{\frac{\varphi\lambda_p'}{\sigma\lambda_0}}\left[1 - \tfrac{1}{2}\frac{\Lambda'}{\varphi^2} - \tfrac{5}{8}\left(\frac{\Lambda'}{\varphi^2}\right)^2 - \tfrac{21}{16}\left(\frac{\Lambda'}{\varphi^2}\right)^3 \cdots \right] \tag{62}$$

and, in particular, for sea level, replacing (27),

$$V_l = \sqrt{\frac{\lambda_p'}{\lambda_0}}\,(1 - \tfrac{1}{2}\Lambda' - \tfrac{5}{8}\Lambda'^2 - \tfrac{21}{16}\Lambda'^3 \cdots) \tag{62'}$$

In Fig. 322 the curve with the ordinates $z'^2 - z'^4$ is shown. Its intersection with a horizontal at the height Λ'/φ^2 gives the solution required. It is seen that for small values of Λ'/φ^2 the right branch of the curve can be gradually approximated by a straight line, a parabola, etc., which is equivalent to stating that the power development (61) can be used. This development, however, loses its validity as one approaches the top of the curve where Λ'/φ^2 equals $\frac{1}{4}$. In this range Table 14, which gives the solution of the equation $z'^2 - z'^4 = Z'$ for all values of the right-hand side, can be used. Once z' is known, the velocity $V_l(h)$ follows from the second equation (58).

TABLE 14.—SOLUTIONS OF $z'^2 - z'^4 = Z'$

$Z' = 0.01$	$z' = 0.995$	$Z' = 0.16$	$z' = 0.894$
02	90	17	85
03	84	18	74
04	79	19	63
05	73	20	51
0.06	0.967	0.21	0.837
07	61	22	21
08	55	23	01
09	49	24	775
10	42	25	06
0.11	0.935		
12	28		
13	20		
14	12		
15	03		

The *speed ratio*

$$v = \frac{V_l(h)}{V_l}$$

is given by the expression

$$v^2 = \frac{\varphi}{\sigma} \frac{1 + \sqrt{1 - 4\Lambda'/\varphi^2}}{1 + \sqrt{1 - 4\Lambda'}} \tag{63}$$

which follows immediately from the second equation (60). In Fig. 323 the values of v are shown for altitudes up to 50,000 ft. and for values of $\Lambda' = 0$, 0.04, 0.08, 0.12, and 0.16. The power-altitude factor $\varphi(h)$ is taken according to Eqs. (10) and (12), Chap. XIII, respectively, with critical altitudes $h_1 = 0$, 10,000, 20,000, and 30,000 ft.

The vertices of the h vs. v-curves correspond to the *ceiling altitudes* h_c. Formulas (60) show that no $V_l(h)$-value exists when φ^2 surpasses $4\Lambda'$. Therefore h_c is determined by

$$\varphi^2(h_c) = 4\Lambda' \qquad \text{or} \qquad \varphi(h_c) = 2\lambda_0 \sqrt{\frac{\lambda_s}{\lambda_p'}} \tag{64}$$

For example, an airplane with the fundamental parameter $\Lambda' = 0.08$ and a nonsupercharged engine would rise to $h_c = 15,000$ ft. since at this altitude the density ratio $\sigma = 0.63$ (see Table, page 10) and, according to Eq. (10), Chap. XIII, $\varphi = 1.76\sigma - 0.176 = 0.565 = 2\sqrt{0.08}$.

The *maximum climbing rate* w_m and the corresponding flying speed V_m can be found from (53') by differentiation. The derivative of the right-hand expression equated to zero gives

$$\frac{\varphi}{\lambda_0} - 3\sigma\frac{V^2}{\lambda_p'} + \frac{1}{\sigma}\frac{\lambda_s}{V^2} = 0 \qquad \text{or} \qquad z'^2 - 3z'^4 + \frac{\Lambda'}{\varphi^2} = 0 \qquad (65)$$

if the transformation (58) is used. The solution of this equation can be written as

$$z_m'^2 = \tfrac{1}{6}\left[1 + \sqrt{1 + \frac{12\Lambda'}{\varphi^2}}\right], \qquad z_m' = \frac{1}{\sqrt{3}}\left[1 + \tfrac{3}{2}\frac{\Lambda'}{\varphi^2} - \tfrac{4.5}{8}\left(\frac{\Lambda'}{\varphi^2}\right)\right] \qquad (66)$$

whence

$$V_m = \sqrt{\frac{\varphi\lambda_p'}{\sigma\lambda_0}}\, z_m' = \sqrt{\frac{\varphi\lambda_p'}{3\sigma\lambda_0}}\left[1 + \tfrac{3}{2}\frac{\Lambda'}{\varphi^2} - \tfrac{4.5}{8}\left(\frac{\Lambda'}{\varphi^2}\right)^2 \cdots\right] \qquad (66')$$

The power development can be used only if Λ'/φ^2 is considerably smaller than $\tfrac{1}{12}$.

The value of w_m equals the right-hand side of (53') with $V = V_m$. If we eliminate, by using (65), the term of third order, we find that

$$w_m = \frac{\varphi}{\lambda_0}V_m - \sigma\frac{V_m^3}{\lambda_p'} - \frac{1}{\sigma}\frac{\lambda_s}{V_m} = \tfrac{2}{3}\frac{\varphi}{\lambda_0}V_m - \tfrac{4}{3}\frac{1}{\sigma}\frac{\lambda_s}{V_m}$$

Upon introducing here the first expression (66') for V_m and rearranging, it follows that

$$w_m = \tfrac{2}{3}\frac{\varphi}{\lambda_0}\sqrt{\frac{\varphi\lambda_p'}{\sigma\lambda_0}}\left(2 - \sqrt{1 + \frac{12\Lambda'}{\varphi^2}}\right)z_m' \qquad (67)$$

Here the first equation (66) must be used for z_m', except for small values of Λ'/φ^2, where the power development still gives a sufficient approximation.

It is seen from (67) that the maximum climbing rate increases with h, if φ is assumed to keep the value 1, *i.e.*, if the brake power is independent of altitude. Simultaneously, V_m increases according to (66') at the same rate. This is due to the fact that if the brake moment of an engine is kept absolutely constant, the indicated velocities keep their values (see Sec. XIV.1). In this respect, the assumption of the ideal airplane is more conservative since with $\chi(h) = 1$ the power is assumed to remain unchanged and thus M_{br} to decrease while the engine speed increases. In fact, with supercharged engines that operate at sea level below the maximum M_{br}-value, an increase of w_m below the critical altitude can be observed. A more detailed investigation of this point would require better experimental data concerning the power-altitude factor. Moreover, in the case where the pilot has a supercharged engine at his disposal, he must, as a rule, start the climb with the engine partly throttled and gradually open the throttle.

The case of a *variable-pitch propeller*, operated at constant engine speed, does not lend itself easily to an analytical form of performance computation since no simple analytical expression, based on the propeller theory, can be derived for the power-available curve. The most reliable manner of computation is then supplied by the graphical method developed in Chap. XIV. If need is felt for having the results in the form of equations, one may use the formula for P_{re}, as has been done throughout this chapter, and then introduce an interpolation formula for the power-available curve, based on the tracing or on the evaluation of a few of its points. How this can be done will be outlined at the end of the next section.

Problem 26. Assume that the three loadings λ_p, λ_s, λ_t are known, and, in addition, the blade setting β', the incidence α', and the parasite area ratio $a_P S_P / W$ of the propeller. How can the P_{av}-curve be found and, from it, the maximum climbing rate?

Problem 27. Develop a chart, analogous to Fig. 316, which shows, for various Λ', the maximum climbing rate as function of altitude, according to (67). Use for $\varphi(h)$ one of the experimental results mentioned in the text.

7. Numerical Discussion. Before giving a numerical illustration of the formulas developed in the preceding section we may discuss the range within which the newly introduced parameters λ_0, λ_p', and Λ' vary and their relationship to the parameters λ_t, λ_p, and Λ used in the theory of the ideal airplane. It was seen in Sec. 4 of this chapter that Λ is a small quantity, theoretically smaller than 0.472 and actually for modern airplanes below 0.19. For an airplane of infinite aspect ratio, the span loading $\lambda_s = 0$ and therefore Λ as well as Λ' vanish. In this case, Eqs. (27) and (62') supply

$$V_t = \sqrt[3]{\frac{\lambda_p}{\lambda_t}} = \sqrt{\frac{\lambda_p'}{\lambda_0}} \qquad (\mathcal{R} = \infty) \quad (68)$$

In the case of actual aspect ratios the third root of λ_p/λ_t and the square root of λ_p'/λ_0 are at least first approximations for V_t and thus are approximately equal. Now, from

$$\lambda_p^{1/3}\lambda_t^{-1/3} \sim \lambda_p'^{1/2}\lambda_0^{-1/2} \qquad \text{or} \qquad \lambda_0 \sim \lambda_p'\lambda_p^{-2/3}\lambda_t^{2/3}$$

we conclude, using the definitions of Λ and Λ',

$$\Lambda' = \lambda_0^2\lambda_s\lambda_p'^{-1} \sim \lambda_p'\lambda_p^{-4/3}\lambda_s\lambda_t^{4/3} = \frac{\lambda_p'}{\lambda_p}\Lambda \quad (69)$$

Since the total parasite loading λ'_p, according to its definition [Eq. (54)], is somewhat smaller than λ_p, we learn from (69) that Λ' will be a little smaller than Λ.   This is the reason why, in Fig. 323, values of Λ' up to 0.16 only were chosen for the representation of the speed ratio v.

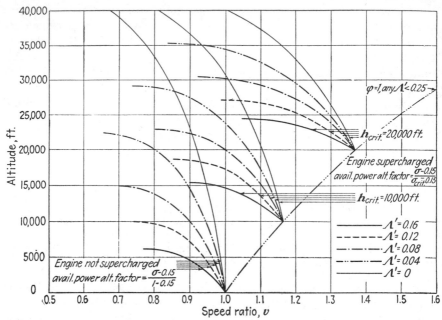

FIG. 323.—Altitude vs. speed ratio for various engine conditions; power available variable with speed.

In no case can Λ' be greater than $\frac{1}{4}$ since, according to (64),

$$4\Lambda' = \varphi^2(h_c) \leqq 1.$$

At sea level P_{av}/W or $\eta_0 P_0/W$ equals $1/\lambda_t$ according to the definition [Eq. (22)] of λ_t.   This combined with the expression (B'') for P_{av}/W gives

$$\frac{1}{\lambda_t} = \frac{V_l}{\lambda_0} - \frac{V_l^3}{\lambda_P} \quad \text{or} \quad \frac{1}{\lambda_0} = \frac{1}{V_l \lambda_t} + \frac{V_l^2}{\lambda_P} \tag{70}$$

The propeller power loading λ_0 as introduced in (13) admits of the following interpretation: Considering that $P_0 = M_0 2\pi n_0$, where n_0 is the engine speed in sea-level flight, and using the angle γ for which $\tan \gamma = V_l/2\pi r n_0$, we have

$$\lambda_0 = \frac{W}{P_0} 2\pi r n_0 \tan \beta' = \frac{W V_l}{P_0} \frac{\tan \beta'}{\tan \gamma} = V_l \eta_0 \lambda_t \frac{\tan \beta'}{\tan \gamma} \tag{71}$$

Now, γ is the angle between the resultant velocity of the representative blade element and the plane of rotation, and $\beta' - \gamma = \alpha'$ is the angle of incidence. As α' is small, the quotient of the two tangents is close to 1, and therefore λ_0 is close to $V_l \eta_0 \lambda_t$. (We cannot conclude safely that the quotient is greater than 1 since quantities of the order α' have been disregarded in deriving the expression of P_{av}.) If we combine $\lambda_0 \sim V_l \eta_0 \lambda_t$ with Eq. (70), we find that

$$\frac{V_l^2}{\lambda_P} \sim \frac{1}{V_l \lambda_t} \left(\frac{1}{\eta_0} - 1 \right)$$

Considering the definition of z in (24), which gives $\lambda_t V_l^3 = \lambda_p z^3$ and the fact that z is close to 1, it follows

$$\frac{1}{\lambda_P} \sim \frac{1}{\lambda_p} \frac{1}{z^3} \left(\frac{1}{\eta_0} - 1 \right) \sim \frac{1}{\lambda_p} \left(\frac{1}{\eta_0} - 1 \right), \qquad \frac{1}{\lambda'_p} = \frac{1}{\lambda_p} + \frac{1}{\lambda_P} \sim \frac{1}{\eta_0 \lambda_p} \qquad (72)$$

This shows how the total parasite loading λ'_p is connected with the parasite loading λ_p of wing and airplane and with the propeller efficiency η_0. The formula, however, as well as $\lambda_0 \sim V_l \lambda_t \eta_0$, gives rough estimates only. The values for establishing a performance computation must be based on specified data or on experience with similar cases.

In the numerical example discussed in Sec. 4 of this chapter the airplane had the weight $W = 8000$ lb., the parasite and span loadings $\lambda_p = 555,600$ and $\lambda_s = 1011$ ft.^{2}/sec.^{2}, and an engine supplying 580 hp. in sea-level flight. With an assumed propeller efficiency $\eta_0 = 0.84$, we computed $\lambda_t = 0.0299$ and the sea-level flight speed $V_l = 254$ ft./sec. This result can be maintained since it is independent of how the available power changes if the conditions vary. Let us now assume that an invariable-pitch propeller is used and that the engine runs at $n_0 = 35$ r.p.s. in sea-level flight and develops a nearly constant brake moment

$$M_0 = \frac{P_0}{2\pi n_0} = \frac{580 \times 550}{70\pi} = 1455 \text{ lb.-ft.}$$

The power available P_{av} will then approximately follow the relation (B), and the results of the preceding section can be applied as soon as the two new loadings λ_0 and λ'_p are known. It may be a fair assumption that $\lambda'_p = 0.80\lambda_p$ in accordance with (72). Then from (54) we find that $\lambda_P = 4\lambda_p = 2,222,400$ and from (70) that

$$\frac{1}{\lambda_0} = \frac{1}{7.60} + \frac{254^2}{2,222,400} = 0.161, \qquad \lambda_0 = 6.23$$

which, in fact, is close to $\eta_0 \lambda_t V_l = 6.38$. The P_{av} vs. V-curve for sea level corresponding to these parameter values is shown in Fig. 324, together with the P_{re} vs. V-curve from Fig. 319. The new fundamental parameter is

$$\Lambda' = \frac{\lambda_0^2 \lambda_s}{\lambda'_p} = 0.0882$$

not much different from $0.80\Lambda = 0.0912$, as was to be expected from (69). With the values of λ'_p, λ_0, Λ' thus found we can check the value of V_l by using either the solution (60) or the power development [Eq. (62')], which gives

$$V_l = 267.2(1 - 0.0441 - 0.0049 - 0.0002) = 254 \text{ ft./sec.}$$

as expected. If the engine is not supercharged and φ is assumed to

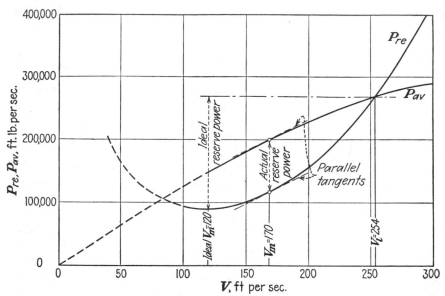

FIG. 324.—Power curves, invariable-pitch propeller.

equal $(\sigma - 0.13)/0.87$, we have, at $h = 10,000$ ft. $(\sigma = 0.738)$,

$$\varphi = 0.699 \qquad \frac{4\Lambda'}{\varphi^2} = 0.722 \qquad V_l(10,000) = 227 \text{ ft./sec.}$$

according to the second formula (60). The ceiling altitude is given by

$$\sigma_c - 0.13 = 0.87 \sqrt{4\Lambda'} = 0.516 \qquad \sigma_c = 0.646 \qquad h_c = 14,300 \text{ ft.}$$

In the case of a supercharged engine with critical altitude $h_1 = 10,000$ ft. $(\sigma_1 = 0.738)$ and the power-altitude factor (beyond h_1) equal to $\varphi = (\sigma - 0.13)/(\sigma_1 - 0.13) = (\sigma - 0.13)/0.608$, the level-flight speed at $h = 20,000$ ft. $(\sigma = 0.533)$ is found as follows:

$$\varphi = 0.663 \qquad \frac{4\Lambda'}{\varphi^2} = 0.803 \qquad V_l(20,000) = 253 \text{ ft./sec.}$$

The ceiling altitude in this case is determined by

$$\sigma_c - 0.13 = 0.608 \sqrt{4\Lambda'} = 0.361 \qquad \sigma_c = 0.491 \qquad h_c = 22,500 \text{ ft.}$$

In order to find the velocity V_m for maximum climbing rate at sea level and the rate w_m itself, we first compute z'_m from (66) with $\varphi = 1$,

$$z'^2_m = \tfrac{1}{6}(1 + \sqrt{1 + 12\Lambda'}) = 0.405, \qquad z'_m = 0.637$$

and then, from (66') and (67),

$$V_m = 267.2 \times 0.637 = 170 \text{ ft./sec.},$$

$$w_m = \frac{2 \times 267.2}{3\lambda_0} (2 - \sqrt{1 + 12\Lambda'})0.637 = 10.3 \text{ ft.}$$

The latter value is considerably smaller than what was found for the ideal airplane (22 ft./sec.), but in better agreement with actual observation. The difference can be seen in Fig. 324, where the two extreme values of $(P_{av} - P_{re})/W$ are shown.

Lastly we discuss a semiempirical way of performance computation, which can be used in all cases, particularly in that of a *controllable-pitch propeller*, if a few points of the power-available curve are known. Let us assume that the three loadings λ_p, λ_s, and λ_t are given so that the P_{re}/W-curve and the level-flight value V_l are determined (Fig. 324a). The ordinate of the point with abscissa V_l is $1/\lambda_t$. Now, some knowledge about the trend of the P_{av}/W-curve is required. If two points (besides V_l, $1/\lambda_t$) or one point and the slope somewhere are given, it is possible to compute two parameters a, b in the equation of an appropriate family of curves. It is proposed to use the form

$$\frac{P_{av}}{W} = \frac{1}{\lambda_t}\left[1 + a + b - a\left(\frac{V}{V_l}\right)^3 - b\left(\frac{V_l}{V}\right)\right] \tag{73}$$

In Fig. 324a several curves of this family, all with a drop of 30 per cent of P_{av} at $V/V_l = \tfrac{1}{2}$, are plotted; these curves show that a sufficiently great variety can be obtained in this way.

The use of (73) in the performance computation is very simple. Introducing the expression for P_{av}/W in the general equation (5'), Chap. XIV, we find that

$$w_m = \left(\frac{dh}{dt}\right)_{\max} = \frac{1}{\lambda_t}(1 + a + b) - \left[\frac{V^3}{\lambda_p}\left(1 + \frac{a\lambda_p}{\lambda_t V_l^3}\right) + \frac{\lambda_s}{V}\left(1 + \frac{bV_l}{\lambda_s\lambda_t}\right)\right]_{\min} \tag{74}$$

This is the same equation as (41) for $\sigma = 1$, $\chi = 1$; that is, the problem, for sea level, is reduced to that of the ideal airplane with changed loading values. In place of λ_p, λ_s, λ_t one has to use λ°_p, λ°_s, λ°_t given by

$$\frac{1}{\lambda^\circ_p} = \frac{1}{\lambda_p}\left(1 + \frac{a\lambda_p}{\lambda_t V_l^3}\right) = \frac{1}{\lambda_p}\left(1 + \frac{a}{z^3}\right)$$

$$\lambda^\circ_s = \lambda_s\left(1 + \frac{bV_l}{\lambda_s\lambda_t}\right) = \lambda_s\left(1 + \frac{bz}{\Lambda}\right) \tag{75}$$

$$\frac{1}{\lambda^\circ_t} = \frac{1}{\lambda_t}(1 + a + b)$$

Then all computations carried out for the ideal airplane in Sec. 3 of this chapter apply to the airplane with P_{av} depending on V, without any other change than the replacement of the loadings by the values (75). By making a, b depend on σ in an appropriate way, one can extend this method to all problems concerning altitude flight, ceiling altitude, etc.

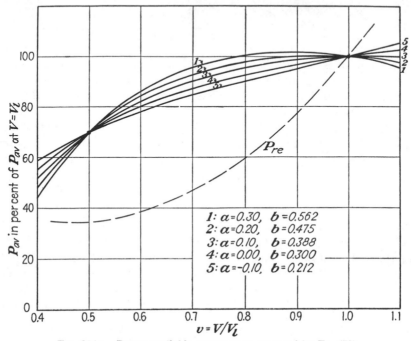

FIG. 324a.—Power-available curves as represented by Eq. (73).

Problem 28. Compute the level-flight velocities and climbing rate at sea level and at 5,000, 10,000, etc., ft. altitude and the ceiling for an airplane with the following loadings:

$$\text{Propeller power loading } \lambda_0 = 7.2$$
$$\text{Span loading } \lambda_s = 1200 \text{ ft.}^2/\text{sec.}^2$$
$$\text{Parasite loading } \lambda_p = 580,000 \text{ ft.}^2/\text{sec.}^2$$
$$\text{Propeller parasite loading } \lambda_P = 3,400,000 \text{ ft.}^2/\text{sec.}^2$$

Assume that the power-altitude factor φ can be taken as $(\sigma - 0.115)/0.885$.

Problem 29. How do the results of the preceding problem change if the engine is supercharged to a critical altitude $h_1 = 18,000$ ft. and, from there on, has the power-altitude factor $\varphi = (\sigma - 0.115)/(\sigma_1 - 0.115)$?

Problem 30. An airplane has the three loading factors $\lambda_p = 720,000$ ft.^{2}/sec.^{2}, $\lambda_s = 1020$ ft.^{2}/sec.^{2}, $\lambda_t = 0.024$ sec./ft. It is known that the power available drops to 90 per cent of its value at $V = V_l$ if $V = 0.80V_l$ and to 75 per cent, if $V = 0.60V_l$. Compute the maximum climbing rate at sea level on the basis of the last paragraph of the preceding text.

***Problem 31.** Give an analysis of the flight at higher altitudes on the basis and with the data of the preceding problem.

CHAPTER XVI

SPECIAL PERFORMANCE PROBLEMS

1. Range and Endurance. In all performance computations discussed in the preceding chapter the gross weight of an airplane was supposed to be constant. This assumption cannot be maintained if a flight over a longer period of time or over a longer distance is considered. In fact, the gross weight diminishes continually according to the fuel consumption of the power plant. If W denotes the gross weight at the time t and c the fuel consumption per unit of power and time (see Sec. XIII.1), a new equation has to be added, which reads

$$\frac{dW}{dt} = -cP \tag{1}$$

where P is the brake power of the engine, equal to the propeller power at the time t, transmission losses, if any, being disregarded. It is true that under the assumption of changing weight the motion of the airplane is no longer uniform (not free from accelerations) and thus the equilibrium equations underlying all our computations are not strictly correct. But it is obvious that the acceleration effects will be so small that we are allowed to consider the entire phenomenon as a continuous succession of uniform motions under gradually changing conditions.

The main questions to be answered in this connection are the following: Given the initial gross weight W_0 and the fuel capacity W_f, so that $W_1 = W_0 - W_f$ is the empty weight, how long is the distance R, *the range*, that can be covered in a straight horizontal flight and what is the time E, the *endurance*, needed for this flight? If we combine Eq. (1) with the definition of the forward velocity V, that is, with $V = ds/dt$, we find

$$ds = V\,dt = -\frac{V}{cP}\,dW, \qquad R = \int_{W_1}^{W_0} \frac{V}{cP}\,dW \tag{2}$$

while

$$dt = -\frac{1}{cP}\,dW, \qquad E = \int_{W_1}^{W_0} \frac{1}{cP}\,dW \tag{3}$$

These formulas answer the two questions if V, c, and P are known as functions of the instantaneous weight W. As to c, it is assumed to be a

461

constant or a given function of P and the engine speed n. Thus, the complete solution of the problems would require us to compute, for each value of W, the level-flight speed V, the engine speed n, and the power P.

According to that which was stated above, the relations between the instantaneous values of W, V, P, n, etc., are supposed to be the same as those for uniform flight. If, for example, an engine with constant throttle opening and with fixed-pitch propeller is used, we have one power-available curve, P_{av} vs. V, independently of W. On the other hand, we know how to draw the set of power-required curves, P_{re} vs. V, for any given value of W (Fig. 325). The respective points of intersection give the successive values of V. To each V, as an abscissa of the

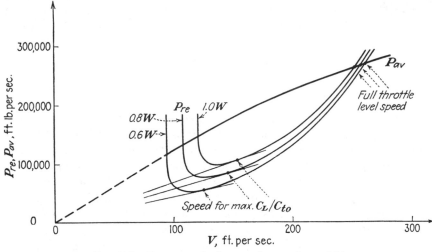

Fig. 325.—Power-required curves for varying weight.

P_{av} vs. V-curve, there corresponds a certain advance ratio J, thus a certain engine speed $n = V/Jd$ and a certain C_P-value so that P can be computed as $C_P \rho n^3 d^5$. If this is carried out for $W = W_0$, for $W = W_1$, and for a number of appropriately chosen intermediate values of W, the integrals of V/cP and of $1/cP$ can be evaluated to any degree of accuracy.

Instead of following this correct but somewhat cumbersome method, rough approximations for both range and endurance can be obtained in the following way, which was first suggested by L. Bréguet: In the expression for R we write $P = P_{re}/\eta = D_{to}V/\eta$ and use the relation $D_{to}/L = D_{to}/W = C_{to}/C_L$ where the total drag coefficient C_{to} is defined as the sum $C_D + (S_p/S)$ with $S_p = $ parasite area of the airplane (except the wing). Thus,

$$\frac{V}{cP} = \frac{V\eta}{cD_{to}V} = \frac{\eta}{c}\frac{W}{D_{to}}\frac{1}{W} = \frac{\eta}{c}\frac{C_L}{C_{to}}\frac{1}{W} \tag{4}$$

Here the propeller efficiency η varies with the advance ratio J, and C_L/C_{to} varies with the angle of incidence. But these variations as well as that of c can be considered as unimportant so that the factor of $1/W$ in (4) is approximately constant and can be replaced by some average value. With this assumption, Eq. (2) supplies

$$R = \int_{W_1}^{W_0} \frac{V}{cP}\,dW = \left(\frac{\eta}{c}\frac{C_L}{C_{to}}\right)_{av} \int_{W_1}^{W_0} \frac{dW}{W} = \left(\frac{\eta}{c}\frac{C_L}{C_{to}}\right)_{av} \log\frac{W_0}{W_1} \quad (5)$$

This formula allows a rapid estimate of R if the expected average value of the parentheses or the average values of η, c, and C_L/C_{to} are known. Take, for example,

$$\eta = 0.70, \qquad c = 0.4 \text{ lb./hp.-hr.} = \frac{0.4}{550 \times 3600} = 0.202 \times 10^{-6}\ \text{ft.}^{-1}$$

and $C_{to} = 0.05 + C_L^2/20$, according to an aspect ratio of about 7 and a total zero lift drag coefficient 0.05. Then the largest value of C_L/C_{to} can be found (see Sec. VII.1) as $\frac{1}{2}\sqrt{20/0.05} = 10.00$, and an average value may be estimated as about 7.5. The greatest possible R, if one-fifth the full initial weight consists of fuel, is thus found to be

$$R_{max} = \frac{0.70}{0.202} \times 10^7 \times \log\tfrac{5}{4} = 7.72 \times 10^6\ \text{ft.} = 1465\ \text{miles.}$$

This would be the correct R if it were possible to maintain during the whole flight the same advance ratio and the angle of incidence that corresponds to the maximum C_L/C_{to}-ratio. In fact, not more than about 75 per cent of R_{max} can be obtained under practical conditions.

Another Bréguet formula can be found for the endurance E by dividing both sides of Eq. (4) by V and substituting for V, according to Eq. (a), Chap. XIV, the expression $\sqrt{2W/\rho S C_L}$.

$$\frac{1}{cP} = \frac{\eta}{c}\frac{C_L}{C_{to}}\frac{1}{WV} = \frac{\eta}{c}\frac{C_L^{3/2}}{C_{to}}\left(\frac{\rho S}{2}\right)^{1/2} W^{-3/2} \quad (6)$$

Then Eq. (3) gives

$$E = \int_{W_1}^{W_0} \frac{dW}{cP} = \left(\frac{\eta}{c}\frac{C_L^{3/2}}{C_{to}}\right)_{av} \sqrt{\frac{\rho S}{2}} \int_{W_1}^{W_0} W^{-3/2}\,dW$$

$$= \left(\frac{\eta}{c}\frac{C_L^{3/2}}{C_{to}}\right)_{av} \sqrt{2\rho S}\left(\frac{1}{\sqrt{W_1}} - \frac{1}{\sqrt{W_0}}\right) \quad (7)$$

if the same assumptions about averaging η, c, C_L, C_{to} are made. Introducing here the initial and the final velocities (under the assumption of a nearly constant angle of incidence), that is $V_0 = \sqrt{2W_0/\rho S C_L}$ and $V_1 = \sqrt{2W_1/\rho S C_L}$, the expression becomes

$$E = 2\left(\frac{\eta}{c}\frac{C_L}{C_{to}}\right)_{av}\left(\frac{1}{V_1} - \frac{1}{V_0}\right) = \left(\frac{\eta}{c}\frac{C_L}{C_{to}}\right)_{av}\frac{2}{V_0}\left(\sqrt{\frac{W_0}{W_1}} - 1\right) \qquad (7')$$

From (5) and (7') the mean velocity during the flight is found to equal

$$\frac{R}{E} = \frac{\log\sqrt{W_0/W_1}}{\sqrt{W_0/W_1} - 1}\,V_0 \qquad (8)$$

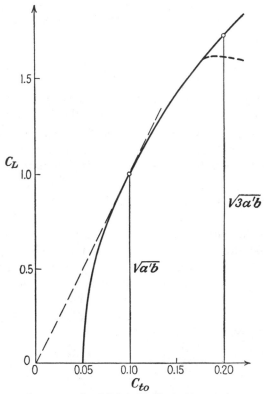

Fig. 326.—Airplane-polar curve, $a' = 0.5$, $b = 20$, illustrating maximum range and maximum endurance conditions.

The factor of V_0 here is smaller than, but very close to, 1. For our example $W_0/W_1 = \frac{5}{4}$, it is 0.945, and for $W_0/W_1 = 2$ we find 0.837.

Sometimes Bréguet's formulas (5) and (7) are used to discuss the question for the *maximum possible range* and the *maximum possible endurance* for given W_0 and W_1. It is obvious that both quantities increase with η and decrease with c. Besides, the range is proportional to C_L/C_{to} and the endurance to $C_L^{3/2}/C_{to}$. If the resultant polar diagram for a plane, with the coordinates C_{to}, C_L, is given (Fig. 326), the point of maximum C_L/C_{to} can be found as the point of contact of the tangent issued from the origin. If C_{to} is assumed in the form $a' + C_L^2/b$, according to the formula just used,

$$\left(\frac{C_L}{C_{to}}\right)_{max} = \tfrac{1}{2}\sqrt{\frac{b}{a'}}, \qquad C_L = \sqrt{a'b} \tag{9}$$

On the other hand, for maximum endurance we have $C_{to}/C_L^{3/2} = $ min.; thus,

$$\frac{a'}{C_L^{3/2}} + \frac{C_L^{1/2}}{b} = \text{min.}, \qquad C_L = \sqrt{3a'b}, \qquad \left(\frac{C_L^{3/2}}{C_{to}}\right)_{max} = \frac{3^{3/4}}{4}\,a'^{-1/4}b^{3/4} \tag{10}$$

This shows that, other things being equal, a higher value of C_L, that is, a smaller velocity (in the ratio $\sqrt[4]{3}:1 = 1.32:1$) has to be used if one wants to fly as long as possible rather than as far as possible. All these conclusions, however, are impaired by the hardly justifiable assumption of a flight at constant angle of attack.

A clear meaning is given to the *problem of finding the greatest possible range* (or endurance) if an engine is assumed to work at full throttle on a propeller with continuously controllable blade setting. Then the answer can be found in the following way.

According to the definitions (2) and (3), the maximum R and the maximum E will be known if for each value of W in the interval

$$W_0 \geqq W \geqq W_1$$

the maximum values f_{max} and g_{max} of the two functions $f(V,W)$ and $g(V,W)$

$$f(V,W) = \frac{V}{cP} = \frac{\eta}{c}\frac{V}{P_{re}}, \qquad g(V,W) = \frac{1}{cP} = \frac{\eta}{c}\frac{1}{P_{re}} \tag{10}$$

have been computed:

$$R_{max} = \int_{W_1}^{W_0} f_{max}(W)\,dW, \qquad E_{max} = \int_{W_1}^{W_0} g_{max}(W)\,dW \tag{10'}$$

To find f and g choose a definite value of W, within the interval W_0, W_1. Then for each V the power required is given, either from the plot of the P_{re} vs. V-curve or from one of the formulas (A), Chap. XV. From the propeller characteristics we can derive a chart that gives the curves $J = $ const. and $\eta = $ const. in a system with C_T/J^2 and C_P/J^2 as coordinates (polar diagram). This is shown in Fig. 327, where the straight lines passing near the origin are the polar curves for the various blade

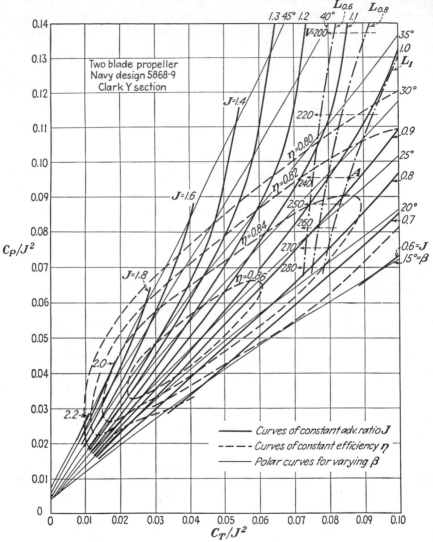

FIG. 327.—Computation of maximum possible range for an airplane with controllable-pitch propeller.

settings β. By definition, the coordinates are

$$x = \frac{C_T}{J^2} = \frac{n^2 d^2}{V^2}\frac{T}{\rho n^2 d^4} = \frac{T}{\rho V^2 d^2} = \frac{P_{av}}{\rho V^3 d^2} = \frac{P_{re}}{\rho V^3 d^2}$$

$$y = \frac{C_P}{J^2} = \frac{n^2 d^2}{V^2}\frac{P}{\rho n^3 d^5} = \frac{P}{n\rho V^2 d^3} = 2\pi\,\frac{M_{br}}{\rho V^2 d^3}$$

(11)

Therefore, if the brake moment M_{br} is supposed to be constant (which is a sufficiently good approximation in almost all cases), a definite point x, y is assigned to each V. For this point, J and η can be read off by means of the curves $J = \text{const.}$ and $\eta = \text{const.}$, and thus $n = V/Jd$ and $P = 2\pi n M_{br}$ are found. As c is a given function of n and P, all quantities occurring in (10) are known for the V under consideration. If V is varied (with the W fixed) one obtains the curves $f(V,W)$ and $g(V,W)$ and can determine their highest points, giving f_{max} and g_{max}. The entire process has to be carried out for a sequence of W-values; then the integrals R_{max} and E_{max} will be calculated by summing up the terms. Only a slight

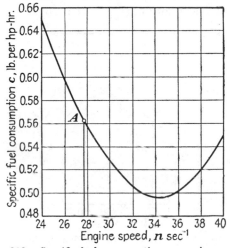

FIG. 328.—Specific fuel consumption vs. engine speed.

complication is involved if the brake moment M_{br} is considered depending on the engine speed n.

In Figs. 327 and 329 the computation of $f(V,W)$ and $f_{max}(W)$ is carried out for the three values $W = W_0$, $0.8W_0$, and $0.6W_0$. The airplane data are those of the example discussed in Sec. XV.3, i.e., gross weight $W_0 = 8000$ lb., $\lambda_s = 1011$, $\lambda_p = 555,600$ ft.^{2}/sec.^{2} The level altitude is assumed as 5000 ft., with $\rho = 0.00205$, $\sigma = 0.86$, and the propeller diameter $d = 9.3$ ft. For $V = 240$ ft./sec. the power required is found from Eq. (A''), Chap. XV, as $P_{re} = 211,600$ ft.-lb./sec. The constant brake moment is estimated as 1455 ft.-lb., corresponding to 500 hp. at $n = 30$. Thus the coordinates x and y follow from (11) as

$$x = \frac{211,600}{0.00205 \times 240^3 \times 9.3^2} = 0.087,$$

$$y = 2\pi \frac{1455}{0.00205 \times 240^2 \times 9.3^2} = 0.096$$

This point is marked A in Fig. 327, and all points computed in the same way for $W = 8000$, but different V, are connected by the curve L_1. Using the curves $J = $ const. and $\eta = $ const. in Fig. 327, we can estimate $\eta = 0.832$ and $J = 0.93$ for A. It follows that $n = V/Jd = 27.7$ sec.$^{-1}$. The fuel consumption is given in Fig. 328 as a function of n in the usual units. At $n = 27.7$ we read off $c = 0.562$ lb./hp.-hr. Thus, finally,

$$f(240, 8000) = \frac{240 \times 0.832}{0.562 \times 211,600} \, 550 \times 3600 = 3320 \text{ ft./lb.}$$

In Fig. 329 the point with abscissa 240 and ordinate 3320 is marked A. All points found in the same way for different values of V give the curve marked W_0. Its highest ordinate is seen to be $f_{max}(8000) = 3440$.

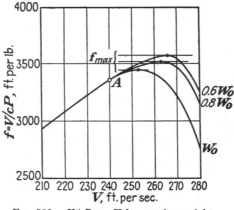

Fig. 329.—V/cP vs. V for varying weight.

Carrying out the same computation for $W = 6400$ and 4800, we find that $f_{max}(6400) = 3520$ and $f_{max}(4800) = 3580$. If the airplane is supposed to have a tank capacity of $8000 - 4800 = 3200$ lb., the maximum range can be found, according to the first formula (10′) (using Simpson's rule for numerical integration), as

$$R_{max} = \tfrac{1}{6}[3440 + (4 \times 3520) + 3580]3200 = 11.25 \times 10^6 \text{ ft.} = 2130 \text{ miles.}$$

The velocities to be used during this flight are seen from Fig. 329 to lie between 253 ft./sec. at the beginning and 266 ft./sec. at the end.

Problem 1. Give an estimate for range and endurance of a plane with 5000 lb. gross weight, 1000 lb. tank capacity, parasite ratio $S_p/S = 0.03$, and aspect ratio 7. Assume $\eta = 0.80$ and $c \sim 0.5$ lb./hp. hr.

*Problem 2. Develop the formulas for range and endurance under the assumption that the functions f and g are proportional to the $(m - 1)$th power of W (and independent of V). Take, in particular, $m = 0.6$, and compare the results with those obtained in the text.

Problem 3. Compute the maximum endurance for the example at the end of the preceding section, using the chart Fig. 327 and the c vs. n-curve Fig. 328.

Problem 4. Develop the propeller chart analogous to that of Fig. 327 for a three-blade propeller with $\lambda = 2.1$, $\mu = 0.023$ (notation of Sec. XI.4), and apply it to the computation of range and endurance for an airplane of gross weight $W_0 = 12,000$ lb., $\lambda_s = 1200$, $\lambda_p = 600,000$ ft.^{2}/sec.^{2}, with a constant brake moment corresponding to 900 hp. at $n = 30$ sec.$^{-1}$.

2. Take-off. The take-off of an airplane begins with a very short period during which the tail is raised from the ground. This is achieved by the moment of the propeller thrust T with respect to the point where the landing gear touches the ground (Fig. 330). The moment Tt acting clockwise (nose down) will outweigh the opposite moment Wl due to the gravity at a comparatively low engine speed. Then, with the tail raised, the propeller axis nearly horizontal, and the engine throttle wide open, the *take-off run* begins; its purpose is to ·accelerate the airplane to a velocity at which climbing is possible.

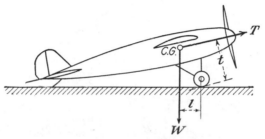

Fig. 330.—Beginning of the take-off run.

The dynamic conditions of the take-off run can be illustrated on a diagram that shows the required power P_{re} or, better, the total drag $D_{to} = P_{re}/V$ as a function of forward speed V. In Fig. 331 the D_{to} vs. V-curve that was discussed in the performance computation (Fig. 290) is reproduced. (If the airplane has retractable landing gear, the P_{re}-curve to be used here must correspond to the drag with the gear extended.)

Any point A on the curve corresponds to definite values of D_{to} and V, say, D_1 and V_1, and to a definite angle of attack $\alpha = \alpha_1$. If the airplane moves on the ground keeping this angle of attack invariable but increasing its velocity from zero up to the value V_1 given by the abscissa of A, the actual drag increases, following the formula

$$D_{to} = \frac{\rho}{2} V^2 C_{to} S = D_1 \frac{V^2}{V_1^2} \tag{12}$$

proportionally to V^2. The drag value is thus represented by the parabola with its vertex at the origin O and passing through A. If the velocity V_1 is reached at the angle α_1, the lift exerted on the airplane will equal the weight W since the points on the D_{to} vs. V-curve are determined by the condition $W = L$. Therefore, during the take-off run the lift will be

$$L = \frac{\rho}{2} V^2 C_L S = L_1 \frac{V^2}{V_1^2} = W \frac{V^2}{V_1^2} \tag{13}$$

The difference between weight and lift at each moment gives the magnitude of the normal pressure between the ground and the landing wheels; thus, if μ denotes the friction factor,

$$\mu(W - L) = \mu W \left(1 - \frac{V^2}{V_1^2}\right) \tag{14}$$

is the friction force acting opposite to the direction of motion.   This

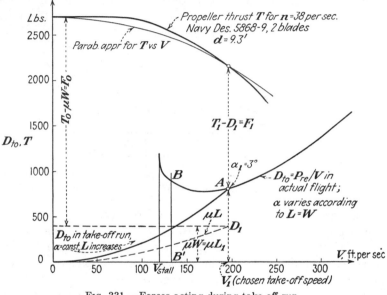

Fig. 331.—Forces acting during take-off run.

force is represented in the figure, for $\mu = 0.7$, by the horizontal line μW and the dotted parabola μL.

During the take-off run the airplane is pulled by the propeller thrust $T = P_{av}/V$, which is decreasing at increasing V. A fair assumption, which leads to a simple method of computation, will be that T diminishes according to the formula $T = T_0 - \text{const. } V^2$ or

$$T = T_0 - (T_0 - T_1) \frac{V^2}{V_1^2} \tag{15}$$

as shown in the upper left of the figure.

Then the equation of motion, stating that mass times acceleration equals the sum of forces, with $dV/dt = V \, dV/ds = \frac{1}{2} d(V^2)/ds$, reads

$$\frac{W}{g}\frac{1}{2}\frac{d(V^2)}{ds} = T - D_{to} - \mu(W - L)$$

$$= T_0 - \mu W - \frac{V^2}{V_1^2}(T_0 - T_1 + D_1 - \mu W) \quad (16)$$

By using the notations F_0 and F_1 for the forces at $V = 0$ and $V = V_1$, respectively,

$$F_0 = T_0 - \mu W, \qquad F_1 = T_1 - D_1 \tag{16'}$$

the right-hand side becomes $F_0 - (F_0 - F_1)V^2/V_1^2$; and Eq. (16) can be written in the form

$$ds = \frac{W}{2g}\frac{d(V^2)}{F_0 - (F_0 - F_1)V^2/V_1^2}$$

This gives, integrated from $V = 0$ up to an arbitrary V,

$$s = \frac{W}{2g}\frac{V_1^2}{F_0 - F_1}\log\frac{F_0}{F} \qquad \text{with } F = F_0 - (F_0 - F_1)\frac{V^2}{V_1^2} \tag{17}$$

If we assume that the pilot leaves the ground only when the velocity $V = V_1$ is reached, expression (17) supplies the *length of the take-off run*

$$S = \frac{W}{2g}\frac{V_1^2}{F_0 - F_1}\log\frac{F_0}{F_1} \tag{18}$$

which, for not too large $(F_0 - F_1)/F_1$, can be replaced approximately by

$$S \sim \frac{WV_1^2}{2gF_0}\left(1 + \tfrac{1}{2}\frac{F_0 - F_1}{F_0}\right) \sim \frac{WV_1^2}{g(F_0 + F_1)} \tag{18'}$$

The analogous approximation for (17) would be

$$s \sim \frac{WV^2}{2gF_0}\left(1 + \tfrac{1}{2}\frac{F_0 - F_1}{F_0}\frac{V^2}{V_1^2}\right) \tag{17'}$$

both approximations being derived by means of the substitution

$$\log(1 + x) \sim x - x^2/2$$

In order to obtain a moderate value for the run S the pilot will elect a comparatively small angle α, that is, a high value of V_1, for example, a V_1 close to the minimum D_{to}, as shown in Fig. 331. This will more or less coincide with the speed of maximum climbing rate or maximum climbing angle. Then, if the run on the ground is continued until $V = V_1$ is reached, a smooth and constant climbing can immediately set in. But the S-value following from (18') seems unnecessarily large. A skilled pilot will pull the elevator stick sometime before reaching V_1, perhaps at a velocity V of about 10 to 20 per cent above stalling speed, and thus, almost instantaneously, increase the angle of attack up to the value that corresponds to the point B at the abscissa V of the D_{to}-curve. If this angle is reached, the airplane will leave the ground and begin a climbing motion at the climbing angle determined by $\overline{BB'}/W$. But here the pilot is in the domain of reversed commands (Sec. XIV.3), and his next movement must be to push the stick slightly forward in order to increase the velocity, to diminish the incidence and thus to change into the regular flight region. The net gain is that the length of the take-off run is decreased, as seen from Eqs. (17') and (18'), essentially at the rate $(V/V_1)^2$. But the flight path does not show the smooth and continuous ascent as in the case of a prolonged run on the ground.

Under these circumstances it is understandable that specifications for the take-off run of an airplane must include some conditions concerning the first phase of the flight path after leaving the ground. A practical way is to require that an obstacle of moderate height, say, 50 ft., shall be passed within a definite distance from the point of departure at $V = 0$. This restricts the choice of the end velocity V of the ground run.

Equations (17') and (18') show that, next to the end velocity V, the take-off run is most influenced by the gross weight W of the airplane. Not only is S proportional to W, but also the force F_0 in the denominator of the formula is decreasing with increasing W, as seen from the definition in (16'). The latter influence depends on the friction coefficient μ, that is, on the quality of the ground. The following experimental values are usually given for regular landing gear with a single axis and normal tires:

$$
\begin{aligned}
\mu &= 0.02 && \text{for concrete or wooden deck} \\
&= 0.04 && \text{for hard turf, level field} \\
&= 0.05 && \text{for average field, short grass} \\
&= 0.10 && \text{for average field, long grass} \\
&= 0.10 \text{ to } 0.30 && \text{for soft ground}
\end{aligned}
$$

Large values of the friction coefficient may prove prohibitive to a take-off operation since the available propeller thrust T_0 at $V = 0$ may be as small as 0.20 to $0.30W$, thus rendering $F_0 = T_0 - \mu W$ very small or zero. There is only one way to improve on T_0 for take-off purposes, *viz.*, the use of variable-pitch propellers. With a pitch much higher than that used in level flight the ratio of T_0/W can readily be increased to double or more of the foregoing values. In general, take-off distances in excess of about 7000 ft. are considered impractical for landplanes.

The influence of altitude over sea level on the take-off run appears mainly in the value of T_0. If the engine output is unaltered, the thrust will be nearly proportional to the air density ρ, that is, to the ratio σ. This may diminish the difference $T_0 - \mu W$ and thus increase the take-off distance considerably.

All this argument presupposes the absence of any wind during the take-off operation. If the atmosphere is not at rest, the airplane always takes off into the wind and then the computation has to be modified in the following way. Let V_w be the wind speed. With respect to a reference system moving with the wind, the airplane at the beginning of the run has the forward speed V_w. At the time t the relative velocity is

$$V' = V + V_w.$$

The equation of motion corresponding to (16) can be written as

$$\frac{W}{g}\frac{dV'}{dt} = \frac{W}{2g}\frac{d(V'^2)}{ds'} = F_0 - (F_0 - F_1)\frac{V'^2}{V_1^2} \tag{19}$$

Here, F_0, F_1, V_1 have the same meaning as before, while s' is the distance covered during the time t in the moving reference system. Introducing the force F_0', which acts at the beginning of the take-off run,

$$F_0' = F_0 - \frac{V_w^2}{V_1^2}(F_0 - F_1) \tag{19'}$$

the integration of (19) over the interval V_w, V' furnishes

$$s' = \frac{W}{2g}\frac{V_1^2}{F_0 - F_1}\log\frac{F_0'}{F''} \qquad \text{with } F'' = F_0 - (F_0 - F_1)\frac{V'^2}{V_1^2} \tag{20}$$

This, however, is not the distance actually covered by the rolling airplane on the ground. To find the correct length s of the take-off run we have to subtract $V_w t$,

$$s = s' - V_w t \tag{21}$$

and therefore must compute *the time t needed for accelerating the airplane* from the (relative) speed V_w to $V_w + V$ or from the (absolute) speed 0 to V. From Eq. (19) we have

$$dt = \frac{W}{g}\frac{dV'}{F_0 - (F_0 - F_1)V'^2/V_1^2} = \frac{W}{gF_0}\frac{dV'}{1 - z^2V'^2} \tag{22}$$

and, by integration,

$$t = \frac{W}{2gF_0}\frac{1}{z}\log\frac{1 + zV'}{1 - zV'}\frac{1 - zV_w}{1 + zV_w} \qquad \text{with } z^2 = \frac{F_0 - F_1}{F_0 V_1^2} \tag{23}$$

Upon combining (20), (21), and (23), the exact length of the take-off run for the (absolute) end velocity V is found. A good approximation can be obtained by considering $(F_0 - F_1)/F_0$ as small and then deriving from (22)

$$ds' = V' dt = \frac{W}{gF_0}\frac{V' dV'}{1 - z^2V'^2} \sim \frac{W}{gF_0}(1 + z^2V'^2)V' dV'$$

This leads to

$$ds = ds' - V_w \, dt = \frac{V' - V_w}{V'} ds' = \frac{W}{gF_0} (V' - V_w)(1 + z^2 V'^2) \, dV'$$

and, upon integration and simple rearrangement, to

$$s = \frac{W}{2gF_0} (V' - V_w)^2 \left[1 + \frac{z^2}{2} (V'^2 + \tfrac{2}{3} V'V_w + \tfrac{1}{3} V_w^2) \right] \qquad (24a)$$

With the same approximation, the integration of (22) yields

$$t = \frac{W}{gF_0} (V' - V_w) \left[1 + \frac{z^2}{3} (V'^2 + V'V_w + V_w^2) \right] \qquad (24b)$$

In the case $V_w = 0$, $V' = V$, Eq. (24a) coincides with (17').

Formula (23) can also be used for finding the time needed in the "normal" take-off run, with $V_w = 0$ and end velocity V_1,

$$T_s = \frac{WV_1}{2gF_0} \frac{1}{\sqrt{1 - F_1/F_0}} \log \frac{1 + \sqrt{1 - F_1/F_0}}{1 - \sqrt{1 - F_1/F_0}} \qquad (25)$$

or, in first approximation,

$$T_s \sim \frac{WV_1}{gF_0} \left(1 + \tfrac{1}{3} \frac{F_0 - F_1}{F_0} \right) \qquad (25')$$

The following example shows how these formulas can be applied.

Let us take the airplane that was discussed in Sec. XV.3, of gross weight $W = 8000$ lb. with the two thrust curves given in Fig. 331. Here the required thrust D_{to} is computed as P_{re}/V from the previously used power-required curve. The available-thrust curve is continued toward the left up to $V = 0$ in accordance with experimental tests on a definite propeller of 9.3 ft. diameter (Fig. 331). If the velocity V_1, which determines the angle of attack during the take-off run, is chosen as $V_1 = 196$ ft./sec., ($\alpha_1 = 3°$), the figure shows $F_1 = T_1 - D_1 = 1360$ lb. and, with $\mu = 0.05$ (average field), $F_0 = T_0 - \mu W = 2330$ lb. Then, for the run continued up to $V = V_1$, the length and time required according to (18) and (25) would be

$$S = \frac{8000}{2 \times 32.2} \frac{196^2}{970} \log \frac{2330}{1360} = 2650 \text{ ft.}, \qquad T_s = 24.8 \text{ sec.}$$

The approximation formulae (18') and (25') would supply

$$S \sim \frac{8000 \times 196^2}{32.2 \times 3690} = 2600 \text{ ft.}, \qquad T_s \sim 23.8 \text{ sec.}$$

Actually, the ground run can be discontinued at a much smaller end velocity V. If V is taken as 140 ft./sec., about 20 per cent above stalling, as indicated in the figure, the length s and the time t can be computed from (17') and from the second equation (24) with $V_w = 0$,

$$s \sim \frac{8000}{2 \times 32.2} \frac{140^2}{2330} (1 + 0.106) = 1150 \text{ ft.}, \qquad t \sim 16 \text{ sec.}$$

The actual take-off distance will correspond to intermediate values, depending on the height of the obstacles that have to be cleared immediately after leaving the airfield.

Problem 5. Compute the take-off distance and the time needed for the ground run, for the airplane discussed at the end of the preceding section, under the assumption of a head wind of 20 m.p.h.

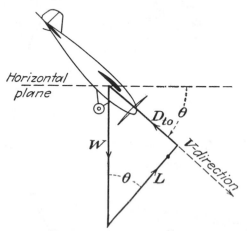

Horizontal plane

Fig. 332.—Attitude in gliding.

Problem 6. How do the results given in the text and those found in Prob. 5 change if the airplane takes off from a practically smooth wooden deck instead of from an airfield?

Problem 7. Find the take-off distance for an airplane with the following specifications: gross weight $W = 5000$ lb., wing loading $W/S = 25$ lb./ft.², effective aspect ratio 6.7, parasite ratio $S_p'/S_p = 0.04$. The power plant supplies $P_{av} = 500$ hp. in sea-level flight and a thrust $T_0 = 1800$ lb. at $V = 0$. The ground run is performed at the angle corresponding to $V_1 = 190$ ft./sec., but the pilot leaves the ground when $V = 160$ ft./sec. is reached. Assume average airfield conditions.

Problem 8. How long is the take-off run in the case of Prob. 7 if the ground is very soft ($\mu = 0.25$) but a 20 m.p.h. wind is blowing?

*Problem 9. Develop the formulas for take-off distance and take-off time under the assumption that the propeller thrust decreases linearly with V. Find, in particular, the first approximations, assuming the ratio $(F_0 - F_1)/F_0$ as small, and compare the results with those given in the text.

3. Steep Gliding and Diving. It was seen in Chap. XIV how gliding angles and gliding velocities can be derived from the power-required curve, provided that the slope of the gliding path is not too steep. In Fig. 291a we saw that the angles, in fact, are small (below about 12°) as long as the forward speed of the airplane is restricted to the range between the low and high speed of level flight.   But when the upper limit of this domain is exceeded, the corresponding gliding angle increases almost suddenly to such high values that the approximation method, which neglects the difference between W and $W \cos \vartheta$, can no longer be used.

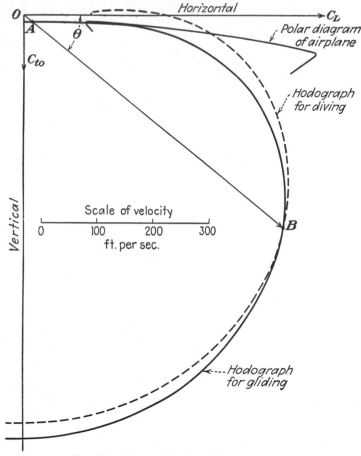

Fig. 333.—Hodographs for gliding and diving.

However, it is not difficult to find the exact answers to all questions concerning uniform gliding without using that simplifying assumption.

Let us now call ϑ the angle of the flight path with the horizontal plane, but counted positive downward (gliding angle), while in Chaps. XIV and XV positive ϑ meant climbing.   Then the equilibrium equations can be written (see Fig. 332) as

$$L = W \cos \vartheta; \qquad D_{to} = W \sin \vartheta; \qquad \tan \vartheta = \frac{D_{to}}{L} = \frac{C_{to}}{C_L} \qquad (26)$$

In Fig. 333 the polar diagram of the airplane, *i.e.*, the polar diagram of the wing with the origin shifted by S'_p/S, is shown with the C_L horizontal and C_{to} vertical downward. According to the last equation (26), the radius vector from the origin O to a point A on the polar curve immediately gives the gliding direction that corresponds to the point A. On

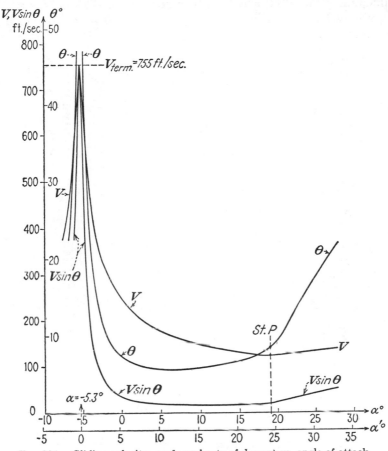

Fig. 334.—Gliding velocity, angle and rate of descent vs. angle of attack.

the other hand, from

$$W = \sqrt{L^2 + D_{to}^2} = \frac{\rho}{2} V^2 S \sqrt{C_L^2 + C_{to}^2} \qquad (27)$$

we learn that the length $\overline{OA}$ of this vector equals $2W/\rho V^2 S$. As long as ρ can be considered as constant, *i.e.*, within moderate level differences, $\overline{OA}$ is a measure of $1/V^2$. That is, the forward velocity V for the angle of incidence represented by the point A is inversely proportional to the square root of $\overline{OA}$. This makes it possible to derive from the polar curve of the airplane the *hodograph*, *i.e.*, the curve connecting the end points of all velocity vectors $\overline{OB}$, the lengths of which equal the square root of $2W/\rho S\ \overline{OA}$. The distances of the points B from the horizontal axis give the sinking velocities $V \sin \vartheta$. In Fig. 334 the magnitudes ϑ, V, and $V \sin \vartheta$ are plotted against the angle of attack α. This supplements the diagram of Fig. 291a, which was computed for small values of ϑ. The maximum velocity that occurs at, or near to, $\vartheta = 90°$ is called the *terminal velocity* of the airplane.

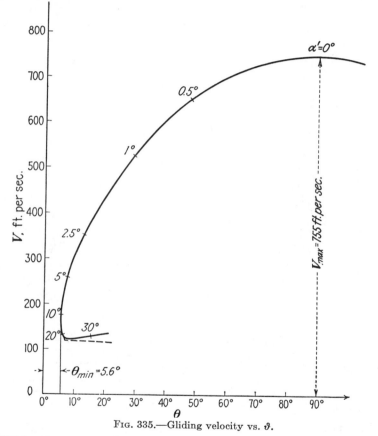

Fɪɢ. 335.—Gliding velocity vs. ϑ.

Within the range of α where the polar curve can be considered a parabola, $C_{t_o} = a' + C_L^2/b$, the relation between V and ϑ can be computed from the equations

$$\tan \vartheta = \frac{a'}{C_L} + \frac{C_L}{b}, \qquad W = \frac{\rho}{2} S V^2 \frac{C_L}{\cos \vartheta} \qquad (28)$$

which follow from (26). Eliminating C_L we find

$$\sin \vartheta = a' \frac{\rho S}{2W} V^2 + \frac{1}{b} \frac{2W}{\rho S} \frac{1}{V^2} \cos^2 \vartheta \qquad (29)$$

and solving this biquadratic equation for V^2 we have

$$V^2 = \frac{W}{\rho S a'} \sin \vartheta \left(1 \pm \sqrt{1 - \frac{4a'}{b} \cot^2 \vartheta} \right) \tag{30}$$

If it is assumed that C_L remains constant beyond the stalling point, we have here $V^2/\cos \vartheta = \text{const.} = V_{st}^2$. The dotted line in Fig. 335 represents V vs. ϑ according to (30) below stalling and according to constant C_L beyond stalling. The solid line gives the values of V derived from the

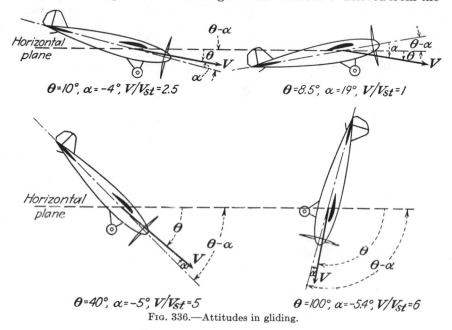

$\theta=10°, \alpha=-4°, V/V_{st}=2.5$ $\theta=8.5°, \alpha=19°, V/V_{st}=1$

$\theta=40°, \alpha=-5°, V/V_{st}=5$ $\theta=100°, \alpha=-5.4°, V/V_{st}=6$

FIG. 336.—Attitudes in gliding.

hodograph (Fig. 334); below stalling, it practically coincides with the curve derived from (30).

The *smallest gliding angle* ϑ_{min} is determined by setting the square root in (30) equal to zero, which supplies

$$\vartheta_{min} = \text{arc tan } 2 \sqrt{\frac{a'}{b}} \tag{31}$$

in accordance with Eq. (8), Chap. VII. There we had a instead of a' since the gliding angle ϵ of the wing alone was computed. The forward speed corresponding to ϑ_{min} follows from (30) as

$$V^2 = \frac{W}{\rho S a'} \sin \vartheta_{min} = \frac{2W}{\rho S \sqrt{a'(4a' + b)}} \tag{31'}$$

Figure 336 shows the attitudes of an airplane in gliding for several values of ϑ. If the angle of attack α is counted from the longitudinal axis on, $\vartheta - \alpha$ is the angle this axis forms with the horizontal plane. For $\vartheta - \alpha > 90°$ the plane is flying upside down. It is seen from Fig. 334 that at angles α higher than some 15° the attitude $\vartheta - \alpha$ and the velocity V remain nearly constant, while the gliding angle ϑ changes rapidly. The result is that in this region the pilot has great difficulty in controlling the angle of gliding. He will do better not to approach too closely the stalling speed.

It is understood that in all these calculations the atmosphere is supposed to be at rest, *i.e.*, no wind present. *In the case of a steady wind*, if the pilot keeps in the wind plane, the vector of wind velocity $\overrightarrow{V_w}$ has to

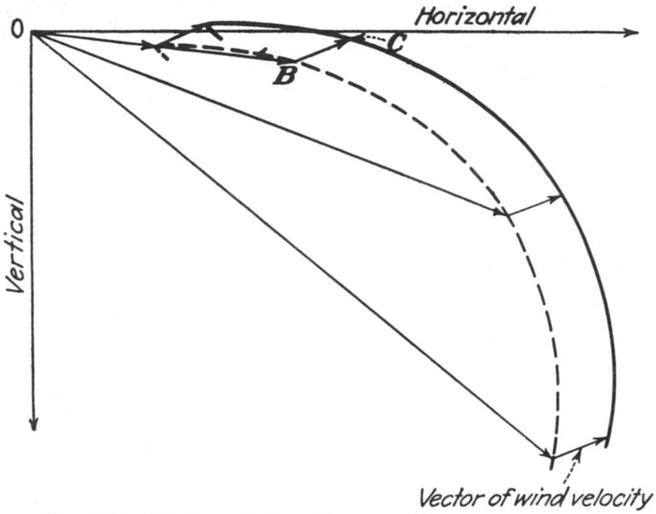

Fig. 337.—Hodograph for gliding and soaring against wind.

be added to the velocities found in the hodograph (Fig. 333). In Fig. 337 the hodograph curve is replotted as a dotted line, and to each vector $\overrightarrow{OB}$ is added the constant vector $\overrightarrow{BC} = \overrightarrow{V_w}$, which is supposed to lie in the plane of flight. This gives the resultant hodograph curve shown as a solid line. It is seen that the presence of wind with an upward component lifts the hodograph curve so that a flight with negative ϑ (climbing) or at least with $\vartheta = 0$ may become possible. These are the conditions under which a *glider plane* performs. The lifting effect depends on the vertical component only. Therefore, if this component has a sufficient magnitude, the airplane can "glide" horizontally in all directions; *i.e.*, it can remain cruising or moving in closed circuits for an indefinite time, at a constant level. Airplanes designed for use without engines have a small wing loading W/S, comparatively small a' (*i.e.*, small parasite ratio S'_p/S), and high value of b (large aspect ratio R). With $W/S = 2.5$

lb./ft.^{2}, $a' = 0.04$, $b = 9\pi = 28.3$, formula (31) gives $\vartheta_{\min} = 0.075 = 4\frac{1}{3}^\circ$, and from Eq. (31') we find, with $\rho = 0.0024$ slug/ft.^{3}, the forward speed $V = 44.2$ ft./sec. Thus an upward wind of $44.2 \times \sin 4\frac{1}{3}^\circ = 3.3$ ft./sec. would be sufficient to maintain this glider in level flight indefinitely. Such a wind component may readily develop in the neighborhood of a hillside toward which a wind blows.

Steep descent under high forward speed is also possible with the engine working. This kind of flight is usually known as *diving*. For small angles ϑ (up to about 15°) between the direction of flight and the horizontal plane the theory of downward flight is included in the general

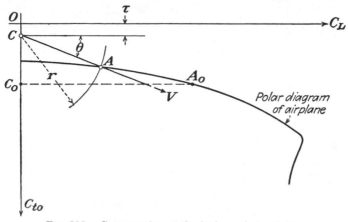

Fig. 338.—Computation of the hodograph for diving.

performance theory developed in Chaps. XIV and XV. In the case of large ϑ, however, the hodograph method just developed for gliding flight can be used. Assuming that the propeller thrust T acts in a direction sufficiently close to the direction of flight, the only modification to be applied to Eqs. (26) and (27) is that D_{to} has to be replaced by $D_{to} - T$. Let us use the abbreviations

$$\frac{2W}{\rho V^2 S} = r, \qquad \frac{2T}{\rho V^2 S} = \frac{2P_{av}}{\rho V^3 S} = \tau \tag{32}$$

Then (26) and (27) supply

$$\tan \vartheta = \frac{C_{to} - \tau}{C_L}, \qquad r^2 = C_L^2 + (C_{to} - \tau)^2, \qquad C_L = r \cos \vartheta \tag{33}$$

The second equation is the equation of a circle in the C_L-C_{to}-plane.

In Fig. 338 the resultant polar diagram is plotted again in the same position as in Fig. 333. The power-available curve P_{av} vs. V is supposed to be known. Then, for any assumed V we can compute r and τ according to (32). The circle of radius r, with its center C on the vertical axis at the distance τ from the origin, intersects the polar curve at the point A and thus determines the direction of flight parallel to $\overrightarrow{CA}$. If the center falls on C_0, the flight path has the horizontal direction C_0A_0. To centers below C_0 belong smaller velocities and negative ϑ, *i.e.*, upward flight. Letting V assume all values within the range of interest, we

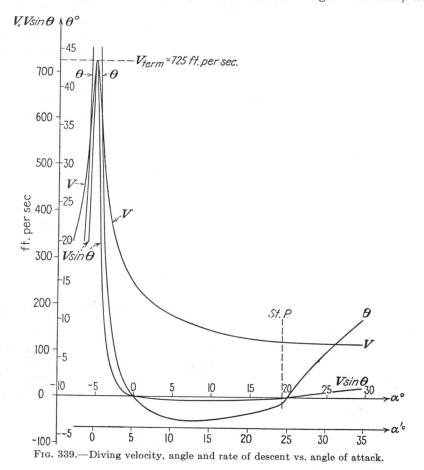

FIG. 339.—Diving velocity, angle and rate of descent vs. angle of attack.

find the correct hodograph as shown in Fig. 333 as a dotted line and can derive from it the graph of Fig. 339, which gives, in analogy to Fig. 334, the magnitudes ϑ, V, and $V \sin \vartheta$ vs. the angle of attack α, for diving. The diagrams of Fig. 339 are plotted under the assumption of a parabolic polar diagram with $a' = 0.04$ and $b = 16.4$ and a power-available function of the form

$$P_{av} = W\left(\frac{V}{\lambda_0} - \frac{V^3}{\lambda_P}\right)$$

[in accordance with Eq. (B''), Chap. XV] with $\lambda_0 = 6.23$, $\lambda_P = 2,222,000$ ft.^{2}/sec.^{2} The gross weight W is assumed to be 8000 lb., the wing area $S = 300$ ft.^{2}, and the air density $\rho = 0.0024$. It is seen from the graph that the terminal velocity, for straight downward flight, is now 725 ft./sec., while without engine it was 755 ft./sec. The difference is explained by the fact that for too high velocities (above about 600 ft./sec.) the engine is driven by the propeller, *i.e.*, the power available is negative.

The assumption made in this argument that the propeller thrust direction coincides with the direction of flight will not tangibly impair the correctness of the results, particularly not within the range of high velocities where only small variations of the angle of attack are involved. On the other hand, the deviation between the directions of flight and the propeller axis may cause a pitching moment, which renders the task of maintaining the moment equilibrium more difficult. The question of balancing the moments with the help of the elevator is decisive for the practical possibility of steep descent and will be discussed in Chap. XVII.

Problem 10. Give the equation of the hodograph for gliding and the expression for the smallest gliding angle in terms of span loading λ_s and parasite loading λ_p. Discuss the influence of altitude.

Problem 11. Plot the gliding hodograph for the airplane discussed in Sec. XV.4. Find its smallest gliding angle and the corresponding forward speed. Compute the velocities for gliding at 30° at 30,000 ft. altitude and at sea level.

*****Problem 12.** Compute the minimum sinking velocity in gliding and the corresponding angle from Eq. (30). Show that under ordinary conditions the method of Chap. XIV gives a sufficiently good approximation.

Problem 13. Plot the hodograph for a glider with 3 lb./ft.^{2} wing loading whose parasite ratio S'_p/S is 0.025 and whose effective aspect ratio is 8.5. Assume that a wind of 8 m.p.h. is present, with an upward angle of 5°, and that the pilot keeps in a vertical plane against the wind.

Problem 14. Explain how the gliding hodograph of an airplane can be used to determine the power required for level flight at a given velocity. In particular, assume that the glider of the preceding problem, with a weight of 500 lb., is towed to a tractor plane flying at 250 ft. in a horizontal direction, and compute the tension in the towing line.

Problem 15. Develop the diving hodograph for an airplane with 28 lb./ft.^{2} wing loading, parasite ratio $S'_p/S = 0.04$, effective aspect ratio 6.8, and available power equal to that used in the example of the text.

4. Landing Operation. Landing Impact. Various questions arise in connection with the landing of an airplane, and a few of them may briefly be discussed here. In general, the pilot will begin the landing operation by shutting off the engine and approaching the ground in a more or less steep glide. Before reaching the ground he must change to a flat glide, for two reasons. (1) The sinking velocity must be reduced in order to avoid, as far as possible, a landing shock. (2) The horizontal velocity must be kept low in order to avoid too long a landing run.

The conditions of steady gliding were discussed in the preceding section. During the last stage of a landing operation, however, both the gliding angle and the gliding velocity continually change under the influence of the pilot's handling of the elevator stick. The theory of such a nonuniform motion, where in addition to the actual forces the so-called "inertia forces" interfere, will be dealt with in Chap. XVIII. This kind of complete analysis will rarely be required in connection with landing problems. If the products of mass times acceleration (inertia forces) are neglected, we can explain the transition, or "flare," flight path in the landing operation by means of the polar diagram and the hodograph curve as given in Fig. 333. The assumption, equivalent to the disregarding of all inertia forces, is that the forward speed V and the

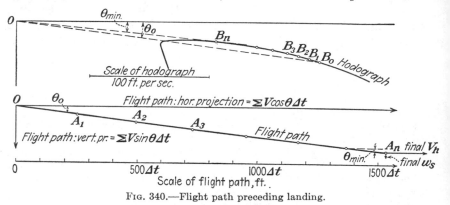

Fig. 340.—Flight path preceding landing.

gliding angle ϑ have at each moment exactly the values which would correspond to a uniform gliding at the respective angle of incidence. The transition period starts at the end of a uniform steep descent represented by the point B_0 of the hodograph (Fig. 340). Then the pilot gradually increases the angle of attack until the representative point falls at B_n, corresponding to the flattest gliding (tangent to the hodograph from O). If the timing for the intermediate stages as represented by B_2, B_3, etc., is known, we can develop by graphical integration an approximate picture of the flight path, as seen in Fig. 340. Each single segment of the flat polygonal line is parallel to one of the rays OB and proportional in length to the product of the time interval by $\overline{OB}$. In our example, with the foregoing values of W, S, ρ it is assumed that the initial gliding angle of 7.4°, corresponding to $V = 254$ ft./sec. (point B_0), is gradually decreased at a uniform rate of 0.3° per time interval Δt, until it reaches the value $\vartheta_{min} = 5.6$ (point B_n). The transition flight extended over 7 time intervals covers a horizontal distance of 1530 Δt and a level difference of 175 Δt. If the pilot begins the transition at an altitude of 600 ft. above the ground, he would have to take Δt equal to $\frac{600}{175} = 3.4$ sec.

that is, to change ϑ at the rate of 0.088°/sec. This whole analysis, of course, gives a very rough approximation only.

It was assumed in our example that the airplane reaches the landing ground in the state of flattest gliding. Instead of this, other points of the diagram could be chosen as terminal points of the landing flight, for example, the point with the smallest sinking rate or the point with the minimum horizontal velocity component. It may also be that by skillfully taking advantage of the inertia forces the pilot succeeds in arriving under more favorable conditions than those corresponding to any state of steady gliding. In addition, modern aircraft often use brakes or flaps (Sec. X.2), which change C_{to} and thus the polar diagram underlying the hodograph. This reduces the landing speed considerably. In any case, some definite values of sinking rate $w_s = V \sin \vartheta$ and of horizontal velocity $V_h = V \cos \vartheta$ will mark the end of the transition flight. While the first component, w_s, determines the *landing shock*, the second, V_h, is the initial velocity of the *landing run*, on the ground.

The conditions of the *landing run* are similar to those of the take-off run as discussed in Sec. 2 of this chapter. If we could assume that the pilot keeps the airplane at a practically constant angle, the air resistance would be proportional to the square of the velocity V. The propeller thrust can be considered as nonexistent since the engine is shut off and the reaction of the still running propeller unimportant. A certain indefiniteness is caused by the fact that after a few seconds the tail skid or the tail wheel will touch the ground. This is connected with a considerable change of the angle of attack for the further run; besides, the friction coefficient for the skid is much affected by its particular shape.

We may use formula (18) or (18'), as given in the foregoing for the length of the take-off run, with $V_1 = V_h$ and $F_0 = \mu W$, $F_1 = D_1$.

$$S = \frac{W V_h^2}{2g(D_1 - \mu W)} \log \frac{D_1}{\mu W} \sim \frac{W V_h^2}{g(D_1 + \mu W)} \tag{34}$$

Here, D_1 should be taken as the drag value for the high angle of attack that corresponds to the position of the airplane with the tail skid or the tail wheel on the ground. When a tail wheel is used, μ may have the same value (or a little higher one) as that indicated in Sec. 2 for various kinds of running ground. For a tail skid the friction coefficient reaches as high a value as 0.50; thus, if one-third the total weight rests on the tail, the former μ should be replaced by $0.17 + \frac{2}{3}\mu$, that is, by 0.27 in the case of $\mu = 0.15$. Under ordinary conditions the ratio $W/(D_1 + \mu W)$ is about 3 to 6, which is equivalent to the statement that gS/V_h^2 lies normally between 3 and 6.

The question of *landing shock*, or landing impact, requires careful consideration in the design of an airplane structure. A certain amount of vertical velocity must be reckoned with at the moment when the ground is reached. This amount may be a little smaller than the theoretical value of $V \sin \vartheta$ at the end of the landing glide, owing to the cushioning effect of the horizontal ground beneath the wing; under particularly favorable conditions, as mentioned in the foregoing, it may be even zero or nearly zero. But no airplane will be designed without providing for some appropriate kind of shock-absorbing device. In principle, a system of elastic springs may be used that has to be inserted between the axles of the main landing wheels and the airplane body. The essential theory of the shock phenomenon can briefly be sketched, as follows.

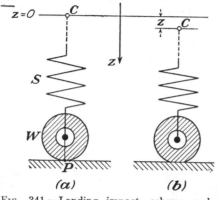

FIG. 341.—Landing impact, scheme and notation.

We denote by z the vertical displacement of the center of mass of the plane, counted positive downward. Let $z = 0$ correspond to the position where the lowest point of the landing wheel just touches the ground while the airplane is suspended, so that no thrust is exerted by the wheel on the ground and no loading is acting on the springs. In the schematic diagram of Fig. 341a, the point C is the center of gravity at $z = 0$, and S represents the spring and W the undistorted wheel with the point of contact P. In this position the shock phenomenon begins. The point C has a downward velocity $dz/dt = w_s$ and, at the first moment, no acceleration since the weight is balanced by the lift. As C continues its downward motion according to the vertical speed w_s, the spring is compressed and the tire flattened (Fig. 341b). This causes a resistance force that may be assumed to be proportional to z.

Thus, the equation of motion for the plane of gross weight W will read

$$\frac{W}{g} \frac{d^2z}{dt^2} = -kz \qquad (35)$$

where k depends on the dimensions of the spring and of the tires. The solution of the differential equation (35) that fulfills the initial conditions

$$z = 0, \qquad \frac{dz}{dt} = w_s \qquad \text{at } t = 0 \qquad (35')$$

is easily seen to be

$$z = w_s \sqrt{\frac{W}{kg}} \sin \sqrt{\frac{kg}{W}} t \qquad (36)$$

This means that a harmonic oscillation sets in, the frequency of which is the square root of kg/W while its amplitude equals

$$z_1 = w_s \sqrt{\frac{W}{kg}} \tag{37}$$

The value of k can be determined by a static experiment, as follows: If the airplane, initially suspended, is let down to the ground until it assumes a position of equilibrium under the influence of its weight, the displacement of the mass center will have a certain magnitude z_0. In this position the weight W is balanced by the elastic resistance kz_0. Thus

$$k = \frac{W}{z_0} \quad \text{and} \quad z_1 = w_s \sqrt{\frac{z_0}{g}}; \quad \frac{z_1}{z_0} = \frac{w_s}{\sqrt{gz_0}} \tag{37'}$$

The ratio of the displacements z_1/z_0 is at the same time the ratio of the forces $kz_1/kz_0 = kz_1/W$. Since z_1 is the maximum displacement (the amplitude), kz_1 is the maximum force, and the result following from (37') is that the greatest load which the landing gear has to stand is

$$F_{\max} = kz_1 = W \frac{w_s}{\sqrt{gz_0}} \tag{38}$$

For a given w_s the spring system must be designed in such a way as to supply a sufficiently large value of the elastic displacement z_0 to keep the load factor $w_s/\sqrt{gz_0}$ within reasonable limits.

In Eq. (35) it was assumed that during the whole process the weight of the airplane is completely balanced by the lift. In reality, the lift force will gradually lessen since the forward speed of the airplane decreases during the landing run. In the extreme case that the lift completely vanishes at once one should write, instead of (35),

$$\frac{W}{g} \frac{d^2z}{dt^2} = W - kz \tag{39}$$

The solution of this differential equation with the initial conditions (35') is

$$z = \frac{W}{k}\left(1 - \cos\sqrt{\frac{kg}{W}}\,t\right) + w_s\sqrt{\frac{W}{kg}}\,\sin\sqrt{\frac{kg}{W}}\,t \tag{39'}$$

as can be seen by differentiating the right-hand side and substituting in (39). The maximum value of z is found from (39') to be

$$z_1 = \frac{W}{k}\left(1 + \sqrt{1 + w_s^2\frac{k}{Wg}}\right) \tag{40}$$

If z_0 is substituted for W/k, the ratio z_1/z_0 and the maximum load kz_1.

respectively, are seen to equal

$$\frac{z_1}{z_0} = 1 + \sqrt{1 + \frac{w_s^2}{gz_0}}, \qquad F_{\text{max}} = kz_1 = W \left(1 + \sqrt{1 + \frac{w_s^2}{gz_0}}\right) \quad (41)$$

The latter value is a little larger than that found in (38).
The actual maximum load can be expected to lie between the two limits. For example, if an airplane comes down at an angle $\vartheta = 5°$ with a forward speed $V = 111$ ft./sec., the sinking velocity is

$$w_s = 111 \sin 5° = 9.7 \text{ ft./sec.}$$

In the case of a static displacement $z_0 = 0.1$ ft., Eq. (38) would give $kz_1/W = 5.4$ and Eq. (41) would yield $kz_1/W = 6.5$. We can conclude that the maximum load to which the landing gear is exposed lies between 5.4 and 6.5 times the gross weight of the airplane. This result is not too much affected by the fact that the elasticity of the system under consideration is not a perfect one. While the oscillations are not permanent but die out sooner or later, the first amplitude will have approximately the computed value, provided, of course, that the corresponding stresses lie within the limits the material can withstand.

Problem 16. Use the hodograph given in Fig. 333 (or that developed according to Prob. 11) to determine the gliding path that starts at the slope of 15° (with the corresponding gliding velocity) and changes at the rate of 0.12°/sec.

Problem 17. The landing gear of an airplane experiences an overload $F_{\text{max}}/W = 5$ at a landing velocity of vertical component $w_s = 10.5$ ft./sec. How large will the overload factor be if the same landing gear is used for an airplane of 20 per cent greater weight, landing with $w_s = 12$ ft./sec.?

5. Seaplane Problems. A very brief account of the conditions under which a seaplane takes off follows. Any seaplane, whether of the flying boat type or essentially a landplane furnished with floats instead of a landing gear, must be capable of floating on the water surface. Before rising into the air the seaplane, starting from a state of rest, performs a take-off run on the water surface during which both hydrodynamic and aerodynamic forces act upon it. The aerodynamic action consisting of drag and lift on the wing, parasite drag on the body, and thrust and torque moments on the propeller blades is subject to the same rules as in the case of a landplane. The interaction between the boat hull or the floats and the water presents a very difficult and interesting problem in the hydrodynamics of shipbuilding. It is not the problem one has to deal with in the design of ordinary ships, which float under the conditions of static buoyancy, but rather the problem of modern racing boats, the so-called "hydroplanes."

Figure 342 shows in a longitudinal section the underwater line of a body that could serve as a hydroplane, a flying boat, or the main float of a seaplane. Its most conspicuous characteristic is the presence of the sharp *step* at about the middle of the length, sometimes called after its inventor the *Ramus step*. The purpose of this design of the hull can be described as follows: As long as the body is at rest or in a sufficiently slow motion, it will be immersed in the water up to a level AA (Fig. 342a) that is determined by Archimedes' principle, or the principle of buoyancy, *i.e.*, by the condition that the weight of the displaced water equals the actual weight of the boat and its load. But with increasing velocities the conditions change. The water below the front part of the hull rises and exerts on the moving body an additional thrust upward which is approximately proportional to the square of the velocity like the lift on an air wing (Fig. 342b). This dynamic lift force adds to the buoyancy, and thus the vertical equilibrium will be maintained with a smaller portion of the body immersed. In fact, with an appropriate provision for the distribution of the masses (position of the center of gravity), the rear part of the hull, behind the step, will emerge more and more from the water as the speed increases. An effective decrease of immersion through the action of dynamic forces seems to occur only with hull forms with a marked step. At least, seaplanes without such a broken underwater line on their floating parts have so far never been satisfactory.

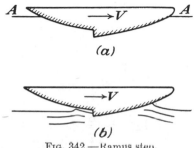

(a)

(b)

FIG. 342.—Ramus step.

A possible explanation of this failure would be as follows: In the last moment before getting into the air an airplane must increase its angle of incidence and therefore make a slight turn, nose up. If it were still in contact with the water along a continuous line (in the longitudinal section), this turn would cause the resultant buoyancy force to shift backward and thus produce a moment that counteracts the lifting of the nose. This is avoided if the last contact occurs just along the edge of the step.

The *take-off run on the water surface* presents a problem of three degrees of freedom. The center of gravity has two velocity components, one horizontal and one vertical, and besides there is the rotation in the vertical plane about the lateral axis through the center. Three dynamic equations, one for each degree of freedom, control the motion. For a first approximation we may simplify the problem by two assumptions, as follows: (1) The inertia force in the vertical direction, *i.e.*, the vertical acceleration and, in a certain way, the vertical velocity component, can be neglected since their amounts are certainly unimportant. (2) We disregard completely the rotation and suppose that the airplane remains parallel to itself during the run; this may alter considerably the forces at the beginning and the end of the run, but seems a fair assumption for its main part.

Let us denote by W the gross weight, by T the propeller thrust, by L and D lift and (total) drag exerted by the air, by L_W and D_W the analogous components due to the dynamic action of the water, and finally by B the buoyancy, *i.e.*, the weight of the displaced water. Then, under the above assumptions, the two equations of motion are (Fig. 343)

$$0 = -W + B + L + L_W, \qquad \frac{W}{g}\frac{dV}{dt} = T - D - D_W \qquad (42)$$

In taking L, L_W and D, D_W as the vertical and horizontal force components, respectively, we make use of the assumption that the vertical velocity component is practically negligible. According to our second assumption the aerodynamic forces L and W will depend on the velocity

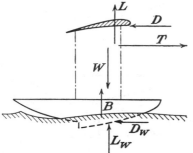

V only, except for the slight increase of the parasite drag due to the emerging parts of the boat. If V_1 is the velocity at which, at the actual angle of incidence, the lift L reaches the value W and the total drag the value D_1, we have, as in Sec. 2,

$$L = W\frac{V^2}{V_1^2}, \qquad D = D_1\frac{V^2}{V_1^2} \qquad (43)$$

Here, V_1, D_1 are the coordinates of that point of the thrust-required curve which corresponds to the state of steady climb-

FIG. 343.—Forces acting on a seaplane during take-off.

ing setting in at the end of the take-off run.

The vertical displacement of the center of gravity or of any point on the body will be called z. At the beginning, when the body is at rest and fully supported by the buoyancy, we set $z = 0$. At the end of the run, when the body has completely emerged and no buoyancy is present, the z-value may be denoted by z_1. Then the buoyancy B is a function of z with

$$B(0) = W, \qquad B(z_1) = 0 \qquad (44)$$

The dynamic action of the water can be expressed in a form similar to that used throughout this volume for the aerodynamic forces,

$$L_W = \frac{\rho_W}{2}V^2 S_L, \qquad D_W = \frac{\rho_W}{2}V^2 S_D \qquad (45)$$

Here, ρ_W is the density of the water and the areas S_L and S_D may be understood as corresponding to the concept of the parasite area of an airplane. Both S_L and S_D must be considered as given functions of z, fulfilling the conditions

$$S_L(z_1) = 0, \qquad S_D(z_1) = 0 \tag{46}$$

For the propeller thrust T we introduce the same expression as in Sec. 2, and for dV/dt the same transformation will be used. Then the two equations (42) can be written as

$$W - B(z) = V^2 \left[\frac{W}{V_1^2} + \frac{\rho_W}{2} S_L(z) \right]$$

$$\frac{W}{g} \frac{dV}{dt} = \frac{W}{2g} \frac{d(V^2)}{ds} = T_0 - V^2 \left[\frac{T_0 - T_1 - D_1}{V_1^2} + \frac{\rho_W}{2} S_D(z) \right] \tag{47}$$

From the first equation we compute V^2 as a function of z:

$$\frac{V^2}{V_1^2} = \frac{1 - B(z)/W}{1 + \rho_W V_1^2 S_L(z)/2W} \tag{48}$$

If this function is reversed, i.e., if z is expressed in terms of V^2 and introduced in the second equation (47), then this equation relates V^2 to its derivative with respect to s and thus allows us to find the length s of the take-off run by integration from $V = 0$ to $V = V_1$.

The relation (48), between z and V^2, can be represented in dimensionless form as shown in Fig. 344, in a coordinate system z/z_1 vs. V^2/V_1^2. It follows from the conditions (44) and (46) that the curve must pass through the origin and through the point 1, 1. Its shape depends entirely on the shape of the boat hull and on the assumed angle of attack, i.e., the angle between boat axis and velocity during the floating. In particular, $B(z)$ is proportional to the volume immersed when the elevation of the boat is z. From (45) and the first equation (47) we can derive

$$\frac{D_W}{W} = \frac{S_D}{S_L} \frac{\rho_W V^2 S_L(z)}{2W} = \frac{S_D}{S_L} \left[1 - \frac{B(z)}{W} - \frac{V^2}{V_1^2} \right] \tag{49}$$

The expression in brackets is zero for $V = 0$ and for $V = V_1$ according to (44). Thus the water resistance D_W vanishes at the beginning and the end of the take-off run, while the friction on the ground in the previous case of take-off from land decreases from a positive initial value to zero. As V/V_1 is known as a function of z from (48), D_W can be computed if an assumption about S_D/S_L is made. This ratio is found by experiments to go down to 0.14 for the best racing boats but to have values up to 0.30 to 0.45 in the case of seaplane hulls that run at high angles of attack.

In order to obtain an idea how the seaplane run proceeds, we may choose for $B(z)$, $S_L(z)$, and S_D/S_L the functions

$$B(z) = W \left(1 - \frac{z^2}{z_1^2} \right), \qquad \frac{\rho_W V_1^2 S_L(z)}{2W} = k \left(1 - \frac{z}{z_1} \right), \qquad \frac{S_D}{S_L} = \text{const.} = 0.4 \tag{50}$$

This may roughly correspond to an underwater line formed by two

straight segments.   The curves in Fig. 344 represent the relation between
z/z_1 and V^2/V_1^2 for various k.   In Fig. 345, D_W/W is plotted according
to (49) against V^2/V_1^2 with $k = 20$.   This graph includes also the straight

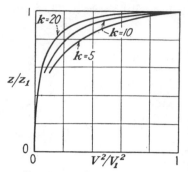

lines representing T/W and D/W according to the assumptions expressed in (47)
and (43).   Figures 346 and 347 give the
resultant force ratio

$$\frac{F}{W} = \frac{(T - D - D_W)}{W}$$

as a function of V^2/V_1^2 and of V/V_1,
respectively.   The integrations required
by the second equation (47) for finding
s and t, distance and time of the run, can
now be carried out in the following way:

Fig. 344.—Float emergence vs.
square of air speed, according to
Eqs. (48) and (50).

For the moment, let us call x and y the
coordinates of either of the curves (Figs.
346 and 347).   Then, starting at the origin,
we draw a sequence of straight lines all under the same small angle with the
y-axis so as to form a sequence of isosceles triangles with their summits on
the curve and their bases coinciding with the x-axis.   It turns out that the
number of such triangles which can be placed on the interval 0 to 1
of the x-axis gives, except for a constant factor, in the first case the length

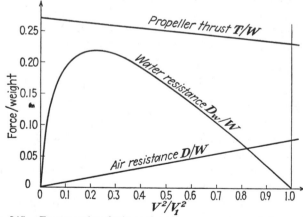

Fig. 345.—Forces acting during take-off run vs. square of air speed.

s, in the second case the time t.   In fact, consider one of the triangles
ABC, say, in the second case (Fig. 347), where $\overline{AC'} = \overline{C'B} = \Delta x = \Delta V/V_1$
and $\overline{C'C} = y = F/W = a/g$ (a = acceleration).   As Δx is supposed to be
small, we may assume that for the small time interval corresponding to
AB, during which V increases by $2V_1 \Delta x$, the acceleration has the

constant value gy. Thus the time interval is

$$\Delta t = \frac{2V_1}{g} \frac{\Delta x}{y}$$

But the ratio $\Delta x/y$ is the same for all triangles and is determined by the initial choice of the first straight line through the origin. In our case this straight line passes through the point $x = 0.05$, $y = 0.256$; thus the ratio $\Delta x/y = 0.195$, and therefore the time interval for each segment $\Delta t = 0.39V_1/g$. Figure 347 shows 45 complete triangles; the remainder on the x-axis may be estimated as nine-tenths of another one. Thus the

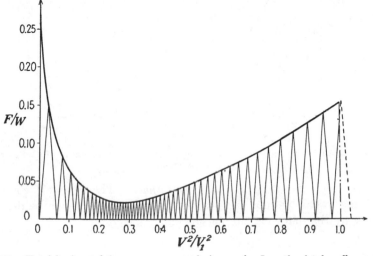

FIG. 346.—Total horizontal force vs. square of air speed. Length of take-off run found by graphic integration.

time needed for accelerating the airplane from $V = 0$ to $V = V_1$ equals $45.9 \times 0.39V_1/g = 17.9V_1/g$ (that is, 67 sec. for $V_1 = 120$ ft./sec.).

The same argument applies to Fig. 346. Here, the increase of V^2 corresponding to a segment AB is $2V_1^2 \Delta x$. As one-half only of the ratio $\Delta V^2/\Delta s$ equals the acceleration $a = gy$, we have

$$\Delta s = \tfrac{1}{2} \frac{\Delta V^2}{gy} = \tfrac{1}{2} \frac{2V_1^2 \Delta x}{gy} = \frac{V_1^2}{g} \frac{\Delta x}{y}$$

With the same slope $\Delta x/y$ as in the foregoing, the 49.4 triangles seen in Fig. 346 give the length s over which the acceleration from 0 to V_1 is achieved as $49.4 \times 0.195V_1^2/g = 9.64V_1^2/g$ (that is, 4320 ft. for $V_1 = 120$ ft./sec.).

For a correct account of the seaplane run the assumptions (50) should be replaced by equations based on the hydrodynamic theory of the interaction between the water and the boat hull.   The elements of such a theory were developed by Herbert Wagner (1930); they lie beyond the scope of this book.   A complete aerodynamic theory would also supply

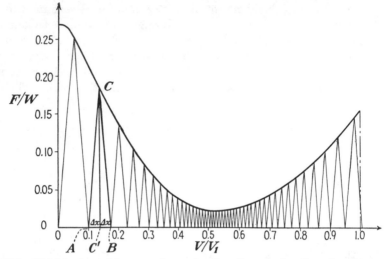

Fig. 347.—Total horizontal force vs. air speed.   Duration of take-off run found by graphic integration.

the basis for computing the *landing impact*.   The dynamic conditions prevailing in the landing, or "watering," of a seaplane can be described roughly by the following simple scheme.

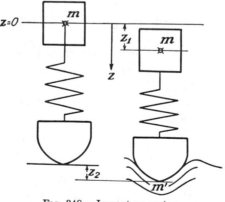

Fig. 348.—Impact on water.

Again as in the foregoing section it is assumed (Fig. 348) that a kind of elastic system is inserted between the main mass of the airplane and the boat or the floats which contact the water.   Let z_1 be the deviation of the center of gravity of the airplane from the position in which the lowest point of the hull just touches the water surface, with the spring unstrained. Then, as z_1 increases from zero on, the floating parts will penetrate the water surface, moving through a distance z_2 and imparting a certain momentum to the surrounding water particles. Let us write $m = W/g$ for the mass of the main body and m' for the variable mass of water on which the velocity dz_2/dt is imposed. As the elastic force tends to restore the initial distance between the main body and the floats and since the change of this distance is $z_2 - z_1$, the z-component of the force on the floats is $k(z_1 - z_2)$ and on the main body $k(z_2 - z_1)$. The simplified equations of motion then are

$$m \frac{d^2 z_1}{dt^2} = k(z_2 - z_1)$$
$$\frac{d}{dt}\left(m' \frac{dz_2}{dt}\right) = k(z_1 - z_2)$$

(51)

and the initial conditions

$$z_1 = z_2 = 0, \qquad \frac{dz_1}{dt} = w_s, \qquad \frac{dz_2}{dt} = 0 \quad \text{at } t = 0 \qquad (51')$$

An indication for m' as a function of z_2 should be supplied by the hydrodynamic theory. For the purpose of preliminary information, let us take for m' a constant average value. Then the first term in the second equation reduces to $m' d^2 z_2/dt^2$. Dividing the two equations by m and m', respectively, and subtracting the second from the first, we find

$$\frac{d^2(z_1 - z_2)}{dt^2} = -k\left(\frac{1}{m} + \frac{1}{m'}\right)(z_1 - z_2) \qquad (52)$$

with the initial conditions $z_1 - z_2 = 0$, $d(z_1 - z_2)/dt = w_s$. The solution can now be found in exactly the same way as was shown for Eq. (35) in the preceding section, except that k/m has to be replaced by

$$k\left(\frac{1}{m} + \frac{1}{m'}\right)$$

Thus, according to (37) the amplitude is

$$|z_1 - z_2|_{\max} = w_s \sqrt{\frac{1}{k\left(\dfrac{1}{m} + \dfrac{1}{m'}\right)}} \qquad (53)$$

and with $W = mg = kz_0$ the load factor becomes

$$\frac{k|z_1 - z_2|_{\max}}{W} = \frac{w_s}{z_0} \sqrt{\frac{1}{k(1/m + 1/m')}} = \frac{w_s}{\sqrt{gz_0}} \frac{1}{\sqrt{1 + W/m'g}} \qquad (54)$$

This coincides with (38) if m' is infinite. For a finite m' the shock is less strong than in the case of landing on solid ground with the same values of sinking velocity w_s and of static displacement z_0. As far as inferences from experimental investigations can be drawn the ratio $W/m'g$ does not exceed about $\frac{1}{2}$. This would mean a reduction of the impact effect by not more than 20 per cent as compared with the landing on solid ground.

Problem 18. Compute the length and the duration of the take-off run for a seaplane, under assumptions (50), with $S_D/S_L = 0.3$ and $k = 25$. How do the results change if the ratio S_D/S_L increases?

Part Five

AIRPLANE CONTROL AND STABILITY

CHAPTER XVII

MOMENT EQUILIBRIUM AND STATIC STABILITY

1. Pitching-moment Equilibrium. If a body moves in a straight path at constant velocity, the sum of forces as well as the sum of moments of all forces acting on it must be zero. In the performance computation of an airplane the first condition only, concerning the resultant force, has been taken into account. Assuming that the airplane is perfectly symmetrical and that it moves parallel to the vertical symmetry plane, the forces perpendicular to this plane cancel each other, and only two components, one parallel to and one perpendicular to the velocity vector, have to be considered. As to the force moments, there is only one component that is not automatically zero in the case of full symmetry, *viz.*, the moment about an axis perpendicular to the symmetry plane. It is known as the *pitching moment* and has already been introduced in Chap. VI in connection with airfoil tests. Note that full symmetry includes the condition that the airplane has two (or, in general, an even number of) propellers, with opposite sense of rotation. The influence of one-sided propeller rotation will be discussed in Sec. 6 of this chapter. At present, one pair of symmetrical propellers will be assumed and the projection of their axes on the symmetry plane will be called the propeller axis.

The concept of total drag D_{to}, which includes the drag of the air wing and all other parts of the airplane, has already been used in the performance computations. Analogically, the *total lift* L_{to} may be introduced as the sum of all thrust components perpendicular to the velocity vector, acting on the wing, the tail surfaces, the fuselage, etc. The weight W acts along a vertical line, lying in the symmetry plane and passing through the center of gravity (c.g.). Symmetrically arranged propellers will supply a resultant thrust T, whose line of action also lies in the symmetry plane. Then the moment condition of equilibrium can be expressed in the following form: *The resultant of total lift and total drag must pass through the point of intersection A of the propeller thrust and the gravity*

497

force. If the propeller thrust line passes through the center of gravity, the resultant of L_{to} and D_{to} must do so, too.

In Fig. 349 the situation of horizontal normal flight is indicated, with the propeller thrust acting parallel to the forward speed in the propeller axis. If we denote by h_l, h_d, h_t the distances from the c.g. of the forces L_{to}, D_{to}, T, the equilibrium condition reads

$$h_l L_{to} - h_d D_{to} + h_t T = 0 \tag{1}$$

The horizontal distances are counted positive toward the right, the vertical ones positive downward.

In normal flight it can be assumed that no lift force acts on the fuselage, etc., so that L_{to} equals $L + L'$, wing lift plus tail lift. If the distances of the respective lines of action from the c.g. are denoted by

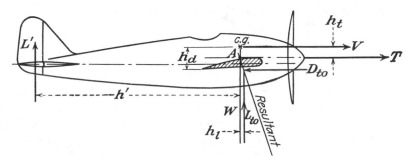

FIG. 349.—Normal flight conditions.

h and h' (h' negative in the usual design), Eq. (1) can be replaced by

$$hL + h'L' - h_d D_{to} + h_t T = 0, \qquad L_{to} = L + L' \tag{2}$$

The ideal case is that $L' = 0$. If a positive L' is required to fulfill Eq. (2), the plane is called *tail-heavy;* in the case of $L' < 0$, *nose-heavy.* The tail lift L' can be supplied by the horizontal stabilizer or by the elevator or by the combination of both. The way in which the tail works will be discussed in Sec. 2 of this chapter.

The normal conditions can prevail only when the airplane flies at a certain incidence. In horizontal flight with the engine throttled or overcharged, and in climbing and descending, the angle of incidence changes and the propeller axis (longitudinal axis of the plane) can no longer be parallel to the direction of flight. In this case Eq. (1) can still be maintained if the terms are adequately interpreted (Fig. 350). The h_l, h_d, h_t are the distances of the respective forces from the c.g. But the line of action of T will not exactly coincide with the propeller axis and the line of action of D_{to} may considerably change, because of a pitching moment experienced by the fuselage. These circumstances will be discussed in Sec. 3 of this chapter.

The condition of pitching-moment equilibrium can be expressed in terms of the *aerodynamic center* (a.c.), which was introduced in Part Two. It has been seen in Sec. VIII.3 that in two-dimensional flow around a body or a set of bodies there always exists one definite point with respect to which the moment of the aerodynamic forces acting on the bodies has a constant value, *i.e.*, a value independent of the direction of flow. This applies, first, to a wing of infinite span or to the combination of a wing and a tail surface of likewise infinite span, when both move in a straight path perpendicular to the span direction. The resultant force in this case is a mere lift, *i.e.*, a force normal to the velocity vector in the plane of flow. All experiments, however, carried out with airfoils or

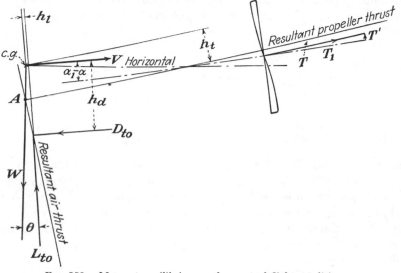

Fig. 350.—Moment equilibrium under general flight conditions.

wing models of finite span confirm that, below stalling, a point of constant pitching moment exists in these cases also, although the resultant force is no longer a mere lift as in the two-dimensional theory but includes a drag component. With the interpretation given to Prandtl's wing theory in this book (Chap. IX), this can easily be explained. In fact, according to this interpretation, the velocity and pressure distribution around a finite wing under the actual incidence coincide with the distribution that would prevail with a wing of infinite span under a smaller incidence (effective incidence). Therefore, the set of all forces, each determined by magnitude and line of action, which correspond to varying angles of incidence is the same in both cases, infinite and finite span, and a point for which the moment is constant in the first case must have this property in the general case, also. Only the relation between the directions of flow and the individual forces is modified.   If two or more profiles (wing and tail) are involved and if each is considered independently as far as the two-dimensional motion is concerned, then, under the influence of some down-wash velocity, the foregoing interpretation leads still to the existence of an a.c., that is, to a point of constant moment.

In this argument no consideration is given to the parasite drag of the wing or the tail.   It can hardly be expected that when the angle of incidence varies the parasite drag has an exactly invariable moment with respect to the a.c.   It rather seems that in the experiments the influence of the parasite drag on the moment is too small to be observed. In fact, compared with the perfect fluid forces, lift and induced drag, the additional drag is small and its line of action must be assumed to pass close by the a.c.   The conclusion is that the moment of the wing forces

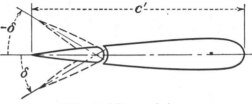

FIG. 351.—Stabilizer and elevator.

(and in the same way for the tail) with respect to center of gravity equals the constant moment M_0 with respect to the aerodynamic center plus the moments of lift and drag as attacking in the aerodynamic center.

Let x_a, z_a be the coordinates of the a.c. of the wing in a coordinate system with the origin in the c.g. and the x-axis in the direction of flight. If accents denote the analogous values for the tail, the condition of moment equilibrium takes the form

$$M_0 + x_a L - z_a D + M_0' + x_a' L' - z_a' D' + h_t T - h_d' D_p = 0 \qquad (3)$$

Here, D_p is the parasite drag on the fuselage, etc. (or, better, the total air reaction on these parts), and h_d' the distance of its line of action from the c.g.

In an airplane of normal design z_a and z_a' are small quantities and so are D and D' when compared with L.   Moreover, the tail with the elevator in normal position has, as a rule, a symmetrical profile; thus, $M_0' = 0$.   Accordingly, Eq. (3) reduces to

$$M_0 + x_a L + x_a' L' + h_t T - h_d' D_p = 0 \qquad (4)$$

In most cases the last term can be disregarded, at least for a first approximation, assuming that the resultant air reaction on fuselage, etc., acts along a line passing close by the c.g.   Equation (4) supplies the following result: If an airplane is correctly balanced, *i.e.*, is neither tail- nor nose-heavy $(L' = 0, L = W)$, the distance x_a between the c.g. and the a.c. of the wing must be

$$x_a = -\frac{M_0}{W} - \epsilon h_t \tag{5}$$

where $\epsilon = T/W$ is the resultant gliding angle of the plane. It was seen in Chap. VII that for the most usual wing profiles M_0 has a negative value. Thus, *with the usual wing profiles* $(M_0 < 0)$ *a correctly balanced airplane has its center of gravity behind the aerodynamic center of the wing, provided that the propeller axis passes not too far below the center of gravity.*

It is sometimes preferable to write these equations in a dimensionless form. With the lift and moment coefficients as defined in Sec. VI.1, Eq. (5) takes the form

$$\frac{x_a}{c} = -\frac{C_{M_0}}{C_L} - \epsilon \frac{h_t}{c} \tag{6}$$

where c is the chord length. For a modern profile with $C_{M_0} = -0.05$ at an average lift coefficient $C_L = 0.4$, the c.g. will lie three-eighths of a chord length aft the leading end of the wing, if the propeller axis passes through the c.g.

The designer has to take care to fulfill condition (5), or a more accurate one, for the average state of flight for which the airplane is intended. Then, if the state of motion changes or the loads are shifted, the airplane will become slightly nose- or tail-heavy so that the pilot has to operate the elevator. If large changes are expected, special devices can be used for fixing the stabilizer in various positions. In cases where considerable parts of the load are dropped during the flight, the performance data are modified and care must be taken to keep the tail lift L', required for balance, within the limits set by the effectiveness of the elevator.

Problem 1. A plane of gross weight $W = 12{,}000$ lb., power loading 30 lb./ft.[2], aspect ratio 7 has in normal flight a wing incidence $\alpha' = 0.09$, with the tail neutral and the propeller axis passing through the c.g. The moment coefficient for the a.c. is $C_{M_0} = -0.06$. Where must the c.g. lie?

Problem 2. If in the case of the preceding problem a load of 900 lb. (included in W) is shifted backward by 3 ft., what force must be supplied by the tail if its distance from the c.g. is 21 ft.? Neglect the small change in incidence due to change of the wing load.

Problem 3. What happens if the load of 900 lb., originally situated 3 ft. behind the c.g., is dropped?

*Problem 4. Describe the change occurring in the position of the c.g. due to the emptying of the gas tank. Assume a prismatic tank, symmetrically arranged, its c.g., when full, having coordinates x, z with respect to the c.g. of the fully loaded plane. The full gasoline weight is given as 18 per cent of the gross weight.

2. The Contribution to the Pitching Moment from the Tail. As long as an airplane moves parallel to its symmetry plane, only the horizontal

tail surfaces, stabilizer and elevator, contribute to the pitching moment. Both surfaces have, in general, symmetrical profiles, which means that the zero lift direction coincides with the symmetry axis of the profile, the aerodynamic center lies on this axis (for these profiles at about one-fourth of a chord length aft the nose), and the constant moment with respect to the a.c. is zero.   The C_L-curve for such a profile passes nearly straight through the origin under a constant slope, up to the stalling points on the positive and negative sides.

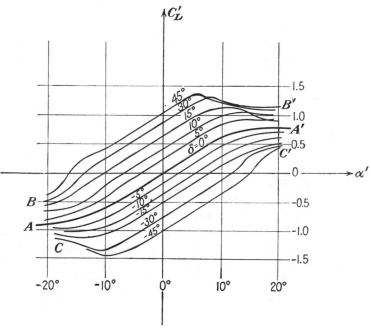

Fig. 352.—Lift coefficient for isolated tail.

If the combination of a stabilizer with an elevator in neutral position is tested, the whole behaves like one symmetrical profile.   In Fig. 352 the line $A'A$ shows the experimental results for the lift coefficient C_L' of a normal tail with the elevator neutral against the angle of incidence α'.   As was to be expected, the curve is antisymmetrical.   Only the section between the two stalling points is of interest.   If the tail combination is tested in an otherwise undisturbed flow, independently of its connection with the airplane, the slope of the straight part $A'A$ will correspond to the actual aspect ratio $R' = B'^2/S'$ where S' is the sum of the areas of the stabilizer and the elevator and B' the span of both surfaces:

$$C_L' = \frac{2\pi\alpha'}{1 + 2/R'} \tag{7}$$

Under the conditions of flight of the complete airplane an allowance must be made for the *influence of the wing on the tail.* As a general rule, the forces exerted by the airflow on a body B are strongly affected by the presence of other bodies ahead of B, while practically independent of what happens behind B. An approximate estimate for the wing influence on the tail can be given on the basis of Prandtl's wing theory in the following way.*

Formula (7) is based on the assumption—prompted by Prandtl's theory (see Chap. IX)—that in the case of a lifting surface of aspect ratio $Æ'$, moving with the forward speed V, a down-wash velocity w' is induced whose magnitude is given by

$$\frac{w'}{V} = \frac{C'_L}{\pi Æ'} \tag{8}$$

If w, C_L, and $Æ$ are the analogous quantities for the wing, the down-wash angle produced by the wing wake at the wing and the value of C_L will be

$$\frac{w}{V} = \frac{C_L}{\pi Æ}, \qquad C_L = \frac{2\pi\alpha}{1 + 2/Æ} \tag{9}$$

Behind the wing the down-wash velocity increases and takes, theoretically, the double value at infinity (see Chap. IX). It may be assumed that, for the usual distances between wing and tail, $2w$ can be considered as an approximate value for the down-wash velocity induced by the wing in the neighborhood of the tail. Thus the down-wash angle for the tail surface will be

$$\frac{w' + 2w}{V} = \frac{C'_L}{\pi Æ'} + \frac{2C_L}{\pi Æ} \tag{10}$$

The zero lift direction of the wing profile will, in general, form an angle α_d with the zero lift direction of the tail. This angle is known as the *tail decalage.* If α and α' are the simultaneous values of incidence for the wing and tail, respectively, we write $\alpha_d = \alpha - \alpha'$. Evidently, α_d is the wing incidence with the tail axis parallel to the direction of flight.

Under the influence of the combined down-wash velocities $w' + 2w$ the effective angle of incidence for the tail is

$$\alpha' - \frac{w' + 2w}{V} = \alpha' - \frac{C'_L}{\pi Æ} - \frac{2C_L}{\pi Æ} \tag{11}$$

and therefore the lift coefficient C'_L for the tail will be determined by

$$C'_L = 2\pi \left(\alpha' - \frac{C'_L}{\pi Æ'} - \frac{2C_L}{\pi Æ} \right) \tag{12}$$

---

* Millikan, C. B. "Aerodynamics of the Airplane," p. 145, New York, 1941.

If (12) is solved for C'_L, then C_L substituted from (9) and $\alpha - \alpha_d$ written for α' we find

$$C'_L = k'\alpha + \text{const. with } k' = \frac{2\pi}{1 + 2/R'} \frac{R - 2}{R + 2} \tag{13}$$

This formula, due to C. Millikan, shows that the slope of the lift coefficient curve for the tail is diminished at the rate $(R - 2)/(R + 2)$ by the influence of the wing. For the usual $R = 7$, this would give k' equal to 55 per cent of its value in the experiments with the isolated tail. Equation (12) leads to the conclusion that C'_L vanishes if $\alpha' = 2C_L/\pi R$. If this is correct, the axis of the tail should make an angle $2C_L/\pi R$ (i.e., about $1.5°$ for $C_L = 0.3$, $R = 7$) with the direction of flight when the tail is supposed to contribute no lift. Actually, an appropriate adjustment has to be made in the process of the so-called "rigging" of the airplane.

So far, elevator and stabilizer have been considered as forming one rigid surface with an essentially straight-lined profile, i.e., the elevator has been supposed to be in a nonoperative position, forming only an extension of the stabilizer. If the elevator is thrown out of the neutral position, as in the execution of a maneuver, the combination of the two surfaces acts essentially as one more or less cambered surface. If the elevator is lowered, as for descending, the camber will be positive; and, according to the general rules (see Chap. VIII), the zero lift angle shifts to the left. In this case, there will be a positive lift at $\alpha' = 0$, that is, when the air speed is parallel to the axis of the stabilizer. In Fig. 352 typical C'_L vs. α'-curves are shown for various settings δ of the elevator, as observed with an isolated tail. The δ-values are counted positive when the elevator is lowered, and α' is the angle between the velocity direction and the axis of the stabilizer, counted in the usual way. The ordinates are force coefficients referred to the factor $\rho S'V^2/2$ ($S' = $ total tail area) and refer to the force component perpendicular to the stabilizer axis (which, for small α', is essentially the same as the component normal to the velocity direction). It will be noticed that in the region below stalling all lines $A'A$, $B'B$, $C'C$, etc., run almost exactly parallel. This is in accordance with the fact that the slope of the lift coefficient curve for thin profiles is very slightly affected by the shape of the profile (Secs. VII.1 and VIII.6). In the stalling region where the lines cease to continue their straight course their behavior depends on the shape of the profiles and also on the details of the arrangement at the hinge.

With the elevator in operative position, the horizontal tail has no longer a symmetrical profile and the constant moment with respect to the a.c. will not be zero. This moment, in general, will have the sign of $-\delta$ and an absolute value increasing with $|\delta|$. However, the variations, occurring in the position of the line of action of the tail force, are very small as compared with the large distance between the tail and the c.g. of the airplane.   They may therefore be neglected in the balance of pitching moments for the airplane as a whole.   It is common practice to consider the value h' occurring in Eq. (2) as a constant, calling its absolute value *tail distance l.*   Practically, l may be taken as the distance between the c.g. and the position of the a.c. of the tail with the elevator neutral.

As long as more detailed experimental data are not available, we must assume that the influence of the wing flow on the tail, also with the

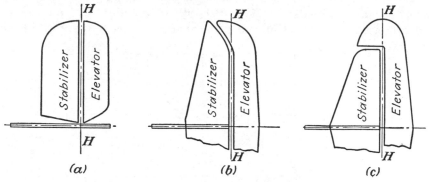

FIG. 353.—Balancing of the hinge moment.

elevator thrown out, can be estimated according to (13).   This means that the moment supplied by the tail can be written as

$$M' = -l\,\frac{\rho}{2}\,S'V^2k'\alpha + \text{const.} \tag{14}$$

with the k' from (13).   The constant depends on the elevator setting but not on α.   If the moment coefficient is computed with respect to wing area S and wing chord c, we have

$$C'_M = \frac{2M'}{\rho S V^2 c} = -\frac{l}{c}\frac{S'}{S}\,k'\alpha + \text{const.} \tag{15}$$

and can formulate the following approximate result: *The horizontal tail supplies a pitching moment whose (negative) derivative with respect to the wing incidence α is proportional to the tail distance l and the tail area S', with a factor k' that depends on the aspect ratios of both tail and wing but is essentially independent of the elevator setting δ.*   This statement, of course, is restricted to tail incidences below stalling and must be understood as a rough approximation only.   With average values $S'/S = \frac{1}{8}$, $l/c = 3$, $k' = 2.5$ the derivative of C'_M with respect to α becomes nearly -1, which gives the order of magnitude in regular cases.

For operating the elevator, the moment with respect to the hinge of that part of the lift force which acts on the elevator itself is decisive. This *hinge moment* must be kept within reasonable limits for all settings that are in use. Evidently M_h will depend on the plane form of the elevator surface and on the position of the hinge relative to it. Some typical arrangements are shown in Fig. 353. In case *a*, the elevator is said to be "completely unbalanced," *i.e.*, the surface elements on which the air pressure acts lie entirely on one side of the hinge axis. The hinge moment coefficient for this case is given in Fig. 354 as a function of the elevator setting δ for various angles of tail incidence α'. The coefficients refer to total tail area and total tail chord. If the hinge axis crosses the elevator area, as in Fig. 353*b* and *c*, the resultant force moves toward the

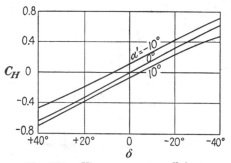

Fig. 354.—Hinge moment coefficient.

axis and the hinge moment becomes smaller. One could make it vanish entirely, in the case of a "completely balanced" elevator, but this would not serve the purpose of effective control. As a rule, a resultant hinge moment of the same sign as that present in the case of an unbalanced surface is used. The same applies also to rudders and ailerons.

Problem 5. An airplane of gross weight 8000 lb. has a horizontal tail whose area is one-eighth of the wing area. The aspect ratio is 7.5 for the wing and 5.5 for the tail. Compute the force that the tail would experience at an incidence equal to 0.6 the actual level-flight incidence of the wing.

Problem 6. An airplane of gross weight 15,000 lb. is flying under the incidence 0.08 in level flight and 0.09 at a certain climbing. Assume that the forward speed is essentially the same in both cases. Compute the tail lift in climbing for neutral elevator position, if the aspect ratio of the wing is 7.5, that of the tail 5, and the total tail area one-eighth of the wing area.

Problem 7. Assume that by setting the elevator at 5° the zero lift angle of the tail shifts by 6°. With the data of the preceding problem compute the tail lift produced by this elevator operation in level flight. Compute the hinge moment if the elevator is completely unbalanced and of rectangular shape and two-fifths of the tail lift is assumed to be uniformly distributed over its area.

Problem 8. How would Eq. (13) change if the down-wash velocity due to the wing, in the neighborhood of the tail, were assumed as $1.5w$ instead of $2w$?

*Problem 9. On the basis of Sec. IX.6 and the preceding section, develop the formula analogous to (13) for the case that the tail is designed as a biplane (twin tail).

3. The Contribution from the Propeller and the Fuselage. Under the supposed symmetrical arrangement of the propellers, with the airplane flying parallel to the symmetry plane, the resultant aerodynamic force acting on the propeller system reduces to a single force (or a couple) in the symmetry plane. In particular, if the airplane moves parallel to the propeller axis, the resultant must lie in this axis, which, in the present argument, will be used as x-axis. If the velocity is inclined against the x-axis (Fig. 355), the line of action of the resultant force will have a different position and will be given if the two components of the force

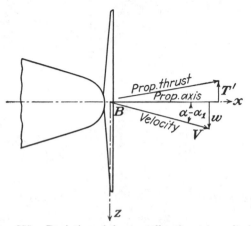

Fig. 355.—Deviation of the propeller thrust in a climb.

and its moment, with respect to one point, are known. For convenience the point B where the propeller plane intersects the axis will be chosen as reference point and as origin of the coordinate system. Taking now the x-axis as horizontal, the z-axis will be drawn vertically downward and the y-axis normal to both, positive toward the right when looked at in the x-direction (starboard).

The situation as shown in Fig. 355 with a positive z-component of the velocity corresponds to the conditions of climbing flight. In fact, it is known from Sec. XIV.4 that during a steady climb the angle of incidence is greater than in level flight. Therefore, if in level flight the velocity has the direction of the propeller axis, in climbing it must be less inclined than the propeller axis (see Fig. 289). This means that its z-component w is positive, supplying a positive angle $w/V = \alpha - \alpha_1$, where α is the actual incidence and α_1 the incidence in normal flight. Observation shows that in such cases an upward component of the propeller force (negative z-direction) occurs. This phenomenon is often described as *apparent increase of lift* in upward flight with the engine on.   An approximate account of the phenomenon can be given by using the concept of a representative blade element as introduced in Sec. XI.4.

In Fig. 356a to d, the blade element, represented by a solid line segment parallel to its zero lift direction (angle β'), is shown together with the two velocity components, the forward speed V in the x-direction and the rotational velocity $u = r\omega$ in four different directions.   If the propeller rotates in the positive sense (from positive y on the shortest way to positive z), the plot (a) with u in y-direction corresponds to the top

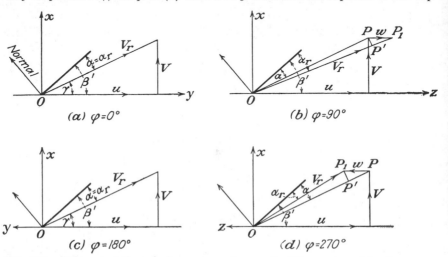

FIG. 356.—Influence of the velocity-component w on the lift experienced by the representative blade element.

position of the element, marked $\varphi = 0$; case (b), marked $\varphi = 90°$, has u parallel to the positive z-axis and corresponds to the horizontal position of the propeller with the blade element on the positive side of the y-axis. Analogically, (c) and (d) show the element in its lowest position and in the horizontal position on the negative side of the y-axis.   The additional velocity w, parallel to the z-axis, is invisible in (a) and (c).   The magnitude of w is assumed to be small, and higher powers of the ratio w/V are consistently disregarded in the present argument.

Considering first the lift force on the propeller element, its magnitude, below stalling, is given by

$$\frac{\rho}{2} A_P V_r^2 C_L = \frac{\rho}{2} A_P V_r^2 k \alpha_r \qquad (16)$$

Its direction is normal to the resultant velocity V_r and lies in the plane subtended by the vector V_r and the normal to the blade element.   In expression (16) A_P is the fictitious area ascribed to the representative blade element, *i.e.*, substantially, the effective blade area (see Sec. XI.4), k is the lift factor, approximately 2π, and α_r is the resultant angle of incidence, that is, $90°$ minus the angle between the velocity direction and the normal of the blade element. It is seen from Fig. 356 that in cases a and c neither α_r nor V_r is affected by the additional velocity w, except for terms of higher order. We have then for $\varphi = 0°$ and $180°$,

$$V_r^2 = V^2 + u^2 + w^2 \sim V^2 + u^2 = \frac{V^2}{\sin^2 \gamma}, \qquad \alpha_r = \beta' - \gamma = \alpha$$

On the other hand, in cases b and d, that is, for $\varphi = 90°$ and $270°$,

$$V_r^2 = V^2 + (u \pm w)^2 \sim V^2 + u^2 \pm 2uw = \frac{V^2}{\sin^2 \gamma} \pm 2uw$$

$$= \frac{V^2}{\sin^2 \gamma}\left(1 \pm 2\frac{w}{V}\sin \gamma \cos \gamma\right)$$

$$\alpha_r \sim \alpha + \frac{\overline{PP'}}{\overline{OP}} = \alpha \pm \frac{\overline{PP}_1 \sin \gamma}{\overline{OP}} = \alpha \pm \frac{w}{V}\sin^2 \gamma \tag{17}$$

At the same time, in cases b and d, the direction of the lift force is rotated through the angle $\sin^2\gamma\, w/V$ so that its angle with the negative z-direction is $90° \mp (\gamma \mp \sin^2\gamma\, w/V)$ and its cosine equals

$$\pm \sin\left(\gamma \mp \frac{w}{V}\sin^2 \gamma\right) = \pm \sin \gamma - \frac{w}{V}\sin^2 \gamma \cos \gamma \tag{18}$$

(The upper sign refers always to $\varphi = 90°$; the lower sign, to $\varphi = 270°$.) Thus, in positions b and d, according to (16), (17), and (18) there is a lift component in the negative z-direction of magnitude

$$T' = \frac{\rho}{2} A_P k \frac{V^2}{\sin^2 \gamma}\left(1 \pm 2\frac{w}{V}\sin \gamma \cos \gamma\right)\left(\alpha \pm \frac{w}{V}\sin^2 \gamma\right)$$

$$\left(\pm \sin \gamma - \frac{w}{V}\sin^2 \gamma \cos \gamma\right)$$

$$= \frac{\rho}{2} A_P k \frac{V^2}{\sin^2 \gamma}\left(\pm \alpha \sin \gamma + \frac{w}{V}\sin^3 \gamma + \alpha\frac{w}{V}\sin^2 \gamma \cos \gamma\right) \tag{19}$$

if terms of higher order in w/V are disregarded.

The first term in the parentheses with the $\pm$ sign gives no resultant contribution in the case of a two-blade propeller, or in any case when the temporary mean value of the force is computed. The two other terms determine a force component in the negative z-direction, due to lift, and equal to

$$T' = \frac{\rho}{2} A_P k V^2 \frac{w}{V}(\sin \gamma + \alpha \cos \gamma) \tag{20}$$

It may be mentioned that in positions a and c, also, the additional velocity w causes a change in the lift direction which leads to a force component parallel to the z-axis. But this contribution can easily be seen to be of the order $\alpha^2 w/V$ which, for small α, is of higher order than the terms kept in (19).

Introducing a coefficient C_T' of the normal force by setting T' equal to $C_T'\rho n^2 d^4$ and using the parameters λ and μ as introduced in Eq. (19), Chap. XI,

$$\lambda = \frac{2\pi r}{d}, \qquad \mu = \frac{A_P}{2d^2}, \qquad \tan \gamma = \frac{1}{\lambda}\frac{V}{nd} = \frac{J}{\lambda}$$

we find that

$$C_T' = \frac{A_P}{2d^2} k \frac{V^2}{n^2 d^2} \frac{w}{V} (\sin \gamma + \alpha \cos \gamma) = \mu k \frac{w}{V} J^2 \frac{J + \alpha\lambda}{\sqrt{J^2 + \lambda^2}} \qquad (21)$$

A quite similar expression can be derived for the normal force due to the drag component of the aerodynamic force. But under the usual assumptions for the drag coefficient below stalling this contribution seems numerically unimportant as compared with the effect of the lift component.

The force in the negative z-direction as computed in (19) and represented by the coefficient C_T' in (21) acts on a two-blade propeller when its blades are in the horizontal position. In general, for an arbitrary φ each of the three factors in the first expression (19) can be expressed as a linear function of $\sin \varphi$ and $\cos \varphi$, and the fact that in the two terms retained in (20) the sign is positive for both $\varphi = 90°$ and $\varphi = 270°$ indicates that these terms are coefficients of $\sin^2 \varphi$, while the term with $\pm$ in (19) has the factor $\sin \varphi$. Now the mean value of $\sin^2 \varphi$ over a whole period is $\frac{1}{2}$. Thus one would expect that only one-half of the computed T' really occurs. However, observed values of C_T' seem to be in good agreement with the expression given in (21). Whether this is due to the nonsteady character of the motion, to the neglected terms of higher order, to the influence of drag, to our disregarding the induced velocities (see Chap. XII), or to other causes cannot be decided at present. The most likely explanation would be that the flow pattern around the blades does not adapt itself to the very fast change of conditions during each period and that, for some reason, the flow which corresponds to the maximum normal force prevails.[1]

Figure 357 shows the results of observations[2] on a propeller with the nominal blade setting $\beta = 28.6°$ tested under 20° deviation between the

---

[1] In a recent investigation by L. B. Rumph, R. J. White, and H. R. Grumann, *Jour. Aeronautical Sci.*, **9**, 465 (1942), the authors come to the conclusion that "the theory of Harris and Glauert . . . predicted only half of the force in yaw as indicated by the test." This theory is also based on the assumption of steady flow conditions.

[2] LESLEY, E. P., G. F. WORLEY, and S. MOY, *NACA Rept.* 597 (1937).

propeller axis and the air speed. Two of the solid lines give the thrust coefficient C_T and the values of another coefficient C_T'', referring to the normal force, both observed for varying advance ratios J. From the C_T vs. J-line sufficiently approximate values for λ, μ, and β' can be derived: $\lambda = 2.2$, $\mu = 0.0145$, $\beta' = 32\frac{1}{2}°$. The value of w/V is given by $\tan 20° = 0.364$; the coefficient k is taken $= 2\pi$. The observed C_T'' corresponds to the force component normal to the air speed rather than

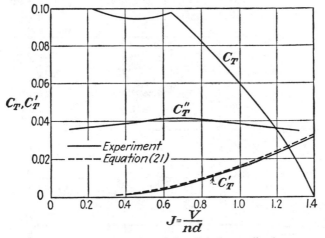

Fig. 357.—Coefficient C_T' of the normal propeller force.

to the propeller axis and thus includes the component of the axial thrust T, normal to the velocity:

$$C_T'' = C_T' + \frac{w}{V} C_T \tag{22}$$

The third solid line in Fig. 357 shows the difference $C_T'' - 0.364C_T$ and can be compared with the dotted line, which represents formula (21). The agreement seems satisfactory within the region below stalling. Other experiments of the same series confirm the proportionality with w/V and the dependence of C_T' on the blade setting, as included in (21). In an earlier experimental investigation[1] a propeller with very small pitch, about $\beta' = 8°$, was tested, so that the region below stalling extended to $J = 0$. Here, the behavior of the expression (21) near the origin $J = 0$ seems to be well confirmed.

Besides the force normal to the axis a propeller in a flight path whose direction deviates from that of the axis develops a moment about a diameter in the propeller plane. If the velocity vector lies in the longi-

[1] FLACHSBART, O., and G. KROEBER, *Z. Flugtechnik u. Motorluftschiffahrt*, **20**, 605 (1929).

tudinal symmetry plane, the moment is a pitching moment. Experiments show that its magnitude is very small and that it changes its sign with varying advance ratio J. The observed order of magnitude of the moment coefficient referred to $\rho n^2 d^5$ is 0.001. For an average value of $C_T \sim 0.05$ this would mean a displacement of the line of action of the propeller thrust by $\frac{1}{50}$ diameter or, with $d = 10$ ft. and $T/W = \frac{1}{8}$, a displacement of the total lift by 0.3 in. At the present time, neither theoretical nor experimental evidence is sufficient to justify any practical application of this result.

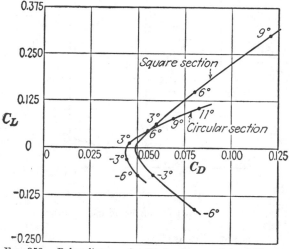

Fig. 358.—Polar diagram for the fuselage models of Fig. 63.

If one wants to take into account the pitching moment supplied by the propeller in a nonnormal state of flight, one has to recur to formula (20) for the normal force due to the deviation w/V. This ratio w/V must be identified with the change of incidence $\alpha - \alpha_1$. Writing now α_P for the propeller incidence and 2π for k and passing to the moment coefficient (referring to wing area and wing chord), we conclude from (20)

$$C_p = \frac{A_P}{S}(\sin \gamma + \alpha_P \cos \gamma)2\pi(\alpha - \alpha_1), \qquad C_{M_p} = \frac{h_p}{c} C_p \qquad (23)$$

where h_p is the abscissa of the propeller plane with respect to the c.g. *The propeller supplies a pitching moment whose (positive) derivative with respect to the wing incidence α is proportional to the distance of the propeller from the c.g. and to the blade area, with a coefficient depending on the working conditions of the propeller, the blade setting, etc.* If in (23) the values $A_P/S = 0.02$, $h_p/c = 1$, $\gamma = 30°$, $\alpha_P = 6°$ are chosen, the moment coefficient derivative becomes 0.075, that is, considerably smaller (and opposite in sign) as compared with the tail influence, Eq. (15).

The aerodynamic action on the *airplane body*, or the fuselage, was considered in Sec. V.6 only for the case in which the motion is parallel to the longitudinal axis of the plane. In this case the total action reduces, essentially, to a drag force with the axis as the line of application. If the direction of flight deviates from the direction of the axis, the body will experience a lift force in addition to the drag. Figure 358 shows the polar diagram for the two ideal fuselage models represented in Fig. 63. The lift coefficient C_L is plotted against the drag coefficient C_D, both coefficients being based on the area of the greatest cross section of the body. The values of the incidence β, that is, of the angle between the velocity direction and the axis, are indicated along the curves. In the case of the circular cross section the lift coefficient at a deviation of 6° is about 0.03. With a wing lift value $C_L = 0.5$ and the ratio $\frac{1}{4}$ of fuselage cross section to wing area, this would mean an additional lift force, due to the fuselage, of 1.5 per cent of the gross weight.

The influence of these forces on the pitching moment could be determined if the line of application of their resultant were known. One can learn from experiments on bodies of somewhat similar forms, like prolate ellipsoids or airship models, that this question is very delicate. Figure 359 shows the lines of action as found on a prolate body of revolution with a ratio 6.5:1 of length to diameter.[1] It is seen that the lines rapidly change as β increases and that the moment rather than the resultant force seems to be of importance. In fact, the perfect-fluid theory leads to the conclusion that a well streamlined body, when moving uniformly through the air, would experience an air reaction equivalent to no resultant force (D'Alembert's paradox; see Chap. IX), but to a couple only. Both lift and drag must be considered, in this case, as effects of viscosity and are presumably of minor importance. As to the moment of the couple, the perfect-fluid problem has been solved for the case of a spheroid (ellipsoid of revolution). If l_f is the length and A_f the cross-sectional area, the moment for a small angle β is found to be[2]

$$M_f = \frac{\rho}{2} V^2 A_f l_f \beta c_f \tag{24}$$

where c_f depends on the ratio length to diameter. For $l_f/d = 2, 3, 4$, the theoretical values of c_f are 0.66, 0.92, 1.04. Various experiments on airships and fuselage models gave c_f between 0.60 and 0.80.

Formula (24) has a twofold use. As long as angular deviations in the longitudinal symmetry plane are considered, β is our former $\alpha - \alpha_1$, and M_f a pitching moment of the same sign as $\alpha - \alpha_1$. Later, in Sec. 6, β

---

[1] FUHRMANN, G., *Z. Flugtechnik u. Motorluftschiffahrt*, **1**, 161 (1921).

[2] See, for example, Horace Lamb, "Hydrodynamics," 5th ed., pp. 146, 156, Cambridge University Press, London, 1924.

will be identified with a sideslip and M_f will be a yawing moment. Here we introduce the coefficient

$$C_f = \frac{2M_f}{\rho V^2 Sc} = \frac{A_f}{S}\frac{l_f}{c} c_f(\alpha - \alpha_1) \tag{25}$$

and state the following result: *The fuselage supplies a pitching moment whose (positive) derivative with respect to the wing incidence is proportional to the cross section and the length of the body, with a coefficient (0.6 to 0.8) depending on its shape.* With $A_f/S = \frac{1}{4}$, and $l_f/c = 2.5$, the derivative takes a value of 0.4 to 0.5, that is, much more than the propeller influence and rather comparable to the tail effect (but of opposite sign).

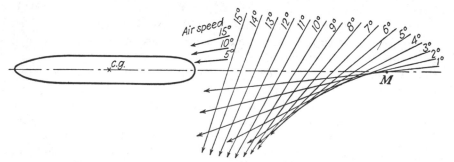

Fig. 359.—Lines of action of the air thrust upon a prolate body.

Problem 10. Give an approximate value for the cross force on a propeller at 5° deviation between propeller axis and direction of flight. The propeller has a blade area of 6 ft.² and a nominal blade setting of 23°, and the velocity of flight is 160 m.p.h.

Problem 11. What amount of tail surface would be necessary to compensate the couple experienced by the fuselage in pitch if the fuselage cross section is one-fifth the wing area? Assume the aspect ratio 6 for both wing and tail.

4. Static Stability and Metacenter. The fundamental concept of *static stability* can be explained in the simplest way by the example of a heavy body subject to the influence of gravity and of a vertical supporting force. In Fig. 360a and b the solid ellipse may be considered as a cross section through a homogeneous cylinder with its generating lines perpendicular to the paper. In both cases a and b the equilibrium conditions are fulfilled. The only forces present, the weight W and the supporting thrust F, are equal in magnitude and opposite in direction and have the same line of action. Nevertheless, there is a considerable difference between these two states of equilibrium. The first is called a *stable equilibrium;* the second is known as *unstable.* In fact, an actual cylinder will never be found to be at rest in position b. How can this phenomenon be accounted for?

The answer will be supplied by considering, in each case, an alternate position of the body, close to the equilibrium position. If, for any reason, the cylinder experiences a slight angular displacement, as shown in the dotted lines of Fig. 360, the state of equilibrium is disturbed in both cases *a* and *b*.   The gravity still acts vertically downward in the center of the ellipse, and the supporting force can still be considered as having the same magnitude and the direction vertically upward.   But its line of action has changed.   In case *a* it shifted to the left; in the case *b*, to the right.   Combined with the weight W the new supporting thrust F' forms, in the first case, a couple that tends to turn the body clockwise, in the second case a couple of opposite sense.   As in both cases the original angular displacement was assumed counterclockwise, we see that in *a*

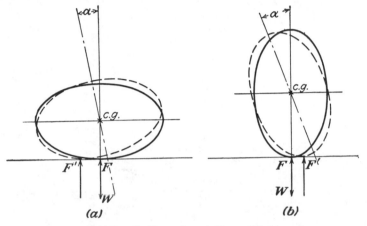

Fig. 360.—Stable and unstable equilibrium.

the couple tends to restore the initial position, while in *b* it tends to increase the displacement.   The couple produced by the displacement is *restoring* in the first case and *deviating* in the second.   Thus the following definition suggests itself: *A state of equilibrium will be called (statically) stable if the system of forces present after a slight displacement tends to restore the initial position and unstable if it tends to increase the disturbance.* There is, obviously, a third intermediate case, when the forces have neither tendency.   This holds, for example, if the ellipse in Fig. 360 is replaced by a circle.   In this case the term "indifferent" state of equilibrium is often used.

The restrictive term "statically" is used in connection with the words "stable" and "unstable" to indicate that only static conditions are taken into consideration.   It may be that the restoring couple in case *a* is insufficient to overcome the inertia forces present in an actual disturbance or so large that it overcompensates the disturbance.   Then the statically stable equilibrium would be dynamically unstable.   The theory of dynamic stability will be discussed in the last chapter of this book. It also follows from the definition that a state of equilibrium may be stable with respect to certain disturbances and unstable with respect to disturbances of a different kind. The elliptic cylinder supported by a horizontal plane is indifferent, for example, with respect to small horizontal translations.

A suitable *stability criterion* for a body resting under the influence of gravity on a plane horizontal support can be found in the following way: Assume that, in Fig. 361, CAB is the contour of a cylinder with its center of gravity marked c.g. and the supporting force passing through A. If the position is altered, another normal of the curve CAB will be

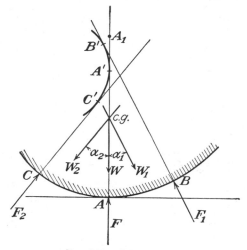

Fig. 361.—Metacenter.

the line of action of the supporting force, for example, the straight line CC', when the body is turned through an angle α_2 counterclockwise. Then the weight will act parallel to CC', in opposite direction, along a line passing through the c.g. It is seen that, in the case of Fig. 361, the couple formed by the two forces acts clockwise and is therefore restoring. If the body were subject to a clockwise displacement, the supporting force would coincide with a normal to the right of A, for example, in B, and again the couple would supply a restoring moment. This is the case for any point of support in the neighborhood of A since the center of gravity in our example lies below each of the normals.

The normals of CAB envelop a curve $C'A'B'$, which is known as the evolute of CAB. The decisive reason for the equilibrium position being stable in Fig. 361 is the fact that the point A' where the equilibrium normal AA' contacts the evolute, *i.e.*, the center of curvature of the contour in A, lies above the c.g. If the mass distribution of the body is changed in such a way that the c.g. is shifted to the point A_1 beyond AA', it is immediately seen that the equilibrium would become unstable.

The point A' where the line of action of the supporting force in the equilibrium position contacts the envelope of all its possible positions is called the *metacenter* for the state of equilibrium under consideration. The stability criterion can thus be formulated in the following simple way: *An equilibrium position for a body, subject to gravity and a supporting force, is statically stable if the metacenter lies above the center of gravity and is unstable in the opposite case.*

To return to the case of Fig. 360, it will be sufficient to consider the evolute of an ellipse (Fig. 362), which is a symmetrical curve with four cusps, A', B', A_1', B_1'; the metacenter for the ellipse supported at A is A',

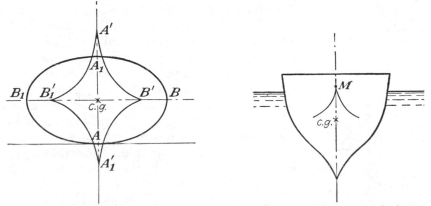

FIG. 362.—Evolute and metacenter.   FIG. 363.—Metacenter of a boat.

above the center of the ellipse.   If the ellipse were supported at B, the corresponding metacenter would be B', a point between B and the c.g. Thus the position shown in Figs. 362 and 360a is stable, while that shown in Fig. 360b is unstable, according to the metacenter criterion.

Exactly the same theory applies when the weight of a body is balanced by any other kind of force instead of the thrust supplied by a horizontal support.   The best known example is that of a body partly immersed in water and balanced by the buoyancy.   Here, in the case of a symmetrical body (Fig. 363), the lines of action of all possible buoyancy forces envelop a curve with the cusp at M, and this point is the metacenter for the equilibrium position with vertical symmetry axis.   The floating body is in stable equilibrium if and only if the center of gravity lies below the metacenter.   An analogous problem presents itself in the case of a symmetrical body falling with uniform velocity under the influence of its weight, balanced by the air resistance (drag).   The curve (Fig. 364) with its cusp at M is here the envelope of all lines of action of the resultant aerodynamic forces that correspond to various angles between the body and the velocity direction. The state of equilibrium (uniform motion) is stable if the center of gravity lies between the metacenter M and the nose of the body.

The problem to be studied in the present chapter is the *stability of an airplane* that flies at constant velocity in a straight path. Let us first consider the case of horizontal flight with the propeller axis in the direction of flight. Here, weight and lift and, on the other hand, propeller thrust and drag balance each other. In a first approximation, the influence of all other forces on the stability, except the total weight and the lift on wing and tail, may be disregarded. The lift can be assumed to follow the laws as derived in the two-dimensional wing theory. Accordingly, the envelope of all lines of action of the lift will be the parabola with its focus in the a.c. and its vertex tangent parallel to the first axis of the *system consisting of the wing and the tail*. It depends on the sign of the constant resultant moment M_0 with respect to the (resultant) a.c. whether the parabola has the position a or b shown in Fig. 365. In each case the metacenter is the point where the vertical through the c.g. contacts the parabola. It is seen in the figure that in case a, that is, for $M_0 < 0$, the metacenter has its position below the a.c., while it lies above the a.c. in case b, where $M_0 > 0$. In general, it is known that the a.c. is somewhere close to the wing chord, in the case of a single wing, and not far away from the straight line connecting wing and tail in the case of this combination. For airplanes of the usual design the c.g. will lie at about the same level. It thus follows that, in general, a positive value of M_0 will lead to a stable equilibrium, *i.e.*, to a c.g. below the metacenter, and a negative M_0 to an unstable situation. But it is by no means a correct statement that the sign of M_0 is a decisive criterion for static stability. It is true only that *in the case of a positive resultant M_0 it is much easier to fulfill the stability condition that the c.g. shall be situated below the metacenter.* This is the reason why sometimes the case of the downward parabola is called stable and that of the upward convex parabola unstable. If the moment with respect to the a.c. is zero, the c.g. must coincide with the a.c. and the equilibrium is indifferent.

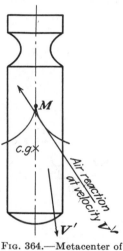

Fig. 364.—Metacenter of a bomb.

It was seen in Chaps. VI to VIII that, for a single airfoil of usual shape, the moment with respect to the a.c. is negative in most cases (acting clockwise, nose down). For a symmetrical profile the moment is zero, and for S-shaped forms it may even have slight positive values. This applies now to the wing as well as to the tail. The question arises how the combination of two airfoils with nonpositive M_0-values can give

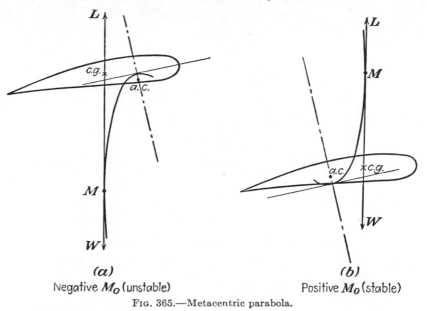

(a)   (b)
Negative M_0 (unstable)   Positive M_0 (stable)

Fig. 365.—Metacentric parabola.

a resultant force distribution with a sufficiently large positive moment at the resultant aerodynamic center. This is answered by the following

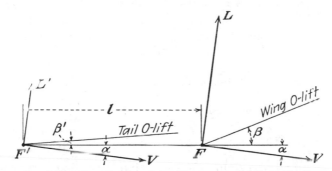

Fig. 366.—Combination of the lift systems of wing and tail.

computation, which will lead us to a decisive condition that must be fulfilled by an effective tail.

In Fig. 366 let F and F' be the a.c. for wing and tail, respectively, their distance $\overline{FF'} = l$, and β, β' the angles that the zero lift directions form with FF'. The constant moments with respect to F and F' will be denoted by M_0, M_0', respectively, the lift magnitudes by L and L', and the angle of attack with respect to the direction FF' by α. Then, in a coordinate system with origin in F and FF' as x-axis, the resultant moment of the lift forces with respect to a point x, y equals

$$M = M_0 + M'_0 - xL \cos \alpha - (x + l)L' \cos \alpha + y(L + L') \sin \alpha \quad (26)$$
$$= M_0 + M'_0 - (L + L')(x \cos \alpha - y \sin \alpha) - lL' \cos \alpha$$

Here the values of L and L' may be introduced from

$$L = \frac{\rho}{2} SV^2 k \sin (\alpha + \beta), \qquad L' = \frac{\rho}{2} S'V^2 k' \sin (\alpha + \beta') \quad (27)$$

Using the abbreviation

$$\frac{k'S'}{kS} = \kappa \quad (28)$$

the variable part of (26), except for the factor $l\rho SV^2 k/2$, can be written

$$-[\sin (\alpha + \beta) + \kappa \sin (\alpha + \beta')](x \cos \alpha - y \sin \alpha) \frac{1}{l}$$
$$- \kappa \sin (\alpha + \beta') \cos \alpha \quad (29)$$

Applying the formula for sine of $\alpha + \beta$ and introducing

$$\kappa \cos \beta' = a, \qquad \kappa \sin \beta' = b; \qquad a + \cos \beta = A, \qquad b + \sin \beta = B$$

expression (29) can be rearranged as

$$- \cos^2 \alpha \left(b + \frac{x}{l} B\right) + \sin^2 \alpha \frac{y}{l} A - \sin \alpha \cos \alpha \left(a + \frac{x}{l} A - \frac{y}{l} B\right) \quad (29')$$

The resultant aerodynamic center is determined by the condition that the total moment with respect to it is independent of α. Therefore, its coordinates x, y must fulfill the condition that in (29') the coefficients of $\sin^2 \alpha$ and $\cos^2 \alpha$ must be equal and the coefficient of $\sin \alpha \cos \alpha$ equal to zero. This supplies the two equations

$$\frac{y}{l} A + \frac{x}{l} B + b = 0, \qquad \frac{x}{l} A - \frac{y}{l} B + a = 0$$

the solution of which can easily be found to be

$$x = -l \frac{aA + bB}{A^2 + B^2}, \qquad y = l \frac{aB - bA}{A^2 + B^2} \quad (30)$$

The factors of l allow a simple geometric interpretation. If, in Fig. 367, the triangle PQR is plotted where $\overline{PQ} = 1$, $\overline{QR} = \kappa$, and the angles of PQ and QR with the x-axis equal β and β', the vector PR has the components A, B and the vector QR the components a, b. It follows, with QS perpendicular to PR,

$$x = -l \frac{\overline{SR}}{\overline{PR}}, \qquad y = l \frac{\overline{QS}}{\overline{PR}} \quad (30')$$

For these values of x, y, expression (29′) reduces simply to the factor of $\sin^2 \alpha$ (or $\cos^2 \alpha$), *i.e.*, to $yA/l = A \overline{QS}/\overline{PR}$. Thus it follows from (26) that the moment with respect to the resultant a.c. has the value

$$M_0 + M_0' + \frac{\rho}{2} SV^2 k l \overline{QS} \frac{A}{PR} = M_0 + M_0' + \frac{\rho}{2} SV^2 k l \overline{QS} \cos \varphi \quad (31)$$

It is seen that the term additional to M_0 and M_0' is positive if and only if $\overline{QS}$ is positive, that is, $\beta > \beta'$. *A tail surface is effective in supplying a positive (stabilizing) moment, only if its zero lift direction forms a smaller angle with the axis FF′ than does the zero lift direction of the wing.* This has been known as an empirical fact for a long time. The angle $\beta - \beta'$ is the tail decalage (Sec. 1 of this chapter).

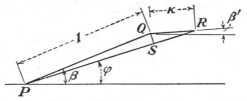

Fig. 367.—Construction of the resultant a.c. and M_0.

In most cases β and β' can be considered as small angles. It is then seen from Fig. 367 that $\overline{SR}/\overline{PR}$ is approximately $\kappa/(1 + \kappa)$ and $\overline{QS}$ approximately $(\beta - \beta')\kappa/(1 + \kappa)$. Thus the coordinates of the resultant a.c. become

$$x \sim -l \frac{\kappa}{1 + \kappa}, \qquad y \sim l \frac{\kappa(\beta - \beta')}{(1 + \kappa)^2} \quad (32)$$

and the resultant moment coefficient, when reduced to wing area S and wing chord c, will equal

$$C_{M_0} + \frac{S'}{S} \frac{c'}{c} C_{M_0}' + k \frac{l}{c} \frac{\kappa}{1 + \kappa} (\beta - \beta') \quad (33)$$

For a symmetric tail profile with $\beta' = 0$, $C_{M_0}' = 0$, this reduces to

$$C_{M_0} + \frac{l}{c} \frac{\kappa}{1 + \kappa} C_{L_0} \quad (34)$$

where $C_{L_0} = k\beta$ denotes the lift coefficient of the wing at the angle of attack $\alpha = 0$. The content of (33) or (34) can also be expressed by stating that *the effect of the tail is proportional to the distance between the two aerodynamic centers and to the angle of decalage, increasing also with the ratio S′/S.*

The conclusions are not much changed if, besides the lift, the induced drag on wing and tail is taken into account. According to Prandtl's wing theory the resultant of lift and induced drag has merely another tangent of the metacentric parabola as its line of action (see Sec. 1 of this chapter). But if all factors interfering with longitudinal stability, like parasite drag and propeller and fuselage influence, should be considered and various states of flight discussed, the metacenter method becomes too complicated. Another approach, based on some simplifying assumptions, will be discussed in the next section.

Problem 12. Compute all data of the resulting metacentric parabola for the combination of two profiles (wing and tail). Given the two aerodynamic centers, the angles β and β', the ratios k'/k, S'/S, and the moment coefficients $2M_0/\rho V^2 Sl$ and $2M_0'/\rho V^2 S'l$. Take, in particular, $k' = k = 2\pi$, $S'/S = \frac{1}{4}$, $M_0' = 0$, and discuss the result for varying M_0.

*Problem 13. Study the resultant metacentric curve in the case of a "duck," *i.e.*, an airplane with stabilizer and elevator ahead of the wing. What becomes of the theorem on decalage?

5. Simplified Stability Discussion.   In order to obtain a rapid survey of the various circumstances that determine the static longitudinal stability of an airplane in straight steady flight, the following simplifications may be introduced:

1. It is assumed that there exists one straight line, called the *mean chord* of the plane, that is parallel or nearly parallel to the propeller axis and contains the center of gravity of the plane and both aerodynamic centers of the wing and the tail, in any state of flight. The variability of the tail a.c. and of the corresponding moment with the elevator setting is disregarded.

2. The angle between the mean chord (or the propeller axis) and the velocity direction is considered as small so that its cosine can be assumed to be 1.

3. It is assumed that the pitching moments of the propeller thrust and of all parasite drag forces cancel each other. Actually, there is almost no information about the lines of action of the main drag contributions, except that they are close to the mean chord. It would not be justified to use assumptions 1 and 2 and to introduce at the same time expressions for the parasite drag moment. The moment of the induced wing and tail drag is included in the constant moments M_0, M_0' with respect to the a.c. The couple M_f, discussed in Sec. 3 of this chapter, which is due rather to lift forces on the fuselage, and the moment of the normal force T' on the propeller, also discussed in Sec. 3, are not included in the moments that are supposed to cancel out.

Under these assumptions the pitching moment M with respect to the c.g. is composed of six terms: the constant moments M_0 and M_0' with respect to the two a.c., the moment M_f and the moment of T', and finally the moments of the lift forces L, L', as attacking in the a.c. Denoting, as in Sec. 1 of this chapter, the abscissas with respect to the c.g. of the two a.c. and the propeller plane by x_a, x'_a, x_p, one has (Fig. 368)

$$M = M_0 + M'_0 + x_a L + x'_a L' + M_p + M_f \tag{35}$$

If the regular lift and moment coefficients for wing and tail are used and C_p, C_f introduced as in (23) and (25),

$$C_p = \frac{2T'}{\rho S V^2}, \qquad C_f = \frac{2M_f}{\rho S V^2 c} \tag{36}$$

Eq. (35) takes the dimensionless form

$$C_M = C_{M_0} + C'_{M_0} \frac{S'c'}{Sc} + \frac{x_a}{c} C_L + \frac{x'_a}{c} C'_L \frac{S'}{S} + \frac{x_p}{c} C_p + C_f \tag{37}$$

The first two terms are absolute constants; the others depend on the wing incidence α and the fourth also on the elevator setting δ. For not

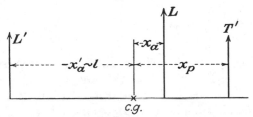

Fig. 368.—Notation for stability discussion.

too large α, all functions can be considered as linear in α, as was seen in the preceding sections. In the C_M vs. α-diagram, Eq. (37) will be represented by a definite straight line for each δ. According to the result obtained in Sec. 2 of this chapter, all these lines can be assumed to be approximately parallel.

The equilibrium condition that must be fulfilled in any state of steady flight requires that

$$M = 0 \qquad \text{or} \qquad C_M = 0 \tag{38}$$

Stability requires, moreover, that

$$\frac{dM}{d\alpha} < 0 \qquad \text{or} \qquad \frac{dC_M}{d\alpha} < 0 \tag{39}$$

In fact, a small disturbance that turns the plane counterclockwise through the angle $d\varphi$ would increase the angle of incidence by $d\alpha = d\varphi$, and this effect would be counteracted only if the increase $d\alpha$ is accompanied by a negative moment dM. *The derivative $dM/d\alpha$ or $dC_M/d\alpha$ must have a negative value for a statically stable state of flight.* The positive quantity $-dC_M/d\alpha$ is often used as a *measure of static stability.*

The straight lines in the C_M vs. α-diagram must have a negative slope, according to (39), and they must intersect the α-axis, in general, on its positive side, since the abscissa of the intersection, according to (38), marks the angle of incidence in the state of flight under consideration. One line corresponds to the flight under normal incidence α_1, for which the last three terms in (35) and (37) vanish. Let us assume $M_0' = 0$ with the usual symmetrical tail profile; then for α_1 the equilibrium condition holds:

$$0 = C_{M_0} + \frac{x_a}{c} C_L, \qquad \frac{x_a}{c} = -\frac{C_{M_0}}{C_L} = -\frac{C_{M_0}}{k\alpha_1} \tag{40}$$

If this is substituted in (37) and then the derivative is taken, we find that

$$-\frac{dC_M}{d\alpha} = \frac{C_{M_0}}{k\alpha_1}\frac{dC_L}{d\alpha} - \frac{x_a'}{c}\frac{S'}{S}\frac{dC_L'}{d\alpha} - \frac{x_p}{c}\frac{dC_p}{d\alpha} - \frac{dC_f}{d\alpha}$$

and if C_L is set $k\alpha$ and $C_L' = k'\alpha' + \text{const.}$ and finally l written for $-x_a'$,

$$-\frac{dC_M}{d\alpha} = \frac{lS'}{cS} k' - \left(-\frac{C_{M_0}}{\alpha_1} + \frac{x_p}{c}\frac{dC_p}{d\alpha} + \frac{dC_f}{d\alpha} \right) \tag{41}$$

With the usual $C_{M_0} < 0$, all terms within the parentheses are positive, according to the result of Sec. 3 of this chapter. *Under normal design conditions (propeller ahead, tail in the rear, $M_0 < 0$) wing, propeller, and fuselage have a destabilizing tendency; the plane can be stabilized by the tail only.* The necessity of a tail for stabilizing a plane was discovered by Alphonse Pénaud about 1870.

The required tail dimensions, expressed in tail area S', tail distance l, and tail-lift factor k', are determined by the stability condition (39) as

$$\frac{lS'}{cS} k' > -\frac{C_{M_0}}{\alpha_1} + \frac{x_p}{c}\frac{dC_p}{d\alpha} + \frac{dC_f}{d\alpha} \tag{42}$$

How the last two derivatives depend on the propeller and fuselage data was shown in Sec. 3 of this chapter. In general, the fuselage term will be comparable to the wing term, while the propeller influence is of smaller order of magnitude. It is seen from (42) that because of the fuselage even a wing with fixed center of pressure ($M_0 = 0$) requires a tail.

A graphical discussion of the equilibrium and stability conditions is given in Fig. 369. Here the straight line (1), intersecting the axis in $\alpha = \alpha_1$, represents the wing terms $M_0 + x_aL$. If the fuselage and propeller terms are added, the resultant of the three destabilizing influences is represented by the straight line (2), which also passes through $\alpha = \alpha_1$ on the axis. The contribution of the tail is shown by one of the lines (3_1), (3_2), etc. The points A_1, A_2, . . . where one of the lines (3) intersects the dotted line (4), which is the reversed line (2), give the values of α for the respective state of flight. If M' is set zero, the points B, where the lines (3) intersect the axis, correspond to the direction of zero lift for the tail. It is seen that B lies to the right of A and therefore has a positive abscissa, as long as that of A is positive. Since the origin O

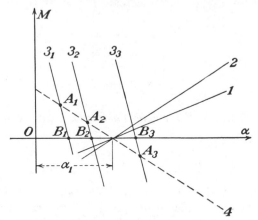

FIG. 369.—Graphical discussion of moment equilibrium and stability.

corresponds to the zero lift direction of the wing, the following conclusion is reached: *In all states of flight with positive wing lift, in a stable plane, the zero lift direction of the combined tail (stabilizer plus elevator) lies below the zero lift direction of the wing, or the decalage is always positive.*

The simplified stability condition (42) can also be interpreted in terms of the *metacenter*, which was introduced in the preceding section. In fact, if the force, supporting the weight W of a body, contacts the envelope of its possible positions at the point M, at a distance h below the c.g. (Fig. 370), the line of action after a small turn $d\alpha$ will intersect the horizontal axis through C at the point C' with $CC' = h \, d\alpha$. Thus a positive moment of magnitude $Wh \, d\alpha$ will be added, and the *metacentric distance* h can be identified as

$$h = \frac{1}{W} \frac{dM}{d\alpha} = \frac{d(M/W)}{d\alpha} \qquad (43)$$

If the moment coefficient C_M is used and W expressed as $\rho C_{L_1} V^2 S/2$ where C_{L_1} means the lift coefficient in level flight, we have $M/W = cC_M/C_{L_1}$ and

FIG. 370.—Metacentric distance.

$$h = c \frac{dC_M}{C_{L_1} \, d\alpha} \qquad (44)$$

The stability condition $dC_M/d\alpha < 0$ is thus seen to be identical with the condition that the metacenter shall lie above the c.g.

It is more usual to apply the concept of metacenter to the forces acting on the airplane except the tail, $i.e.$, on wing, propeller, and fuselage only. In this case, Eq. (41) would read

$$- \frac{dC_M}{d\alpha} = \frac{lS'}{cS} k' - \frac{h_1}{c} C_{L_1} \tag{41'}$$

and the inequality (42)

$$\frac{lS'}{cS} k' > \frac{h_1}{c} C_{L_1} \tag{42'}$$

Here h_1 is the *metacentric distance for the airplane without tail* or simply an abbreviation for

$$h_1 = \frac{c}{C_{L_1}} \left(- \frac{C_{M_0}}{\alpha_1} + \frac{x_p}{c} \frac{dC_p}{d\alpha} + \frac{dC_f}{d\alpha} \right) \tag{45}$$

Although all airplanes in practical use today are stabilized by tail surfaces, it is not impossible to design a *tailless stable airplane.* If the

wing profile has a sufficiently large positive C_{M_0}, condition (42) can be satisfied with $S' = 0$. This agrees, of course, with the fact that in this case the metacentric parabola has the upper position b in Fig. 364, so that the c.g., when supposed to lie on the mean chord,

FIG. 371.—Wing + tail and reflexed wing.

is certainly below the metacenter. It was stressed in Chap. VI that positive C_{M_0} can be obtained if the mean camber line of the profile is strongly reflexed toward the trailing end. This can now be understood in connection with the theorem on the decalage. In fact, the combination of a wing ahead, with a less inclined tail section behind, is, in a certain sense, equivalent to one reflexed profile (Fig. 371). The advantage in using the tail is that its lift appears in expression (42) multiplied by the large distance, while it is difficult to place the reflexed part of a single profile so far away from the c.g. as to achieve a sufficiently large stability value. A solution that has been tried consists in giving the wing a considerable sweepback so that its plan form looks like an arrowhead (Dunne plane, Fig. 372). Such airplanes have been built and successfully flown. It is obvious that here the wing tips, if set at a small angle (zero angle in horizontal flight), have a similar effect as has an ordinary tail under positive decalage.

It should be remembered that in all computations of this section the wing, as well as the plane, was supposed to work at angles of incidence *below stalling.* If, for example, the tail approaches stalling conditions. the derivative $dC'_L/d\alpha$ will decrease and become zero and eventually negative. Then the tail no longer can serve the purpose of stabilizing. In designing an airplane, therefore, one should carefully check the angles of incidence of the tail that may occur under various flying conditions.

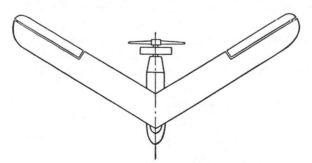

Fig. 372.—Dunne airplane with pusher propeller.

Problem 14. Compute the necessary tail surface for an airplane of gross weight 12,000 lb., wing loading 28 lb./ft.², and cruising speed at sea level 140 m.p.h. The wing profile has a moment coefficient with respect to the a.c. $C_{M_0} = -0.06$. Assume the aspect ratio 6 for the wing, 5 for the tail. The tail distance and fuselage length are about three times the wing chord. The cross section of the fuselage is one-fourth the wing area. The propeller is of the usual size, its nominal blade setting 30°, its plane of rotation ahead of the c.g. by 1½ chord lengths.

Problem 15. Discuss the result of the preceding problem with respect to varying C_{M_0}.

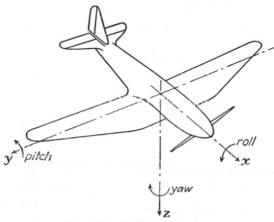

Fig. 373.—Roll, pitch, and yaw.

6. Lateral Moments. Figure 373 shows a suitable system of orthogonal axes through the center of gravity in an airplane. The longitudinal direction (propeller axis, mean chord, level flight) is taken for the *x*-axis, the normal to it, in the symmetry plane, pointing downward, is the z-axis, and the perpendicular to both, pointing to starboard (right wing in the direction of flight), is the y-axis.   The moment components parallel to x, y, z are called *rolling*, *pitching*, and *yawing* moments.   The terms roll, pitch, and yaw will be used for the respective components of rotation. The positive sense of a moment or a rotation is counterclockwise when seen from the positive side of the respective axis.   That is, the positive x-rotation turns the y-direction on the shortest way into the z-direction; the positive y-rotation likewise turns z into x; and the positive z-rotation, x into y.   Thus it is seen that a positive pitching moment turns the nose

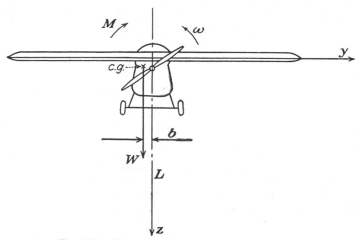

Fig. 374.—Compensation of the propeller moment.

of the airplane upward, in accordance with the assumption used in the earlier sections of this book.

If a completely symmetrical aircraft moves parallel to its symmetry plane, no rolling or yawing moments with respect to the center of gravity are caused by the air reactions on wing, tail, fuselage, etc.   It follows that, *if the aircraft has one engine with a single propeller, the equilibrium will be disturbed*, since the air reactions on the propeller supply a moment about the axis of the propeller, opposite to its sense of rotation.   This moment, which was called Q or P/ω in the discussion of the propeller (Chaps. XI and XII), is a rolling moment, according to the notation just introduced.   When the brothers Wright designed their first airplane, they were anxious to avoid this inconvenience and used two propellers, rotating in opposite directions, symmetrically arranged, and driven by means of chains from the engine shaft in the center.   It was learned later that this precaution was superfluous since *the amount of rolling moment* that must be balanced by a deviation from complete symmetry *is practically unimportant*.

Assume that the propeller is a left-hand screw; then the propeller moment will be a positive rolling moment of magnitude

$$M = \frac{P}{\omega} = \frac{P}{2n\pi} \tag{46}$$

where n is the number of r.p.s. and P the engine power (except for losses in the transmission) in foot-pounds per second. One way to compensate this moment would be to arrange the loads inside the fuselage in such a way that the c.g. is shifted in the negative y-direction to a distance b from the symmetry plane with (Fig. 374)

$$b = \frac{M}{W} = \frac{P}{W}\frac{1}{2n\pi} \tag{47}$$

In this case the lift acting in the symmetry plane would supply a negative rolling moment with respect to the c.g., of magnitude $bL = bW = M$. With the usual values of power loading $W/P = 7$ to 14 lb./hp., the quotient P/W is $\frac{550}{7}$ to $\frac{550}{14}$ ft./sec. This gives for $n = 25$ the displacement $b = 0.25$ to 0.50 ft. This amount is unimportant, at least for larger airplanes.

Another way to balance the rolling moment of the propeller is to make the incidence on the starboard wing a little larger than that on the port wing. The resultant lift on each half wing will act at about one-fourth the span distance from the center. If the lift force is $(1 + \kappa)W/2$ on the right and $(1 - \kappa)W/2$ on the left wing, a negative rolling moment with respect to the center will arise, of magnitude

$$\frac{1+\kappa}{2}W\frac{B}{4} - \frac{1-\kappa}{2}W\frac{B}{4} = \kappa\frac{WB}{4} \tag{48}$$

In order to compensate in this manner the moment (46) one has to make

$$\kappa = \frac{4}{WB}\frac{P}{2n\pi} = \frac{4b}{B} \tag{49}$$

This gives for a span $B = 50$ ft. the value $\kappa = 0.02$ to 0.04. If the mean incidence in this case is $6°$, one should have about $6°12'$ on the starboard and about $5°48'$ on the port wing. This can easily be reached in the process of "rigging" the airplane. It is seen that with increasing size of the airplane the task of balancing the propeller moment becomes easier.

The problem of *static stability against lateral disturbances* plays a much less important part than that for longitudinal motion. With respect to roll an airplane is in a certain way statically indifferent: no moments of air reactions arise when the airplane turns around the longitudinal axis, as long as the velocity changes due to roll are disregarded.

In yaw the vertical control surfaces, fin and rudder, act similarly to the longitudinal tail in longitudinal motion.   If (Fig. 375) the velocity vector deviates from the x-axis by an angle β to port, the vertical tail, whose area may be denoted by S'', experiences a force, nearly parallel to the y-axis, which can be written as

$$F_y = \frac{\rho}{2} V^2 S'' k'' \beta \tag{50}$$

This gives a negative (restoring) yawing moment $-lF_y$.   On the other hand, the fuselage will cause a positive (destabilizing) moment, according

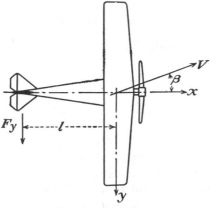

Fig. 375.—Sideslip.

to Eqs. (24) and (25).   Thus the total yawing moment due to the *sideslip* β will be

$$M = \left(-\frac{\rho}{2} V^2 S'' l k'' + \frac{\rho}{2} V^2 A_f l_f c_f \right) \beta \tag{51}$$

and the moment coefficient, based on wing area S and on span B as reference length,

$$C_M = -\left(k'' \frac{S'' l}{SB} - c_f \frac{A_f l_f}{SB} \right) \beta \tag{52}$$

The expression in the parentheses, *i.e.*, the negative derivative $-dC_M/d\beta$, is sometimes called the *static directional stability* of the plane. In these equations no yawing moment produced by sideslip β on the wing itself has been taken into account.   This is justified as long as the wing is essentially plane (no dihedral angle present).   The influence of the dihedral will be discussed in Sec. XIX.5.

The factor k'' in formulas (50) to (52) can hardly be estimated on the basis of experiments on the isolated control surfaces.   If one could assume that the surfaces act like an airfoil of aspect ratio 1, the factor should be $2\pi/3$. But the interference between the fuselage and these surfaces diminishes the force considerably. Some experiments carried out[1] on complete airplane models allow the conclusion that k'' lies between 1 and 1.8. With $S''/S \sim 0.1$, $l/B \sim 0.5$, the first term in (49) would then give for $-dC_M/d\beta$ about 0.05 to 0.09. If the second term is taken into account—the influence of the fuselage—with the c_f-value given in Sec. 3 of this chapter, it would reduce the stability practically to zero or even to negative values. According to Millikan,[2] observations in the California Aeronautical Institute have shown that a satisfactory value of static directional stability amounts to 0.04 to 0.08. But it is certain that airplanes with much lower and even with negative values of $-dC_M/d\beta$ can be operated. More information to this effect will be found in Sec. XX.4, where the dynamic stability theory is discussed. It may be remarked, once more, that with respect to lateral disturbances the static approach is rather inadequate.

By operating the elevator in straight flight the pilot is always in a position to produce an additional pitching moment, whether he wants to turn the longitudinal axis of the plane up or down or has to counterbalance a longitudinal disturbance. The analogous role with respect to yawing moments is played by the *rudder*, *i.e.*, the movable part of the vertical tail whose area is included in what was called S'' in the foregoing argument. Even in the[,] earliest and most primitive designs of aircraft the rudder appears as a vital implement. Its necessity was obviously prompted by the use rudders have found for centuries in shipbuilding. But ships have no device for the control of rolling moments. Their stability in roll is guaranteed by a correct position of the metacenter (Sec. 4 of this chapter; Fig. 363). It seems that the decisive step that made flying practical was taken when Wilbur and Orville Wright (1903) introduced a workable control of rolling moments for the airplane. They followed the pattern of bird flight and used wings that could be distorted during the flight in such a manner that a part of the lift force is shifted from the right wing to the left, or vice versa. This particular method of roll control was soon abandoned and replaced by the introduction of the well-known *ailerons*, *i.e.*, of a pair of wing flaps turnable about an axis of spanwise direction and connected crosswise with each other. In neutral position the ailerons form a part of the wing, and their area is included in what we call the wing area S. When thrown out of this position in straight flight the ailerons supply a rolling moment that may be used, for example, to balance the propeller moment or to counteract any disturbance in the roll. The moment produced by the ailerons is proportional, other things being equal, to their distance from the symmetry plane.

[1] *Brit. Aer. Res. Com., R. & M.* 965 (1925).
[2] *Op. cit.*, p. 158.

Many factors, like plan-form, profile, and twist of the wing influence the effectiveness of aileron operation. The air reactions on the ailerons, and therefore the required area S''', cannot be computed, even approximately, from airfoil data provided by experiments on isolated flaps. The three-dimensional wing theory applied to a wing with thrown-out flap supplies a method of computation. In conventional airplanes the area ratio S'''/S is about 0.08 to 0.12, and the maximum rolling moment has the order of magnitude 0.1 WB (weight × span).

Problem 16. The actual thrust of a propeller of 10 ft. diameter working at the advance ratio $J = 0.8$ with an efficiency $\eta = 0.78$ is $T = 1200$ lb. The wing span B is 40 ft. What forces on the ailerons would be necessary to compensate the rolling moment of the propeller if these forces are supposed to act at a distance $0.8B$ from each other?

Problem 17. Estimate the yawing moment at a sideslip $\beta = 5°$ for an airplane of gross weight 8000 lb., flying at a lift coefficient $C_{L_1} = 0.45$, if the tail distance and fuselage length are 32 ft., the ratio $S''/S = 0.12$, and $A_f/S = 0.22$.

CHAPTER XVIII

NONUNIFORM FLIGHT

1. Introduction. Elementary Results. So far in this book the airplane in flight has been considered as performing a uniform motion in a straight path. No accelerations or, what is the same, inertia forces have been taken into account. However, if the pilot has to change his route or if he wants to change from a state of level flight to climbing, etc., curved pathways and accelerations occur. The general mechanical problem can then be stated as a problem of three-dimensional motion of a rigid body under the influence of variable forces.

The assumption that the airplane behaves like a *rigid body*, *i.e.*, as though it were invariable in shape, is a sufficiently good approximation in almost all cases. But even under this restriction one encounters two big difficulties. First, the equations of motion for a rigid body with all the six degrees of freedom can be integrated only very seldom, *i.e.*, for very specified forces only. In all other cases nothing other than a laborious step-by-step procedure is possible. But a much more important inconvenience is that we really do not know what aerodynamic forces act on a body in nonuniform motion. To find them, it would be necessary to solve a hydrodynamic problem that is far beyond our present limits. It is usual to assume that the forces at each moment are just the same as if the instantaneous state of velocity were a permanent one. This is theoretically justifiable to a certain extent as long as the surrounding air can be supposed to perform an irrotational continuous perfect fluid motion, since a solution of the potential equation is uniquely determined by the instantaneous velocity values at the boundaries. Nothing of this kind can be said when drag forces due to viscosity are decisive. Besides, the experimental data so far available even for the steady air-flow reactions on moving bodies do not cover the most general case of rigid body motion, and it cannot be expected that they will be supplemented in a sufficient way in the near future.

Under these circumstances an analysis of nonuniform flight must content itself with a rather rough approximation, either dealing with the simplest cases of motion only or using extensive idealizations. In the present section some elementary results concerning a curved path in a vertical and in a horizontal plane, respectively, will be deduced. In Secs. 2 to 4 of this chapter some types of motion in the vertical symmetry plane will be discussed in greater detail under certain simplifying assump-

tions. The last section deals with the simplest questions concerning
the asymmetric motion. Finally, in the next chapter, the complete set
of differential equations that control the most general type of airplane
motion and that are needed in the stability theory (Chap. XX) will be
presented.

Let us assume that an airplane is flying straight and steadily and that,
at a given moment, the pilot, in order to perform a *vertical turn*, operates
the elevator in such a way as to increase the angle of incidence. This
change of incidence will produce a sudden increase of lift and drag, provided that the steady flight was below stalling. Let C_L and C'_L be the
lift coefficients before and after the change. Then the equilibrium
condition for the steady flight supplies

$$W = C_L \frac{\rho}{2} V^2 S \tag{1}$$

where W is the gross weight, S the wing area, and V the steady flight
velocity. If R denotes the radius of curvature of the curved path at
the first instant after the change of incidence occurred, the equation of
motion for the direction normal to the path reads (Fig. 376)

$$\frac{W}{g} \frac{V^2}{R} = C'_L \frac{\rho}{2} V^2 S - W = (C'_L - C_L) \frac{\rho}{2} V^2 S = W \frac{C'_L - C_L}{C_L} \tag{2}$$

Actually, V^2/R is the centripetal acceleration; and, assuming that the
angle ϑ between the tangent of the path and the horizontal plane is
small, we need not make any distinction between the vertical direction
and the normal to the path. From Eq. (2) we derive

$$R = \frac{2}{g\rho} \frac{W}{S} \frac{1}{C'_L - C_L} = \frac{V^2}{g} \frac{C_L}{C'_L - C_L} \tag{3}$$

The first expression (3) shows that the radius of curvature is proportional to the wing loading W/S and inversely proportional to the
density ρ and the difference $C'_L - C_L$. This difference reaches its maximum if C'_L is made to equal the stalling value C_{st}. Then its order
of magnitude is about 1 or, for modern airplanes, generally greater than 1.
As a pressure value divided by $\gamma = g\rho$ gives the corresponding pressure
head, the result can be stated as follows: *The smallest radius of curvature
for an overhead turn from a state of horizontal or nearly horizontal steady
flight is approximately twice the pressure head that corresponds to the wing
loading—more exactly, the pressure head multiplied by $2/(C_{st} - C_L)$.*

The magnitude $1/R_{\min}$ can be considered as a measure of *vertical
maneuverability*. The foregoing statement implies that the vertical
maneuverability decreases with increasing wing loading and altitude. A
possibly high value of C_{st} is advantageous. If the stalling speed V_{st} is

introduced by

$$W = C_{st} \frac{\rho}{2} V_{st}^2 S \tag{4}$$

the comparison with (1) shows that $C_L V^2 = C_{st} V_{st}^2$. Thus the second expression (3) leads to

$$R = \frac{V_{st}^2}{g} \frac{C_{st}}{C_L' - C_L} \tag{5}$$

If C_L' equals C_{st}, the second fraction is not much greater than 1 and for modern, fast airplanes will hardly exceed 2. Thus the following state-

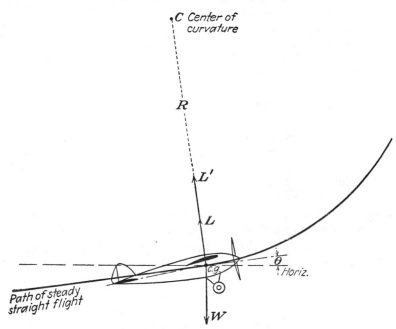

Fig. 376.—Turn in a vertical plane.

ment is justified: *The smallest radius of curvature is about four times the velocity head that corresponds to the stalling speed at the respective altitude—more exactly, the velocity head times* $2C_{st}/(C_{st} - C_L)$.

Similarly, slightly more favorable conditions prevail in the case of downward curves in a vertical plane. Here, $C_L' - C_L$ has to be replaced by $C_L - C_L'$; and since C_L' can take negative values, the maximum can eventually be greater than in the former case.

In connection with the occurrence of vertical accelerations the phenomenon of *apparent weight increase* should be mentioned. If a wing is carrying a mass of weight W in uniform flight, it has to transmit on this mass a thrust of magnitude W. If, however, the same mass is to be accelerated at the rate d^2h/dt^2 (h = elevation), the equation of motion supplies for the vertical thrust F the relation

$$m\frac{d^2h}{dt^2} = F - W, \qquad F = W + m\frac{d^2h}{dt^2} = W\left(1 + \frac{1}{g}\frac{d^2h}{dt^2}\right) \qquad (6)$$

This F is, for example, the actual load to which the wing structure is subjected when the airplane performs the overhead turn just considered. The ratio F/W is known as the *load factor*. If we identify the acceleration d^2h/dt^2 with the foregoing V^2/R, it is seen from (2) that F equals $C'_L \rho V^2 S/2$, and thus the load factor l becomes

$$l = \frac{F}{W} = \frac{C'_L}{C_L} \qquad (7)$$

The maximum load factor that can occur in horizontal or nearly horizontal flight has the value

$$l_{max} = \frac{C_{st}}{C_L} = \left(\frac{V}{V_{st}}\right)^2 \qquad (8)$$

according to (1) and (4). Commercial airplanes are usually designed to withstand a maximum load factor of 4 to 6. More efficient airplanes, like fighters, which operate under small incidence (small C_L, high velocity) and have a high stalling value of C_L must be built for much higher load factors, up to 15 and more.

The conditions for downward turns again are a little more favorable since here the difference rather than the sum of weight and inertia force counts. At the same value of acceleration the load factor will be smaller in the case the pilot turns downward than it is in an upward turn.

Let us now consider an airplane making a *turn in a horizontal plane* so that its pathway is a circle of radius R with a vertical axis (Fig. 377). It may be assumed, first, that the motion is a so-called "flat turn," *i.e.*, that the symmetry plane is kept vertical and the moments generated by this circling motion are fully balanced by an appropriate setting of the control surfaces, elevator, rudder, and ailerons. Then it is not possible to assume that the instantaneous direction of flight coincides with the longitudinal axis of the airplane since in this case no horizontal force normal to the path would be developed. Such a force is needed to produce the centripetal acceleration V^2/R. The only possible hypothesis is that a deviation β between the velocity and the axis exists. Such a deviation, known as the *sideslip*, would produce a cross component of the aerodynamic forces that may be called *side force* or transversal force.

In the case of Fig. 377, which represents a right turn, the force is directed to starboard (right wing, seen in direction of flight). There is no extensive experimental evidence about this kind of aerodynamic forces; but it can be said that, for small sideslips β, the side force may be considered as proportional to β.   Thus, upon introducing the side-force coefficient C_y

$$F_y = C_y \frac{\rho}{2} V^2 S \tag{9}$$

C_y can be assumed as $C_y = k_y\beta$ where k_y is of the order of magnitude 0.4.

The equation of motion for the direction normal to the path supplies

$$\frac{W}{g} \frac{V^2}{R} = (T - F_x) \sin \beta + F_y \cos \beta \tag{10}$$

where T denotes the propeller thrust and F_x the component of aero-

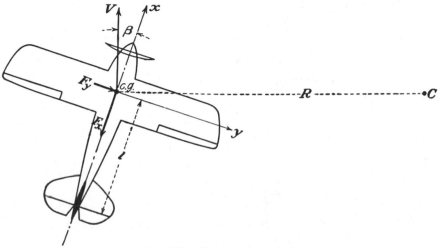

Fig. 377.—Flat turn (top view).

dynamic forces opposite to the x-axis (practically, the drag).   For the first moment we may assume that T and F_x cancel each other.   Then, combining (10) and (9) with the assumption for C_y, we find that

$$R = \frac{WV^2}{F_y g \cos \beta} = \frac{2W}{C_y \rho V^2 S} \frac{V^2}{g \cos \beta}$$

and with $\cos \beta \sim 1$ this can be written in either of the forms

$$R = \frac{W}{g\rho S} \frac{2}{k_y\beta} \quad \text{or} \quad R = \frac{V^2}{g} \frac{C_L}{C_y} = \frac{V^2}{g} \frac{C_L}{k_y\beta} \tag{11}$$

Both formulas show that in a flat turn the radius is inversely proportional to the sideslip β.   Highest suitable values of β may be close to $15° = 0.26$ rad.   Then with the foregoing value $k_y = 0.4$ it is seen from the first equation (11) that the *smallest radius of curvature in a flat turn is about twenty times the pressure head that corresponds to the wing loading.* For a wing loading of 30 lb./ft.^{2} or a pressure head of about 400 ft. at sea level, this would mean a minimum radius of 8000 ft. The pilot will make such a flat turn only when he wants to change his course slowly, but not in a timed maneuver.

Fortunately, there exists another way to turn the path of a flying airplane in the horizontal plane. The maneuver used in all regular cases is not the flat but the *banked turn* as indicated in Fig. 378. By operating the ailerons the pilot brings the airplane into an inclined position, rolling it about the longitudinal axis. Then, with the velocity still in the symmetry plane and parallel to the axis, the lift supplies a cross

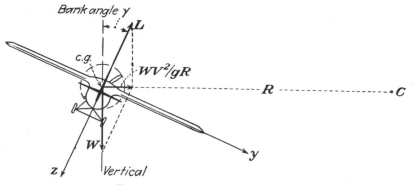

Fig. 378.—True-banked turn.

force of magnitude $L \sin \gamma$, where γ, known as the *bank angle* (or simply the *bank*) is the angle between the normal to the mean wing plane and the vertical direction. The equation of motion for the centripetal direction is now

$$\frac{W}{g}\frac{V^2}{R} = L \sin \gamma = C_L \frac{\rho}{2} V^2 S \sin \gamma \qquad (12)$$

and this combined with the condition for vertical equilibrium,

$$W = L \cos \gamma,$$

gives the well-known formula

$$\frac{V^2}{gR} = \tan \gamma \qquad (13)$$

which determines the correct bank angle for given V and R. It is the same formula that applies to the bank in the curve of a railroad track or a highway for automobiles. From (12) we derive immediately

$$R = \frac{1}{g\rho} \frac{W}{S} \frac{2}{C_L \sin \gamma} \qquad (12')$$

This gives the radius of curvature for any value of γ in the "true-banked" turn. The radius will be a minimum if γ approaches 90° and if stalling speed is used. Thus it is seen that *the lower limit for the radius in a true-banked turn is* $(2/C_{st})$ *times the pressure head corresponding to the wing loading, i.e., in all practical cases smaller than twice the pressure head.* Thus, a fighter plane even with 40 lb./ft.² wing loading can easily reach radii of 800 to 900 ft.

The term "true-banked" refers to the obvious fact that the effects of banking and of sideslip can be combined. If an airplane is not sufficiently banked, *i.e.*, if for a given V and R the banking angle is smaller than it should be according to Eq. (13), the missing part of the cross force will be supplied by sideslip. A more detailed discussion of the horizontal turn follows in Sec. 5 of this chapter.

In all computations of this section it is understood that the controls are supposed to be sufficiently powerful to supply the moments necessary for changing into and then maintaining the required attitude of the airplane. Moreover, it must be emphasized once more that these computations are meant to give only a rough survey and first information in a field where all problems are largely complicated and delicate.

Problem 1. If an airplane is flying at a level flight speed of 300 ft./sec. with a wing loading of 35 lb./ft.² at 15,000 ft. altitude, what is the smallest radius for an overhead turn if the maximum value of C_L is 1.4? What does the apparent weight increase amount to in this case?

Problem 2. Compute the bank angle for a true-banked horizontal turn of radius 800 ft. at 10,000 ft. altitude. The wing loading is 20 lb./ft.², the angle of incidence in level flight $\alpha' = 0.2$, and the aspect ratio 6.8.

Problem 3. How large is the radius of a flat turn in the case of Prob. 2 if the sideslip amounts to 12 to 14° and the side-force coefficient is 0.38?

2. Lanchester's Phugoid Theory. As early as 1908, W. F. Lanchester made a first successful approach to describing the pathways at large that an airplane, moving in its symmetry plane, can perform. In the present section the complete system of equations that control the longitudinal motion of a plane will be set up and then the simplifications introduced that led Lanchester to his interesting results.

With the restriction to symmetric flight, *i.e.*, flight without bank, yaw, and sideslip, etc., the airplane is a body with *three degrees of freedom.* There are two translational components, say, in the horizontal and vertical direction or parallel and perpendicular to the velocity vector, and one rotation about the transversal axis. In the following, φ will be used for the angle of the longitudinal axis (propeller axis, mean chord direction) with the horizontal plane and ϑ for the angle that the velocity of the mass center (c.g.) forms with the same plane, both angles counted positive upward (Fig. 379). Then the c.g. acceleration in the direction of V or of the tangent to the path is dV/dt or VdV/ds and in the direction of the upward normal $V\,d\vartheta/dt$. If the definitions of the radius of curvature, $R = ds/d\vartheta$, and of the velocity $V = ds/dt$ are used (ds = element of arc along the path), the latter expression is seen to be the same as V^2/R or as $V^2\,d\vartheta/ds$.

The aerodynamic force normal to V will be called, as usual, the lift L; the component in the direction of V will be written as $T - D$ where T means the propeller thrust and D the total drag. If, owing to the deviation $\varphi - \vartheta$, there is some normal force T' acting on the propeller (see

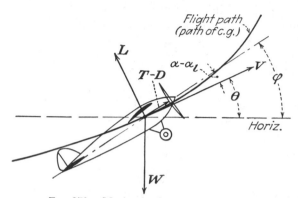

Fig. 379.—Motion in the symmetry plane.

Sec. XVII.3), this may be included in L. With the gross weight W, the two equations of motion for the translational components now read

$$\frac{W}{g}\frac{dV}{dt} = \frac{W}{g}V\frac{dV}{ds} = T - D - W \sin \vartheta \tag{14}$$

$$\frac{W}{g}V\frac{d\vartheta}{dt} = \frac{W}{g}V^2\frac{d\vartheta}{ds} = L - W \cos \vartheta \tag{15}$$

The third equation of motion, for the rotational component, includes the moment of inertia (moment of gyration) of the body with respect to the transversal axis passing through the center of gravity. It will be called J; its dimension is mass times length squared. The resultant pitching moment with respect to the c.g. will be denoted by M. Then, according to a well-known statement of rigid-body dynamics the relation

$$J\frac{d^2\varphi}{dt^2} = M \tag{16}$$

holds.

The three differential equations (14), (15), *and* (16) *determine the motion in the symmetry plane* in the following way: If the initial values of V, ϑ, φ, and $d\varphi/dt$ are given

$$V = V_0, \quad \vartheta = \vartheta_0, \quad \varphi = \varphi_0, \quad \frac{d\varphi}{dt} = \dot\varphi_0 \quad \text{for } t = 0 \qquad (17)$$

integration of (14) to (16) supplies the subsequent values of these variables for all values of time t. In particular, V and ϑ considered as corresponding polar coordinates supply the *hodograph* of the translational motion. Then, in a Cartesian system with the x horizontal and z vertical downward, from

$$\frac{dx}{dt} = V \cos \vartheta, \quad - \frac{dz}{dt} = V \sin \vartheta$$

the x- and z-values can be found by quadratures,

$$x = \int_0^t V \cos \vartheta \, dt, \quad z = - \int_0^t V \sin \vartheta \, dt \qquad (18)$$

The difficulty of the problem lies, of course, in the fact that the force components T, L, D as well as the moment M depend on all three variables V, ϑ, φ.

In order to obtain a rapid survey of the possible solutions Lanchester introduced the following simplifying assumptions:

1. It is assumed that the propeller thrust and the total drag balance each other at any instant so that $T - D$ can be omitted in (14).

2. It is assumed that the moment of inertia is exceedingly small so that with finite M the angular acceleration becomes infinite. As M is supposed to counteract any change of $\varphi - \vartheta$ (stability condition), it follows that the direction φ of the body will adapt itself instantaneously to the position in which the moment is zero, *i.e.*, to a constant value of $\varphi - \vartheta$.

The motion satisfying Eqs. (14) to (16) on these conditions was named by Lanchester *phugoid motion.*[1] Its theory is particularly simplified by the fact that the third equation can now be entirely disregarded since ϑ and φ, with their difference $\varphi - \vartheta = \text{const.}$, count for one variable only. Equations (14) and (15) become

$$V \frac{dV}{ds} = -g \sin \vartheta, \quad V^2 \frac{d\vartheta}{ds} = g \left(\frac{L}{W} - \cos \vartheta \right) \qquad (19)$$

The lift is supposed, in the usual way, to be proportional to V^2, with a coefficient depending on the incidence only. If $\varphi - \vartheta$ does not change, the incidence remains constant and so does the lift coefficient C_L. Calling V_l the level-flight velocity of the plane we have

$$L = C_L \frac{\rho}{2} V^2 S, \quad W = C_L \frac{\rho}{2} V_l^2 S, \quad \frac{L}{W} = \frac{V^2}{V_l^2} \qquad (20)$$

---

[1] It seems that the term phugoid, which is generally adopted today, originates in a linguistic slip. Lanchester sought a Greek equivalent for flight, or flying, and he found the word φυγή, which means flight in the sense of fleeing.

and thus, with $-dz = ds \sin \vartheta$, Eqs. (19) take the form

$$\frac{V \, dV}{dz} = g, \qquad V^2 \frac{d\vartheta}{dz} \sin \vartheta = g \left(-\frac{V^2}{V_i^2} + \cos \vartheta \right) \qquad (21)$$

The first of these equations can immediately be integrated and gives

$$V^2 = 2gz \qquad \text{or} \qquad z = \frac{V^2}{2g}, \qquad V = \sqrt{2gz} \qquad (22)$$

where the integration constant is (arbitrarily) taken as zero. This means that the level $z = 0$ is assumed as the upper level limit, which the airplane can reach only at velocity zero. Relation (22) expresses the fact that the body behaves like a conservative system keeping the sum of kinetic and potential energy invariable. This fact is a consequence of the assumption $T - D = 0$.

Dividing the second equation (21) by the first we find that

$$V \sin \vartheta \frac{d\vartheta}{dV} = -\frac{V^2}{V_i^2} + \cos \vartheta \qquad \text{or} \qquad \cos \vartheta - V \sin \vartheta \frac{d\vartheta}{dV} = \frac{V^2}{V_i^2}$$

The left-hand side of the last equation is obviously the derivative of $V \cos \vartheta$ with respect to V. Therefore, by integration

$$V \cos \vartheta = \int \frac{V^2}{V_i^2} \, dV = \tfrac{1}{3} \frac{V^3}{V_i^2} + \text{const.} \qquad (23)$$

This is the *polar equation of the hodograph.* To each value of the constant there corresponds a particular solution, *i.e.*, a particular pathway of the airplane. If at the beginning $V = V_i$ and $\vartheta = 0$, the constant must have the value $\tfrac{2}{3} V_i$. The corresponding z follows from (22) as $V_i^2/2g$. Since $z = \text{const.} = z_0$ is compatible with $\vartheta = \text{const.} = 0$, it is seen that the horizontal flight at the level $z_0 = V_i^2/2g$ with the constant speed V_i is one solution of the problem—as was to be expected.

In order to find the other solutions, *i.e.*, the other possible pathways of the idealized airplane, it is convenient to introduce in (23) the value for V from (22). This gives

$$\cos \vartheta = \tfrac{1}{3} \frac{2gz}{V_i^2} + \frac{\text{const.}}{\sqrt{2gz}}$$

and, upon using the dimensionless variable

$$\zeta = \frac{2gz}{V_i^2} = \frac{V^2}{V_i^2} \qquad (24)$$

instead of z, the equation can be written in the form

$$\cos \vartheta = \tfrac{1}{3}\zeta + \frac{A}{\sqrt{\zeta}} \qquad (25)$$

where A stands for const./V_l.   In Fig. 380 cos ϑ is plotted against ζ according to (25) with ζ downward and cos ϑ on the horizontal axis positive toward the right.   Only cos ϑ-values between -1 and $+1$ can occur; the half circle between these two points indicates the ϑ-direction corresponding to each abscissa.

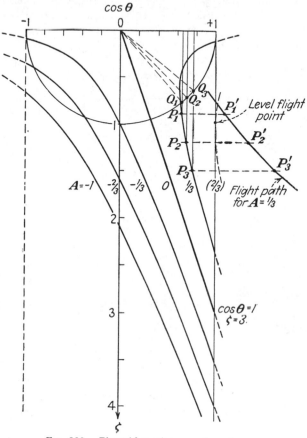

Fig. 380.—Phugoid motion: cos ϑ vs. ζ.

For $A = 0$ the cos ϑ vs. ζ-line goes straight from the origin to the point $\zeta = 3$, cos $\vartheta = 1$.   This straight line separates the curves $A > 0$, which lie to the right, each beginning and ending at cos $\vartheta = 1$, from the curves with $A < 0$, which all cross from cos $\vartheta = -1$ to cos $\vartheta = 1$. For $A = \frac{2}{3}$ just the point $\zeta = 1$, cos $\vartheta = 1$ (level-flight solution) falls into the region $|\cos \vartheta| \leqq 1$, and greater values of A supply no points at all.

The pathway corresponding to any of the curves can be found by the following step-by-step procedure: Through any point P of one of the curves, say, for $A = \frac{1}{3}$, draw the vertical that will intersect the half circle in Q. Then OQ is the direction of the path in the point P' where it crosses the level of P. Thus the path can be pieced together of small straight segments. In Fig. 381 this is carried out for one positive value, $A = \frac{1}{3}$, and one negative value, $A = -1$.

The different character of the path curves for $A > 0$ and $A < 0$ can be seen immediately from Fig. 380. In the first case positive values of $\cos \vartheta$ only occur and ζ is restricted to a small range including $\zeta = 1$: the path will be a wave line winding about the horizontal straight line

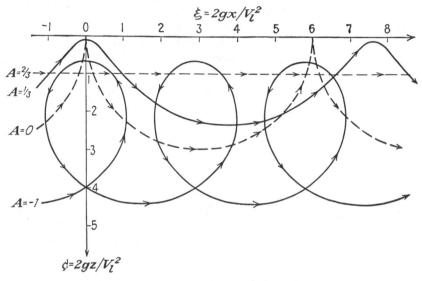

FIG. 381.—Phugoid motion: typical path curves.

that represents the level-flight solution. In the second case, however, $\cos \vartheta$ and thus the horizontal velocity component change from negative values at higher elevation to positive values below: the path curve is a loop with backward velocity at the top and forward velocity at the bottom. Which of the various motions actually occurs in a given case depends on the initial conditions, *i.e.*, magnitude and direction of the launching velocity. Note that in this theory, with $\varphi - \vartheta = \text{const.}$, the orientation of the body follows exactly the rotation of the velocity vector. This means that at the top of the loop the airplane is turned upside down.

The dotted line in Fig. 381 shows the limiting solution that corresponds to $A = 0$. In this case the body reaches the highest possible level, $z = 0$, with vertical tangent, $\cos \vartheta = 0$, $\vartheta = 90°$. The shape of the curve can be found by analytic integration as follows. Noting that, with

$$\zeta = \frac{2gz}{V_l^2}, \qquad \xi = \frac{2gx}{V_l^2}, \qquad \frac{dz}{dx} = \frac{d\zeta}{d\xi} = -\tan\vartheta \qquad (26)$$

the equation of the path curve in dimensionless coordinates ξ, ζ is in all cases given by

$$\xi = -\int \cot\vartheta \, d\zeta \qquad (27)$$

we find, in the case $\cos\vartheta = \zeta/3$, that

$$\xi = \mp \tfrac{1}{3} \int_0^\zeta \frac{\zeta}{\sqrt{1 - \zeta^2/9}} \, d\zeta = \pm 3 \left(1 - \sqrt{1 - \zeta^2/9}\right) \qquad (28)$$

This is the equation of two circles of radius 3 with centers in $\xi = \pm 3$, $\zeta = 0$. As only positive values of ζ count, the pathway at the limit between the sinuous and the looped curves consists of a sequence of half circles as seen in Fig. 381.

It should be remembered that the curves are given here in a dimensionless form, with the reduction factor $V_l^2/2g$. Each unit represents a length equal to the velocity head corresponding to the level-flight velocity. The radii of the circles just discussed are three times $V_l^2/2g$, for example, three times 620 ft. for an airplane with the level-flight velocity $V_l = 200$ ft./sec., etc. Light cardboard models that can be made to fly (nearly) horizontally at 15 ft./sec. will show radii of about 10.5 ft.

It is well known from model experiments and from observations on actual airplanes that the two types of motion, along the wavy path with horizontal trend and the loopings (and also, to a certain extent, the intermediate case of half circles) can be realized approximately. In spite of the far-reaching simplifications introduced in the phugoid theory this theory succeeds in giving an excellent general idea of the possible types of motion of a free flying body. It is not sufficient, however, for a quantitative analysis of the flight path.

Problem 4. An airplane model has a level-flight velocity of 35 ft./sec. Find its phugoid path if it is launched in the horizontal direction at an initial speed (a) of $V_0 = 42$ ft./sec, (b) of $V_0 = 70$ ft./sec. At what value of V_0 will the path approximately correspond to the half circle?

Problem 5. Develop the relation between the constant A in Eq. (25) and the velocity value V_0 at $\vartheta = 0$. Discuss the various types of motion in terms of V_0.

*Problem 6.** Assume that the velocity V during the flight is very close to the level-flight value V_l. Neglecting terms of higher order in $V - V_l$, develop the equations of motion. Show that the paths are sine waves, and compute the wave length, the amplitude, and the period.

3. Longitudinal Flight along a Given Path.

If an airplane actually moves in a curved path, the pilot has to operate the elevator during the flight. Thus the airplane is not an invariable body as is supposed in the phugoid theory. The problem, which one can expect to be solved by means of theoretical mechanics, would consist in describing the flight for a given sequence of elevator operations. The equations of motion, as far as they can be integrated for given initial conditions, should supply this answer. From a practical standpoint, however, a different form of the problem seems to be more appropriate. If a pathway in a vertical plane is given, one wants to know whether an aircraft of given dimensions can fly along this path and under what conditions it can do this.

In both problems the main difficulty lies in finding the right expressions for the forces acting on the moving body. As has been pointed out already, our present knowledge of the air reactions in any general kind of airplane motion is very incomplete. Even if it is admitted that the forces depend on the velocities only (not on the accelerations), one would need much more data for the case of combined rotational and translational motion. The only way the problems can be handled at the present stage of information is by introducing certain very simple hypotheses about the forces, which may be more or less near to the truth. The following assumptions will underlie the computations in this section:

1. The propeller thrust T acts in the direction of the velocity V of the center of gravity, and its magnitude depends only on V or on V and on the density ρ.

2. The components of the resultant air reaction on the craft, in the directions perpendicular and parallel to V, respectively, are of the form

$$L = C_L \frac{\rho}{2} V^2 S, \qquad D = C_D \frac{\rho}{2} V^2 S \tag{29}$$

where C_L and C_D depend on the angle of incidence only.

3. The tail supplies a pitching moment (a couple) to be specified later.

Assumptions 1 and 2 are essentially the same as those used in the performance computation. The influence of the rotational velocity on the resultant force is neglected. A normal force on the propeller due to the deviation of its axis from the velocity direction (Sec. XVII.3) can be added to L as an "apparent lift." In general, however, one will have to use for C_L and C_D the values as given in the scaled resultant polar diagram of the plane. No restriction is made so far concerning the pitching moments due to the forces T, L, D and about a possible additional couple connected with the rotational velocity.

In the problem to be studied here the flight path is considered as being given. This implies that the angle ϑ which the velocity V forms with the x-axis is a known function of the arc length s and that the derivative $d\vartheta/ds = 1/R$ is known, also. Then the first two equations of motion, (14) and (15), include only two unknowns V and φ (the latter determining α) and therefore can be dealt with independently of the third equation (16).

For the sake of brevity a dimensionless form of Eqs. (14) and (15) will be used, by reducing all lengths with respect to the velocity head of the level flight, $V_l^2/2g$, where

$$W = C_{Ll} \frac{\rho}{2} V_l^2 S \qquad (30)$$

For example, let us introduce

$$\sigma = \frac{2g}{V_l^2} s, \qquad \xi = \frac{2g}{V_l^2} x, \qquad \zeta = \frac{2g}{V_l^2} z, \qquad \kappa = \frac{V_l^2}{2g} \frac{1}{R} \qquad (31)$$

Obviously, σ is the arc length and κ the curvature (reciprocal value of the radius of curvature) in the reduced path curves. Then, upon introducing the ratios

$$u = \frac{V^2}{V_l^2}, \qquad c = \frac{C_L}{C_{Ll}}, \qquad \epsilon = \frac{D}{L} \qquad (32)$$

the two equations (14) and (15) when divided by W yield ($\epsilon = D/L$)

$$\frac{du}{d\sigma} = \frac{T}{W} - \epsilon c u - \sin \vartheta \qquad (33)$$

$$2\kappa u = cu - \cos \vartheta \qquad (34)$$

Here, ϑ, κ are known functions of σ, and T/W is a function of u, while the lift ratio c and the gliding angle ϵ are given functions of the incidence, *i.e.*, of $\alpha = \alpha_l + \varphi - \vartheta$. The second equation must serve to express α in terms of σ and u. Then this expression substituted on the right-hand side of (33) makes this side a function of σ and u so that a differential equation of first order between u and σ has to be integrated. From the u-values found in this way, α and thus φ can be computed by means of (34). If we are interested only in the velocity ratio u, we need

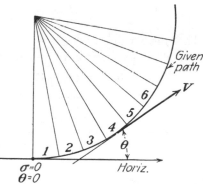

FIG. 382.—Motion along a vertical circle.

the relation between c and ϵ only in the form supplied by a nonscaled polar diagram; then c is computed from (34), etc. To show how this method works, the following numerical example may be considered.

The pathway is given as an arc of circle of radius 2 ($\kappa = 0.5$), starting at the lowest point $\vartheta = 0$ and bent upward (Fig. 382) so as to have $\vartheta = \kappa\sigma$. The initial velocity may be the level-flight speed; thus, $u = 1$ at $\sigma = 0$. The relation between c and ϵ will be assumed in the usual form, corresponding to a parabolic polar curve

$$\epsilon = \frac{a}{C_L} + \frac{C_L}{b} = \frac{a}{C_{Ll}}\frac{1}{c} + \frac{C_{Ll}}{b}c \tag{35}$$

with the parasite-drag coefficient $a = 0.03$ and $b = 3\!R = 20$ corresponding to an aspect ratio of about 7. The level-flight value of C_L is determined according to (30) by ρ, V_l and the wing loading W/S. Taking $V_l = 300$ ft./sec., $W/S = 20$ lb./ft.[2], and, for an altitude of about 10,000 ft., $\rho = 0.0018$ slug/ft.[3], we find from (30) that $C_{Ll} = 0.247$; thus,

$$\epsilon c = 0.121 + 0.0124c^2 \tag{35'}$$

The propeller thrust will conveniently be assumed as decreasing with increasing speed at a rate proportional to V^2 or to u. The level flight value of the ratio T/W must equal the value of ϵ for $c = 1$, that is, 0.133; and, if T/W is assumed 30 per cent larger at zero velocity,

$$\frac{T}{W} = \frac{T_l}{W}[1 + 0.3(1 - u)] = 0.173 - 0.040u = \frac{T_0}{W} - \tau u \tag{36}$$

Thus with the value given already for κ the two equations to be handled in a step-by-step procedure are

$$c = 1 + \frac{\cos \vartheta}{u}, \qquad du = d\sigma[0.173 - u(0.161 + 0.0124c^2) - \sin \vartheta] \tag{37}$$

Let us choose steps of $10° = 0.1745$ rad., which means

$$d\sigma = \frac{d\vartheta}{\kappa} = 0.349.$$

The mean value of ϑ in the first step is $5°$; the initial value of u is 1. Therefore, as a first approximation, $c = 1 + \cos 5° = 1.996$ and

$$du = 0.349(0.173 - 0.209 - 0.087) = -0.043,$$

which gives the end value of u as 0.957. Now, we can improve on this

TABLE 15.—MOTION IN A VERTICAL CIRCLE

	Step					
	0–10°	10–20°	20–30°	30–40°	40–50°	50–60°
Initial u	1.000	0.958	0.860	0.713	0.524	0.301
Mean u	0.98	0.91	0.79	0.62	0.41	0.17
Mean ϑ	5°	15°	25°	35°	45°	55°
$c = 1 + \cos \vartheta/u$	2.016	2.062	2.149	2.320	2.725	4.37
ϵc	0.171	0.174	0.178	0.188	0.213	0.358
$u(\epsilon c + 0.04)$	0.207	0.195	0.172	0.141	0.104	0.068
$\sin \vartheta$	0.087	0.259	0.423	0.573	0.707	0.819
$-du$	0.042	0.098	0.147	0.189	0.223	0.249
Final u	0.958	0.860	0.713	0.524	0.301	0.052
$V/V_l = \sqrt{u}$	0.979	0.927	0.844	0.724	0.549	0.23

approximation by using a mean value between 1 and 0.957 instead of 1 in the computation of du; this changes the increment du to -0.042. In this way, with estimated mean values for each segment, Table 15 has been computed.

The results are represented in Fig. 383. It is seen that the forward speed V drops to about one-quarter the level-flight value at the point $\vartheta = 60°$. In the next step the formulas would give a c-value beyond 8, that is, a lift coefficient beyond 1.9, which cannot be expected to occur under ordinary circumstances. The result is that the airplane under consideration cannot rise higher than to about 60° along the circle of

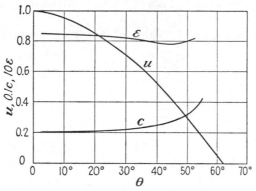

FIG. 383.—Distribution of u, c, and ϵ along the path.

radius $2 \times V_i^2/2g = 2800$ ft. Note that the climb was initiated with $c \sim 2$, that is, at an incidence twice that of level flight. This increased incidence must be provided by an application of the elevator.

In order to compare this motion with the conditions of steady climbing we have to omit the acceleration terms in (33) and (34). These equations then read

$$c = \frac{\cos \vartheta}{u}, \qquad \frac{T}{W} = \epsilon c u + \sin \vartheta = \epsilon \cos \vartheta + \sin \vartheta \qquad (38')$$

If $\cos \vartheta$ is here identified with 1 and consequently L with W, the second equation (38) coincides with Eq. (3), Chap. XIV, stating that $\sin \vartheta$ equals $(T - D)/W = (T - \epsilon W)/W$. The maximum angle for steady climbing is then found as the maximum of

$$0.173 - u\left(0.161 + 0.0124 \frac{1}{u^2}\right)$$

which is about $\vartheta = 0.084 \sim 5°$.

An inspection of Table 15 shows that the value of ϵ changes very little throughout the motion. For the six segments from 0 to 60° the ϵ

are 0.085, 0.084, 0.083, 0.081, 0.078, 0.082. This suggests trying an approximate analytical solution by *assuming a constant average value for* ϵ in (33). Substituting c from (34) and using for T/W the last expression (36), we find the equation

$$\frac{du}{d\sigma} = \frac{T_0}{W} - \sin \vartheta - \epsilon \cos \vartheta - \lambda u, \qquad \text{with } \lambda = 2\kappa\epsilon + \tau \quad (38')$$

which for a constant κ and $\vartheta = \kappa\sigma$ can be integrated analytically. We can verify by differentiation that the function

$$u = Ae^{-\lambda\sigma} + \frac{T_0}{\lambda W} + \sqrt{\frac{1 + \epsilon^2}{\kappa^2 + \lambda^2}} \cos (\vartheta + \vartheta_0), \quad \text{with } \tan \vartheta_0 = \frac{\kappa\epsilon + \lambda}{\kappa - \epsilon\lambda} \quad (39)$$

satisfies Eq. (38) whatever the value of A. This constant is determined

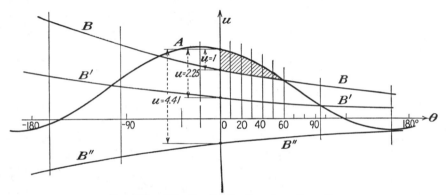

Fig. 384.—Velocity distribution along the path.

by the initial conditions. In Fig. 384 the curve A representing the second and third term of u is plotted for the values of the parameters corresponding to the foregoing example: $\kappa = 0.5$, $\tau = 0.04$, $\epsilon = 0.082$, $\lambda = 0.122$, $T_0/W = 0.173$. Its ordinates are

$$1.42 + 1.95 \cos (\vartheta + 18.42°).$$

The curve B has ordinates proportional to $-e^{-0.244\vartheta}$, with the factor $A = -2.28$ determined in such a way as to make the resultant u-value (the distance between A and B) at $\vartheta = 0$ equal to 1. It is seen that near to $\vartheta = 60°$ the two curves intersect, giving here $u = 0$. The intermediate values (shadowed area) are likewise in good agreement with the results of the step-by-step procedure.

Figure 384 shows two more curves, B' and B'', with ordinates proportional to those of the curve B. For B' the factor $A = -1.03$ is chosen so as to make $u = 2.25$ at $\vartheta = 0$, and for B'' the factor is $A = 1.13$, corresponding to $u = 4.41$ at $\vartheta = 0$. It is seen from the figure that in the first case u becomes zero a little beyond $\vartheta = 90°$, while in the second u remains positive over the whole range of ϑ from -180 to $180°$. This means that, if the airplane passes through the lowest point of the circular path with a velocity equal to 1.5 times the level-flight speed, it can rise to about 90°, *i.e.*, to the level of the center of the circular path; and if its velocity at $\vartheta = 0$ is 2.1 times V, a full turn (looping) would be possible. It must be checked, however, whether the c-values involved do not exceed the stalling limit (there is no danger of this at high velocities) and whether the assumption of a nearly constant ϵ is still justified.

An airplane flight of the type described in this section can be performed only if the airplane is provided with an elevator of sufficient dimensions and if this elevator is appropriately operated. This will be discussed in the following section.

Problem 7. Assume that the airplane discussed in the text ($V_l = 300$ ft./sec., $C_{Ll} = 0.247$, etc.) moves in a circular path of radius $R = 1000$ ft. Compute the point to which it can rise if its initial velocity, at the bottom of the circular arc, is $1.2V_l$.

Problem 8. Discuss the relation between the constant A in Eq. (39) and the initial velocity at $\vartheta = 0$ for varying value of the radius of curvature.

Problem 9. Use the analytical method to compute the motion along a circular path of radius $R = 2500$ ft. for an airplane of $V_l = 250$ ft./sec. level-flight velocity, constant $\epsilon = 0.090$, with $T = T_l(1.3 - 0.3u)$. Under what conditions can this airplane perform a complete looping?

4. Effect of Elevator Operation. Let us first consider the case that an elevator is thrown out of its neutral position while the airplane is moving horizontally at its level-flight velocity V_l. The very beginning of the nonuniform motion which sets in at this instant is to be studied. The angles ϑ and φ are both zero at $t = 0$ and will remain small so that their cosine can be identified with 1. Moreover, the assumption may be made that during the short period under consideration the forward speed V does not sensibly change. That is, it is assumed that the expression on the right side of the first equation of motion, *viz.*,

$$T - D - W \sin \vartheta$$

remains comparatively small. Then, with $V = V_l$, the second and third equations of motion, (14) and (15), give a sufficient basis for the analysis

$$\frac{W}{g} V_l \frac{d\vartheta}{dt} = L - W = W\left(\frac{L}{W} - 1\right) \quad \text{and} \quad J \frac{d^2\varphi}{dt^2} = M \quad (40)$$

In the first of these equations, with $V = V_l = \text{const.}$, the ratio L/W equals the ratio of the lift coefficient C_L at the time t to the level-flight value C_{Ll}. Assuming conditions below stalling, the quotient C_L/C_{Ll} can also be replaced by α/α_l where α is the actual and α_l the level-flight incidence.   Then the first equation (40) can be written as

$$\frac{d\vartheta}{dt} = n(\alpha - \alpha_l) \qquad \text{with } n = \frac{g}{V_l \alpha_l} \quad (41)$$

The displacement of the elevator from its neutral position will cause a pitching moment which, for the sake of first information, can be considered as a constant M'.   It is at least not impossible to operate the elevator in such a way that its influence on the airplane remains constant for a short while.   The airplane may be assumed to be statically neutral (Sec. XVII.4)—the usual amount of restoring moment due to the change in attitude would not much influence the result of the analysis.   But there is another, essential contribution to M besides M'.   The rotational velocity $d\varphi/dt$ that sets in when the elevator leaves its horizontal path will be met by a *damping moment* that is proportional to $d\varphi/dt$ and opposite in direction to the rotation.   If the proportionality factor is called K_m, the second equation (40) can be written in the form

$$\frac{d^2\varphi}{dt^2} = \frac{M'}{J} - m\frac{d\varphi}{dt} \qquad \text{with } m = \frac{K_m}{J} \quad (42)$$

Both coefficients, n and m, in Eqs. (41) and (42) have the dimension t^{-1}.   The value of n corresponding to the numerical example in the preceding section would be about 2 per second.   In general, it may lie between 1 and 3.   A typical value for m is given[1] as 2.5 per second; it will hardly drop below 1.

Since Eq. (42) includes one unknown only, it can be integrated independently of (41).   It can easily be verified by differentiation that

$$\frac{d\varphi}{dt} = \frac{M'}{mJ} + Ae^{-mt} \tag{43}$$

is the general integral.   The initial condition that $d\varphi/dt$ vanishes at $t = 0$ will be fulfilled with $A = -M'/mJ$, and thus

$$\frac{d\varphi}{dt} = \frac{M'}{mJ}(1 - e^{-mt}) \tag{43'}$$

is the definite form of Eq. (43).

In (41) the difference $\alpha - \alpha_l$ equals, by definition, the difference $\varphi - \vartheta$.   Taking the derivative of $\varphi - \vartheta = \alpha - \alpha_l$, we find that

$$\frac{d\vartheta}{dt} = \frac{d\varphi}{dt} - \frac{d\alpha}{dt} \tag{44}$$

[1] DURAND, W. F., "Aerodynamic Theory," vol. 5, p. 99 (B. M. Jones).   See also Eq. (35), Sec. XX.2.

and thus (41) is equivalent to

$$\frac{d\alpha}{dt} + n\alpha = n\alpha_l + \frac{d\varphi}{dt} = n\alpha_l + \frac{M'}{mJ}(1 - e^{-mt}) \tag{45}$$

This differential equation for α has a known integral. Again, it can be verified by differentiation that the general integral, with the integration constant A, reads

$$\alpha = Ae^{-nt} + \alpha_l + \frac{M'}{nmJ}\left(1 - \frac{n}{n-m}e^{-mt}\right) \tag{46}$$

In order to fulfill the condition $\alpha = \alpha_l$ at $t = 0$, the constant A must be taken as $M'/n(n-m)J$, thus supplying

$$\alpha = \alpha_l + \frac{M'}{nmJ}\left(1 + \frac{me^{-nt} - ne^{-mt}}{n-m}\right) \tag{46'}$$

From (43') and (46') the derivative $d\vartheta/dt$ can be found according to (44)

$$\frac{d\vartheta}{dt} = \frac{M'}{mJ}\left(1 + \frac{me^{-nt} - ne^{-mt}}{n-m}\right) \tag{47}$$

which also follows from (41) and (46').

The constants m and n are essentially positive quantities. Therefore, all exponential terms go to zero when t increases indefinitely. With an average value $m \sim n \sim 2.5$ the factor e^{-mt} drops from 1 to 0.01 within 1.8 sec. If n equals m, the second term in the parentheses appears in the indefinite form $0/0$. But it can be seen by the usual method that its "true value" in this case is $-(1 + mt)e^{-mt}$. Thus the main result included in (43'), (46'), and (47) can be stated as follows: *Within a very short time the angular velocity* $d\varphi/dt$ *and the derivative* $d\vartheta/dt$ *approach the same constant value* M'/mJ *or* M'/K_m, *while the angle of incidence* α *approaches the constant* $\alpha_l + M'/nmJ$ *or* $\alpha_l + M'/nK_m$.

If the pilot wants to change from level flight to climbing, he will pull the elevator stick and thus produce a positive moment M'. Then, within 1 to 2 sec. the incidence will settle down to a new value higher than the incidence at level flight, the increase depending on the moment M', the damping capacity K_m, and the factor n. At the same time the airplane has acquired an angular velocity that remains constant as long as the elevator is in operation; and, in addition, the slope of the flight path keeps increasing at a constant rate. The pilot has to push back the elevator stick as soon as he reaches the new direction he wants to take. Now, a state of steady climbing can follow *if and only if the equilibrium conditions allow a climb at higher incidence.* This is the case, as was seen in the performance discussion in Sec. XIV.2, if the airplane was originally flown at the so-called "high speed" of level flight.

This whole argument can easily be extended to the case where the steady flight preceding the application of the elevator is a climbing or flat descending rather than a level flight. The result, already anticipated in Chap. XIV, is that the elevator works normally only when applied in a state of steady flight which does not belong to the region called the region of "reversed commands." A positive elevator moment then leads always to an increase of slope of path and to an increase of incidence, and a negative moment to a decrease of both.

If the airplane with the elevator in neutral position is not statically indifferent, each state of steady flight other than the normal level flight requires a certain amount of elevator moment. This means that in the case of a statically stable airplane the pilot, after reaching the intended slope, must not return the elevator completely to its original position. In a not too fast maneuver it will be possible to limit the rotation of the elevator to the amount required in the new state of steady flight.

The results obtained here also have a bearing on the type of nonuniform motion discussed in the preceding section. Assume that in an actual case Eqs. (33) and (34) have been solved. Then not only is V found as function of s and (upon integration of $ds/dt = V$) as function of t, but also c and thus the angle of incidence α for each time are known. From the given ϑ and the computed α we can derive $\varphi = \vartheta + \alpha - \alpha_l$ and by differentiation $d\varphi/dt$ and $d^2\varphi/dt^2$. The latter quantity multiplied by the moment of gyration J gives the left side of the third equation of motion (16). On the right-hand side the contributions to the moment due to the propeller, the wing reactions, the damping, etc., must be considered as known, and the only unknown in the equation is the moment contributed by the elevator. In this way Eq. (16) makes it possible to compute the elevator moment that, at any instant, would be necessary for performing the flight along the given pathway. This can be used to check whether an airplane of given design is capable of performing a motion represented by a particular solution of (33) and (34) or, as the case may be, whether the dimensions of the elevator can be adapted to the requirements of those types of motion which it is desired to have performed. Attention must be paid to the fact that the effect of the elevator is proportional to the square of the velocity V, that is, to the speed ratio u, thus decreasing considerably at the points where u is small.

Fortunately, this elaborate procedure is made unnecessary by the results already established, except when a very high grade of maneuverability is demanded. It can be assumed that, under normal conditions, the elevator acts practically instantaneously and that, accordingly, the moment to be supplied equals approximately that required for compensating the counteracting static moments of the air reactions, etc. In fact, other circumstances that cannot be taken into account in this roughly approximating theory—like the influence of change in down-wash as induced by the wing on the tail—may be of greater importance than the inaccuracy incurred by neglecting the dynamic terms in the moment equation.

Problem 10.   An airplane of gross weight W has a tail distance $l = 20$ ft. and a radius of gyration equal to $0.3l$. Compute the motion of this airplane under the influence of a tail moment $M' = 0.02lW$. Take $m = 3$ per second and $n = 2$ per second. Find, in particular, the terminal increase of incidence and the time needed to obtain 95 per cent of this increase.

*Problem 11.   Find, with the data of Prob. 10, the time needed for changing the direction of flight by $\vartheta = 15°$, and compute the angle of incidence reached at this moment.

5. Asymmetric Motion.   In Secs. 2 to 4 of this chapter only motions of an airplane have been considered in which its symmetry plane per-

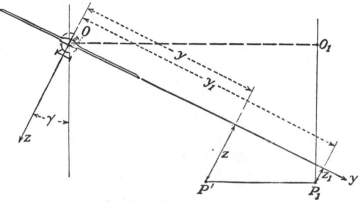

Fig. 385.—True-banked turn.

manently coincides with one and the same vertical plane.   If this restriction is eliminated, a much more complicated problem arises whose complete discussion is beyond the scope of this book.   The general equations for this motion of six degrees of freedom will be given in the next chapter.   At present, only the simplest case of an asymmetric motion, the *steady rotation about a vertical axis*, will be briefly discussed. In all these problems the main difficulty lies, as already mentioned above, in the correct analysis of the air reactions on a body that moves through the air in a general manner.

Let us first take up the case of a *true-banked turn*, already dealt with in Sec. 1.   It follows from the absence of sideslip that the x-axis (longitudinal axis) coincides with the tangent of the circular path.   It is normal to the paper, pointing toward the rear in Fig. 385.   If P' is the projection upon the y-z-plane of a point P of the plane (or a point rigidly connected with the plane) and P_1 the foot of the perpendicular from P'

on the axis of rotation, the mass element dm at P has an acceleration $\omega^2 \overline{PP}_1$ in the direction $\overrightarrow{PP}_1$, ω being the angular velocity. It is well known that the vector sum (integral) of all the products $\omega^2 \overrightarrow{PP}_1 \, dm$ equals a vector $\omega^2 \overrightarrow{OO}_1 m$ where O is the center of gravity, O_1 its projection on the axis, $\overline{OO}_1 = R$, and m the total mass W/g. But applying a force of this magnitude and direction in the c.g. is not sufficient to maintain the steady rotation, as the following analysis shows.

Let x, y, z be the coordinates of P, and 0, y_1, z_1 the coordinates of P_1. Then it is easily seen from the figure that

$$\overline{OO}_1 = R = y_1 \cos \gamma - z_1 \sin \gamma \quad \text{and}$$
$$\overline{O_1 P}_1 = y \sin \gamma + z \cos \gamma = y_1 \sin \gamma + z_1 \cos \gamma$$

Solving these two equations for y_1 and z_1 we find the coordinates of P_1:

$$x_1 = 0, \qquad y_1 = y \sin^2 \gamma + z \sin \gamma \cos \gamma + R \cos \gamma,$$
$$z_1 = z \cos^2 \gamma + y \sin \gamma \cos \gamma - R \sin \gamma \tag{48}$$

The components of the vector $\overrightarrow{PP}_1$ are $x_1 - x$, $y_1 - y$, $z_1 - z$ and the components of the mass $\times$ acceleration product are $\omega^2(x_1 - x) \, dm$, etc. According to (48) the latter components are

$$dI_x = -\omega^2 x \, dm$$
$$dI_y = -\omega^2 \cos \gamma (y \cos \gamma - z \sin \gamma - R) \, dm \tag{49}$$
$$dI_z = -\omega^2 \sin \gamma (z \sin \gamma - y \cos \gamma + R) \, dm$$

In taking the integrals over the total mass, all terms with x, y, z drop out since $\int x \, dm = \int y \, dm = \int z \, dm = 0$, as O is the center of gravity. What is left is

$$I_x = 0, \qquad I_y = m\omega^2 R \cos \gamma, \qquad I_z = -m\omega^2 R \sin \gamma$$

and these are, in fact, the components of the anticipated vector $m\omega^2 \overrightarrow{OO}_1$.

But the elements (49) give rise to moment components, also. For example, the x-component of the moment equals, by definition,

$$y \, dI_z - z \, dI_y,$$

that is, according to (49),

$$\omega^2[yz(\cos^2 \gamma - \sin^2 \gamma) + (y^2 - z^2) \sin \gamma \cos \gamma - R(y \sin \gamma + z \cos \gamma)] \, dm$$

If this is integrated, the linear terms again drop out. The integral over $yz \, dm$ also vanishes since the airplane is supposed to be symmetric so that equal elements dm are found at the same z and opposite y. Thus, only the term $(y^2 - z^2)$ gives a contribution. If the same computation is carried out for the other two moment components also, and the usual notations for inertia moments

$$\int (x^2 + z^2) \, dm = J_y, \qquad \int (y^2 + x^2) \, dm = J_z, \qquad \int xz \, dm = J_{xz} \tag{50}$$

are introduced, the three moment components become

$$\omega^2(J_z - J_y) \sin \gamma \cos \gamma, \qquad -\omega^2 J_{xz} \cos^2 \gamma, \qquad \omega^2 J_{xz} \sin \gamma \cos \gamma \quad (51)$$

Thus it is seen that *in order to keep an airplane in a steady, true-banked, horizontal turn the air reactions must supply, in addition to the centripetal force $R\omega^2 W/g$, a rolling, a pitching, and a yawing moment according to* (51).

The inertia product J_{xz} is comparatively small since the z-values for almost all points of the airplane are not large and are partly positive, partly negative. But also the rolling moment, given by the first expression (51), becomes unimportant if the radius R of the turn is large as compared with the dimensions of the airplane, owing to the factor $\omega^2 = V^2/R^2$. For an airplane of ordinary design $(J_z - J_y)/mB^2$ is of the order of magnitude 0.015 $(B = \text{span})$. Suppose that the moment is to be balanced by forces $\pm \kappa W$ acting on the ailerons at a distance $0.8B$ from each other. Then, with the expression $\tan \gamma = V^2/gR$ for the true-bank angle,

$$0.8\kappa W B = \frac{V^2}{R^2} (J_z - J_y) \sin \gamma \cos \gamma,$$

$$\kappa = \frac{0.015}{0.8} \frac{B}{R} \frac{V^2}{gR} \sin \gamma \cos \gamma \sim 0.02 \frac{B}{R} \sin^2 \gamma \quad (52)$$

This shows that, for $\gamma = 56°$ and even such a small radius as $10B$, the force on each aileron would be 0.1 per cent of the gross weight W.

If the turn is not true-banked but a sideslip is present, the computation becomes more complicated but the numerical result will be almost the same. In either case, *the moments required for maintaining the steady rotation are unimportant under normal conditions.*

We turn now to an analysis of the forces and moments that are actually present when an airplane moves along a circular path at a bank angle γ and a sideslip β. It is seen from Fig. 385 that the rotation ω about the vertical axis has a pitch component $\omega_y = \omega \sin \gamma$ and a yaw component $\omega_z = \omega \cos \gamma$. The presence of a sideslip β can be interpreted as the existence of an additional velocity component of the c.g., in the y-direction and of magnitude $v_y = -V \tan \beta$.

The effect of a pitch upon the airplane has already been discussed in the preceding sections. In the following the influence of a cross velocity v_y (or a sideslip β) and of a yaw ω_z will be studied, to which will be added the influence of a roll ω_x. Cross speed, roll, and yaw $(v_y, \omega_x, \omega_z)$ are the three velocity components characteristic for the lateral motion, as are v_x, v_z, ω_y, forward and vertical speed and pitch, for the longitudinal motion. Of course, separating those influences as though they were completely independent is a somewhat artificial expedient, but it is at present the only way available to provide at least preliminary information. Moreover, in the computation of the forces and moments the so-called "strip hypothesis" will be applied; *i.e.*, infinitesimal elements of the surfaces will be considered as independent from each other, similarly to what was done in the blade-element theory of the propeller.

In Fig. 386 an airplane is shown in top view and the *influence of a (positive) yaw* ω_z on the velocity of a wing element dS indicated. On starboard the forward speed is decreased, on the port increased; in both cases it can be written as $V - y\omega_z$. The elements of lift and drag force are therefore

$$dL = C_L \frac{\rho}{2} (V - y\omega_z)^2 \, dS, \qquad dD = C_D \frac{\rho}{2} (V - y\omega_z)^2 \, dS$$

where C_L and C_D are values of the local lift and drag coefficients. If ω_z is considered as small and ω_z^2 neglected, the integral values L and D are unaffected by ω_z since positive and negative contributions cancel each other. But the dL-forces cause rolling moments $-y \, dL$, and the dD cause yawing moments $+y \, dD$. By integration,

$$M_x = -\frac{\rho}{2} \int_{-B/2}^{B/2} C_L y (V - y\omega_z)^2 \, dS \sim \rho V \omega_z \int_{-B/2}^{B/2} C_L y^2 \, dS \qquad (53)$$

$$M_z = \frac{\rho}{2} \int_{-B/2}^{B/2} C_D y (V - y\omega_z)^2 \, dS \sim -\rho V \omega_z \int_{-B/2}^{B/2} C_D y^2 \, dS \qquad (54)$$

Here, the ω_z^2-terms are again disregarded and the terms linear in y omitted since they vanish owing to the symmetry. As long as no special information is given about the distribution of the C_L- and C_D-values, only a mean value for each coefficient can be used. Then, calling y_* the radius of inertia of the wing area we can write the two integrals as $y_*^2 S C_L$ and $y_* S C_D$.

It will be useful to introduce moment coefficients C_M referring to wing area S and to a reference length l, which will be specified later:

$$C_M = \frac{2M}{\rho V^2 S l} \qquad (55)$$

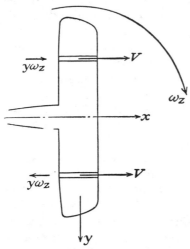

Fig. 386.—Change of forward speed due to yaw.

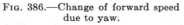

Moreover, the derivatives of the moments with respect to the variable velocity component rather than the moments themselves are of importance. Thus, the result included in (53) and (54) will be represented as

$$\frac{\partial C_{M_z}}{\partial \omega_z} = 2C_L \frac{y_*^2}{Vl}, \qquad \frac{\partial C_{M_s}}{\partial \omega_z} = -2C_D \frac{y_*^2}{Vl} \tag{56}$$

Here y_* is a geometrical constant, depending on the wing form and in usual cases not much different from $B/4$. As to C_L and C_D, one must bear in mind that these are average values, essentially defined by Eqs. (53) and (54), and that they should be determined by special experiments whenever higher accuracy is required. For a first approximation the level-flight values may be used.

In almost the same way the moments due to *roll* can be computed. It is seen in Fig. 387 that a positive roll ω_x adds a downward velocity $y\omega_x$ at a point at the distance y on the starboard side. Thus it increases

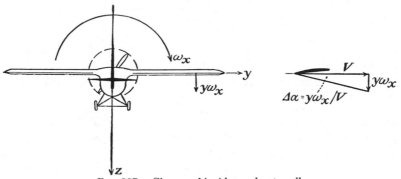

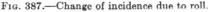

Fig. 387.—Change of incidence due to roll.

the incidence by $y\omega_x/V$. If ω_x is considered as small and C'_L and C'_D denote the derivatives of C_L and C_D with respect to the incidence, lift and drag forces on the wing element dS at y are increased by

$$C'_L \frac{\rho}{2} V^2 \frac{y\omega_x}{V} dS \qquad \text{and} \qquad C'_D \frac{\rho}{2} V^2 \frac{y\omega_x}{V} dS \tag{57}$$

Lift on the starboard side gives a negative rolling moment and drag on the starboard a positive yawing moment. Thus the new values are

$$M_x = -C'_L \frac{\rho}{2} V\omega_x \int_{-B/2}^{B/2} y^2 \, dS, \qquad M_z = C'_D \frac{\rho}{2} V\omega_x \int_{-B/2}^{B/2} y^2 \, dS \tag{58}$$

Upon introducing moment coefficients according to (55) and taking the derivatives, it follows that

$$\frac{\partial C_{M_x}}{\partial \omega_x} = -C'_L \frac{y_*^2}{Vl}, \qquad \frac{\partial C_{M_s}}{\partial \omega_x} = C'_D \frac{y_*^2}{Vl} \tag{59}$$

supplementing formulas (56).

Finally, the influence of a *sideslip* β or a *cross velocity* v_y has to be studied. If β is counted positive as indicated in Fig. 377, the sideslip turns the direction of the drag force so as to add a positive cross force of magnitude $D\beta$. Accordingly,

$$F_y = C_D \frac{\rho}{2} V^2 S\beta, \qquad \frac{\partial C_y}{\partial \beta} = C_D \tag{60}$$

If the wing is supposed to be a plane parallel to the x- and y-axes, no further influence would be observed. It is different for a V-shaped

FIG. 388.—Dihedral angle δ.

wing with a *dihedral angle* δ (Fig. 388; for the definition of dihedral, see Sec. VI.4).

As will be seen, the positive sideslip β increases the incidence at all points on the left wing and reduces the incidence on the starboard side.

In Fig. 389, A is a point on the left wing, the vector $\overrightarrow{AB}$ of magnitude V is plotted in the direction of the longitudinal axis, and AC is the zero

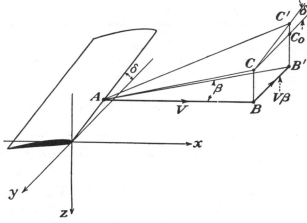

FIG. 389.—Sideslip and dihedral combined.

lift direction for the wing profile through A. Thus $\angle CAB$ is the incidence as long as no sideslip exists. The vector $\overrightarrow{BB'}$ in the negative y-direction is made to equal $V\beta$ so that $\angle B'AB = \beta$ and $\overrightarrow{AB'}$ is the velocity vector in the presence of sideslip. The parallel to the z-axis through B' intersects the plane that includes the zero lift directions of all profiles at a point C', which—in the case of positive dihedral—lies higher than C. The difference between $\overline{B'C'}$ and $\overline{BC}$ equals $\overline{C_0C'} = \overline{BB'}\delta = V\beta\delta$. The new incidence $\angle C'AB'$ is therefore greater than the original $\angle CAB$ by $\overline{C_0C'}/\overline{AB} = V\beta\delta/V = \beta\delta$. The opposite is true for points on the starboard side. *The effect of a dihedral angle δ, when a sideslip β occurs, consists in reducing the incidence on the starboard side and raising it on the port by $\beta\delta$.*

The influence of these changes in incidence is obviously nil for the total lift and total drag. But moments are produced. An area element dS on either half wing contributes a rolling and a yawing moment whose magnitudes, in analogy to (57), are

$$|y|C_L' \frac{\rho}{2} V^2\beta \, \delta \, dS \qquad \text{and} \qquad -|y|C_D' \frac{\rho}{2} V^2\beta \, \delta \, dS \qquad (61)$$

The resultant moments are

$$M_x = C_L' \frac{\rho}{2} V^2\beta\delta \int_{-B/2}^{B/2} |y| \, dS, \qquad M_z = -C_D' \frac{\rho}{2} V^2\beta\delta \int_{-B/2}^{B/2} |y| \, dS \qquad (62)$$

If the distance y^* of the centroid of each half wing area is introduced, the derivatives of the respective moment coefficients are

$$\frac{\partial C_{M_x}}{\partial \beta} = C_L'\delta \frac{y^*}{l}, \qquad \frac{\partial C_{M_z}}{\partial \beta} = -C_D'\delta \frac{y^*}{l} \qquad (63)$$

All these derivatives (56), (59), (60), and (63) will later be used in the discussion of lateral stability (Sec. XX.4). They include the influence of β, ω_x, and ω_z on the wing only.

Returning now to the problem of a turn in a horizontal plane, we may omit the terms due to the moments of inertia as given in (51). The only inertia effect, then, is the vector W/g times V^2/R in the horizontal direction (Fig. 385) with the components WV^2/gR times $\cos\gamma$ in the y-direction and times $-\sin\gamma$ in the direction of z. If we call F_x', F_y', F_z', the components of the air reactions on the tail, the fuselage, etc., the three equations of motion for the center of gravity will be

$$0 = T - D + F_x'$$
$$\frac{W}{g} \frac{V^2}{R} \cos\gamma = W \sin\gamma + \beta D + F_y' \qquad (64)$$
$$-\frac{W}{g} \frac{V^2}{R} \sin\gamma = W \cos\gamma - L + F_z'$$

Here, $-F_x'$ is essentially the parasite drag of the airplane (except the wing) so that the first equation means $T = D_{to}$, propeller thrust equal to total drag. The second component F_y' will be increasing with β. It includes the cross force on fuselage and vertical tail surface due to sideslip. As mentioned in Sec. 1 of this chapter, the coefficient for the sum $\beta D + F'_y$ is about 0.4 in normal conditions, *i.e.*,

$$\beta D + F'_y = \frac{\rho}{2} V^2 S \times 0.4\beta \tag{65}$$

At any rate, the second equation (64) shows how an underbanked turn (γ smaller than the true bank angle) can be carried out with positive sideslip. In the third equation, F'_z will be unimportant. If this equation is multiplied by $\cos \gamma$ and the second by $\sin \gamma$, the sum gives

$$L \cos \gamma = W + (\beta D + F'_y) \sin \gamma \tag{66}$$

This determines the incidence required in the curved path. Dividing by $\rho V^2 S/2$ and using (65), we find that

$$C_L = \frac{C_{Ll}}{\cos \gamma} + 0.4\beta \tan \gamma \tag{66'}$$

which shows that C_L and therefore *the incidence in the turn must always be greater than the values C_{Ll} and α_l for straightforward flight.*

The three equations for moment equilibrium, divided by $\rho V^2 Sl/2$ and with C'_{M_x}, etc., for the coefficients regarding the moments on tail and airplane body, read, since $\omega_x = 0$ and $\omega_z = \cos \gamma \, V/R$:

$$\begin{aligned}
0 &= \frac{\partial C_{M_x}}{\partial \beta} \beta + \frac{\partial C_{M_x}}{\partial \omega_z} \frac{V}{R} \cos \gamma + C'_{M_x} \\
0 &= C'_{M_y} \\
0 &= \frac{\partial C_{M_z}}{\partial \beta} \beta + \frac{\partial C_{M_z}}{\partial \omega_z} \frac{V}{R} \cos \gamma + C'_{M_z}
\end{aligned} \tag{67}$$

Here, the four derivatives are those given in (56) and (63). The second equation, or $M'_y = 0$, does not mean that the elevator can be kept in neutral position. On the contrary, the tail must be used to compensate both the damping moment, which is proportional and opposite in direction to the pitch $\omega_y = V \sin \gamma/R$, and also the static moment due to the increased incidence. Both contributions are acting in the same sense in the case of a statically stable plane. The conclusion is that *in a horizontal turn the elevator must be operated in the same sense as in climbing and the stronger, the higher the static stability of the plane.*

The moments M'_x and M'_z must be supplied by the ailerons and the rudder, respectively. Their amounts can be computed by solving the first and the third equation (67) with the derivatives given in (56) and (63). It can happen that M'_x and M'_z both become zero. This, again, does not exactly mean that the turn is made with the ailerons and the rudder neutral, but it may be that the additions which should be made to the moment derivatives when the influence of ω_z and β on the fin and the rudder in neutral position is taken into account are comparatively unimportant. (A discussion of these terms will be given in Sec. XX.4.) If it is assumed that these additions do not change the signs of the derivatives as computed from (56) and (63), Eqs. (67) show that M'_x must be negative and M'_z positive. This means that *in normal conditions the ailerons must be operated in the sense of counteracting the bank and the rudder in the sense of going into the turn.*

Problem 12. Compute the rolling and yawing moments due to dihedral for an airplane moving at $V = 280$ ft./sec. at sea level, with a slideslip β. The dihedral angle is $\delta = 4°$. The wing has a span $B = 24$ ft. and consists of two trapezoids of taper ratio $c_0/c_1 = 1.8$ (see Sec. VI.4). The total parasite-drag coefficient is 0.04, the actual incidence 6.5°. Give also the moment coefficients referred to wing area and tail distance $l = 3$ times mean chord length.

Problem 13. Find the derivatives of the moment coefficients due to yaw for the airplane determined by the data of Prob. 12.

Problem 14. Solve the same problem for the moments due to roll.

*Problem 15. Give the formula for the rolling moment due to yaw, replacing the assumption of constant C_L by the hypothesis of an elliptic lift distribution.

CHAPTER XIX

GENERAL THEORY OF MOTION AND STABILITY

1. The General Equations of Motion of an Airplane. The theory of dynamics of a rigid body allows one to set up the equations for the most general motion of an airplane in various forms. The way that seems best suited, considering the special kind of forces present in the case of airplane flight, consists in using a *reference system rigidly connected with the airplane* and of taking the six velocity components with respect to this system as the unknowns. The origin of the coordinate system may coincide with the center of gravity (c.g.); the axes 1 and 3 may lie in the symmetry plane and the axis 2 perpendicular to it. Thus 2 is the lateral, or cross, axis, used as the y-axis in the preceding chapters, while the 1- and 3-axes in the symmetry plane are not specified. For convenience, however, we shall sometimes call the axis 3 the down axis.

The six equations of motion will express the fact that the integral of the products acceleration times mass element equals the sum of forces and that the integral of the moments of acceleration multiplied by mass element equals the sum of the force moments. If a_1, a_2, a_3 denote the components of acceleration; x, y, z the coordinates of the mass element dm; F_1, F_2, F_3 the components of the resultant air reactions; W_1, W_2, W_3 the components of gravity; and M_1, M_2, M_3 the components of the resultant moment, the equations are

$$\int a_1\, dm = F_1 + W_1, \qquad \int (ya_3 - za_2)\, dm = M_1,$$
$$\int a_2\, dm = F_2 + W_2, \quad (1) \qquad \int (za_1 - xa_3)\, dm = M_2, \quad (2)$$
$$\int a_3\, dm = F_3 + W_3 \qquad \int (xa_2 - ya_1)\, dm = M_3$$

The first task is to express the integrals on the left-hand side in terms of velocities and their derivatives.

Let V_1, V_2, V_3 be the three components of the velocity of the c.g. and ω_1, ω_2, ω_3 the components of the angular velocity, all, of course, referring to the directions of the coordinate axes as already described above. The velocity components v_1, v_2, v_3 for the point with the coordinates x, y, z will then be

$$v_1 = V_1 + \omega_2 z - \omega_3 y, \quad v_2 = V_2 + \omega_3 x - \omega_1 z, \quad v_3 = V_3 + \omega_1 y - \omega_2 x \quad (3)$$

according to the well-known formulas for the components of a vector product. The acceleration component a_1 will not coincide with the derivative dv_1/dt since the direction of the 1-axis is changing in time. In fact, a vector OA with constant components v_1, v_2, v_3 changes its position in space (Fig. 390) because of the rotation of the reference system. During the time dt the point A moves to A', and the components of $\overrightarrow{AA'}$ can be computed by the same formulas that give the displacement of the point with coordinates v_1, v_2, v_3 owing to the rotation ω_1, ω_2, ω_3. The first component, according to the first equation (3), will be $(\omega_2 v_3 - \omega_3 v_2)\,dt$, etc. This gives the acceleration components, if the differentiation with respect to time is denoted by dots:

$$a_1 = \dot{v}_1 + \omega_2 v_3 - \omega_3 v_2, \; a_2 = \dot{v}_2 + \omega_3 v_1 - \omega_1 v_3, \; a_3 = \dot{v}_3 + \omega_1 v_2 - \omega_2 v_1 \quad (4)$$

Introducing here the values of v_1, v_2, v_3 from (3), we obtain

$$\begin{aligned}
a_1 &= \dot{V}_1 + \omega_2 V_3 - \omega_3 V_2 - x(\omega_2^2 + \omega_3^2) - y(\dot{\omega}_3 - \omega_1\omega_2) + z(\dot{\omega}_2 + \omega_3\omega_1) \\
a_2 &= \dot{V}_2 + \omega_3 V_1 - \omega_1 V_3 - y(\omega_3^2 + \omega_1^2) - z(\dot{\omega}_1 - \omega_2\omega_3) + x(\dot{\omega}_3 + \omega_1\omega_2) \quad (5) \\
a_3 &= \dot{V}_3 + \omega_1 V_2 - \omega_2 V_1 - z(\omega_1^2 + \omega_2^2) - x(\dot{\omega}_2 - \omega_3\omega_1) + y(\dot{\omega}_1 + \omega_2\omega_3)
\end{aligned}$$

In forming the integral of $a_1\,dm$ it should be noted that the c.g. is defined by the conditions that the integrals of $x\,dm$, $y\,dm$, $z\,dm$ vanish. Thus, only the constant terms (those independent of x, y, z) in the expressions (5) contribute to $\int a_1\,dm$, etc. Upon denoting the total mass $\int dm$ by W/g, the left-hand sides of Eqs. (1) become

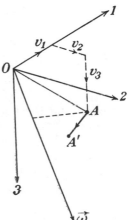

FIG. 390.—Change of a vector with constant components.

$$\frac{W}{g}(\dot{V}_1 + \omega_2 V_3 - \omega_3 V_2),$$

$$\frac{W}{g}(\dot{V}_2 + \omega_3 V_1 - \omega_1 V_3), \quad (6)$$

$$\frac{W}{g}(\dot{V}_3 + \omega_1 V_2 - \omega_2 V_1)$$

In computing the left-hand sides of (2) we can omit the constant terms in the expressions for the accelerations since they lead to linear terms that vanish after integration over dm. The linear terms in (5) lead to terms of second order in the moments; for example, in $ya_3 - za_2$ we have

$$yz(\omega_3^2 - \omega_2^2) + z^2(\dot{\omega}_1 - \omega_2\omega_3) + y^2(\dot{\omega}_1 + \omega_2\omega_3) - xy(\dot{\omega}_2 - \omega_3\omega_1)$$
$$- zx(\dot{\omega}_3 + \omega_1\omega_2)$$

or, by simple rearranging,

$$(y^2 + z^2)\dot{\omega}_1 + (y^2 - z^2)\omega_2\omega_3 - yz(\omega_2^2 - \omega_3^2) - zx(\dot{\omega}_3 + \omega_1\omega_2)$$
$$- xy(\dot{\omega}_2 - \omega_3\omega_1) \quad (7)$$

with two analogous formulas for the second and the third components of the moment of acceleration. In integrating over dm the following quantities evolve: (1) the three *moments of gyration*

$$J_1 = \int (y^2 + z^2)\, dm, \qquad J_2 = \int (z^2 + x^2)\, dm, \qquad J_3 = \int (x^2 + y^2)\, dm \qquad (8)$$

(2) the *moments of deviation*, or inertia products of second order,

$$J_{12} = \int xy\, dm, \qquad J_{23} = \int yz\, dm, \qquad J_{31} = \int zx\, dm \qquad (9)$$

As the body is supposed to be symmetric with respect to the xz-plane, the magnitudes J_{12} and J_{23} must be zero since to each mass element with a positive y there corresponds another with negative y, the same x, z, and the same dm. The complete expression (7) when integrated over dm supplies

$$J_1\dot\omega_1 + (J_3 - J_2)\omega_2\omega_3 - J_{31}(\dot\omega_3 + \omega_1\omega_2) - J_{12}(\dot\omega_2 - \omega_3\omega_1) - J_{23}(\omega_2^2 - \omega_3^2)$$

and, by replacing the subscripts 1, 2, 3 by 2, 3, 1 and 3, 1, 2, the analogous values for the y- and z-directions are found. With $J_{12} = J_{23} = 0$ the final expressions for the left sides of Eqs. (2) are

$$\begin{aligned}
J_1\dot\omega_1 + (J_3 - J_2)\omega_2\omega_3 - J_{31}(\dot\omega_3 + \omega_1\omega_2) \\
J_2\dot\omega_2 + (J_1 - J_3)\omega_3\omega_1 - J_{31}(\omega_3^2 - \omega_1^2) \\
J_3\dot\omega_3 + (J_2 - J_1)\omega_1\omega_2 - J_{31}(\dot\omega_1 - \omega_2\omega_3)
\end{aligned} \qquad (10)$$

By an appropriate choice of the 1- and 3-axes in the symmetry plane one can make J_{31} vanish, also. The three axes are then called the *principal axes* of the body. Under normal conditions the mean chord direction and the normal to it will not be far away from the principal directions. This means that for this choice of the 1- and 3-axes J_{31} will have a comparatively small value.

The forces appearing on the right-hand side of Eqs. (1) and (2) are the gravity with the resultant components W_1, W_2, W_3 acting in the c.g. and the air reactions on all parts of the airplane including the propeller, with the resultants F_1, F_2, F_3 and the moments M_1, M_2, M_3 with respect to the c.g. It is the basic assumption in the dynamics of an airplane that, at each moment, the air reactions on a moving body are completely determined if the instantaneous velocities of all parts of the body are known. (For a criticism of this hypothesis, see Sec. 2 of this chapter.) Now, for each part of the airplane body proper, *i.e.*, except the propeller, the velocity vector is given, according to (3), by the six quantities V_1, V_2, V_3, ω_1, ω_2, ω_3. All experimental and theoretical data concerning the air reactions can be considered as equations giving some of the components F_1, F_2, F_3, M_1, M_2, M_3 as functions of some of the six velocity components. If, for example, the wing moves parallel to its symmetry plane at a velocity V and an angle of attack α, then to give lift L and drag D as functions of V and α (Fig. 391) is the same as to give F_1 and F_3 as functions of V_1 and V_3. A similar consideration applies to any force or moment component, provided that the geometry of the airplane body, *i.e.*, the relative position of all its points with respect to the coordinate system, is considered as given. The same holds true for the propeller if the propeller speed n is a known constant. One may separate the air reactions on the propeller and consider them as an additional system of forces acting on the airplane body. At any rate, the contributions of the propeller to the right-hand sides of Eqs. (1) and (2) are determined by the six velocity parameters V_1, . . . , ω_3 and the propeller speed n. *If the geometry of the airplane and the propeller speed are known, the six components F_1, F_2, F_3, M_1, M_2, M_3 of the air reactions are known functions of the six velocity components V_1, V_2, V_3, ω_1, ω_2, ω_3.*

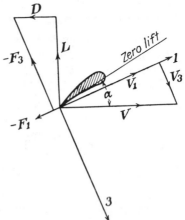

Note that this statement would not be correct if a different kind of coordinate system were used, for example, a system of axes fixed in space. In this case the components of the air reactions would depend on the velocity components and on additional variables that determine the instantaneous position of the body with respect to the coordinate axes. This is the reason why the coordinate system as described in the foregoing has been chosen.

The gravity forces, however, behave differently. With respect to a system fixed in space the components of gravity would be constants, but they enter

Fig. 391.—Velocity and force components for motion in the symmetry plane.

as variables in Eqs. (1) and (2). In fact, the gravity vector W is a constant in space. It was seen in Eq. (4) and Fig. 391 how the components of the vector $\overrightarrow{a} = \overrightarrow{dv}/dt$ depend on the components v_1, v_2, v_3 of $\overrightarrow{v}$ and their derivatives. If on the right-hand sides of (4) the v_1, v_2, v_3 are replaced by W_1, W_2, W_3, we find the components of dW/dt. Since W is a constant vector in space, all these expressions must be zero. This supplies the relations

$$0 = \dot{W}_1 + \omega_2 W_3 - \omega_3 W_2, \qquad 0 = \dot{W}_2 + \omega_3 W_1 - \omega_1 W_3,$$
$$0 = \dot{W}_3 + \omega_1 W_2 - \omega_2 W_1 \tag{11}$$

If the first is multiplied by W_1, the second by W_2, and the third by W_3, the sum of the equations is

$$W_1 \dot{W}_1 + W_2 \dot{W}_2 + W_3 \dot{W}_3 = 0 \text{ or } W_1^2 + W_2^2 + W_3^2 = \text{const.} = W^2 \tag{12}$$

The *complete system of equations of motion* can now be written as follows:

$$\frac{W}{g}(\dot{V}_1 + \omega_2 V_3 - \omega_3 V_2) = F_1 + W_1$$

$$\frac{W}{g}(\dot{V}_2 + \omega_3 V_1 - \omega_1 V_3) = F_2 + W_2$$

$$\frac{W}{g}(\dot{V}_3 + \omega_1 V_2 - \omega_2 V_1) = F_3 + W_3 \tag{13}$$

$$J_1\dot{\omega}_1 + (J_3 - J_2)\omega_2\omega_3 - J_{31}(\dot{\omega}_3 + \omega_1\omega_2) = M_1$$

$$J_2\dot{\omega}_2 + (J_1 - J_3)\omega_3\omega_1 - J_{31}(\omega_3^2 - \omega_1^2) = M_2$$

$$J_3\dot{\omega}_3 + (J_2 - J_1)\omega_1\omega_2 - J_{31}(\dot{\omega}_1 - \omega_2\omega_3) = M_3$$

$$\dot{W}_1 = \omega_3 W_2 - \omega_2 W_3, \qquad \dot{W}_2 = \omega_1 W_3 - \omega_3 W_1, \qquad \dot{W}_3 = \omega_2 W_1 - \omega_1 W_2$$

This is a system of nine simultaneous differential equations, each of first order, for the nine unknowns V_1, V_2, V_3; ω_1, ω_2, ω_3; W_1, W_2, W_3. As one of the last three equations can be replaced by an equation without derivatives ($W_1^2 + W_2^2 + W_3^2 = W^2$), *the integration problem is of the eighth order.* Solution of this problem would supply the six velocity values V_1, . . . , ω_3 and the three components W_1, W_2, W_3 as functions of time. Now, if the angles that the three axes 1, 2, 3 form with the vertical (direction of gravity) are called ψ_1, ψ_2, ψ_3, we have

$$W_1 = W \cos \psi_1, \qquad W_2 = W \cos \psi_2, \qquad W_3 = W \cos \psi_3 \tag{14}$$

Thus knowing W_1, W_2, W_3 is the same as knowing the attitude of the airplane body toward the vertical. Since

$$\cos^2 \psi_1 + \cos^2 \psi_2 + \cos^2 \psi_3 = 1,$$

it is seen again that W_1, W_2, W_3 count for two independent variables only. A complete integral of (13) will be determined by eight constants or eight "initial conditions," for example, by the initial values of V_1, . . . , ω_3 and of two out of W_1, W_2, W_3 or ψ_1, ψ_2, ψ_3. To summarize:

If for any instant $t = 0$ the state of velocity, i.e., the values of V_1, V_2, V_3, ω_1, ω_2, ω_3 and the attitude of the airplane toward the vertical, i.e., two of the values ψ_1, ψ_2, ψ_3 or W_1, W_2, W_3, are given, the differential equations (13) determine both the state of velocity and the attitude toward the vertical, for any subsequent instant t.

To find the path of the c.g. and the complete attitude of the airplane (one angle more, in addition to the ψ), further integrations and further initial conditions are necessary. This problem is of minor importance and will not be treated here in general form. Special cases have been discussed in Chap. 18.

For some purposes it will be useful to transform Eqs. (13) into a *dimensionless form.* If the similarity laws derived in Sec. IV.2 are applied to the components of the air reactions, each of them can be expressed by a dimensionless coefficient C:

$$F_1 = \frac{\rho}{2} V^2 S C_1, \qquad F_2 = \frac{\rho}{2} V^2 S C_2, \qquad F_3 = \frac{\rho}{2} V^2 S C_3,$$

$$M_1 = \frac{\rho}{2} V^2 S l C_4, \qquad M_2 = \frac{\rho}{2} V^2 S l C_5, \qquad M_3 = \frac{\rho}{2} V^2 S l C_6 \tag{15}$$

Here S may be the wing area and l some suitably chosen reference length, for example, $\sqrt{S}$, or the span B, or the tail distance, etc. The coefficients C_1, etc., depend on the ratios $V_1 : V_2 : V_3$; $\omega_1 : \omega_2 : \omega_3$; on the geometry of the airplane, and on the Reynolds number. Let us introduce the level-flight velocity V_l at a density ρ_l with the lift coefficient C_{Ll} by

$$W = \frac{\rho_l}{2} C_{Ll} V_l^2 S \tag{15'}$$

The velocity components V_1, V_2, V_3, ω_1, ω_2, ω_3 will be represented by the ratios

$$u_1 = \frac{V_1}{V_l}, \quad u_2 = \frac{V_2}{V_l}, \quad u_3 = \frac{V_3}{V_l}, \quad \sigma_1 = \frac{\omega_1 V_l}{g}, \quad \sigma_2 = \frac{\omega_2 V_l}{g}, \quad \sigma_3 = \frac{\omega_3 V_l}{g} \tag{16}$$

and u^2 will be written for $u_1^2 + u_2^2 + u_3^2$. For the moments of gyration and deviation we use

$$j_1^2 = \frac{gJ_1}{Wl^2}, \qquad j_2^2 = \frac{gJ_2}{Wl^2}, \qquad j_3^2 = \frac{gJ_3}{Wl^2}, \qquad j_{31} = \frac{gJ_{31}}{Wl^2} \tag{16'}$$

where evidently j_1 is the ratio of the first radius of gyration to l, etc. Finally, a dimensionless time scale may be defined by

$$\tau = \frac{gt}{V_l} \tag{17}$$

and the derivatives with respect to τ may now be designated by dots. Omitting, then, for the sake of brevity, the terms due to J_{13}, we find that Eqs. (13) take the form

$$\dot{u}_1 + \sigma_2 u_3 - \sigma_3 u_2 = \cos\psi_1 + \frac{\rho}{\rho_l}\frac{C_1}{C_{Ll}} u^2$$

$$\dot{u}_2 + \sigma_3 u_1 - \sigma_1 u_3 = \cos\psi_2 + \frac{\rho}{\rho_l}\frac{C_2}{C_{Ll}} u^2$$

$$\dot{u}_3 + \sigma_1 u_2 - \sigma_2 u_1 = \cos\psi_3 + \frac{\rho}{\rho_l}\frac{C_3}{C_{Ll}} u^2$$

$$j_1^2 \dot{\sigma}_1 + (j_3^2 - j_2^2)\sigma_2\sigma_3 = \frac{\rho}{\rho_l}\frac{V_l^2}{gl}\frac{C_4}{C_{Ll}} u^2 \tag{18}$$

$$j_2^2 \dot{\sigma}_2 + (j_1^2 - j_3^2)\sigma_3\sigma_1 = \frac{\rho}{\rho_l}\frac{V_l^2}{gl}\frac{C_5}{C_{Ll}} u^2$$

$$j_3^2 \dot{\sigma}_3 + (j_2^2 - j_1^2)\sigma_1\sigma_2 = \frac{\rho}{\rho_l}\frac{V_l^2}{gl}\frac{C_6}{C_{Ll}} u^2$$

$$\sin\psi_1 \dot{\psi}_1 + \sigma_3 \cos\psi_2 - \sigma_2 \cos\psi_3 = 0,$$

$$\sin\psi_2 \dot{\psi}_2 + \sigma_1 \cos\psi_3 - \sigma_3 \cos\psi_1 = 0$$

Here, the coefficients C_1, C_2, . . . , C_6 are functions of u_1, u_2, u_3, σ_1, σ_2, σ_3. Thus the eight simultaneous differential equations (18) of first order determine the velocity ratios u_1, u_2, u_3, σ_1, σ_2, σ_3 and the angles ψ_1, ψ_2 (with ψ_3 from $\cos^2 \psi_1 + \cos^2 \psi_2 + \cos^2 \psi_3 = 1$) as functions of the dimensionless time variable. It is seen that, besides j_1, j_2, j_3 (and j_{13}), three parameters are essential: the density ratio ρ/ρ_l, the lift coefficient C_{Ll} of level flight, and the ratio V_l^2/gl, that is, velocity head of level flight to size of the airplane.

Problem 1. Develop the system of dimensionless equations that replaces (18) if instead of (16) and (17) the time scale is taken as

$$\tau' = \frac{tV_l}{l}$$

and the angular speed components are reduced to

$$\sigma' = \frac{l\omega}{V_l}$$

Problem 2. Calling m the mass of the airplane and S the wing area, derive the system of dimensionless equations of motion with the following units of time, velocity, and angular speed:

$$\text{Time} = \frac{m}{\rho V_l S} \qquad \text{Velocity} = \frac{\rho V_l S l}{m} \qquad \text{Angular speed} = \frac{\rho S V_l}{m}$$

*Problem 3. Show that in the case of a merely longitudinal motion ($V_2 = \omega_1 = \omega_3 = 0$) the first, third, fifth, seventh, and ninth of Eqs. (13) are equivalent to Eqs. (14) to (16), Chap. XVIII.

2. Steady Motion. Specification of Forces.

The motion of an airplane will properly be called a steady motion if the time derivatives of the nine variables involved, that is, V_1, V_2, V_3; ω_1, ω_2, ω_3; ψ_1, ψ_2, ψ_3 permanently vanish. If $\dot{\psi}_1 = \dot{\psi}_2 = \dot{\psi}_3 = 0$ is introduced in the last two equations (18) and the analogous equation for ψ_3, we find

$$\sigma_3 \cos \psi_2 - \sigma_2 \cos \psi_3 = \sigma_1 \cos \psi_3 - \sigma_3 \cos \psi_1 = \sigma_2 \cos \psi_1 - \sigma_1 \cos \psi_2 = 0$$

This is equivalent to
$$(19)$$

$$\sigma_1 : \sigma_2 : \sigma_3 = \omega_1 : \omega_2 : \omega_3 = \cos \psi_1 : \cos \psi_2 : \cos \psi_3 \qquad (20)$$

and the latter equation expresses the condition that the resultant rotation vector is vertical. As the derivatives of V_1, V_2, V_3 are zero, the velocity vector $\overrightarrow{V}$ moves invariably with the airplane, *i.e.*, participates in the rotation about a vertical axis. Therefore the direction of V, *i.e.*, the direction of the path of the c.g. forms a constant angle with the vertical, which we may call $90 - \vartheta$.

With $\dot{u}_1 = \dot{u}_2 = \dot{u}_3 = 0$ the left-hand sides of the first three equations (18) include the terms $\sigma_2 u_3 - \sigma_3 u_2$, etc., only. These are the

components of the (dimensionless) acceleration of the c.g.   If the first
of these terms is multiplied by σ_1, the second by σ_2, and the third by σ_3,
the sum is zero.   This shows that the acceleration is perpendicular to
the rotation, *i.e.*, is directed horizontally.   On the other hand, the same
three terms, when multiplied by u_1, u_2, u_3, give, again, the sum zero;
and this means that the acceleration is normal to the velocity vector.
Moreover, the acceleration vector appears as the vector product of
$\overrightarrow{u}$ and $\overrightarrow{\sigma}$, the (dimensionless) velocity and rotation.   Since the magni-

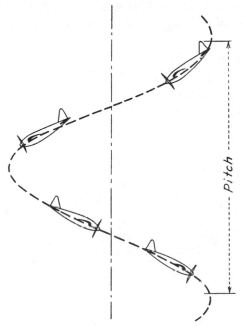

Fig. 392.—Helicoidal motion.

tudes of these vectors and their angle $90 - \vartheta$ are constant, the acceleration vector has a constant magnitude, also.   Now, the acceleration was
seen to be normal to the velocity; it has therefore the magnitude V^2/R
where R is the radius of curvature of the c.g. path.   The conclusion is
that R must be constant.   The only curve with constant slope against
the horizontal plane and constant curvature is the helix.   Thus it is
proved that *the most general steady motion of an airplane is a helicoidal
motion with vertical axis at constant velocity and constant rotational speed;
all points of the body travel along helical curves with the same pitch.*

This general type of motion includes as special cases the rotation in a
horizontal plane ($\vartheta = 0$) and the straight motion at an angle ϑ
($R = \infty$).

The airplane will perform a steady helicoidal motion if the air reactions, including the propeller, supply a vertical force component to balance the weight, a constant horizontal force equal to WV^2/gR, and, moreover, the moments determined by the left-hand sides of the fourth, fifth, and sixth of Eqs. (18). These conditions have been discussed for the case $\vartheta = 0$, horizontal turn, in Sec. XVIII.5. The case $R = \infty$, straight flight, was the subject matter of Chaps. XIV and XV, which dealt with the performance computation. Some remarks concerning the general case of helicoidal motion are given in Chap. XX.

Any further conclusions can be drawn from the equations of motion only on the basis of a detailed analysis of their right-hand sides, *i.e.*, of

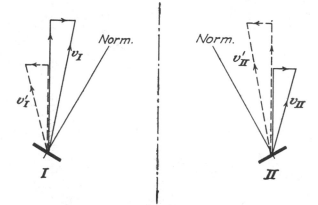

FIG. 393.—Two symmetrical elements.

the force and moment components, considered as functions of the velocity parameters $V_1, \ldots, \omega_3$. Such an analysis in general form is beyond the scope of this book. As a preliminary step to the applications that will be the subject of the last chapter, only two decisive properties of the air reactions in the case of a *completely symmetric airplane* must be discussed here.

The first statement, as follows, is immediately obvious and needs no proof: *If a completely symmetric airplane performs a "longitudinal" motion, i.e., a motion with $V_2 = 0$, $\omega_1 = 0$, $\omega_3 = 0$, the three lateral components F_2, M_1, M_3, that is, cross force and rolling and yawing moments, vanish.* This theorem has already been used in Sec. XVIII.2 and later.

The analogous statement with respect to lateral motion and longitudinal forces would not be correct. There is no reason why, for example, in pure yaw the fuselage walls should not experience a longitudinal force, due to friction. There exists, however, a more general theorem that expresses the consequences of symmetry for the relation between the six force components and the velocities V_2, ω_1, ω_3.

In Fig. 393 two symmetric surface elements, I and II, of the airplane are shown with their respective normals n_I and n_{II}. The two elements have the same x- and z- but opposite y-values. In any state of motion, i.e., for any values of V_1, V_2, V_3, ω_1, ω_2, ω_3, the velocities of I and II have the same component v_2, as seen from the second equation (3), which does not include y. The two resultant velocity vectors are drawn as solid lines and denoted v_I' and v_{II}. Now, if V_1, V_3, ω_2 are kept unchanged and the signs of V_2, ω_1, ω_3 reversed, formulas (3) show that the 2-component of both velocities changes its sign while the 1-components as well as the 3-components mutually interchange, because the respective formulas (3) include the products $\omega_3 y$ and $\omega_1 y$, respectively. The new velocity vectors are denoted by v_I' and v_{II}' and shown in Fig. 393 as dotted lines. It is seen that v_I' is symmetric to v_{II} and v_{II}' to v_I. The relative position of the velocity vectors with respect to the surface elements is the same in both states of motion, except for the reflection with respect to the symmetry plane. In other words, if the original force components were

$$F_1^I, \, F_2^I, \, F_3^I \text{ in } x, \, y, \, z; \text{ and } F_1^{II}, \, F_2^{II}, \, F_3^{II} \text{ in } x, \, -y, \, z$$

the new components (second state of motion) are

$$F_1^{II}, \, -F_2^{II}, \, F_3^{II} \text{ in } x, \, y, \, z; \text{ and } F_1^I, \, -F_2^I, \, F_3^I \text{ in } x, \, -y, \, z$$

The sum of forces on the two elements has consequently the same 1- and 3-components and opposite 2-components in the two different states of motion.

If the moments are computed and summed up, just the opposite is found. For example, the 3-component of the moments is

$$(xF_2^I - yF_1^I) + (xF_2^{II} + yF_1^{II}) \quad \text{in the first motion}$$
$$(-xF_2^{II} - yF_1^{II}) + (-xF_2^I + yF_1^I) \quad \text{in the second motion}$$

These two values have opposite signs. The same is true for the first component of the moment, while the 2-component remains unchanged.

The result of this argument can now be stated: *If in any state of motion of a completely symmetric airplane the lateral velocities V_2, ω_1, ω_3 are simultaneously reversed, the longitudinal forces F_1, F_3, M_2 stay unchanged, while the lateral forces F_2, M_1, M_3 change their sign.* In a shorter form one could say that F_1, F_3, M_2 are even functions and F_2, M_1, M_3 are odd functions of the group V_2, ω_1, ω_3 of variables.

It will be seen in Sec. 4 of the present chapter that this property of the air reactions plays an important part in the stability theory.

In this argument, as well as in the preceding section, it has been assumed that the air reactions on a moving body *depend on the instantaneous state of velocity only*, not on the accelerations (and higher derivatives). It is obvious that this can be only an approximation and that some influence of the acceleration must exist.   The theory of irrotational flow of a perfect fluid gives a certain answer to this question.   According to this theory, a body moving in a fluid originally at rest behaves like a body of *increased inertia:* there is a term of apparent mass to be added to its real mass W/g, and certain additions to its moments and products of inertia, all depending on the shape of the body.   For example, in the case of a sphere of radius a and mass W/g, the reaction of the surrounding fluid would be taken into account when the sphere is assumed to have the mass

$$\frac{W}{g} + \tfrac{2}{3}\pi\rho a^3 \tag{21}$$

and to move under the influence of the other forces (weight, etc.) alone. This includes (for the case of a sphere) the result, expressed in D'Alembert's paradox (Sec. IX.3) that no reaction exists if the motion is uniform.

To apply this theory to the motion of an airplane meets with essential difficulties.   (1) The terms of apparent mass, etc., have so far not been elaborated for bodies of such complicated shape as an aircraft body. (2) It was seen in Chap. IX that, as far as perfect-fluid theory can be used, the flow around an air wing is a discontinuous irrotational motion. It is doubtful whether the classical results that refer to continuous flow patterns can be applied under these conditions.   (3) In all dynamic problems dealt with in the present part of this book, the nonconservative (drag) forces seem to play a decisive part.   Thus it must be concluded that at the present stage of research no sufficient basis exists for taking into account the influence of acceleration terms on the forces exerted by the air on a moving airplane.   Throughout the book this simplifying assumption will be maintained.

3. Theory of Dynamic Stability.   The theory of dynamic stability, or *stability of a given state of motion*, is due to E. J. Routh (1877).   In a certain way, it is a counterpart or a generalization of the theory of static stability, which was outlined in Sec. XVII.4 in a nonanalytic form. Here, in the broader problem, the analytic method must be followed. To explain its principal idea a simple example may be taken up first.

Let m be the mass of a particle that moves in the x-y-plane under the influence of force components F_x and F_y, which may be given as functions of the coordinates x, y and the velocity components $\dot{x}$, $\dot{y}$.   The equations of motion, then (dots denoting time derivatives), are

$$m\ddot{x} = F_x(x,y,\dot{x},\dot{y}), \qquad m\ddot{y} = F_y(x,y,\dot{x},\dot{y}) \tag{22}$$

By a set of initial values x_0, y_0, $\dot{x}_0$, $\dot{y}_0$ a particular solution of (22) is singled out.   Such a solution, which can be represented by two functions of t,

$$x = X(t) \qquad y = Y(t) \tag{23}$$

describes one possible motion of the particle. The problem is whether this particular motion can be classified as a stable or an unstable one. In order to secure a decision Routh considers a "neighboring" motion, setting

$$x = X(t) + x', \qquad y = Y(t) + y' \tag{24}$$

where x' and y' are supposed to be small quantities. If (24) is substituted on the right-hand side of (22) the power development of F_x supplies

$$F_x(x,y,\dot{x},\dot{y}) = F_x(X,Y,\dot{X},\dot{Y}) + \frac{\partial F_x}{\partial x} x' + \frac{\partial F_x}{\partial y} y' + \frac{\partial F_x}{\partial \dot{x}} \dot{x}' + \frac{\partial F_x}{\partial \dot{y}} \dot{y}' \tag{25}$$

and an analogous expression for F_y, if all terms of higher than first order in x', y', $\dot{x}'$, $\dot{y}'$ are neglected. Introducing these expressions in (22) the terms $m\ddot{X}$ and $m\ddot{Y}$ to the left cancel out against the first terms $F_x(X,Y,\dot{X},\dot{Y})$ and $F_y(X,Y,\dot{X},\dot{Y})$ to the right, since (23) is supposed to be an exact solution of Eqs. (22). What is left are two differential equations for x' and y',

$$\begin{aligned}
m\ddot{x}' &= \frac{\partial F_x}{\partial x} x' + \frac{\partial F_x}{\partial y} y' + \frac{\partial F_x}{\partial \dot{x}} \dot{x}' + \frac{\partial F_x}{\partial \dot{y}} \dot{y}' \\
m\ddot{y}' &= \frac{\partial F_y}{\partial x} x' + \frac{\partial F_y}{\partial y} y' + \frac{\partial F_y}{\partial \dot{x}} \dot{x}' + \frac{\partial F_y}{\partial \dot{y}} \dot{y}'
\end{aligned} \tag{26}$$

Here, the differential quotients $\partial F_x/\partial x$, etc., are obviously the derivatives of F_x or F_y at the point $x = X$, $y = Y$, $\dot{x} = \dot{X}$, $\dot{y} = \dot{Y}$ and therefore known functions of t. The set (26) accordingly is a set of *linear homogeneous differential equations* for x', y'. Its general solution must be found, *i.e.*, the functions x', y' of t satisfying (25) for an arbitrary set of initial conditions x_0', y_0', $\dot{x}_0'$, $\dot{y}_0'$. If this general integral shows that, whatever is taken for x_0', y_0', $\dot{x}_0'$, $\dot{y}_0'$, the functions $x'(t)$ and $y'(t)$ tend toward zero, when t increases indefinitely, the motion (23) will properly be called a *stable state of motion*. In fact, this motion then has the property that if the initial values X_0, Y_0, $\dot{X}_0$, $\dot{Y}_0$ are slightly modified into $X_0 + x_0'$, $Y_0 + y_0'$, $\dot{X}_0 + \dot{x}_0'$, $\dot{Y}_0 + \dot{y}_0'$, the influence of this change vanishes more and more as time goes on. If for any definite set of initial values x_0', y_0', ... either x' or y' or both are not going toward zero, the state of motion given by $X(t)$ and $Y(t)$ will be *unstable* with respect to this particular type of disturbance x_0', y_0', ... Note that, since Eqs. (26) are homogeneous, multiplying the initial values x_0', y_0', ... by a common constant factor means multiplying the solution x', y' by the same factor.

This method is particularly workable if the force derivatives $\partial F_x/\partial x$, etc., become independent of time when X and Y from (23) are introduced for x and y (and $\dot{X}$, $\dot{Y}$ for $\dot{x}$, $\dot{y}$). Then Eqs. (26) become linear homogeneous equations *with constant coefficients*, and such a system can easily be integrated.   Let us write (26) in this case as

$$\ddot{x}' = a_1 x' + a_2 y' + a_3 \dot{x}' + a_4 \dot{y}'$$
$$\ddot{y}' = b_1 x' + b_2 y' + b_3 \dot{x}' + b_4 \dot{y}' \tag{27}$$

Here ma_1 obviously equals the value of $\partial F_x / \partial x$ when X, Y are substituted for x, y, etc.   To integrate the system (27) one makes the assumption

$$x' = A e^{\lambda t}, \quad y' = B e^{\lambda t} \tag{28}$$

with A, B, and λ constant.   Then, since $\dot{x}' = \lambda A e^{\lambda t}$, $\ddot{x}' = \lambda^2 A e^{\lambda t}$, . . . introducing (28) in (27) supplies

$$\lambda^2 A = a_1 A + a_2 B + \lambda a_3 A + \lambda a_4 B$$
$$\lambda^2 B = b_1 A + b_2 B + \lambda b_3 A + \lambda b_4 B \tag{29}$$

These are two linear homogeneous equations for A and B.   One can solve each of them for the quotient A/B.

$$\frac{A}{B} = \frac{a_2 + \lambda a_4}{\lambda^2 - a_1 - \lambda a_3} = \frac{\lambda^2 - b_2 - \lambda b_4}{b_1 + \lambda b_3} \tag{30}$$

It is obvious that this is consistent only if the two quotients to the right are equal, *i.e.*, if λ satisfies the condition

$$(\lambda^2 - a_1 - \lambda a_3)(\lambda^2 - b_2 - \lambda b_4) = (a_2 + \lambda a_4)(b_1 + \lambda b_2) \tag{31}$$

Upon carrying out the multiplications and rearranging, (31) takes the form

$$\lambda^4 + a\lambda^3 + b\lambda^2 + c\lambda + d = 0 \tag{31'}$$

where the a, b, c, d are certain expressions in a_1, b_1, a_2, b_2, . . . , a_4, b_4. In a more general case of more than two unknowns one would have more factors A, B, C . . . and the corresponding number of linear homogeneous equations, like (29), for them.   Then the condition of their being compatible will be that the determinant of their coefficients vanishes.   It is seen that (31) is nothing else than the determinant (of second order) formed by the coefficients of (29), equaled to zero.   In any case, one obtains an algebraic equation for λ and definite expressions for the ratios $A:B:C$ . . . in terms of λ.

Let us now assume that there exists a real number λ_1 which fulfills Eq. (31) or (31').   Then, introducing this value in (30) we can compute the ratio $A:B$, using either of the two quotients to the right.   Let A_1, B_1 be two values that have this particular ratio.   Then,

$$x' = A_1 e^{\lambda_1 t}, \qquad y' = B_1 e^{\lambda_1 t} \tag{32}$$

is a solution of the differential equations (27).   It does not matter that

A_1 and B_1 are determined except for a common factor only, since Eqs. (27) are homogeneous and thus any multiple of a solution is a solution, too.

The simplest case presents itself if the algebraic equation (31) or (31') has four different real roots, λ_1, λ_2, λ_3, λ_4. Then for each of these λ-values one obtains from (30) a ratio $A:B$ and thus four solutions of the form (32), for example, $A_2 e^{\lambda_2 t}$ and $B_2 e^{\lambda_2 t}$ for the root λ_2, etc. The fact that the set of differential equations (27) is linear and homogeneous leads to the conclusion that any linear combination of solutions is a solution, also. Thus we can choose four arbitrary constants K_1, K_2, K_3, K_4 and write the general integral in the form

$$x' = K_1 A_1 e^{\lambda_1 t} + K_2 A_2 e^{\lambda_2 t} + K_3 A_3 e^{\lambda_3 t} + K_4 A_4 e^{\lambda_4 t}$$
$$y' = K_1 B_1 e^{\lambda_1 t} + K_2 B_2 e^{\lambda_2 t} + K_3 B_3 e^{\lambda_3 t} + K_4 B_4 e^{\lambda_4 t} \tag{33}$$

Here the eight factors A_1, B_1, A_2, . . . , B_4 are determined by (30), while the four integration constants K_1, K_2, K_3, K_4 depend on the initial conditions, *i.e.*, on the initial values of x', y', $\dot{x}'$, $\dot{y}'$ or on the nature of the disturbance applied to the original motion as given by (23).

But, without being concerned with the 12 constants, one can immediately decide whether or not the state of motion under consideration is stable. If all numbers λ_1, λ_2, λ_3, λ_4 are negative, each of the terms in (33), and therefore the two sums, will approach zero when t increases indefinitely, whatever the values of the 12 constants. Thus the result is ascertained that *the state of motion described by (23) is stable if the algebraic equation (31) has four negative real roots*. If one of the roots, say, λ_1, is positive or zero, it still may happen that x' and y' according to (33) tend toward zero, *viz.*, if the initial conditions are such as to supply $K_1 = 0$. This means that an unstable motion can behave like a stable one with respect to some particular types of initial disturbance.

Two more cases have to be studied, (1) if the roots of (31) are not all real and (2) if they are not all different from each other. If there exists one complex root $\lambda_1 = \lambda' + \lambda''i$, there must necessarily be the conjugate root $\lambda_2 = \lambda' - \lambda''i$, also. If the constants A_1, B_1 and A_2, B_2 are computed from (30), as already explained above (except for insignificant factors), by setting once $\lambda = \lambda_1$ and once $\lambda = \lambda_2$, one will find conjugate complex values for them.

$$\begin{matrix} A_1 = A' + A''i, \\ B_1 = B' + B''i \end{matrix} \quad \text{and} \quad \begin{matrix} A_2 = A' - A''i, \\ B_2 = B' - B''i \end{matrix} \tag{34}$$

The corresponding solutions of the differential equations (27) are

$$x' = (A' + A''i)e^{(\lambda' + \lambda''i)t}, \qquad y' = (B' + B''i)e^{(\lambda' + \lambda''i)t}$$

and
$$x' = (A' - A''i)e^{(\lambda' - \lambda''i)t}, \qquad y' = (B' - B''i)e^{(\lambda' - \lambda''i)t} \tag{35}$$

Now, the sum and the difference of two possible solutions are solutions, too. Using the well-known formulas

$$\cos \alpha = \frac{1}{2}(e^{\alpha i} + e^{-\alpha i}), \qquad \sin \alpha = -\frac{i}{2}(e^{\alpha i} - e^{-\alpha i}) \qquad (36)$$

the sum of the expressions (35) is

$$\begin{aligned} x' &= 2e^{\lambda' t}(A' \cos \lambda'' t - A'' \sin \lambda'' t), \\ y' &= 2e^{\lambda' t}(B' \cos \lambda'' t - B'' \sin \lambda'' t) \end{aligned} \qquad (37)$$

and the difference multiplied by i

$$\begin{aligned} x' &= 2e^{\lambda' t}(-A' \sin \lambda'' t - A'' \cos \lambda'' t), \\ y' &= 2e^{\lambda' t}(-B' \sin \lambda'' t - B'' \cos \lambda'' t) \end{aligned} \qquad (37')$$

If $|A|$ is the absolute value and α the argument of the complex number A and $|B|$, β the same for B, one can also replace these expressions, omitting the factor 2 in the first, and -2 in the second case, by

$$\begin{aligned} x' &= |A|e^{\lambda' t} \cos (\lambda'' t + \alpha) & y' &= |B|e^{\lambda' t} \cos (\lambda'' t + \beta) \\ x' &= |A|e^{\lambda' t} \sin (\lambda'' t + \alpha) & y' &= |B|e^{\lambda' t} \sin (\lambda'' t + \beta) \end{aligned} \qquad (37'')$$

Thus two different, real integrals of the differential equations (27) are found, and they can be used with arbitrary integration constants K_1, K_2 in the place of the first two terms of the general solution (33). As the sine and cosine, whatever the arguments are, do not exceed the range ± 1, it is seen that *in the case of complex roots the motion will be stable, if and only if, the real parts λ' of such roots have the negative sign.* The difference between this case and the preceding one is only that there each single term represented a monotonously (aperiodically) vanishing disturbance, while here each pair of conjugate roots gives two damped oscillations with the frequency λ'' and the damping factor λ' and different phases. This will be seen more clearly in the examples worked out in Chap. XX.

If two or more roots of (31) coincide, formula (33) will not represent, in general, the complete solution of the integration problem since it is of no avail to use two coefficients, K_1 and K_2, for the same terms $A_1 e^{\lambda_1 t} = A_2 e^{\lambda_2 t}$. In this case one can conclude as follows: Assume that $\lambda_1 = \lambda_2$ equals a definite negative number. Then it is possible to modify very slightly the coefficients $a_1, b_1, \ldots, a_4, b_4$ of (31) in such a way that the modified equation has two different real roots, very close to the original one. The solution of the modified problem can be written in the form (33), which supplies; for definite initial conditions and a definite large value of t, values of x' and y' that are practically zero. Now, the integral of a set of linear differential equations is known to vary continuously with its coefficients. Thus, going back from the modified to the original problem one must find x'- and y'-values that are still practically zero. This does not apply to infinite t, but the real use that can be made of a mechanical theory is restricted to finite values of time. Besides, it can be proved by certain analytical methods that even for $t = \infty$ a double real root behaves like a single one, leading to vanishing disturbance when it is negative. The same holds true for complex roots with negative real parts. *The necessary and sufficient condition for stability is that all roots of* (31) *are either negative real numbers or complex numbers with negative real parts.*

Routh's theory as developed here, supplies more than a mere stability criterion. It leads to a full description of the motion that follows a given initial disturbance. If one wants to know only whether a state of motion is stable or not, the procedure can be simplified. It is not necessary, as the following argument will show, to compute the roots by solving Eq. (31).

Let us denote by $f(\lambda)$ the polynomial on the left-hand side of (31'). It is known that $f(\lambda)$ must be identically equal to the product of the so-called "root factors" $\lambda - \lambda_1$, $\lambda - \lambda_2$, etc. If all roots are real, we have

$$f(\lambda) = (\lambda - \lambda_1)(\lambda - \lambda_2)(\lambda - \lambda_3)(\lambda - \lambda_4) \tag{38}$$

If λ_1, λ_2 are conjugate complex numbers $\lambda' \pm \lambda''i$, this becomes

$$f(\lambda) = [\lambda^2 - 2\lambda'\lambda + (\lambda'^2 + \lambda''^2)](\lambda - \lambda_3)(\lambda - \lambda_4) \tag{39}$$

In a stable motion the λ_1, λ_2, λ_3, λ_4 in (38) or the λ', λ_3, λ_4 in (38') are negative. In both cases (and also in the case of two pairs of complex roots), when the numerical values are introduced $f(\lambda)$ appears as a product of only positive terms. Thus *the four coefficients a, b, c, d of the polynomial* (31') *must all be positive in the stable case.*

Now, assume that the coefficients a and c have been given definite positive values, while b and d are allowed to vary within the first quadrant of a b-d-plane (Fig. 394). There will be certain points S in the plane that correspond to stable motions and other points U corresponding to instability.

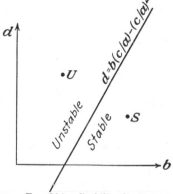

Fig. 394.—Stability limit.

Between the two categories of points there must exist some borderline. As d is supposed to be positive, no root can be zero. Therefore, the transition from stable to unstable roots can occur only when the real part of a complex root passes from a negative to a positive value. At the border line the real part must be zero, *i.e.*, there must exist a pure imaginary root. For an imaginary λ the terms $c\lambda + a\lambda^3$

in the equation are imaginary and the terms $\lambda^4 + b\lambda^2 + d$ are real. Consequently, each of these sums must vanish separately if $f(\lambda)$ should vanish:

$$c\lambda + a\lambda^3 = 0, \qquad \lambda^4 + b\lambda^2 + d = 0 \qquad (\lambda \neq 0) \qquad (40)$$

Computing $\lambda^2 = -c/a$ from the first and substituting in the second, we find that

$$\frac{c^2}{a^2} - \frac{cb}{a} + d = 0 \qquad (41)$$

as the condition for the existence of an imaginary root. This represents, for given a and c, a straight line in the b-d-plane as shown in Fig. 394. On one side of this line lie the points for stable motions; on the other side, the points corresponding to unstable motions. The two sides are also characterized by the fact that for one of them the expression (41) is positive and for the other negative. To decide which is which, let us take the case of all roots equaling -1. Then, according to (38), the polynomial is $(\lambda + 1)^4 = \lambda^4 + 4\lambda^3 + 6\lambda^2 + 4\lambda + 1$. In this case, $a = c = 4$, $b = 6$, $d = 1$, the left-hand side of (41) is $1 - 6 + 1 = -4$. Thus all stable points must lie on the side where

$$\frac{c^2}{a^2} - \frac{cb}{a} + d < 0 \qquad \text{or} \qquad d < \frac{c}{a}\left(b - \frac{c}{a}\right) \qquad (42)$$

The result can be summarized in the following statement: *The state of motion given by* (23) *is stable if, and only if, the four coefficients a, b, c, d, which depend on the force derivatives a_1, b_1, . . . , a_4, b_4, are positive and fulfill the inequality* (42).

In case the algebraic equation (31) is of higher than fourth order, this analysis becomes more complicated. It will be seen in Chap. XX that the stability of an airplane in level flight depends on problems of fourth order only.

Problem 4. Prove, starting from the general stability theory, that the uniform, straight motion of a body that is subjected to no force is an unstable state of motion.

Problem 5. A mass point under the influence of gravity is in unstable position on the top of a curved surface under all conditions and in stable position at the bottom if any kind of friction is assumed. These statements should be proved on the basis of the general stability theory.

Problem 6. A mass point P which is movable in the x-y-plane and of which the equilibrium position coincides with the origin O is linked to three fixed points A, B, C in the plane by elastic springs PA, PB, PC. Prove that this equilibrium is stable if any amount of damping exists.

4. Application to the Airplane. The first to apply Routh's theory of dynamic stability to airplane motion were G. H. Bryan and W. E. Williams (1904). They studied mainly the stability of gliding flight. In the present section the state of motion whose stability is discussed will be the uniform level flight. The investigation starts from the general equations of motion (13) and follows closely the lines of the stability theory as given in Sec. 3 of this chapter.

The variables that were x and y in the example of Sec. 3 are now the nine quantities V_1, V_2, V_3, ω_1, ω_2, ω_3, W_1, W_2, W_3. In a state of uniform level flight at the velocity V_l the 1-axis may form the constant angle ψ with the velocity direction (Fig. 395). It is then immediately seen that the nine variables in this state of motion have the following values:

$$V_1 = V_l \cos \psi, \qquad V_2 = 0, \qquad V_3 = V_l \sin \psi; \qquad \omega_1 = \omega_2 = \omega_3 = 0$$
$$W_1 = -W \sin \psi, \qquad W_2 = 0, \qquad W_3 = W \cos \psi \qquad (43)$$

Equations (43) can take the place of (23) since they describe a possible motion, *i.e.*, a particular solution of the equations of motion. In fact, it is known that in the uniform level flight of a completely symmetric plane the air reactions can be reduced to three forces, the lift L, the drag D, and the propeller thrust T, all acting in the symmetry plane. Therefore, F_2 (cross force), M_1 and M_3 (rolling and yawing moments) vanish. The left-hand sides of all equations (13) vanish if (43) is substituted for the variables. On the right-hand side only the first, third, and fifth equations include terms that are not identically zero. These equations now read:

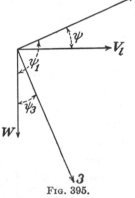

Fig. 395.

$$0 = -W \sin \psi + F_1, \qquad 0 = W \cos \psi + F_3, \qquad 0 = M_2 \qquad (44)$$

The first two equations, in a slightly modified form, are the two conditions that underlie the performance computation and that express the fact that the vector sum of weight, lift, drag, and propeller thrust must be zero (Sec. XIV.1). The last equation is the condition of pitching moment equilibrium, discussed in Chap. XVII.

The next step is to introduce the small additions to the values (43) corresponding to what was called x' and y' in (24). The following notation is self-explanatory in connection with (43):

$$V_1 = V_l \cos \psi + v_1, \qquad V_2 = v_2, \qquad V_3 = V_l \sin \psi + v_3;$$
$$\omega_1 = \omega_1, \qquad \omega_2 = \omega_2, \qquad \omega_3 = \omega_3 \qquad (45)$$
$$W_1 = -W \sin \psi + w_1, \qquad W_2 = w_2, \qquad W_3 = W \cos \psi + w_3$$

The new variables that have to be considered as small quantities are v_1, v_2, v_3; ω_1, ω_2, ω_3; w_1, w_2, w_3. If the expressions (45) are substituted in (13) and all terms of higher than first order in the small quantities neglected, these equations take the form

$$\frac{W}{g}(\dot{v}_1 + \omega_2 V_l \sin \psi) = -W \sin \psi + F_1 + w_1$$

$$\frac{W}{g}[\dot{v}_2 + (\omega_3 \cos \psi - \omega_1 \sin \psi)V_l] = F_2 + w_2 \qquad (46)$$

$$\frac{W}{g}(\dot{v}_3 - \omega_2 V_l \cos \psi) = W \cos \psi + F_3 + w_3$$

$$J_1\dot{\omega}_1 - J_{31}\dot{\omega}_3 = M_1, \qquad J_2\dot{\omega}_2 = M_2, \qquad J_3\dot{\omega}_3 - J_{31}\dot{\omega}_1 = M_3$$

$$w_1 = -W\omega_2 \cos \psi, \qquad w_2 = W(\omega_1 \cos \psi + \omega_3 \sin \psi), \qquad w_3 = -W\omega_2 \sin \psi$$

Here, the force and moment components F_1, F_2, F_3; M_1, M_2, M_3 must be replaced by their developments, according to Eqs. (25) and (26) in the general theory. Each component depends, in general, on the six velocities V_1, V_2, V_3, ω_1, ω_2, ω_3. As the level flight value of F_1 is $W \sin \psi$ [first equation (44)], we can write

$$F_1 = W \sin \psi + \frac{\partial F_1}{\partial V_1} v_1 + \frac{\partial F_1}{\partial V_2} v_2 + \frac{\partial F_1}{\partial V_3} v_3 + \frac{\partial F_1}{\partial \omega_1} \omega_1 + \frac{\partial F_1}{\partial \omega_2} \omega_2 + \frac{\partial F_1}{\partial \omega_3} \omega_3 \quad (47)$$

In the derivatives the level-flight values (43) have to be introduced for the variables, so that they become constants. If we call these constants A_1, A_2, . . . , A_6, Eq. (47) will read

$$-W \sin \psi + F_1 = A_1v_1 + A_2v_2 + A_3v_3 + A_4\omega_1 + A_5\omega_2 + A_6\omega_3 \quad (47')$$

and analogous expressions hold for the other components. The derivatives may be called B and C for F_2 and F_3, respectively, and D, E, G for the three moments M_1, M_2, M_3. For example,

$$M_3 = G_1v_1 + G_2v_2 + G_3v_3 + G_4\omega_1 + G_5\omega_2 + G_6\omega_3 \qquad (47'')$$

since the level-flight value of M_3 is zero. Only F_1 and F_3 include absolute terms.

However, all the formulas (47), (47'), etc., simplify considerably if the symmetry of the plane is taken into account. It was seen in Sec. 2 of this chapter that the lateral force components F_2, M_1, M_3 vanish identically if the lateral velocities V_2, ω_1, ω_3 are zero. Therefore, the nine coefficients B_1, B_3, B_5, D_1, D_3, D_5, G_1, G_3, G_5 must be omitted. The second theorem in Sec. 2 stated that the longitudinal components F_1, F_3, M_2 remain unchanged when, starting from any state of motion, the V_2, ω_1, ω_3 are simultaneously reversed. Now, one may start from a motion with $\omega_1 = \omega_3 = 0$ and reverse $V_2 = v_2$. Then the coefficients of v_2 in the expressions for F_1, F_3, M_2 must vanish. These are A_2, C_2, E_2, and in exactly the same way it can be seen that A_4, C_4, E_4 and A_6, C_6, E_6 disappear, also. Thus the original number 36 of force derivatives reduces to 18, because of the symmetry. In A, C, E only the odd subscripts remain, and in B, D, G only the even subscripts.

More important than the reduction in the number of coefficients is it to learn that now the first, third, and fifth equations (46) include only the longitudinal variables v_1, v_3, ω_2 while in the second, fourth, and sixth only the lateral variables v_2, ω_1, ω_3 appear.   Thus the problem indicated by the nine equations (46) splits into two independent problems.   The first concerning the so-called "longitudinal stability" is stated in the following five equations:

$$\frac{W}{g}(\dot{v}_1 + \omega_2 V_l \sin \psi) - A_1 v_1 + A_3 v_3 + A_5 \omega_2 + w_1$$

$$\frac{W}{g}(\dot{v}_3 - \omega_2 V_l \cos \psi) = C_1 v_1 + C_3 v_3 + C'_5 \omega_2 + w_3 \qquad (48)$$

$$J_2 \dot{\omega}_2 = E_1 v_1 + E_3 v_3 + E_5 \omega_2$$

$$\dot{w}_1 = -W \omega_2 \cos \psi, \qquad \dot{w}_3 = -W \omega_2 \sin \psi$$

These are five (linear and homogeneous) differential equations, each of first order, for the five variables v_1, v_3, ω_2, w_1, and w_3.   But the last two equations supply the relation $\dot{w}_1 \sin \psi - \dot{w}_3 \cos \psi = 0$, which gives, integrated, $w_1 \sin \psi - w_3 \cos \psi = $ const.   Therefore, one of the variables w_1, w_3 can be eliminated, and the remaining integration problem is thus seen to be *of fourth order.*

The second set of equations, which concerns the so-called "lateral stability," consists of the following:

$$\frac{W}{g}[\dot{v}_2 + (\omega_3 \cos \psi - \omega_1 \sin \psi)V_l] = B_2 v_2 + B_4 \omega_1 + B_6 \omega_3 + w_2$$

$$J_1 \dot{\omega}_1 - J_{31} \dot{\omega}_3 = D_2 v_2 + D_4 \omega_1 + D_6 \omega_3 \qquad (49)$$

$$J_3 \dot{\omega}_3 - J_{31} \dot{\omega}_1 = G_2 v_2 + G_4 \omega_1 + G_6 \omega_3$$

$$\dot{w}_2 = W(\omega_1 \cos \psi + \omega_3 \sin \psi)$$

Here it is immediately seen that the integration problem is of the order 4, since we have four equations each of first order, for the four unknowns v_2, ω_1, ω_3, w_2.   The main theorem, due to Bryan and Williams, can be stated as follows: *In the case of a completely symmetric airplane the stability problem for the straight level flight divides into two independent problems, one concerning the longitudinal and one concerning the lateral variables.   Each of them is an integration problem of fourth order.*

It can easily be seen that the same is true if another state of steady straight flight, climbing or diving, is studied.   The independence of lateral and longitudinal variables, however, does not subsist in the cases of curved steady flight, *e.g.*, a horizontal turn or spiral motion.   It was seen in Sec. XVIII.4 how in the case of a horizontal turn, even in the steady flight conditions, both types of variables interfere.   Some remarks about the helicoidal motion will be made in Sec. XX.5.

The two sets (48) and (49) can be brought into a *dimensionless form,* according to the method used in Sec. 1 of this chapter.   Let us again take V_l/g as the time unit, setting

$$\tau = \frac{gt}{V_l}; \qquad u_1 = \frac{v_1}{V_l}, \qquad u_2 = \frac{v_2}{V_l}, \qquad u_3 = \frac{v_3}{V_l};$$

$$\sigma_1 = \frac{V_l\omega_1}{g}, \qquad \sigma_2 = \frac{V_l\omega_2}{g}, \qquad \sigma_3 = \frac{V_l\omega_3}{g} \tag{50}$$

Instead of the quantities w_1, w_2, w_3 the angles ψ_1, ψ_2, ψ_3 formed by the three coordinate axes with the vertical may be introduced. In the undisturbed level flight we have $\psi_1 = 90° + \psi$, $\psi_2 = 90°$, $\psi_3 = \psi$. Therefore we write

$$\psi_1 = 90° + \psi + \epsilon_1, \qquad \psi_2 = 90° + \epsilon_2, \qquad \psi_3 = \psi + \epsilon_3 \tag{51}$$

From the last three equations (45) then follows, since ϵ_1, ϵ_2, ϵ_3 are small,

$$\frac{w_1}{W} = \cos \psi_1 + \sin \psi = -\epsilon_1 \cos \psi, \qquad \frac{w_2}{W} = \cos \psi_2 = -\epsilon_2,$$

$$\frac{w_3}{W} = \cos \psi_3 - \cos \psi = -\epsilon_3 \sin \psi \tag{51'}$$

Thus the last three equations (46) take the form

$$\dot{\epsilon}_1 = \omega_2, \qquad -\dot{\epsilon}_2 = \omega_1 \cos \psi + \omega_3 \sin \psi, \qquad \dot{\epsilon}_3 = \omega_2 \tag{51''}$$

the first and third appearing in (48), the second in (49).

The set (48) of differential equations for the longitudinal disturbance can now be written as

$$\dot{u}_1 + \sigma_2 \sin \psi = a_1u_1 + a_3u_3 + a_5\sigma_2 - \epsilon_1 \cos \psi$$
$$\dot{u}_3 - \sigma_2 \cos \psi = c_1u_1 + c_3u_3 + c_5\sigma_2 - \epsilon_3 \sin \psi$$
$$\dot{\sigma}_2 = e_1u_1 + e_3u_3 + e_5\sigma_2 \tag{52}$$
$$\dot{\epsilon}_1 = \sigma_2, \quad \dot{\epsilon}_3 = \sigma_2$$

Here the dots mean differentiation with respect to τ (not t). The factors a_1, a_3, . . . , e_5 are multiples of the A_1, A_3, . . . , E_5. They can be expressed in terms of derivatives of force and moment coefficients that refer to the level-flight velocity V_l and to the actual air density $\rho = \rho_l$. Using the wing area S and an appropriately chosen reference length l (chord, tail distance, span, etc.), we define the coefficients.

$$K_1 = \frac{2F_1}{\rho V_l^2 S}, \qquad K_2 = \frac{2F_2}{\rho V_l^2 S}, \qquad K_3 = \frac{2F_3}{\rho V_l^2 S};$$

$$K_4 = \frac{2M_1}{\rho V_l^2 Sl}, \qquad K_5 = \frac{2M_2}{\rho V_l^2 Sl}, \qquad K_6 = \frac{2M_3}{\rho V_l^2 Sl} \tag{53}$$

Besides, we have to use the lift coefficient of level flight C_{Ll} given by

$$W = C_{Ll} \frac{\rho}{2} V_l^2 S \tag{54}$$

In the transition from (48) to (52) the first two equations had to be divided by W and the third by Wj_2^2 and gl^2/V_l^2. The K_1, K_2, ..., K_6 depend on the velocity ratios V_1/V_l ... $V_l\omega_1/g$ ... and can be considered as functions of the dimensionless variables u_1, u_2 ... σ_1, σ_2 .... Thus the coefficients a_1, a_3, ..., e_5 are

$$a_1 = \frac{1}{C_{Ll}}\frac{\partial K_1}{\partial u_1} \qquad a_3 = \frac{1}{C_{Ll}}\frac{\partial K_1}{\partial u_3} \qquad a_5 = \frac{1}{C_{Ll}}\frac{\partial K_1}{\partial \sigma_2}$$

$$c_1 = \frac{1}{C_{Ll}}\frac{\partial K_3}{\partial u_1} \qquad c_3 = \frac{1}{C_{Ll}}\frac{\partial K_3}{\partial u_3} \qquad c_5 = \frac{1}{C_{Ll}}\frac{\partial K_3}{\partial \sigma_2} \qquad (55)$$

$$e_1 = \frac{1}{C_{Ll}j_2^2}\frac{V_l^2}{gl}\frac{\partial K_5}{\partial u_1} \qquad e_3 = \frac{1}{C_{Ll}j_2^2}\frac{V_l^2}{gl}\frac{\partial K_5}{\partial u_3} \qquad e_5 = \frac{1}{C_{Ll}j_2^2}\frac{V_l^2}{gl}\frac{\partial K_5}{\partial \sigma_2}$$

It will be seen later how these nine derivatives, which determine the longitudinal stability of level flight, can be computed from the airplane data.

An analogous transformation applies to the set (49). We omit the terms depending on J_{31}, that is, we assume that the 1- and 3-axes are the principal axes of the body or close to them. Then, upon using (51) and (51'), Eqs. (49) take the form

$$\dot{u}_2 - \sigma_1 \sin\psi + \sigma_3 \cos\psi = b_2u_2 + b_4\sigma_1 + b_6\sigma_3 - \epsilon_2$$

$$\dot{\sigma}_1 = d_2u_2 + d_4\sigma_1 + d_6\sigma_3 \qquad (56)$$

$$\dot{\sigma}_3 = g_2u_2 + g_4\sigma_1 + g_6\sigma_3$$

$$-\dot{\epsilon}_2 = \sigma_1 \cos\psi + \sigma_3 \sin\psi$$

Here dots refer to differentiation with respect to τ and the coefficients b_2, b_4, ..., g_6 correspond to those given in (55). They may be written in the following way:

$$b_2 = \frac{1}{C_{Ll}}\frac{\partial K_2}{\partial u_2} \qquad b_4 = \frac{1}{C_{Ll}}\frac{\partial K_2}{\partial \sigma_1} \qquad b_6 = \frac{1}{C_{Ll}}\frac{\partial K_2}{\partial \sigma_3}$$

$$d_2 = \frac{1}{C_{Ll}j_1^2}\frac{V_l^2}{gl}\frac{\partial K_4}{\partial u_2} \qquad d_4 = \frac{1}{C_{Ll}j_1^2}\frac{V_l^2}{gl}\frac{\partial K_4}{\partial \sigma_1} \qquad d_6 = \frac{1}{C_{Ll}j_1^2}\frac{V_l^2}{gl}\frac{\partial K_4}{\partial \sigma_3} \qquad (57)$$

$$g_2 = \frac{1}{C_{Ll}j_3^2}\frac{V_l^2}{gl}\frac{\partial K_6}{\partial u_2} \qquad g_4 = \frac{1}{C_{Ll}j_3^2}\frac{V_l^2}{gl}\frac{\partial K_6}{\partial \sigma_1} \qquad g_6 = \frac{1}{C_{Ll}j_3^2}\frac{V_l^2}{gl}\frac{\partial K_6}{\partial \sigma_3}$$

These nine coefficients determine the lateral stability of level flight and will be discussed in Sec. XX.3.

Problem 7. Derive the equations for the longitudinal disturbance of the steady level flight by applying the theory developed in Sec. 3 of this chapter to Eqs. (14) to (16), Chap. XVIII, and show that the set of equations obtained in this way is equivalent to (52).

*Problem 8. Develop, in analogy to (52), the disturbance equations for the case of a bomb falling vertically at its terminal velocity (weight = drag). Assume the bomb to be a body of revolution with its axis vertical in the undisturbed motion.

*Problem 9. What modifications have to take place in Eqs. (52) and (56) if a state of steady climbing or gliding at a small angle is considered, instead of the straight level flight?

CHAPTER XX

DYNAMIC STABILITY OF AN AIRPLANE

1. Longitudinal Stability of Level Flight. The equations that describe the motion of an airplane following a longitudinal disturbance in level flight have been given in Sec. XIX.4, first in their original form (48) and then using dimensionless variables in (52). In both cases ψ was the angle between the 1-axis and that direction in the airplane body which coincides with the velocity vector in the state of level flight under consideration. An axis of the latter direction passing through the c.g. may be called the *longitudinal axis of the airplane.* (It depends slightly on which state of level flight is studied, whether full power, normal load, etc.) Now, in the equations the inertia product J_{31} does not appear. It therefore makes no difference whether or not the 1-axis is a principal axis of the system. We are completely free to assume $\psi = 0$, that is, to make *the 1-axis coincide with the longitudinal axis of the airplane.*

With $\psi = 0$ the angles ψ_1 and $\psi_1 - 90°$ are the angles that, after the disturbance occurred, the 1-axis makes with the vertical direction and with the horizontal plane, respectively. The latter angle was called φ in all discussions of Chap. XVIII. From Eq. (51), Chap. XIX, it follows that $\epsilon_1 = \psi_1 - 90° - \psi = \psi_1 - 90° = \varphi$. We thus can replace ϵ_1 by φ in (52); and since the last but one equation of the set gives

$$\dot{\epsilon}_1 = \dot{\varphi} = \sigma_2,$$

the set (52) can be written, in the case $\psi = 0$ (the dots denoting the differentiation with respect to τ),

$$\begin{aligned}
\dot{u}_1 &= a_1 u_1 + a_3 u_3 + a_5 \dot{\varphi} - \varphi \\
\dot{u}_3 &= c_1 u_1 + c_3 u_3 + c_5 \dot{\varphi} + \dot{\varphi} \\
\ddot{\varphi} &= e_1 u_1 + e_3 u_3 + e_5 \dot{\varphi}
\end{aligned} \tag{1}$$

The angle between the instantaneous velocity vector and the horizontal direction of the undisturbed flight was called ϑ in Chap. XVIII. It is seen from Fig. 396 that $u_3 = v_3/V_l$ equals the angle $\varphi - \vartheta$ between the 1-axis and the instantaneous velocity vector, *i.e.*, the angle of attack referring to the longitudinal axis. If α_l is the angle of incidence (angle between velocity and zero lift direction of the plane) in level flight and α the actual incidence, we have

$$\alpha - \alpha_l = \varphi - \vartheta = u_3 \tag{2}$$

as in Chap. XVIII.   Thus differentiation with respect to u_3 is the same as with respect to α.   For the resultant velocity V one has

$$V^2 = (V_l + v_1)^2 + v_3^2 \sim V_l^2 + 2V_l v_1, \qquad \frac{V^2}{V_l^2} = 1 + 2\frac{v_1}{V_l} \qquad (3)$$

Therefore, the derivative of any function with respect to $u_1 = v_1/V_l$ is the derivative with respect to V^2/V_l^2 multiplied by 2.

The following assumptions about the forces and moments in longitudinal motion are usually made: (1) The resultant force (components F_1 and F_3) does not depend on the rotational velocity $\sigma_2 = \dot{\varphi}$; it equals

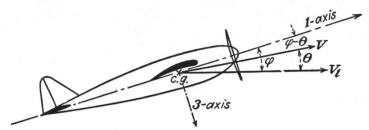

Fig. 396.—Disturbance of level-flight attitude.

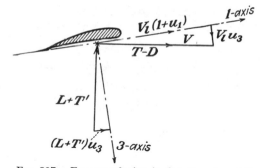

Fig. 397.—Force resolution in directions 1 and 3.

the product of V^2 and a function of α.   (2) The pitching moment M_2 consists of one term that is the product of V^2 and a function of α and another term (damping) that is proportional to σ_2.

The forces are usually given by their components parallel and normal to the velocity.   We call L and D lift and drag and denote by T and T' the propeller force in the direction of flight and normal to it.   Then (Fig. 397) the components in the 1- and 3-directions are

$$F_1 = T - D + (L + T')u_3 \sim T - D + L_l u_3$$
$$F_3 = -L - T' + (T - D)u_3 \sim -L - T' \qquad (4)$$

if terms of higher order in $u_3 = \varphi - \vartheta$ are neglected.   In (4) the level-flight values of $T - D$ and of T' are assumed to be zero, while the level-flight value of L is $L_l = W$. If C_L and C_D are the usual lift and (total) drag coefficients and C_T, $C_{T'}$ analogous coefficients (referring to wing area) of the two components of the propeller force, the coefficients K_1 and K_3 as defined in Eq. (53), Chap. XIX, corresponding to the first of the foregoing assumptions are

$$K_1 = \frac{V^2}{V_l^2}(C_T - C_D) + C_{Ll}u_3, \qquad K_3 = -\frac{V^2}{V_l^2}(C_L + C_{T'}) \qquad (5)$$

The derivatives of both expressions with respect to $\dot{\varphi}$ are zero. The derivatives with respect to u_1 and u_3 are, for level flight conditions,

$$\frac{\partial K_1}{\partial u_1} = 2(C_T - C_D) = 0, \qquad \frac{\partial K_1}{\partial u_3} = \frac{\partial(C_T - C_D)}{\partial \alpha} + C_{Ll}$$

$$\frac{\partial K_3}{\partial u_1} = -2(C_L + C_{T'}) = -2C_{Ll}, \qquad \frac{\partial K_3}{\partial u_3} = -\frac{\partial(C_L + C_{T'})}{\partial \alpha} \qquad (6)$$

The moment M_2 may be written, according to the second of the above assumptions, as

$$M_2 = \frac{\rho}{2}SV_l^2 l K_5 = \frac{\rho}{2}SV^2\left[cC_M(\alpha) - lk_M\frac{d\varphi}{dt}\right] \qquad (7)$$

Here C_M is the pitching moment coefficient as used in Chap. XVII, and k_M a damping factor. It follows, by differentiation, that

$$\frac{\partial K_5}{\partial u_1} = 2C_M(\alpha_l)\frac{c}{l} = 0, \qquad \frac{\partial K_5}{\partial u_3} = \frac{c}{l}\frac{\partial C_M}{\partial \alpha}, \qquad \frac{\partial K_5}{\partial \sigma_2} = -k_M\frac{g}{V_l} \qquad (8)$$

Thus the following values for the coefficients in (1) result from the definitions in Eq. (55), Chap. XIX:

$$a_1 = 0, \qquad a_3 = 1 + \frac{\partial(C_T - C_D)}{C_{Ll}\,\partial\alpha}, \qquad a_5 = 0$$

$$c_1 = -2, \qquad c_3 = -\frac{\partial(C_L + C_{T'})}{C_{Ll}\,\partial\alpha}, \qquad c_5 = 0 \qquad (9)$$

$$e_1 = 0, \qquad e_3 = \frac{1}{C_{Ll}j_2^2}\frac{V_l^2}{gl}\frac{c}{l}\frac{\partial C_M}{\partial\alpha}, \qquad e_5 = -\frac{1}{C_{Ll}j_2^2}\frac{V_l}{l}k_M$$

The principal coefficients are e_3 and e_5, the first proportional to the static stability, which was defined in Chap. XVII as $-dC_M/d\alpha$, the second proportional to the damping factor k_M. In a_3 the second term, in general, will be small as compared with 1, and so will be the term $C_{T'}$ in c_3 when compared with C_L. Thus a_3 is positive and c_3 negative, at least below stalling. The last coefficient e_5 is certainly negative. Upon omitting the terms that vanish and rearranging, Eqs. (1) become

$$\dot{u}_1 - a_3u_3 + \varphi = 0$$
$$2u_1 + \dot{u}_3 - c_3u_3 - \dot{\varphi} = 0 \qquad (10)$$
$$-e_3u_3 + \ddot{\varphi} - e_5\dot{\varphi} = 0$$

Here, according to the method developed in Sec. XIX.3, we introduce the solution in the form

$$u_1 = A e^{\lambda \tau}, \qquad u_3 = B e^{\lambda \tau}, \qquad \varphi = C e^{\lambda \tau} \tag{11}$$

Then the A, B, C must satisfy the equations

$$
\begin{aligned}
\lambda A - a_3 B + C &= 0 \\
2A + (\lambda - c_3)B - \lambda C &= 0 \\
-e_3 B + \lambda(\lambda - e_5)C &= 0 \; .
\end{aligned} \tag{12}
$$

The condition of compatibility of these three linear homogeneous equations for the three unknowns A, B, C can be found by eliminating A from the first two and then comparing the ratio B/C with that given by the last equation or simply by equaling the determinant to zero.   In either case the condition reads

$$\lambda^4 - \lambda^3(c_3 + e_5) + \lambda^2(c_3 e_5 + 2a_3 - e_3) - 2a_3 e_5 \lambda - 2e_3 = 0 \tag{13}$$

The first conditions of stability are that the coefficients of all powers of λ must be positive:

$$
\begin{aligned}
a = -c_3 - e_5 > 0, \qquad b &= c_3 e_5 + 2a_3 - e_3 > 0, \\
c = -2a_3 e_5 > 0, \qquad d &= -2e_3 > 0
\end{aligned} \tag{14}
$$

For a and c the inequalities are satisfied, as it has already been stated that $c_3 < 0$, $e_5 < 0$, $a_3 > 0$.   The last inequality asks for $e_3 < 0$; and if this is fulfilled, the second will be fulfilled too.   As $-e_3$ is proportional to what was called "static stability" in Chap. XVII, the following result is reached: *A necessary condition for dynamic longitudinal stability is that the airplane is statically stable with respect to a longitudinal disturbance.*

The further stability condition as developed in Eq. (42), Chap. XIX, can be written in the form

$$a^2 d < abc - c^2 \tag{15}$$

If the values for a, b, c, d are substituted from (14) and then the inequality solved for e_3, one finds that

$$-e_3 \left[1 + \frac{e_5}{c_3} (1 - a_3) \right] < a_3 e_5^2 + \frac{2a_3^2 e_5}{c_3 + e_5} \tag{16}$$

The expression to the right and the term in the bracket are positive, according to what has already been seen.   Thus (16) gives an upper limit for the coefficient $-e_3$ which, essentially, measures the static stability. In connection with the result previously reached we now can state: *An airplane is stable in level flight with respect to longitudinal disturbances if it is statically stable and if the static stability does not surpass a limit determined by the damping factor and the other force derivatives.*

One might have expected that the complete stability discussion as worked out in this chapter would lead to a lower limit of static stability rather than to an upper limit. It can be understood why this is not the case. In fact, a too strong restoring moment can produce an overcompensation of the initial disturbance if it is allowed to exert its influence unopposed. The presence of a sufficient damping moment is necessary to restrain the tendency toward overcompensation.

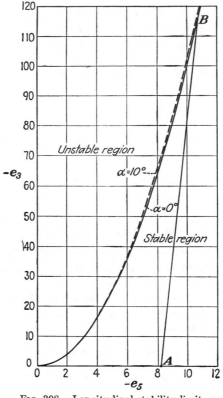

FIG. 398.—Longitudinal stability limit.

For a *numerical discussion* of the stability conditions we may omit two quantities of minor importance: the derivatives of $(C_T - C_D)$ and of $C_{T'}$ with respect to α. If the data are available, there is no difficulty in using the complete expressions for a_3 and c_3 in (9). Here it will be sufficient to take

$$a_3 = 1, \qquad c_3 = -\frac{1}{C_L}\frac{\partial C_L}{\partial \alpha} = -\frac{1}{k\alpha_l}\,k = -\frac{1}{\alpha_l} \qquad (17)$$

Under these assumptions, (16) simplifies considerably, and the complete stability conditions can be written in the form

$$0 < -e_3 < e_5^2 - \frac{2\alpha_l e_5}{1 - \alpha_l e_5} \tag{18}$$

Thus the whole problem is reduced to a *relation between the factor e_3 representing the static stability and the factor e_5 representing the damping, a relation that depends on the level-flight incidence α_l only.*

In Fig. 398 the curves according to (18), limiting the region of stable values e_3, e_5, are shown for incidences $\alpha_l = 0°$ and $10°$ in a coordinate system with the abscissas $-e_5$ and the ordinates $-e_3$. The curves differ very little from each other within the usual range of α_l.

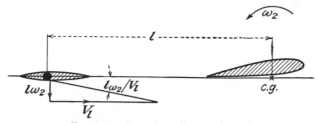

Fig. 399.—Damping effect of the tail.

The *static stability* was fully discussed in Chap. XVII. We may take the approximation given in Eq. (41), omitting also the less important propeller term, so as to have, with α_l replacing α_1,

$$-\frac{\partial C_M}{\partial \alpha} = \frac{lS'}{cS} k' - \left(-\frac{C_{M_0}}{\alpha_l} + \frac{dC_f}{d\alpha} \right) \tag{19}$$

Here, S' is the tail area, l the tail distance, which we also choose for reference length, k' the lift factor for the tail, C_{M_0} the moment coefficient for the a.c. of the wing, and $dC_f/d\alpha$ a positive constant referring to the fuselage moment (Sec. XVII.3). This substituted in the formula for e_3 in (9) gives, with $C_{Ll} = k\alpha_l$,

$$-e_3 = \frac{1}{j_z^2 k \alpha_l} \frac{V_l^2}{gl} \left[\frac{S'}{S} k' - \frac{c}{l} \left(-\frac{C_{M_0}}{\alpha_l} + \frac{dC_f}{d\alpha} \right) \right] \tag{20}$$

In order to obtain a rough conservative estimate for the *damping moment* e_5, one may assume that only the tail surface contributes to the damping of the pitch. It is seen in Fig. 399 how a positive pitch ω_2 adds a downward component $l\omega_2$ to the forward speed V_l at the points on the tail and thus increases the angle of incidence by $l\omega_2/V_l$. With the lift factor k' and $\omega_2 = d\varphi/dt$, the corresponding increase of lift will be

$$\frac{\rho}{2} S' V_l^2 k' \frac{l\omega_2}{V_l} = \frac{\rho}{2} S' V_l k' l \frac{d\varphi}{dt}$$

This force, multiplied by the distance l, gives the damping moment that was assumed in (7) in the form $(\rho/2)SV_l^2 k_M l d\varphi/dt$. Therefore, the damping coefficient k_M must be

$$k_M = \frac{S'}{S} \frac{l}{V_l} k'$$ (21)

and the factor e_5, according to the last formula (9),

$$e_5 = - \frac{1}{j_2^2 k \alpha_l} \frac{S'}{S} k'$$ (22)

It is, of course, a rough approximation only that the lift factor k' in (22) is identified with the factor occurring in (19). Both quantities should be determined by independent experiments.

Eliminating the terms referring to the tail from (20) and (22), we obtain

$$-e_3 = \frac{V_l^2}{gl} \left[-e_5 - \frac{c}{l} \frac{1}{j_2^2 k \alpha_l} \left(-\frac{C_{M_0}}{\alpha_l} + \frac{dC_f}{d\alpha} \right) \right]$$ (23)

This is a linear relation between e_3 and e_5 that will be represented by a straight line AB in Fig. 398. The straight line is determined by airplane data, independently of the tail-area ratio S'/S. Any point between A and B gives a pair of values e_3, e_5 that fulfills the stability conditions. The corresponding value of S'/S is proportional to the abscissa $-e_5$.

Let us take for an example a big transport plane of 90,000 lb. gross weight, flying at 10,000 ft. altitude ($\rho = 0.0018$ slug/ft.^{3}) with a forward speed $V = 300$ ft./sec. If the wing loading is 35 lb./ft.^{2}, an area $S = 2571$ ft.^{2} and a lift coefficient $C_L = 0.432$ are required. With a lift factor $k = 4.8$ (corresponding to an aspect ratio of about 7) the level-flight incidence will be $\alpha_l = 0.090 = 5°$ and the mean wing chord $c = 19.17$ ft. Assume the tail distance (which is also chosen as reference length) $l = 60$ ft.; thus $V_l^2/gl = 46.6$. In general, the radius of gyration is of the same order of magnitude as the chord or as one-third the tail distance. In our case, let j_2^2 be 0.08. The wing profile of modern type may have a C_{M_0}-value of -0.045. As to the fuselage influence, Eq. (25), Chap. XVII, with the assumptions $A_f/S = \frac{1}{6}$, $l_f/c = 3.0$, and $c_f = 0.78$, supplies the derivative $dC_f/d\alpha$ equal to 0.39. Thus, Eq. (23) reads

$$-e_3 = 46.6[-e_5 - 9.24(0.5 + 0.39)] = 46.6(-e_5 - 8.22)$$

In Fig. 398 this straight line is drawn. It intersects the e_5-axis at the abscissa 8.22 (point A), and the curve corresponding to the right-hand side of (18) with $\alpha_l = 0.09$ at B. The coordinates of B, found by solving the two equations, are $-e_5 = 10.70$ and $-e_3 = 115.6$. Only in this comparatively narrow range, between 8.22 and 10.70, is $-e_5$ allowed to vary. The limits for the admissible tail ratio are then found from (22):

$$\frac{S'}{S} = - \frac{j_2^2 k \alpha_l}{k'} e_5$$

With $k' = 2.5$ (see Sec. XVII.2) this gives S/S' between 6.8 and 8.8 for the tail-area ratio consistent with longitudinal stability.

This example shows how the practical questions arising in connection with the longitudinal stability of the steady level flight can be solved without much computational work. If sufficiently detailed data are available, the more precise formula (16) instead of (18) may be used, which does not change the procedure considerably. The only difficulty lies in the fact that really reliable and complete values for all the coefficients involved will be at the disposal of the designer only in very rare cases.

Problem 1. Prove that an airplane of 11,250 lb. gross weight, geometrically similar to the airplane discussed in the numerical example of the text, would behave in exactly the same way with respect to stability if its wing loading were 17.5 lb./ft.[2] and its velocity 212 ft./sec. Compute the ratio of the linear dimensions of the two airplanes.

Problem 2. If all linear dimensions of an airplane are multiplied by a factor Λ while the air density changes at the rate σ, how must weight, wing loading, and level-flight velocity be varied if the stability properties should remain unchanged?

Problem 3. Compute the admissible limits of the tail-area ratio S'/S for the following specifications: gross weight $W = 1650$ lb., wing loading $W/S = 10.5$ lb./ft.[2], aspect ratio 6, level-flight velocity at 15,000 ft. altitude $V_l = 160$ ft./sec., tail distance $l = 3c$, radius of gyration $0.3l$. Assume that the wing has a moment coefficient with respect to the a.c. of $C_{M_0} = -0.06$ and that the influence of the airplane body, etc., doubles the static instability due to the wing.

*Problem 4. Develop the stability theory, as given in the preceding section, for a straight flight at a small inclination ϑ to the horizon.

*Problem 5. What has to be changed in the stability argument, if it is assumed that the tail surface contributes to the lift in the state of steady flight under consideration?

2. The Small Oscillations Following a Disturbance.

From Eq. (1) or (10) set up in the foregoing section much more information can be obtained about the behavior of an airplane in longitudinal motion than merely the stability criterion. Referring to the general form of the solution, given in Sec. XIX.3, for the equations of motion in the case of small disturbances, one can analyze in detail the kind of motion that sets in after a disturbance in the steady level flight of the airplane has occurred. A brief account of such a discussion follows, under the simplifying assumptions that allowed the expressions for a_3 and c_3 to reduce to those given in Eq. (17).

Equation (13) for λ may be rewritten in the form

$$\lambda^4 + a\lambda^3 + b\lambda^2 + c\lambda + d = 0 \tag{24}$$

where, according to (14) and (17),

$$a = \frac{1}{\alpha_l} - e_5, \qquad b = 2 - \frac{e_5}{\alpha_l} - e_3, \qquad c = -2e_5, \qquad d = -2e_3 \quad (25)$$

The two coefficients e_3 and e_5 are connected by the relation (23), which now will be written in the shorter form

$$-e_3 = \frac{V_l^2}{gl}\left(-e_5 - \frac{h}{j_{2l}^2}\right) \quad (26)$$

where h is the metacentric distance for the airplane without tail, as introduced in Eq. (45), Chap. XVII. For a stable airplane, e_3 and e_5 moreover, are subject to the inequalities (18).

In the numerical example discussed at the end of Sec. 1 of this chapter we had $V_l^2/g = 46.6$ and $h/j_{2l}^2 = 8.22$. These values determine the straight line (26) in the e_3 vs. e_5-diagram, and the extreme values for e_3, e_5, which fulfill the stability condition, were found (see Fig. 400) as

$$\begin{array}{lll} -e_5 = 8.22, & -e_3 = 0 & \text{(point } A\text{);} \\ -e_5 = 10.70, & -e_3 = 116 & \text{(point } B\text{)} \end{array} \quad (27)$$

For the discussion of the motion following a disturbance, an intermediate point C may be chosen corresponding to a tail-area ratio $S'/S = 0.131$. The coordinates of C, if k' is assumed as 2.5, are, according to (22) and (26),

$$-e_5 = 9.50, \qquad -e_3 = 59.6 \qquad \text{(point } C\text{)} \quad (28)$$

With these values, $\alpha_l = 0.09$ and Eq. (13) becomes, since $a_3 = 1$ and $c_3 = -1/\alpha_l = -11.1$,

$$\lambda^4 + 20.6\lambda^3 + 167\lambda^2 + 19.0\lambda + 119 = 0 \quad (29)$$

The four roots of this algebraic equation determine, as was shown in Sec. XIX.4, the motion of the airplane caused by any initial disturbance.

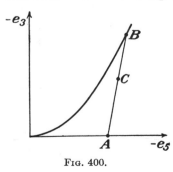

FIG. 400.

In order to discuss in a more general way what types of motion are possible, one may start with considering the two extreme cases represented by points A and B in the graph (Fig. 400). Point A corresponds to the minimum, point B to the maximum admissible tail area. Mathematically speaking, as was seen in Sec. XIX.4, in case A one root must be zero (*i.e.*, a disturbance may stay unchanged), while in case B one pair of imaginary roots exists (*i.e.*, an oscillation with constant amplitude may occur).

In A, with $e_3 = 0$, the last term in (13) vanishes and the polynomial has the factor λ. It is then easily seen to have the factor $\lambda - e_5$, also. Equation (13), in the general case, with $a_3 = 1$, $c_3 = -1/\alpha_l$, can be written as

$$\lambda(\lambda - e_5)\left(\lambda^2 + \frac{\lambda}{\alpha_l} + 2\right) = e_3(\lambda^2 + 2) \tag{30}$$

where the right-hand side is now zero. Then one root is $\lambda = 0$, one is $\lambda = e_5$, and, since α_l is small, another root will be of the order $-2\alpha_l$ and the last one of the order $-1/\alpha_l$. The exact solutions can be written as

$$\lambda_1 = 0, \qquad \lambda_3 = e_5, \qquad \lambda_2, \lambda_4 = -\frac{1}{2\alpha_l}(1 \mp \sqrt{1 - 8\alpha_l^2}) \tag{31}$$

and the numerical values in our example are

$$\lambda_1 = 0, \qquad \lambda_2 = -0.183, \qquad \lambda_3 = -8.22, \qquad \lambda_4 = -10.9$$

Under the conditions of point B, that is, maximum tail, one pair of roots is imaginary; thus, for it, the second and fourth terms in (24) must vanish for themselves. This means that the polynomial must be divisible by the factor

$$\lambda^2 + \frac{c}{a} = \lambda^2 - \frac{2e_5}{1/\alpha_l - e_5}$$

The other factor of second order is then easily found since the product of the absolute terms is $-2e_3$, and the linear term must give $-2e_5$ when multiplied by the absolute term in the first factor. This leads to the following transformation of (24) with $a_3 = 1$, $c_3 = -1/\alpha_l$:

$$\left(\lambda^2 - \frac{2e_5}{1/\alpha_l - e_5}\right)\left[\lambda^2 + \left(\frac{1}{\alpha_l} - e_5\right)\lambda + \frac{e_3}{e_5}\left(\frac{1}{\alpha_l} - e_5\right)\right]$$

$$= \frac{\lambda^2}{\alpha_l e_5}\left(c_3 + e_5^2 - \frac{2\alpha_l e_5}{1 - \alpha_l e_5}\right) \tag{32}$$

The right hand side is indeed zero, if the condition (18) is fulfilled at its limit. The roots are now

$$\lambda_1, \lambda_2 = \pm i\sqrt{\frac{-2\alpha_l e_5}{1 - \alpha_l e_5}};$$

$$\lambda_3, \lambda_4 = \frac{1 - \alpha_l e_5}{2\alpha_l}\left(1 \pm i\sqrt{\frac{4\alpha_l e_3}{e_5(1 - \alpha_l e_5)} - 1}\right) \tag{33}$$

and in the numerical example

$$\lambda_1, \lambda_2 = \pm 0.99i; \qquad \lambda_3, \lambda_4 = -10.90 \pm 10.88i$$

Note that the amount of the first two roots can in no case surpass $\sqrt{2}$.

In Fig. 401 the four roots corresponding to point A are marked A_1, A_2, A_3, and A_4, and those corresponding to B are B_1 to B_4. The diagram represents a λ-plane with the real part of each λ plotted toward the right and the imaginary parts upward. When the point $-e_3$, $-e_5$ in Fig. 400 moves from A through C to B, the points in Fig. 401 representing the four roots must change from their positions at A_1, . . . , A_4 to B_1, . . . , B_4 along certain curves. First, the points of each of the pairs

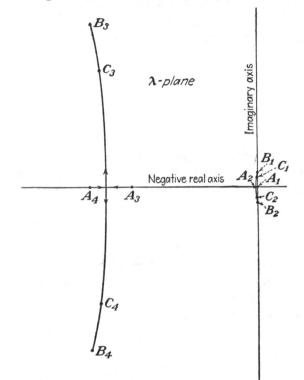

Fig. 401.—Roots of the characteristic equation.

A_1A_2 and A_3A_4 will approach each other along the real axis. Then, when they have met, they separate again and move upward and downward, along symmetrical curves, to join the points B_1, B_2 on the imaginary axis and B_3, B_4 in the upper and lower half plane. The roots C_1, C_2, C_3, C_4, which correspond to the point C in Fig. 400, must lie somewhere on these curves. It has already been seen that the ordinate of B_1 cannot be greater than $\sqrt{2}$ and that the abscissa of A_2 is about $-2\alpha_l$. On the other hand, the coordinates of the points B_3, B_4 and the abscissas of A_3, A_4 are comparatively large. It may therefore be concluded that *the motion which sets in after a disturbance in level flight consists of two damped oscillations superimposed on the uniform forward speed—one very slightly damped with a long period and one strongly damped with a short period.*

The numerical relation between the real parts of the roots λ and the damping and that between the imaginary parts and the frequency of the oscillations can easily be determined. Each root $\lambda = \lambda' + \lambda''i$ leads to an integral with the factor $e^{\lambda'\tau}$ and the factor sin (or cos) $\lambda''\tau$. A *full period* is achieved when $\lambda''\tau = 2\pi$. According to the relation between τ and t, as given in Eq. (50), Chap. XIX, the length of a period in seconds is

$$t_1 = \frac{V_l}{g}\tau_1 = \frac{V_l}{g}\frac{2\pi}{\lambda''} \tag{34}$$

This gives in our case ($V_l = 300$ ft./sec.) about 60 sec. for $\lambda'' \sim 1$ (points B_1, B_2) and about 6 sec. for $\lambda'' \sim 10$ (points B_3, B_4).

The *damping intensity* is usually measured by the time needed to reduce the amplitude to one-half its initial value. This means $e^{\lambda'\tau} = \frac{1}{2}$ or $\lambda'\tau = -\log 2$. Thus the *half time* t' of an oscillation is

$$t' = \frac{V_l}{g}\tau' = -\frac{V_l}{g}\frac{\log 2}{\lambda'} = -\frac{V_l}{g}\frac{0.693}{\lambda'} \tag{35}$$

With $V_l = 300$ ft./sec. this gives t' about 0.7 sec. for $\lambda' \sim -10$ and 70 sec. for $\lambda' \sim -0.1$.

The actual *problem of finding the roots* of Eq. (29), *i.e.*, of (24) with the given values of e_3, e_5 (point C), requires only simple numerical computations, once approximate values for one pair of roots have been found. Such an approximate solution is immediately supplied by the foregoing statements and the information offered by Fig. 401. There must exist one pair of roots C_1, C_2 close to the imaginary axis and of the order of magnitude 1. For a λ of amount 1 or near to it, the terms with the highest coefficients in (29) will prevail. Thus, an appropriate guess will be to assume $\lambda^2 \sim -\frac{119}{167}$, that is, $\lambda = \pm 0.84i$, as a first approximation. Then, there are various methods to improve on a first assumption. One of them is the procedure of successive approximations. One writes the prevailing terms on one side,

$$167\lambda^2 + 119 = -\lambda^4 - 20.6\lambda^3 - 19.0\lambda$$

and introduces on the other side the value of the first choice. This gives

$$167\lambda^2 + 119 = -0.498 \mp 3.75i$$

and thus

$$167\lambda^2 = -119.50(1 \pm 0.0314i), \qquad \lambda = \pm 0.846i - 0.013$$

Repeating the same procedure once more, one finds

$$\lambda_1, \lambda_2 = -0.0129 \pm 0.847i \tag{36}$$

Once one pair of roots is known, the other can be easily computed since the coefficient 20.6 in Eq. (29) is the negative sum of all roots and the absolute term 119 their product. Thus, the real part of λ_3, λ_4 is $-\frac{1}{2}(20.6 - 0.026) = -10.3$, and the square of the absolute value $\frac{119}{0.718} = 165.8$. Therefore,

$$\lambda_3, \lambda_4 = -10.3 \pm 7.7i \tag{36'}$$

The exact values of period and half time are $t_1 = 69$ sec. and $t' = 500$ sec. for the first and $t_1 = 7.6$ sec. and $t' = 0.63$ sec. for the second type of oscillation.

To find the actual *components of the motion* in the first and the second case we must recur to Eqs. (12), which determine the ratios $A:B:C$, once the characteristic λ-roots are known. From the first two equations (12) we draw, with $a_3 = 1$, $c_3 = -1/\alpha_l$,

$$A:B:C: = -\frac{1}{\alpha_l}:(2 + \lambda^2):\left(2 + \lambda^2 + \frac{\lambda}{\alpha_l}\right)$$

Omitting here an arbitrary factor, one may write

$$A = 1, \qquad B = -\alpha_l(2 + \lambda^2), \qquad C = -\alpha_l(2 + \lambda^2) - \lambda \tag{37}$$

If λ is introduced from (36) and formulas (37''), Chap. XIX, are used, the final result for the first type of oscillation becomes

$$u_1 = e^{-0.013\tau} \begin{Bmatrix} \cos \\ \sin \end{Bmatrix} 0.847\tau, \qquad u_3 = -0.115e^{-0.013\tau} \begin{Bmatrix} \cos \\ \sin \end{Bmatrix} (0.847\tau - 0.017)$$

$$\varphi = -0.855e^{-0.013\tau} \begin{Bmatrix} \cos \\ \sin \end{Bmatrix} (0.847\tau + 1.42) \tag{38}$$

In the same way, with the values of (36') the second type of motion is given by

$$u_1 = e^{-10.3\tau} \begin{Bmatrix} \cos \\ \sin \end{Bmatrix} 7.7\tau, \qquad u_3 = 14.9e^{-10.3\tau} \begin{Bmatrix} \cos \\ \sin \end{Bmatrix} (7.7\tau + 1.87)$$

$$\varphi = 8.86e^{-10.3\tau} \begin{Bmatrix} \cos \\ \sin \end{Bmatrix} (7.7\tau + 0.84) \tag{38'}$$

Formulas (38) and (38') represent four particular solutions of the equations of motion. Each of them can be multiplied by a factor (K_1, K_2, K_3, K_4); and from a complete set of initial conditions, i.e., values of u_1, u_3, φ and $\dot{\varphi}$ at $t = 0$, the four constants can be determined. As the second type of motion (38') dies away very quickly, the really observable motion will be of the form (38). The ratio of the three factors of u_1, u_3, φ is $1:0.115:0.855$, and their absolute values as well as the phases are determined by the initial conditions.

Problem 6. Compute the characteristic roots and the corresponding frequency and half-time values for the example of the text if the tail-area ratio is chosen as $\frac{1}{4}$.

Problem 7. Discuss the position of the roots in the λ-plane for the specifications given in Prob. 3. Assume an intermediate value for the tail-area ratio S'/S, and find the frequencies and damping factors.

Problem 8. Compute, for the example of the text, the actual motion that follows a sudden change in the attitude of the airplane of 1°. That is, assume the initial conditions $u_1 = u_3 = \phi = 0$, $\varphi = 1°$.

Problem 9. Compute, for the same example, the motion that follows a sudden upward gust of 10 ft./sec.

3. Lateral Stability.

The problem of stability against lateral disturbances can be dealt with, mathematically, in exactly the same way as has been done in the last two sections for a disturbance in the longitudinal plane. The only difference is that now a great number of poorly known force and moment coefficients interfere, which makes it difficult to reach simple conclusions of general significance.

We start from Eqs. (56), Chap. XIX, which were set up under the assumption that the 1- and 3-axes, forming the angle ψ with the horizontal and vertical directions, respectively, are the principal axes of the body. The last of the four equations, $\dot{\epsilon}_2 = -\sigma_1 \cos \psi - \sigma_3 \sin \psi$, can be integrated and gives

$$\epsilon_2 = - \cos \psi \int \sigma_1 \, d\tau - \sin \psi \int \sigma_3 \, d\tau \tag{39}$$

If this is substituted in the first equation, the set of equations for the three unknowns u_2, σ_1, σ_3 reads

$$\dot{u}_2 = b_2 u_2 + (b_4 + \sin \psi)\sigma_1 + \cos \psi \int \sigma_1 \, d\tau + (b_6 - \cos \psi)\sigma_3 + \sin \psi \int \sigma_3 \, d\tau$$
$$\dot{\sigma}_1 = d_2 u_2 + d_4 \sigma_1 + d_6 \sigma_3 \tag{40}$$
$$\dot{\sigma}_3 = g_2 u_2 + g_4 \sigma_1 + g_6 \sigma_3$$

According to the general rules developed in Sec. XIX.3, we introduce

$$u_2 = A e^{\lambda \tau}, \qquad \sigma_1 = B e^{\lambda \tau}, \qquad \sigma_3 = C e^{\lambda \tau} \tag{41}$$

and note that $\int \sigma_1 \, d\tau = \sigma_1/\lambda$, $\int \sigma_3 \, d\tau = \sigma_3/\lambda$. (Actually, it would be exactly the same to retain ϵ_2 as a fourth unknown in the equations and to use a fourth constant D setting $\epsilon_2 = D e^{\lambda \tau}$.) If (41) is introduced in (40) and the factor $e^{\lambda \tau}$ in all equations canceled, the following three linear algebraic equations for A, B, C, result:

$$A(b_2 - \lambda) + B\left(b_4 + \sin \psi + \frac{\cos \psi}{\lambda}\right) + C\left(b_6 - \cos \psi + \frac{\sin \psi}{\lambda}\right) = 0$$
$$A d_2 + B(d_4 - \lambda) + C d_6 = \tag{42}$$
$$A g_2 + B g_4 + C(g_6 - \lambda) = 0$$

Either by computing the determinant or by successive elimination of the unknowns one finds the condition of compatibility of these three equations in the form of an equation of fourth order for λ,

$$\lambda^4 \div a\lambda^3 + b\lambda^2 + c\lambda + d = 0 \tag{43}$$

where the a, b, c, d depend on the force and moment derivatives b_2, b_4, . . . , g_6 and the angle ψ. If b_4 and b_6 are assumed to be zero see page 604), one has

$$a = -(b_2 + d_4 + g_6),$$
$$b = d_4 g_6 - d_6 g_4 + b_2(d_4 + g_6) + g_2 \cos \psi - d_2 \sin \psi$$
$$c = -b_2(d_4 g_6 - d_6 g_4) - \sin \psi (g_2 + d_6 g_2 - d_2 g_6)$$
$$- \cos \psi (d_2 + d_4 g_2 - d_2 g_4) \tag{44}$$
$$d = \cos \psi (d_2 g_6 - d_6 g_2) + \sin \psi (d_4 g_2 - d_2 g_4)$$

The stability conditions are the same as those repeatedly discussed in Secs. XIX.3 and XX.1,

$$a > 0, \quad b > 0, \quad c > 0, \quad d > 0; \quad \frac{c}{a} < b - \frac{ad}{c} \tag{45}$$

The only difficulty of the present problem as compared with that of longitudinal stability lies in the complication caused by the great number of equally important force and moment derivatives. Only two out of the nine can fairly be neglected, and among the others no particular group has such a prevailing influence, as was the case with e_3 and e_5 in the longitudinal-stability problem. Sufficiently reliable values for the nine coefficients b_2, b_4, . . . , g_6 could be derived only from special experiments carried out with a complete model of the airplane—and even this procedure is precarious since models with too much detail are likely to behave differently from real airplanes. The only basis for a theoretical discussion of the various influences is to separate the contributions to the force and moment coefficients and their derivatives as originating in the wing, in the tail, and in the fuselage. One may assume that the mutual interference of these parts has a secondary effect only.

For the sake of simplicity it will also be assumed that the angle between the 1-axis and the velocity direction is exactly or nearly zero. There is of course no objection against setting $\psi = 0$ if in the second and third of Eqs. (40) the terms due to the inertia product J_{31} are not omitted. But, owing to the fact that J_{31} is certainly small and that precise values for it are hardly available, one may use Eqs. (40) as they stand and introduce here $\sin \psi = 0$, $\cos \psi = 1$. The advantage is that one then can identify the 1- and 3-directions with what is usually called the longitudinal and the down axes (x- and z-axes). Note that in this case the dimensionless quantity u_2 is nothing else than the angular deviation between the longitudinal axis and the projection of the velocity vector on the x-y-plane. This angle, with the opposite sign, was called sideslip β in previous arguments. Thus we have $u_2 = -\beta$.

The *air reactions on the wing* have been computed in Sec. XVIII.5; *i.e.*, certain roughly approximating expressions have been derived, essentially based on the assumption that each wing element can be considered as a separate airfoil. The cross force F_2 can now be identified with the y- (or 2-) component of the wing drag D. Its magnitude is consequently $D\beta = -Du_2$ and its coefficient $K_2 = -C_D u_2$, in accordance with Eq. (60), Chap. XVIII. On the other hand, one may assume that no considerable force in cross direction is produced by roll and yaw. Thus the wing contributions to the derivatives of K_2 are

$$\frac{\partial K_2}{\partial u_2} = -C_D, \qquad \frac{\partial K_2}{\partial \omega_1} = 0, \qquad \frac{\partial K_2}{\partial \omega_3} = 0 \qquad (46)$$

For the influence of sideslip, roll, and yaw on the moments, formulas (63), (59), and (56), Chap. XVIII, can be used with slight modifications. As the x- and z-axes now coincide with the 1- and 3-axes, respectively, the former C_{M_x} and C_{M_z} are now K_4 and K_6. Besides, β has to be replaced by $-u_2$. Upon using the same notations C_L, C_D, C'_L, C'_D, y^*, y_*, δ (dihedral), the six derivatives become

$$\frac{\partial K_4}{\partial u_2} = -C'_L \delta \frac{y^*}{l}, \qquad \frac{\partial K_4}{\partial \omega_1} = -C'_L \frac{y_*^2}{V_l l}, \qquad \frac{\partial K_4}{\partial \omega_3} = 2C_L \frac{y_*^2}{V_l l}$$

$$\frac{\partial K_6}{\partial u_2} = C'_D \delta \frac{y^*}{l}, \qquad \frac{\partial K_6}{\partial \omega_1} = C'_D \frac{y_*^2}{V_l l}, \qquad \frac{\partial K_6}{\partial \omega_3} = -2C_D \frac{y_*^2}{V_l l} \qquad (47)$$

For the air reactions *on the tail* simple approximations can be found in a similar way by considering a single vertical tail surface (fin plus rudder) at a distance l from the c.g. If k'' is the lift factor for this surface (corresponding to $k = 2\pi$ for an isolated airfoil of infinite span), a sideslip β would produce a cross force $\rho S'' V_l^2 k'' \beta/2$ in the positive y-direction, with its line of action at the distance l from the c.g., thus supplying

$$\frac{\partial K_2}{\partial u_2} = -k'' \frac{S''}{S}, \qquad \frac{\partial K_6}{\partial u_2} = k'' \frac{S''}{S} \qquad (48)$$

A yaw of magnitude ω_3 is equivalent to a sideslip $-l\omega_3/V_l$ as far as the tail is concerned, as seen in Fig. 402. This gives

$$\frac{\partial K_2}{\partial \omega_3} = k'' \frac{l}{V_l} \frac{S''}{S}, \qquad \frac{\partial K_6}{\partial \omega_3} = -k'' \frac{l}{V_l} \frac{S''}{S} \qquad (49)$$

The rolling moment caused by these cross forces will be proportional to the distance z^* of the centroid of the tail from the 1-2-plane (counted positive downward); thus (Fig. 403),

$$\frac{\partial K_4}{\partial u_2} = k'' \frac{z^*}{l} \frac{S''}{S}, \qquad \frac{\partial K_4}{\partial \omega_3} = -k'' \frac{z^*}{V_l} \frac{S''}{S} \qquad (50)$$

Finally, a roll of magnitude ω_1 adds to the velocity V_l a cross component $-z\omega_1$ for the tail element dS'' at the distance z (Fig. 403). This produces a cross force and a yawing moment for each element, which sum up to a force and a moment proportional to $z*$.

$$\frac{\partial K_2}{\partial \omega_1} = k'' \frac{z*}{V_l} \frac{S''}{S}, \qquad \frac{\partial K_6}{\partial \omega_1} = -k'' \frac{z*}{V_l} \frac{S''}{S} \qquad (51)$$

The rolling moment due to these forces will be proportional to z^2 for each element and will sum up to

$$\frac{\partial K_4}{\partial \omega_1} = -k'' \frac{z_*^2}{V_l l} \frac{S''}{S} \qquad (52)$$

if z_* is the radius of gyration for the tail surface. It is understood that all these formulas (48) to (52) can be considered as giving the order of

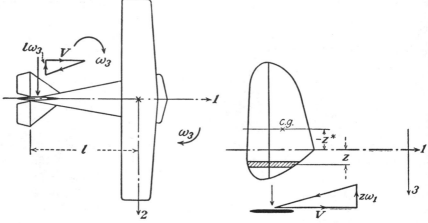

FIG. 402.—Reaction of the vertical tail to yaw.   FIG. 403.—Reaction of the vertical tail to roll.

magnitude of the respective derivatives only. In general, a more or less varying factor instead of k'' has to be used.

It is still more difficult to take into account the air reactions, due to the lateral disturbance, *on the airplane body*, fuselage, etc., and on the *propeller*. The propeller influences may be considered as negligible (see Sec. XVII.3). As to the fuselage, we may restrict ourselves to two terms: (1) In the expression (46), which refers to the cross force caused by sideslip, the wing-drag coefficient C_D should be replaced by the coefficient C_{to} of total drag (Sec. XIV.1). (2) A yawing moment on the fuselage, due to sideslip, like that considered in the discussion of static stability in Sec. XVII.3, may be introduced, with

$$\frac{\partial K_6}{\partial u_2} = -c_f \frac{A_f}{S} \qquad (53)$$

where A_f is the fuselage cross section and c_f the coefficient defined in Sec. XVII.3.

Combining now all the formulas (46) to (53) and the definitions given in (57) of Chap. XIX, the nine coefficients in the set (40) are found.

$$b_2 = -\frac{1}{C_{Ll}}\left(k''\frac{S''}{S} + C_{to}\right), \qquad b_4 = \frac{gz^*}{V_l^2}\frac{k''}{C_{Ll}}\frac{S''}{S}, \qquad b_6 = \frac{gl}{V_l^2}\frac{k''}{C_{Ll}}\frac{S''}{S}$$

$$d_2 = \frac{V_l^2}{j_1^2 gl}\frac{1}{C_{Ll}}\left(k''\frac{z^*}{l}\frac{S''}{S} - C'_L\delta\frac{y^*}{l}\right), d_4 = -\frac{1}{j_1^2 C_{Ll}}\left(k''\frac{z^2_*}{l^2}\frac{S''}{S} + C'_L\frac{y^2_*}{l^2}\right)$$

$$d_6 = -\frac{1}{j_1^2 C_{Ll}}\left(k''\frac{z^*}{l}\frac{S''}{S} - 2C_L\frac{y^2_*}{l^2}\right) \tag{54}$$

$$g_2 = \frac{V_l^2}{j_3^2 gl}\frac{1}{C_{Ll}}\left(k''\frac{S''}{S} - c_f\frac{A_f}{S} + C'_D\delta\frac{y^*}{l}\right),$$

$$g_4 = -\frac{1}{j_3^2 C_{Ll}}\left(k''\frac{z^*}{l}\frac{S''}{S} - C'_D\frac{y^2_*}{l^2}\right), \qquad g_6 = -\frac{1}{j_3^2 C_{Ll}}\left(k''\frac{S''}{S} + 2C_D\frac{y^2_*}{l^2}\right)$$

The further discussion of these formulas must be based on a numerical example.

Problem 10. What influence would it have on Eqs. (40) if instead of the level flight a steady flight in an inclined straight path were studied?

Problem 11. Give the complete form of Eqs. (40), taking into account that the 1- and 3-axes are not the principal axes of the airplane body.

4. Numerical Discussion. Let us take up the same airplane and the same state of level flight whose longitudinal stability was analyzed in Secs. 1 and 2 of this chapter. The main characteristics involved in both stability investigations are the ratio V_l^2/gl and the level-flight value C_{Ll} (now simply written C_L) of the lift coefficient. As reference length l we choose, as in the preceding sections, the tail distance, *i.e.*, usually the distance between the c.g. and the elevator and rudder axles. The values of these parameters in our example are

$$\frac{V_l^2}{gl} = 46.6 \qquad C_L = 0.432$$

Of the other parameters already used in the preceding sections we shall need the wing-drag and the total-drag coefficients, $C_D = 0.03$ and $C_{to} = 0.06$, and the derivatives with respect to α of both lift and drag coefficients, $C'_L = k = 4.8$, and $C'_D = 2kC_L/\pi R \sim 0.21$. As to the horizontal tail, it would supply only a numerically unimportant addition to the damping rolling moment, due to roll, which may be neglected.

More important parameters in the problem of lateral stability are the dihedral angle δ, the area S'' of the vertical tail, and the two radii of gyration for roll and yaw, j_1l and j_3l. The following assumptions correspond, at least in the order of magnitude, to normal conditions

$$\frac{S''}{S} = 0.1, \qquad j_1^2 = 0.07, \qquad j_3^2 = 0.11$$

For the dihedral angle we choose first $\delta = 0.07$ (about 4°) and shall later discuss the influence of varying δ. Furthermore, the two mean-span distances y_* and y^* and the mean distances for the vertical tail z_* and z^* are involved in the present argument. As correct in order of magnitude one may assume $y^*/l = 0.4$, $y_*^2/l^2 = 0.25$, while z^* and z_* are much smaller. It may be assumed $z^* = 0$ (*i.e.*, the centroid of the vertical tail on the longitudinal axis) and $z_*^2/l^2 = 0.01$. Then, except for the destabilizing contribution to g_2, due to the airplane body, the nine coefficients as defined in (54) can be computed. The contributions from the tail and from the wing are separately indicated in the formulas:

$$b_2 = -\frac{0.18 + 0.06}{0.432} \sim -0.6, \qquad b_4 = 0 \qquad b_6 = 0.009 \sim 0$$

$$d_2 = \frac{46.6(0 - 0.134)}{0.07 \times 0.432} \sim -210, \qquad d_4 = -\frac{0.00 + 1.20}{0.030} = -40,$$

$$d_6 = -\frac{0 - 0.216}{0.030} \sim 7$$

$$g_2 = \frac{46.6(0.18 - \cdots + 0.01)}{0.11 \times 0.432} \qquad g_4 = -\frac{0 - 0.05}{0.047} \sim 1$$

$$g_6 = -\frac{0.18 + 0.015}{0.047} \sim -4$$

The factor g_2 determines the static directional stability, as discussed in Sec. XVII.5. Using the moment coefficient C_M introduced in Sec. XVII.6, one can write g_2 in the form

$$g_2 = \frac{1}{C_{Ll}} \frac{V_l^2}{gl} \frac{1}{j_3^2} \frac{B}{l} \frac{dC_M}{du_2} \tag{55}$$

where B is the wing span and dC_M/du_2 is written for $-dC_M/d\beta$. In the foregoing numerical expression, the two contributions from tail and wing would yield $g_2 \sim 190$, while subtracting the destabilizing influence of the fuselage, under the usual assumptions, would render g_2 much smaller. As was mentioned in Sec. XVII.6, a value $-dC_M/d\beta$ of about 0.04 gives a satisfactory degree of static stability. With this figure and $B/l \sim 1.8$, Eq. (55) would give $g_2 \sim 70$. In the following computation half this value is first assumed to be valid, and the influence of changing g_2 is discussed later.

From Eqs. (44) with $\psi = 0$ and the values for $b_2, b_4, \ldots, g_6$ just derived we find, slightly abbreviating,

$$a = 45, \qquad b = 214, \qquad c = 1492, \qquad d = 595 \tag{56}$$

and the algebraic equation for λ,

$$\lambda^4 + 45\lambda^3 + 214\lambda^2 + 1492\lambda + 595 = 0 \tag{57}$$

It is seen that all coefficients are positive, as required for stability. Moreover, the two sides of the last inequality (45) are

$$\frac{c}{a} = 33.16 \qquad b - \frac{d}{c/a} = 214 - \frac{595}{33.16} = 196$$

Thus, all stability conditions are fulfilled.

The types of motion that can set in after a lateral disturbance has occurred depend on the roots of (57). It is easily seen that two real roots exist, one very small, determined by the last two terms

$$\lambda_1 \sim -\tfrac{595}{1492} = -0.399$$

and a large one determined by the first and second term

$$\lambda_2 \sim -45$$

In order to improve on λ_1, we write (57) in the form

$$\lambda = -\tfrac{595}{1492} - \tfrac{1}{1492}(214\lambda^2 + 45\lambda^3 + \lambda^4)$$

and introduce on the right-hand side the approximation -0.4 for λ. This yields $\lambda_1 = -0.421$; and if the same procedure is repeated once more (i.e., -0.421 substituted on the right side), one finds -0.422. Thus $\lambda_1 = -0.422$ can be considered as a sufficiently exact solution.

The second root can be improved by rearranging (57) in the form

$$\lambda = -45 - \frac{214}{\lambda} - \frac{1492}{\lambda^2} - \frac{595}{\lambda^3}$$

If, here, -45 is substituted to the right, one obtains $\lambda = -41$; and if the procedure is repeated, $\lambda_2 = -40.6$ can be secured as a good approximation of the second root.

The remaining two roots are found by using the facts that the sum of all four roots must be -45 and their product 595. This gives

$$\lambda_3 + \lambda_4 = -45 + 40.6 + 0.4 = -4.0, \qquad \lambda_3\lambda_4 = \frac{595}{0.422 \times 40.6} = 34.7$$

These two relations allow us to compute λ_3 and λ_4 in the well-known way,

$$\lambda_3, \lambda_4 = -2.0 \pm \sqrt{4.00 - 34.7} = -2.0 \pm 5.54i$$

Thus, according to (34) and (35), the possible motions are two aperiodically damped modes with the half-time values

$$t' = \frac{V_l}{g}\frac{0.693}{\lambda'} = 15 \text{ sec. and } 0.16 \text{ sec.}$$

and one damped oscillation with the half-time $t' = 3.2$ sec. and the period

$$t_1 = \frac{V_l}{g} \frac{2\pi}{\lambda''} = 10.6 \text{ sec.}$$

In Fig. 404 the four types of motion are indicated. The abscissas are the values of τ, and the ordinates represent, on an arbitrary scale, any of the variable u_2, σ_1, σ_3. The actual values of u_2, σ_1, σ_3, as they occur following a disturbance of the steady level flight, can be determined

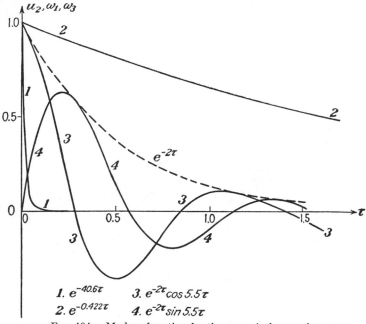

1. $e^{-40.6\tau}$ 3. $e^{-2\tau}\cos 5.5\tau$
2. $e^{-0.422\tau}$ 4. $e^{-2\tau}\sin 5.5\tau$

Fig. 404.—Modes of motion for the numerical example.

in the same way as was seen in Sec. 2 of this chapter in the case of longitudinal motion.

In discussing the influence of the various factors on the lateral stability one may first vary the coefficient g_2, which determines the static stability. If the eight other coefficients b_2, b_4, . . . , etc., are kept unchanged, formulas (44) with $\psi = 0$ yield

$$a = 45, \qquad b = 179 + g_2, \qquad c = 92 + 40g_2, \qquad d = 840 - 7g_2$$

The a, b, c, d will be positive if g_2 is smaller than $\frac{840}{7} = 120$ and greater than $-\frac{92}{40} = -2.3$. The last inequality (45) now reads

$$\frac{92 + 40g_2}{45} < 179 + g_2 - \frac{45(840 - 7g_2)}{92 + 40g_2}$$

It can easily be computed that this is fulfilled only if g_2 is greater than 2.9. Thus, with

$$2.9 < g_2 < 120$$

an upper and lower limit for the static stability is found, similar to what was seen in the problem of longitudinal stability. The only difference is that the lower limit is not zero but, in the present case, a positive value. That even slight negative values are admissible in other cases can be seen from the following argument.

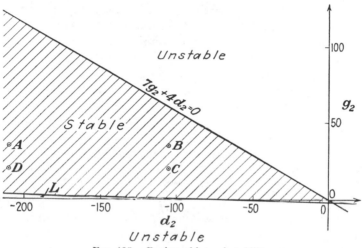

Fig. 405.—Region of lateral stability.

Let us vary, besides g_2, the dihedral angle δ. The value of δ influences d_2 and, to a slight extent, g_2. If the other coefficients remain unchanged, formulas (44) give

$$a = 45, \qquad b = 179 + g_2, \qquad c = 92 + 40g_2, \qquad d = -7g_2 - 4d_2$$

Thus the upper bound for g_2 will be $-4d_2/7$, that is, decreasing proportional to d_2 or to δ. The lower limit is determined by the inequality

$$\frac{92 + 40g_2}{45} < 179 + g_2 + \frac{45(7g_2 + 4d_2)}{92 + 40g_2}$$

The limiting curve in the d_2-g_2-plane is seen to be a parabola whose essential part is nearly a straight line, L in Fig. 405. It passes through the points $d_2 = 0$, $g_2 = -2.2$; $d_2 = -90.5$, $g_2 = 0$; $d_2 = -210$, $g_2 = 2.9$, thus nearly coinciding with the d_2-axis. The region of stability, hatched in Fig. 405, is confined between this parabola and the straight line $7g_2 + 4d_2 = 0$ and also includes points with negative g_2.

The roots λ turn out to be only slightly sensitive to a change of d_2 and g_2 within reasonable limits. In our first argument point A of Fig. 405

with the coordinates -210, 35 was used. If only half the original dihedral $\delta = 4°$ is applied, d_2 reduces to -105. With the original $g_2 = 35$ (point B), the roots are

$$\lambda_1 = -0.119, \qquad \lambda_2 = -40.6, \qquad \lambda_3, \lambda_4 = -2.15 \pm 5.62i$$

and if g_2 is reduced to 20 (point C), the roots become

$$\lambda_1 = -0.337, \qquad \lambda_2 = -40.6, \qquad \lambda_3, \lambda_4 = -2.05 \pm 4.03i$$

In neither case is the behavior of the airplane much altered.

The main results of this argument can be roughly summarized as follows: *Airplanes of conventional design, with a small amount of dihedral, are dynamically stable against lateral disturbances, independently of their being statically stable. Exceedingly high static stability (large fins and rudders) destroys the dynamic stability.* Slight violations of the dynamic stability conditions are not dangerous since they lead to modes of motion that can easily be controlled by rudders and ailerons.

Problem 12. Compute the minimum value of dihedral angle required for lateral stability, in the case of the airplane discussed in the text, with $g_2 = 35$. How does the minimum required dihedral angle change if the static stability factor g_2 varies?

Problem 13. Discuss the lateral stability and the modes of motion following a disturbance, for the example of the text, with the only change that the area ratio S''/S is reduced to 0.08.

5. Final Remarks. Autorotation. Spinning. The stability investigation in the preceding sections covers only the case of steady level flight. But the argument can easily be extended to any other state of steady motion in a straight path, climbing, diving, or gliding. It is much more difficult to analyze the stability conditions of the more general types of steady motion mentioned in Sec. XIX.2, horizontal turn and helicoidal motion with vertical axis. In these cases the problem cannot be divided into two independent problems, each concerning only one group of the variables. Here the stability problem is of the type discussed in Sec. XIX.3, but of the eighth order. The characteristic λ-equation is of the eighth degree, and its eight roots must be examined. In addition to all the parameters involved in the longitudinal and lateral stability investigation there appear the parameters that determine the state of motion under consideration, radius of curvature, angular velocity, and, in the second case, the pitch. Such a complete analysis has so far not yet been worked out, and it seems reasonable to postpone it until more reliable and more detailed data on force and moment derivatives are available.

From a practical point of view it may be expected that the conventional airplane in steady flight with the usual large radii of curvature will behave approximately in the same manner as it does in the case of straight flight. Thus fulfillment of the stability conditions for straight flight may be considered sufficient for all practical purposes. This refers to horizontal turns as well as to climbing or gliding in flat spirals.

There is, however, another restriction included in the theory as developed in Secs. 1 to 4 of this chapter. In all formulas for force and moment derivatives we assumed that all airfoil elements, in steady flight as well as during the motion that follows a disturbance, work under angles of incidence *below stalling*, where lift and drag coefficients follow the linear and the parabolic laws, respectively. With a view to practical airplane conditions it is necessary to study certain cases of motion under higher incidence, also. The most important phenomenon of this kind is a downward flight in a flat or steep spiral at angles of incidence going as high as 70°. This type of motion is generally known as the *spinning of an airplane*. It is not intended to give here an exhaustive theory of

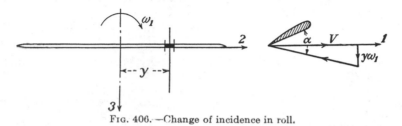

Fig. 406.—Change of incidence in roll.

the spinning phenomenon. Only a few of its characteristics will be briefly discussed.

Let us first consider an airplane moving in straight level flight at the wing incidence α_m. For the sake of simplicity it will be assumed that the airplane is exposed to one certain type of disturbance only, *viz.*, to an additional roll ω_1. This condition can easily be realized in a wind-tunnel experiment. The airplane model has to be constrained to rotate about a fixed axis parallel to the 1-direction and passing through the c.g. In this case, with a constant wind speed V, we have a problem of one degree of freedom and one single equation of motion

$$J_1 \frac{d\omega_1}{dt} = M_1 \tag{58}$$

The rolling moment M_1 is zero for $\omega_1 = 0$ because of symmetry, and the type of motion following an initial value of ω_1 depends on how M_1 is connected with ω_1-values different from zero.

In order to obtain information about the function $M_1(\omega_1)$ consider a wing of any plan-form (Fig. 406), twisted in such a way that the effective angle of incidence in the undisturbed state of motion equals α_m throughout the span. (If α_m is large, the twist will be negligible.) Now, if a roll of amount ω_1 occurs, the resultant velocity vector at the distance y from the axis will consist of the forward speed V and a component $y\omega_1$ normal to it. Setting

$$\frac{\omega_1 y}{V} = \eta, \qquad \frac{\omega_1 B}{2V} = \eta_1 \tag{59}$$

the angle of incidence for the wing element under consideration will be $\alpha_m + \text{arc tan } \eta$ and the resulting velocity equal to $V_r = V \sqrt{1 + \eta^2}$. Using the "strip hypothesis" introduced in Sec. XVIII.5, we may assume that a force of magnitude

$$dF = C_r \frac{\rho}{2} V_r^2 \, dS \tag{60}$$

is acting on the wing element of area $dS = c \, dy$. The direction of this force will be approximately perpendicular to the axis, thus supplying a rolling moment equal to $-y \, dF$.

The coefficient C_r in (60) depends on the angle of incidence. In Fig. 407 a typical curve giving $C_r = \sqrt{C_L^2 + C_D^2}$ vs. α is shown for α-values up to $70°$ (see Fig. 187). One may combine two symmetric elements dS and introduce the difference

$$\Delta C_r = C_r(\alpha_m + \text{arc tan } \eta) - C_r(\alpha_m - \text{arc tan } \eta) \tag{61}$$

Then the rolling moment, according to the strip hypothesis, will be

$$M_1 = -\frac{\rho}{2} V^2 \int \Delta C_r(\eta)(1 + \eta^2) cy \, dy$$

$$= -\frac{\rho}{2} \frac{V^4}{\omega_1^2} \int_0^{\eta_1} (1 + \eta^2) \, \Delta C_r(\eta) c\eta \, d\eta \tag{62}$$

If all angles of incidence involved fall in the range below stalling, the difference ΔC_r is throughout positive, and therefore the rolling moment, due to a positive roll, is negative; the roll is automatically damped and dies away, practically within a finite time. It is different when the stalling limit is exceeded.

In Fig. 407 a definite value $\alpha_m = 24°$ is indicated. The dotted line is the image, reflected with respect to the line $\alpha_m = 24°$, of the C_r vs. α-curve. Thus the ordinate difference between the two curves gives the values of ΔC_r for all η. It is seen that in the present example ΔC_r is negative for small η and becomes positive later. Thus, if the *roll rate* $\eta_1 = \omega_1 B / 2V$ is small, the rolling moment, according to (62), will have a positive value; it will tend to increase an initial roll. As the usual stability definition refers to small disturbances only, the state of flight under the incidence $\alpha_m = 24°$ must be called *unstable*. If the same reasoning is applied to $\alpha_m = 34°$ where the C_r vs. α-curve is again rising, one finds that this angle of incidence leads to a stable flight.

It is, however, more important in the present problem to study the moment M_1 as function of ω_1 or η_1 for larger values of roll. If the planform of the wing is given by $c(y)$ and the angle α_m chosen, the Eq. (62) allows us to find, by graphical or numerical integration, the rolling moment M_1 for each roll rate η_1. Typical results are shown in Fig. 408. Curve 1 corresponds to a small value of α_m (below stalling conditions); it shows only negative values of M_1 and represents the case of complete stability. For $\alpha_m = 24°$, curve 2, the moment is first positive, then negative. The curve intersects the η_1-axis at a point A. Whatever small initial roll occurs, the angular velocity will increase up to the value that corresponds to the point A. This phenomenon or the η_1-range

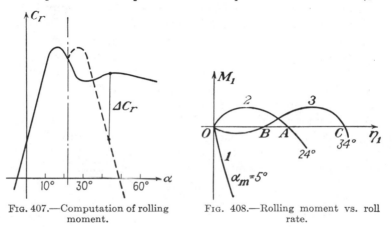

FIG. 407.—Computation of rolling moment.

FIG. 408.—Rolling moment vs. roll rate.

within which it occurs is known as the *autorotation of an airplane*. When the ω_1 corresponding to A is reached, the roll speed will stay unchanged.

For still higher values of η_1 (curve 3), there is first a region of negative M_1 up to the point B, followed by the interval BC with positive values of the moment. This case is known as *latent autorotation*. A sufficiently small initial roll will die away; but if a certain narrow limit (represented by the abscissa of B) is exceeded, the initial angular speed will automatically increase up to a higher value corresponding to the point C and will stay there unchanged.

Autorotation is probably the beginning of each spinning motion. When an airplane is flying under too high an incidence a gust or some deviation from full symmetry (for example, an unbalanced propeller moment) may produce an initial roll that then increases automatically up to a considerable value. A bank angle corresponding to a right turn is connected with a positive roll. The bank brings about a horizontal component of the air reaction, directed starboard. This force component causes the airplane to rotate about a vertical axis with the rotation vector $\vec{\omega}$ pointing downward. At the same time the nose of the airplane must go down since the 1-component of $\vec{\omega}$, that is, the roll, is positive. Yawing and pitching moments seem to be of minor importance. Experience shows that in this way airplanes of conventional design very soon reach a state of flight in which all equilibrium conditions are fulfilled. The center of gravity of the airplane (and every other point) moves downward in a helical path with vertical axis, at a constant velocity and constant inclination to the vertical (Sec. XIX.2). Values of $\omega B/2V$ (where V is the c.g. speed) from 0.3 up to 1.2 have been observed with an inclination of the c.g. path ranging from about 30 to 60°.

In the beginning of flying practice, until about 1914, the spinning flight used to end fatally since the pilot was unable to change his steady state of flight into conditions suitable for landing. It was then learned that, with the high angle of incidence, the airplane is in the regime of *reversed commands* (Sec. XIV.3) so that maneuvers contrary to the natural reaction of the pilot are necessary. In order to change from the steep to a flatter descent the elevator stick must be pushed forward as though steeper diving were intended. If at the same time the lateral controls are operated in the sense opposite to the rotation, the transition to a normal flight with an incidence in the domain of direct commands can be performed in most cases. There exists, however, a dangerous type of flat spin with exceedingly high value of the rotation rate $\omega B/2V$ that often cannot be checked by such maneuvers. It seems that the pitching moment due to centrifugal forces which cannot be overcome by the elevator moment is partly responsible for it. Besides, the lateral controls may refuse to act in the required sense, the ailerons because their rolling moment is accompanied by an adverse yawing moment, and the rudder—which should be the most effective control for checking the spin—because it is shielded by the elevator from the action of the air. The most effective protection against catastrophic spin is a well-considered tail design that keeps the control surfaces in action in all conditions that may occur in actual flight. The danger of spin is greater for smaller airplanes than for larger ones. This indicates that the size factor V^2/gl, which is involved in all stability questions, plays an important role in the spinning problem, also.

These remarks can give only a remote idea of the difficulties that are encountered in the dynamics of the airplane when other than the simplest types of motion are considered. It must be left to the reader to complete his information by studying the currently published research papers and by his own efforts in elaborating the arguments presented in the text. In this last chapter, as well as in all the preceding, only the basic facts could be mentioned and only the fundamentals of the theoretical approach have been discussed.   For some topics, like the wing theory, the use of *more advanced mathematical methods*, for others a broader foundation gained through *observation and experiment* will provide a deeper comprehension of all the diversified phenomena involved in the flight of an airplane than this textbook could offer.

BIBLIOGRAPHICAL AND HISTORICAL NOTES

For a brief introduction to the more elementary topics the reader may consult Clark B. Millikan, "Aerodynamics of the Airplane," New York, 1941. A comprehensive text, dealing with all problems of flight theory is "Aerodynamic Theory, A General Review of Progress," William Frederick Durand, editor in chief, 6 vols., Berlin, 1934–1936 (in English); reprinted 1943 by the Durand Reprinting Committee, California Institute of Technology, Pasadena, Calif.[1]

Most of the information concerning observation data will be found in the following publications:

Advisory Committee for Aeronautics (since 1920, *Aeronautical Research Committee*), *Reports and Memoranda (R. & M.)* 1, (1909)—(Great Britain).
National Advisory Committee for Aeronautics (NACA) Tech. Repts. (Repts.) 1, 1915; *Tech. Notes* 1, (1920)—(United States).
"Ergebnisse der Aerodynamischen Versuchsanstalt zu Göttingen," herausgegeben von L. Prandtl, München-Berlin, 1923—— (Germany).

The most important periodicals, covering both the experimental and the theoretical field are:

Jour. Roy. Aeronautical Soc., **1** (1897)——.
Zeitschrift für Flugtechnik und Motorluftschiffahrt **1**, (1910)—— **24** (1933) and *Luftfahrtforschung*, **1** (1928)——.
Jour. Aeronautical Sci., **1** (1934)——.

Chapter I

The principles of hydrostatics and aerostatics were established in the seventeenth century by Galilei, Torricelli, and Pascal. The latter was the first to recommend the use of pressure measurements for the determination of altitude. The equation of state for a perfect gas, usually connected with the names of Boyle, Mariotte, Gay-Lussac, and Charles, has been known since about 1800.

The need for a "standard atmosphere" made itself felt in all countries from the beginning of aviation practice. The standard values now generally adopted in this country are given in *NACA Rept.* 147 (1922); for tables see also *Repts.* 218 (1925) and 246 (1926). The international standard, which is only slightly different, is defined in *British Air Pub.* 1173 (September, 1925), H. M. Stationery Office, London, Regulation Governing the Graduation of Altimeters Consequent upon the Adoption of an International Standard Atmosphere.

Information about the physics of the atmosphere can be found in most textbooks on meteorology.

Chapter II

The first substantial result in the theory of fluid motion was Bernoulli's equation, given by Daniel Bernoulli (1700–1782) in his "Hydraulico-statica" of 1738. Later, Leonhard Euler (1707–1783), combining Newton's second law with the assumption of

[1] Hereafter referred to as W. F. Durand, "Aerodynamic Theory."

a mere normal stress at all points, developed the complete system of partial differential equations that control the motion of a "perfect" fluid. Our equation (16) or (19) is only a very special consequence of Euler's equations, valid for two-dimensional steady flow. New forms were given to Euler's theory by J. L. Lagrange (1736–1813), by H. v. Helmholtz (1821–1894), and by Lord Kelvin (William Thomson) (1824–1907). The interpretations given in Secs. 4 and 5 are based on the results of Helmholtz, who introduced the notion of vorticity, and of Kelvin, who mainly used the notion of circulation. The papers, still worth reading, are: Helmholtz, Über Integrale der hydrodynamischen Gleichungen, die den Wirbelbewegungen entsprechen, *Crelle's Jour.*, **55** (1858); Kelvin, On Vortex Motion, *Trans. Roy. Soc. Edinborough*, **25** (1869). The first is reprinted in Ostwald's "Klassiker der exakten Wissenschaften 79," Leipzig, 1896; the second in Kelvin's "Mathematical and Physical Papers," Cambridge, 1882.

A complete survey of the classical theory of perfect fluids can be found in H. Lamb, "Hydrodynamics," 5th ed., Cambridge, 1924; a shorter introduction in L. M. Milne-Thomson, "Theoretical Hydrodynamics," London, 1938. For those parts of the theory which are more closely connected with aviation problems, see also the mimeographed lectures quoted in the preface, p. vii.

Chapter III

The earliest investigation of the reaction of moving fluid upon rigid bodies is due to Daniel Bernoulli (1738). The moment of momentum equation in form (14) was given by Euler (1754). A complete and correct discussion of the problems connected with the momentum and energy equations can hardly be found in the existing literature. Our text attempts only a brief introduction. For a discussion of the incompressible case, see also R. v. Mises, "Theorie der Wasserräder," Leipzig, 1908.

Chapter IV

The concept of fluid friction being proportional to the relative "gliding" of neighboring layers is due to Newton. L. Navier developed (1827) a molecular theory that led to a definite system of partial differential equations, and G. G. Stokes showed later (1845) that the same equations can be derived on the basis of the general concept of the stress tensor in a continually distributed mass. The main facts of hydraulics, known since about the beginning of the nineteenth century, were in strong contradiction with the theory of viscous fluids. I. Boussinesq (1877) was the first to explain the discrepancies by ascribing them to turbulence, and Osborne Reynolds (1883) gave the full experimental evidence by discovering the critical limit that separates laminar and turbulent flow. Reynolds also initiated the dimensional analysis as outlined in Sec. 2. The hydraulic hypothesis, generally used for a long time, was explicitly formulated by R. v. Mises (1908).

The discovery of the discontinuous type of perfect fluid motion is one of the great achievements of H. v. Helmholtz (1868). The theory has been extended by T. Levi-Civita (1907) to include the formation of a wake behind rounded obstacles. Special questions of great importance have been studied by C. Schmieden [*Ingenieur Arch.* **3** (1932) and **5** (1934)], by St. Bergman [*Z. angew. Mathem. Mechanik*, **12** (1932)], and by I. Leroy [*Commentarii Helvetici*, **8** (1935–1936)] and others. The theory of the vortex street was given by Th. v. Kármán in *Physik. Z.*, **13** (1912).

L. Prandtl found (1904) an approximate solution of the Navier-Stokes equations that represents a laminar boundary layer. Many examples have been studied later in connection with the problem of separation. A bibliography is given in the reprint of Prandtl's original paper:" Vier Abhandlungen zur Hydrodynamik," Göttingen,

1927. Later R. v. Mises [*Z. angew. Mathem. Mechanik*, **8** (1928)] gave the theory a new form, which was used by Th. v. Kármán and C. B. Millikan [*NACA Tech. Rept.* 504 (1934)] to solve a problem connected with the flow past an airplane wing. The idea of a turbulent boundary layer was developed by Th. v. Kármán in *Z. angew. Mathem. Mechanik*, **1**, 233 (1921).

More information about the subject of this chapter may be found in the books: Prandtl-Tietjens, "Fundamentals of Hydro- and Aeromechanics," New York, 1934; "Modern Developments in Fluid Dynamics" (composed by the Fluid Motion Panel of the Aeronautical Research Committee and others and edited by S. Goldstein) 2 vols., Oxford, 1938.

Chapter V

The fact that fluid resistance of bluff bodies is approximately proportional to the square of the velocity has been known since the late eighteenth century. The first theoretical account of this fact was supplied by Helmholtz's investigation on discontinuous flow (see Chap. IV). The subsequent experimental research was mainly promoted by the use of Reynolds's dimensional analysis, F. W. Lanchester's notion of streamlining, and Prandtl's boundary-layer theory. Actual sources for information on experimental results are the book by G. Eiffel, "La Résistance de l'air," 2d ed., Paris, 1911; and the current reports of the Aerodynamische Versuchsanstalt Göttingen, the NACA, and the (British) Aeronautical Research Committee. A collection of data can also be found in W. S. Diehl, "Engineering Aerodynamics," New York, 1936, and much useful information, particularly concerning Sec. 6, in the "Handbook of Aeronautics" (published under the authority of the Royal Aeronautical Society), Vol. I, London, 1938. For Kármán's theory leading to the logarithmic form of the skin-friction relation see his paper, *Nachr. Ges. Wiss. Göttingen*, 1930, pp. 50–76.

Chapter VI

For a more recent system of profiles, similar to that of the four- and five-digit sets, see H. B. Helmbold and F. Keune, *Luftfahrtforschung*, **20**, 81–96 (1943). The three types of theoretically defined profiles (Joukowski, Kármán-Trefftz, Mises) are discussed in most treatises on aerodynamics, *e.g.*, in Vol. III of W. F. Durand, "Aerodynamic Theory."

Chapter VII

The systematic study of airfoil characteristics has been considered as one of the main tasks of all aerodynamic research institutions during the last 35 years. The results can be found in the current publications of the institutions; in addition to those already cited (Chap. V), see also "Allegato ai rendiconti tecnici del ministero dell' aeronautica," Roma.

Chapter VIII

The two-dimensional wing theory started with a paper by W. M. Kutta (only partly published in *Illustrierte aeronautische Mitteilungen*, 1902, p. 133) in which the author, at the instigation of S. Finsterwalder, studied the lift force produced on an airfoil by the irrotational flow of a perfect fluid. Independently, N. Joukowski (1847–1921) in a Russian paper of 1906 arrived at the same formula $L' = \rho \Gamma V$, but he also showed how the circulation Γ is uniquely determined for a profile with one sharp end. In his paper of 1910 [*Z. Flugtech. u. Motorluftschiffahrt*, **1**, 280 (1910)] he gave the first correct lift formula, valid for the so-called "Joukowski profile." The general theory, supplying lift and moment of lift for arbitrary profiles and leading to the statements concerning the aerodynamic center, was developed by R. v. Mises [*Z. Flugtech. u. Motorluftschiffahrt*, **8**, 157 (1917); **11**, 68, 87 (1920)]. The theory is presented in most textbooks, *e.g.*, H. Glauert, "The Elements of Aerofoil and Airscrew Theory," Cambridge, 1926; Harry Schmidt, "Aerodynamik des Fluges," Berlin-Leipzig, 1929; W. F. Durand, "Aerodynamic Theory," Vol. II (Th. v. Kármán). Among numerous papers that supply various contributions to the theory may be quoted: W. Müller, *Z. angew. Mathem. Mechanik*, **3** (1923), **4** (1924), **5** (1925); W. Birnbaum, *ibid.*, **3** (1923); F. Höhndor , *ibid.* **6** (1926); Th. Theodorsen, *NACA Repts.* 383 (1931) and 411 (1932).

The thin-wing theory was developed by M. Munk [*NACA Tech. Rept.* 142 (1922)]. It was shown later by R. v. Mises [*Jour. Aeronautical Sci.*, **7**, 290 (1940)] how this method can also be used as the first step in a sequence of successive approximations that leads to the complete solution of the two-dimensional airfoil problem.

Chapter IX

The theory of wings with finite span was inaugurated by F. W. Lanchester, who in a series of research papers, beginning in the year 1894, clearly recognized the importance of the trailing vortices behind a supporting surface. He summed up his main results in two books, "Aerodynamics" and "Aerodonetics," London, 1907 and 1908, and supplemented them in a paper The Airofoil in the Light of Theory and Experiments, *Proc. Inst. Automobile Eng.*, **9**, 169 (1915). Meanwhile, since 1910, L. Prandtl studied the same problem and arrived at similar conclusions. He gave a complete presentation of his farther reaching theory in two papers "Tragflügeltheorie" in *Nachr. Ges. Wiss. Göttingen*, 1918 and 1919 (reprinted in "Vier Abhandlungen zur Hydrodynamik und Aerodynamik," Göttingen, 1927). By incorporating the results of the two-dimensional wing theory (see Sec. 4 of this chapter) Prandtl succeeded in supplying a definite formula for computing lift and drag of wings and systems of wings (biplanes) of arbitrary shape. In numerous papers by Prandtl's collaborators and others the theory has been improved and supplemented in various directions. In particular, E. Trefftz by introducing the Fourier analysis, as used in Sec. 7 of this chapter, showed a convenient way for arriving at numerical results [*Z. angew. Mathem. Mechanik*, **1**, 206 (1921)]. A brief presentation of Prandtl's theory can be found in the books of Glauert and Harry Schmidt already cited (Chap. VIII); some attempts at generalization are indicated in the article of J. M. Burgers in Vol. II of W. F. Durand, "Aerodynamic Theory." Here also a long list of papers dealing with the subject can be found.

Chapter X

For the theory of stalling consult the paper by Th. v. Kármán and C. B. Millikan, *Jour. Applied Mech.*, **57**, A-21 (1935). Experimental studies on stalling conditions and maximum lift can be found in many NACA reports, in most cases in connection with a discussion of high-lift devices; see *e.g.*, Nos. 664, 668, 677, and 679. The boundary-layer control by suction, originally suggested by L. Prandtl, has been carefully investigated, for example, by A. Gerber in *Mitt. Inst. Aerodynamik Tech. Hochschule Zürich*, No. 6 (Zürich, 1938). For experiments on pressure distribution over airfoils with flaps, see *NACA Tech. Repts.* 574, 614, 620, and 633.

The chordwise pressure distribution for Joukowski profiles, typical for all kinds of profile, was first computed by O. Blumenthal, *Z. Flugtech. u. Motorluftschiffahrt*, **4**, 125 (1913). The computation of the spanwise distribution is an essential part of Prandtl's wing theory, and various procedures are suggested in treatises and papers dealing with this theory; see, *e.g.*, I. Lotz, *Z. Flugtech. u. Motorluftschiffahrt*, **22,** 189 (1931).

The influence of the compressibility of the air on airfoil characteristics is now being studied thoroughly in all countries. A review of the main problems and a mathematical approach, not too complicated for the mature reader, based on the fundamental ideas of the Russian mathematician A. Chaplygin (1904), were given by Th. v. Kármán, *Jour. Aeronautical Sci.*, **8,** 337 (1940). The Mach number is named for Ernst Mach (1838–1916), famous philosopher of science, who made the first fundamental observations on fast-moving projectiles.

Chapters XI and XII

The study of the action of screw propellers dates back to the work of W. J. M. Rankine (1820–1872) and W. Froude, who both had in mind marine propulsion. The geometry of an airplane screw is somewhat different from that of a ship screw; thus, here, the blade-element theory, inaugurated by Froude [*Trans. Inst. Naval Architects*, **30,** 390 (1889)], wins higher importance. This concept, based on the analogy between a propeller element and an airfoil, was first applied to the computation of airplane propellers by S. Drzewiecky (1900). In his book, "Des hélices aériennes, théorie générale des propulseurs," Paris, 1909, the blade-element theory is developed in its original form (Sec. XII.1), which leaves the question undecided which aspect ratio of the airfoil characteristics should be used in the computation.

The momentum theory was initiated by Rankine [*Trans. Inst. Naval Architects*, **6,** 13 (1865)] and later developed by many writers. In the earlier papers no circumferential velocity in the slip stream was taken into account. A. Betz [*Z. Flugtech. u. Motorluftschiffahrt*, **11,** 105 (1920)] considered this rotation, and H. Glauert gave a complete set of equations for the general momentum theory (see W. F. Durand, "Aerodynamic Theory," Vol. IV, p. 191). The first who tried to combine the notions of the momentum theory with those of the blade-element theory was H. Reissner [*Z. Flugtech. u. Motorluftschiffahrt*, **1,** 257, 309 (1910)]. Later Th. Bienen and Th.v. Kármán developed the modified momentum theory and its combination with blade-element conceptions along the lines shown in Secs. XII.5 and 5 [*Z. Ver. deutsch. Ing.*, **68,** 1237 (1924)]. They also gave an interpretation of their results from the standpoint of the Lanchester-Prandtl wing theory. Much work has been spent on attempts to utilize the wing theory for the purpose of propeller computation; see particularly L. Goldstein, On the Vortex Theory of Screw Propellers, *Proc. Roy. Soc.*, **A 123,** 440 (1929). But it seems that, except for the question of blade interference, not much more can be reached than a plausible interpretation of the results otherwise obtained. A bibliography and a review of these attempts can be found in Glauert's article in Vol. IV of W. F. Durand, "Aerodynamic Theory." For a comprehensive study of propeller problems and a more complete bibliography, see also F. Weinig, "Aerodynamik der Luftschraube," Berlin, 1940.

Chapter XIII

More detailed information about the principles underlying the operation of aircraft engines can be found in any of the numerous books on this subject, *e.g.*, A. W. Judge, "Aircraft Engines," Vol. I, New York, 1941; or "Handbook of Aeronautics" (see Chap. V), 3d ed., Vol. II, London, 1938. For the early development see L. S. Marks, "The Airplane Engine," New York, 1922.

The foundation of the dynamics of reciprocating engines was laid by J. v. Radinger in his book "Über Dampfmaschinen mit hoher Kolbengeschwindigkeit," 3d ed., Vienna, 1892. A modern text, including the theory of crankshaft vibrations, is given by D. L. Thornton, "Mechanics Applied to Vibrations and Balancing," New York, 1941. The tuned pendulum was introduced by E. S. Taylor, *Jour. S.A.E.* **38**, 81 (1936). For the theory see also the articles of J. P. Den Hartog in "Contributions to the Mechanics of Solids, Dedicated to Stephen Timoshenko," New York, 1938, and of V. Moore in *Jour. Aeronautical Sci.*, **9**, 229 (1942).

Chapters XIV and XV

The main problems of performance computation are solved by a discussion of the simplest equilibrium conditions for a solid body. Thus there was no need here to develop some new theory. However, a considerable time was needed before the ideas about the conditions of airplane operation were clarified. The earliest writers, up to about 1908, erroneously searched for the amount of work needed to keep a body of weight W suspended in the air. In one of the first competent texts, written by two distinguished mathematicians, E. Borel and P. Painlevé ("L'Aviation," Paris, 1910), no clear statement of the performance problem can be found. The correct solution, based on the discussion of the two power curves (Sec. XIV.2), developed in the years 1910 to 1915. Among the earlier books that give a correct discussion may be quoted: Capitaine Duchêne, "L'Aéroplane étudié et calculé par les mathématiques élémentaires," 3d ed., Paris, 1913 (first published 1911); R. v. Mises, "Fluglehre," 5th ed. by K. Hohenemser, Berlin, 1936 (first published 1915); L. Bairstow, "Applied Aerodynamics," 2d ed., New York, 1939 (first edition 1920).

As long as wing characteristics and/or propeller-engine characteristics are given as graphs, only a graphical performance computation is possible. There is, however, general agreement today that the polar diagram of a wing can be represented, below stalling, as a parabola. On this basis and with the assumption of a constant power-available value W. B. Oswald [*NACA Tech. Rept.* 408 (1932)] developed the procedure referred to in Secs. XV.2 to 5. It is obvious that various different forms of diagrams (*e.g.*, logarithmic scales) can be used without affecting the essence of the method. Real progress in the performance computation would require the adoption of simple and sufficiently exact expressions for the propeller-engine characteristics and the power-altitude factor. A tentative approach is presented in Secs. XV.6 and 7.

Chapter XVI

The first to study systematically the problem of range and endurance was Louis Bréguet, one of the French pioneer aircraft designers. His formulas, already widely adopted at that time, are published in *L'Aérophile*, **29**, 271 (1921). A more elaborate discussion, similar to that given in Sec. XVI.1, can be found in the paper by A. B. Scoles and W. A. Schoech, *Jour. Aeronautical Sci.*, **5**, 436 (1938). This paper deals also with take-off computation. For more empirical data about the take-off (and similar questions) see W. S. Diehl, "Engineering Aerodynamics," pp. 435–446, New York, 1936. An interesting study on the landing of airplanes is given by H. Glauert in *R. & M.* 666 (1920).

Seaplane problems are very involved, and most of them are not yet sufficiently studied. A valuable review and a bibliography up to 1936 are given by E. G. Barrillon in Vol. VI of W. F. Durand, "Aerodynamic Theory," pp. 134–222. For an attempt at a hydrodynamic theory of watering, see Herbert Wagner, *Z. angew. Mathem. Mechanik*, **12**, 193 (1932), and *Proc. Fourth Intern. Congress Applied Mech.*, Cambridge, 1934, p. 126.

Chapter XVII

The practical problem of maintaining the equilibrium of pitching moments and the longitudinal stability of an airplane has been solved by Alphonse Pénaud, who introduced the horizontal tail surface as a constituent part of a model airplane; see *L'Aéronaute, Paris,* January, 1872. The elementary computations connected with this problem are now given in almost all textbooks. For the theorem on decalage as derived in Sec. 4 of this chapter from the combination of two aerodynamic centers, see the author's paper, *Jour. Aeronautical Sci.*, **7**, 303 (1940). The influence of the fuselage upon the static stability was recently stressed by H. P. Liepmann, *Jour. Aeronautical Sci.*, **9**, 181 (1942). A brief representation of the longitudinal-control and static-stability theory was recently published by Hans Reissner, *Trans. A.S.M.E.*, **65**, 625 (1943).

Chapter XVIII

The first to study the pathway at large of an airplane model was F. W. Lanchester; see his book "Aerodonetics," London, 1908 (2d ed. 1910). Under more general conditions the airplane motion was studied, for example, by W. Mueller, *Z. angew. Mathem. Mechanik*, **19**, 193 (1939), *Ingenieur Arch.*, **9**, 63 (1938), **10**, 63 (1939), **11**, 99 (1940). The results of these computations are in general agreement with those of the simpler approach as developed in Secs. 3 and 4 of this chapter. The more elementary relations of airplane dynamics as discussed in Sec. 1 of this chapter and later are presented in many texts on the theory of flight.

Chapters XIX and XX

The general equations for the motion of an airplane (Secs. XIX.1 and 2) were discussed, including the gyroscopic effect of the propeller rotation, by the author in "Encyclopaedie der mathematischen Wissenschaften," Vol. IV, 2, pp. 343–352 (1911). The general theory of stability has been developed by E. J. Routh, "Essay on Stability of a Given State of Motion," London, 1877. The application to airplane stability was first made by G. H. Bryan and W. E. Williams, *Proc. Roy. Soc. London*, **73**, 1904, p. 100, and G. H. Bryan, "Stability in Aviation," London, 1911. A simplified discussion of longitudinal stability is due to Th. v. Kármán and E. Trefftz, *Jahr. Wiss. Ges. Luftfahrt*, **3**, 116–138 (1914–1915). Both the longitudinal and the lateral stability were studied theoretically and experimentally by J. C. Hunsaker, Dynamical Stability of Airplanes, *Smithsonian Misc. Collections*, **62**, No. 5 (Washington, 1916). For a recent attempt to adapt the longitudinal-stability theory to practical requirements see M. Munk, *Aero Digest*, **42**, Nos. 5 and 6 (1934), **43**, Nos. 1 and 2 (1944). The stability of the steady motion in a horizontal turn and related questions of lateral stability have been studied by H. Reissner, *Z. Flugtech. u. Motorluftschiffahrt* **1**, 101 (1910). No definite results have as yet been reached in the problem of lateral stability. An exhaustive numerical discussion is given in the contribution of B. M. Jones in Vol. V, W. F. Durand, "Aerodynamic Theory," pp. 121–221; here also more indications about spin and autorotation can be found.

INDEX

A

Actuator disk, 327*ff*.
Advance ratio (def.), 286
Aerodynamic center, 124, (def.) 144, 187
  existence in finite wing, 499
  of wing and tail system, 519*ff*.
  general existence, two-dimensional theory, 185*f*.
Aileron, 531*f*.
Aileron operation in a turn, 563
Air, constituents of, 18
  to gasoline, ratio of, 357
Air reactions determined by instantaneous velocities, 566*f*., 573*f*.
Air resistance, 95*ff*.
  (*See also* Drag)
Airfoil (def.), 113
  thin (*see* Thin airfoil)
Airfoil geometry, 115–121
Airfoils, empirical data on, 139–169
Airplane components, major, drag of, 107–111
Airship, offsets of, 103
Airship body, drag of, 103
Altitude, critical in supercharging (def.), 368
  true, 13*f*.
Altitude flight, 409–418, 431*ff*., 438*f*.
  and change of weight, 418
Altitude performance, short-cut method, 412*ff*.
  of supercharged engine, 413*f*., 433, 435*f*., 439
Angle of attack, 114, (def.) 140
Angle of incidence (def.), 140
  effective (def.), 234
Area, equivalent frontal (def.), 95
  parasite (def.), 96
    of airplane, 385, 421, 444
    of propeller, 426, 450
    total, 451
Aspect ratio (def.), 113, 134
  in elliptic wing theory, 243*f*.

Aspect ratio, influence on wing characteristics (empirical), 148–156
Atmosphere at rest, 1–21
  isothermal, 5
  polytropic, 6*f*.
  standard, 8–13
    table, 10
Attitudes in gliding, 479
Autorotation, 611
  latent, 611
Axes, of airplane, 527, 586
  principal, (def.) 566, 586, 599

B

Bank angle (def.), 538
Barometric formula, 5*f*.
Bénard, H., 89
Bernoulli's equation, 26–30, 32*ff*.
  applied in propeller theory, 330*f*.
  and energy equation, 69*f*.
Best performance curve (prop. charts), 315*f*.
Bicirculation vector, 49*ff*.
  (*See also* Motion, bicirculating)
Biot-Savart formula, 216*f*.
Biplane decalage, 138
Biplane geometry, 137*f*.
Biplane interference factor, 247*ff*.
Biplane theory, 244–249
Bluff bodies, drag of, 96–98
Body forces, 3
Boundary conditions, similar, 81
Boundary layer, 90–94
  control, 265*ff*.
  and stalling, 260*ff*.
  thickness of, 91
  turbulent, 92*f*.
Brake, 271
Brake moment (def.), 361
Brake moment constant, altitude flight with, 382
  assumption of, 396
Brake power (def.), 359

Brake power-altitude factor, 364*ff.*, 412
  (*See also* Power-altitude factor)
Bréguet, L., 134, 462
  formula for endurance, 463
  formula for range, 463
Bryan, G. H., 580
Buoyancy, moment of, 173
Buoyancy term in momentum equation, 172

C

Camber (def.), 116
  influence of, 163*f.*, 166, 207
Ceiling altitude, (def.) 383, 432*ff.*, 453
  vs. fundamental performance parameter, 433
Center, aerodynamic (*see* Aerodynamic center)
Chord (def.), 115*f.*
Chordwise distribution of pressure and lift, 272
Circulating motion (def.), 39*f.*
  (*See also* Motion, circulating)
Circulation (def.), 40
  in the field of a vortex line (three-dimensional), 219
  and lift force, two-dimensional, 176*ff.*
  in two-dimensional irrotational field of flow, 44*f.*
  and vorticity, 221*f.*
Climb, flat, balance of forces in, 384
Climb curves, 416*ff.*, 440
Climbing rate, maximum, 402, 438, 454
Combustion energy, 356
Compressibility, influence on airfoil characteristics, 280*ff.*
Compression ratio (def.), 358
Continuity condition, 24
  for compressible fluid, 277, 283
Contour map (prop. charts), 314*f.*
Crank angle, 372
Cross force due to sideslip, 536*f.*, 560
Cup anemometer, 98*f.*
Cylinder, circular, flow around, boundary layer theory, 92*ff.*
  two-dimensional theory, 177*f.*, 198

D

D'Alembert's paradox, 225
  and finite wing theory, 227

Damping intensity of oscillation, 597
Damping moment, 552
  estimate of, 591*f.*
Dead water, in perfect fluid theory, 86
  underpressure in, 88, 97, 101
Density (def.), 1
Density altitude (def.), 11*f.*
Diesel engine in airplanes, 356
Dihedral angle (def.), 132
  influence on stability, 560*f.*, 607
Discontinuity surface, equivalent with vortex sheet, 223
Discontinuous flow, 86*ff.*, 200
Disks, pair of, drag of, 97*f.*
Diving, 481–483
Diving hodograph, 481
Down-wash velocity, (def.) 233, 237*ff.*
  constant, 240*ff.*
  at the tail, 503
Drag, induced, 142, (def.) 238
  in biplane theory, 246*ff.*
  parasite, 95–111
  of rectangular plate (surface drag), 106
  smallest for given lift, 253
Drag coefficient, parasite (def.), 95
  surface, 106
  of wing, empirical, 141*ff.*
Drag interference, 109
Dunne airplane, 526

E

Efficiency of engine, mechanical (def.), 359
  thermal (def.), 359
  volumetric, (def.) 358, 366
Eiffel, G., 96, 97, 100, 142
Elevator, 502*ff.*
Elevator operation, effect of, 551–555
  in a turn, 562
Ellipsoid, drag of, 100
Endurance, 461–468
  maximum, 464*ff.*
Energy, total, of a fluid, 67*f.*, 72
Energy equation, 67–73
Engine, 356*ff.*
  at altitude, 363–370
  at sea level, 356–363
Engine vibrations, 371–380
Equation, characteristic, approximate solution, 597*f.*

Equation, complex roots of, 578
  real roots of, 577
  situation of roots, 596
Equation of state, 3
Equations of motion, complete system
  of, 568
  dimensionless, 569
  general, 564–570
Euler's rule of differentation, for nonsteady flow, 64
  for relative steady flow, 66
  for steady flow, 25*f*.

F

Fineness ratio, streamlined bodies, 102*ff*
First axis, (def.) 124, 139
  (*See also* Zero lift direction)
Five-digit series, NACA, 119*f*.
Flap, split, 268
  trailing-end, 268*ff*.
Flap and slot in combination, 270
Flight, at altitude, 409–418, 431*ff*., 438*f*.
  along a given path, longitudinal, 545–
    551
  nonuniform, elementary results, 533–
    539
  at sea level, 398–409
Flight conditions, normal, 498
Floats, drag of, 111
Flow, discontinuous, 86*ff*., 200
  inverse, (def.) 23, 66
  laminar, 81*ff*.
  quasi-steady, 65*ff*.
  spanwise, 229
  turbulent, 83*ff*.
  two-dimensional (def.), 24
    around airfoil, 44, 171*ff*., 189, 199*ff*.
    around circular cylinder, 178
    experimental realization of, 158
    irrotational, 38
    as superposition of uniform flow and
      flow induced by vortex sheet,
      195*ff*.
Fluid, compressible, in two-dimensional
  steady irrotational flow, 277*ff*.
  perfect (def.), 2, 26
    interference with rigid bodies, 72
  real, 77
  viscous, 74*ff*.
Flux (def.), 24

Flying boats, drag of, 109
"Flying Fortress," 133
Focus of a profile, 144*f*.
Force coefficient (def.), 78*ff*.
Force derivatives in stability theory, 584*f*.
Four-digit series, NACA, 118*f*.
Fowler flap, 270
Friction, of engine piston, 359, 382
  in take-off, 472
Friction factor in pipe flow, 82
Fuel consumption, (def.) 362, 461, 467
Fuselage, drag of, 108*f*.

G

Γ-distribution (def.) 233
  elliptic, 239*ff*.
  represented by trigonometric series,
    250*ff*.
Gasoline, 356
Glaisher's factor, 20
Glauert, H., 252, 281, 355
Gliding, attitudes in, 479
  flat, 427
  at smallest angle, 404, 479
  steep, 475–481
Gliding angle (def.), 143*f*.
Gliding hodograph, 476*f*.
Goldstein, S., 350

H

Handley-Page slot, 266*f*.
Head, total (def.), 31
  change of, and power transfer, 72
Helmholtz, H. v., 38
Helmholtz flow, 86*ff*., 200
Hemisphere, drag of, 98
Hexane, 356
High-lift devices, 264–271
Hinge moment, 506
Horseshoe vortex, 235*ff*., 238
Humidity, atmospheric, 20*f*.
Hydraulic hypothesis, 84*f*.
Hydroplane, 488

I

Ideal airplane, 437–442
  altitude performance, 439
  maximum climbing rate, 438, 440

Ideal airplane, service ceiling, 439
Incidence (*see* Angle of incidence)
Indicator diagram, 357*f*.
Induced drag (def.), 238
　(*See also* Drag, induced)
Induced velocity (def.), 190
　(*See also* Velocity, induced)
Inflow factor, propeller, 346
Isothermal level, standard atmosphere, 9

J

Joukowski, N., 122, 157, 201, 350
Joukowski profile, 122*ff*.
Joukowski's airfoil theory, 177*ff*.

K

Kármán, Th. v., 89, 106, 126, 337
Kármán vortex street, 191
Kutta, W. M., 201
Kutta-Joukowski theorem, 176, 183, 202, 237
Kutta's condition, 201

L

Lachmann, G., 266
Lanchester, W. F., 226, 539
Lanchester-Prandtl, wing theory of, 224–239
Landing, 483–488
　flight path preceding, 484*f*.
Landing gear, drag of, 110
Landing impact, 485*ff*.
　for seaplane, 494*ff*.
Landing run, 485
Level flight, balance of forces in, 384
　with given power, 428–436
　high speed of, 399*ff*., 429
　low speed of, 400*f*.
"Liberator," 133
Lift, of airfoil of infinite span, 174–181
　apparent increase of, 508*ff*., 546
　total (def.), 497
Lift coefficient, empirical, 139*ff*.
　for extended angular range, 258
　of theoretical profiles, 124, 180
　in three-dimensional theory, 241, 243, 255*f*.
　in two-dimensional theory, 180

Lift distribution, 233
　chordwise, 271*ff*.
　elliptic, 239–244
　general, 250–257
　rectangular wing, 254*f*., 274
Lift factor, in compressible flow, 281*f*.
　in thin airfoil theory, 207*f*.
Lifting line (three-dimensional theory), 232*ff*.
Lilienthal, O., 157
Load factor (def.), 536
Loadings, range of, 444

M

Mach number (def.), 278
Maneuverability, vertical, 534
Mapping, conformal, 121, 180
Mass, apparent, 574
Mean camber line (def.), 116
Mean rotation, 212*ff*.
Metacenter (def.), 517
Metacentric distance, 525*f*., 594
Metacentric parabola, 186*ff*., 519
　and stability, 188*f*.
Millikan, C. B., 504
Mises, R. v., 128, 158, 185
Moment, pitching (*see* Pitching moment)
　resultant, on immersed body, 61*ff*.
Moment of deviation, 566, 569
Moment of gyration, 566, 569
Moment of momentum, flux of, 60
　theorem, 58*ff*.
　　applied in airfoil theory, 173*f*., 181*ff*.
　　applied in propeller theory, 329
Moment coefficient, 80, 114, 124*ff*., 185
　empirical, 144*ff*.
Moment derivatives in stability theory, 585
Moment equilibrium in terms of aerodynamic center, 499*f*.
Moments, lateral, 527–532
Momentum, flux of, 52*ff*.
Momentum theorem, 55*ff*.
　applied in airfoil theory, 170*ff*.
　applied in propeller theory, 331
Motion, bicirculating, 45–51
　applied in airfoil theory, 182
　circulating, (def.) 39*f*.
　　applied in airfoil theory, 176
　　induced by vortex, 190
　　with solid rotating core, 189

Motion, helical (propeller), 290*f.*
  helicoidal, of airplane, 571
  steady (*see* Steady motion)
Motion of airplane, asymmetric, 555–563
  tail reactions in, 601
  in the symmetry plane, 540*ff.*
  along a vertical circle, 547*ff.*
Munk, M., 206
Munk's span factor, 166, 248*f.*, 420, 422

N

Nacelles, drag of, 109
Navier-Stokes equations, 76
Nose-heavy condition, 498, 501

O

Octane rating, 360
Order of firing, 373
Oscillations following disturbance, 593–599
Oswald, W. B., 421, 444

P

Parachute, drag of, 9
Parasite area (*see* Area, parasite)
Parasite drag (*see* Drag, parasite)
Parasite loading (def.), 421
  total (def.), 450*f.*
Peak-efficiency propeller, 449
Pénaud, A., 524
Pendulum, tuned, 374*ff.*
Perfect fluid (def.), 2
Perfect gas, equilibrium under gravity, 4–8
Performance, of airplane, 381*ff.*
  fundamental parameter of, (def.) 428, 451
  for various types of airplanes, 444
Performance analysis, dimensionless, 394–398
Performance computation, influence of small variations of data, 447*f.*
  numerical example, 442–447
Phugoid motion, theory of, 539–545
Piston displacement, 357
Pitch (airplane motion) (def.), 528
Pitch of propeller (def.), 292

Pitching moment, of airplane, 497
  of wing, two-dimensional theory, 183*ff.*
Pitching-moment contribution, from fuselage, 513*f.*
  from propeller, 507–512
  from tail, 501–506
Pitching-moment derivative, 523*ff.*
Pitching-moment equilibrium, 497–501
Pitot tube, 33
  correction for adiabatic flow, 34
Plate, rectangular, drag of, 97
Plate flow, boundary layer theory, 91
Poiseuille's laws, 82
Polar diagram (def.), 142*ff.*
  in airplane performance, 395, 419*f.*
  elliptic wing theory, 241
  of propeller, 308*f.*, 398, 423*f.*
Power, per cylinder, 359
Power-altitude factor, (def.) 364*f.*, 412, 431, 438, 451
Power available (def.), 382
  for constant-speed propeller, 392*f.*
  varying with speed, 450–460
  for varying brake moment, 407
  for varying propeller diameter, 408
  for varying throttle setting, 402*f.*
Power available-altitude factor, 431
Power available curve, analytical, 425
  point by point, 389*ff.*
Power curves analytic, 419–426
  construction of, 383, 385–394
Power-drop factor, 361
Power indicated, 358
Power indicated-altitude factor, 364
Power loading (def.), 428
Power required (def.), 382
  at altitude, 411
  indicated (def.), 410
  for varying weight, 405, 462
  for varying wing area, 406
Power-required curve, analytical, 422
  point by point, 386*ff.*
Prandtl, L., 90, 231, 350
Prandtl's hypothesis, 234*f.*, 245
Pressure (def.), 1*ff.*
  brake mean effective (def.), 359
  dynamic, (def.), 31–34
    table, 32
  indicated mean effective (def.) 358
  reduced (humidity correction), 21

Pressure altitude, 11
center of (def.), 146
Pressure distribution, chordwise, 271*ff.*
spanwise, 273*f.*
Pressure head (def.), 28
in isothermal and polytropic flow, 29
Pressure slot, 266
Profile, Antoinette (1907), 157
Clark Y (1922), 160
parasite drag, 142
Farman (1906), 156
Göttingen 360 (1915), 159
influence of the shape of, 161–167
on stalling, 261*ff.*
Joukowski, 122*ff.*, 163
Kármán-Trefftz, 126*f.*
Mises, theoretical, 128*ff.*
1915, 159
NACA 2409, 161
NACA 23009, 162
principal, 113
thickness function (NACA), 117*f.*
Wright (1903), 155
Profile variation along wing, 135*f.*
Profiles, historical development of, 157*ff.*
theoretically developed, 121–131
Propeller, 285–355
adjustable-pitch, 310
at advance ratio zero, 353*ff.*
attempts of three-dimensional theory, 350*ff.*
best performance, 449
blade-element efficiency, 324*f.*
blade-element theory, 317–326
choice of, 448–450
constant-speed, 392*f.*, 408, 455
controllable-pitch, (def.) 294, 310, 459
figure of merit, 354
high-speed, 299
with infinite number of blades, 328
interference with fuselage, 300
multiple interference, 351*ff.*
representative blade element method, 302
variable pitch (*see* Constant-speed propeller)
Propeller advance ratio (def.), 286
Propeller blade setting (def.), 286
nominal (def.), 293
Propeller characteristics, 296–301
Propeller charts, 311*ff.*

Propeller efficiency, (def.) 289, 298
Propeller geometry, 290–296
Propeller grading curves, 347*ff.*
Propeller induced efficiency (def.), 335
Propeller kinematics, 285*f.*
Propeller loading, parasite (def.), 425*f.*
Propeller moment, compensation of, 528*f.*
Propeller momentum theory, basic relations, 326–334
conclusions, 334–339
modified, 339–345
Propeller pitch (def.), 292
distribution, 292*f.*
effective (def.), 298
nominal (def.), 293
Propeller plan-form, 294
Propeller polar diagram, 308*f.*, 398, 423*f.*
Propeller power, coefficient (def.), 287, 298
from energy equation, 70*ff.*
Propeller power loading, in airplane performance, (def.) 425, 451*ff.*
Propeller power-loading factor (def.), 335
Propeller power-speed coefficient (def.), 312
Propeller sets, 310*ff.*
Propeller thrust coefficient (def.), 287
Propeller thrust-loading factor (def.), 335
Propeller torque (def.), 288
Propeller slip, 298
Propeller solidity, 303, 306, (def.) 346

Q

Quarter chord point (def.), 145*f.*
Quasi-steady flow, (def.) 65*ff.*, 70*ff.*

R

Radius of gyration, 569, 592, 604
Ram effect, 367*f.*
Ramus step, 489
Range, 461–468
maximum, 464*ff.*
Rankine, W. J. M., 337
Rate of climb curves, 411, 413, 417, 440, 446
graphical integration, 415
Rayleigh, Lord, 75
Reciprocal integration, method of, 493
Reduction of a climb, 14*ff.*

Region of transition of drag coefficient, 99
Reverse of commands, 401f., 612
Reynolds number, (def.) 78ff.
  and boundary layer, 92
  critical, for pipe flow, 82
  and drag of round bodies, 99ff.
  and drag of streamlined bodies, 105
  effective, 169
  influence of, on airfoil characteristics,
    167ff., 262ff.
  and surface drag, 106f.
Roll (def.), 528
  change of incidence in, 609
Rolling moment, due to dihedral, 561
  due to roll, 559
  due to yaw, 558
  vs. roll rate, 611
Rotation, 35ff., (def.) 37, 212
  conservation of, 38
  steady, about a vertical axis, 555ff.
Rotation vector (def.), 213
Roughness, relative, 107
Round bodies, drag of, 99ff.
Routh, E. J., 574
Routh's stability theory, 574–580
  applied to the airplane, 580–585
Rudder, 531
  operation in a turn, 563

S

Schmeidler, W., 268
Seaplane problems, 488–496
Second axis (def.), 129
Separation, 86ff.
Separation point in boundary-layer
  theory, 93f., 260, 264
Service ceiling (def.), 416
Sharp trailing edge, in three-dimensional
  theory, 225ff.
  in two-dimensional theory, 177ff.
Side force due to sideslip, 536f., 560
Sideslip (def.), 530
  and dihedral, 560f.
Similitude, law of, 77–81
Sink, 46
Sinking rate, minimum, 404, 427
Skin friction, 105ff.
Soaring, 480
Sound velocity, 282ff.
  local, 30, 277ff.

Sound velocity, in standard atmosphere,
  10
Source, 46
Span loading, (def.) 422, 451
Spanwise flow, 229
Spanwise lift distribution, (def.) 233, 273
Speed ratio vs. altitude, 430ff., 453
Sphere, drag of, 100
Spinning, 609ff.
Spoiler, 271
Stability, dynamic (Routh's theory),
  574–580
  condition of, 579
  independence of longitudinal and
    lateral variables, 583
  lateral, 599–603
    force derivatives in terms of air-
      plane data, 603
    numerical discussion, 603–608
  longitudinal, force derivatives in
    terms of airplane data, 588
    of level flight, 586–593
    limit of, 590
  static, 514ff.
    and sign of M_0, 518f.
    simplified discussion, 522–527
  static directional, 529ff, 602
Stabilizer, 502ff.
Stagnation pressure tables, 32, 34
Stagnation point (def.), 31
Stalling, 258–264, 609f.
Stalling point (def.), 141
Steady motion (def.), 22
  of airplane (general case), 570–574
Steady rotation of airplane, moments
  required in, 556f.
Stratosphere, 19
Stream filament (def.), 24
Stream tube (def.), 24
Streamline (def.), 23f.
Streamlined bodies, drag of, 102ff.
Streamlined strut, offsets of, 103
Stresses, viscous, 76
Strip hypothesis, 558, 610
Stokes formula, 217f.
Suction slit, 265
Supercharger, 366ff.
  power consumption of, 368
Superposition, principle of, 50, 177ff., 275
Surface drag (*see* Skin friction)
Sweepback, 135

Symmetry of airplane, influence of, on air reactions, 572*ff*., 583

T

Tail, damping effect of, 591
  influence of the wing on, 503
Tail angle of a profile, 127
Tail area, admissible maximum, 594
Tail decalage, 503
  theorem, 521
Tail distance (def.), 505
Tail-heavy condition, 498, 501
Tail lift for varying elevator setting, 504
Tail surfaces, drag of, 109
Tailless airplane, 526
Take-off, 469–475
  on water, 489*ff*.
  against wind, 472*f*.
Take-off run, length and duration of, 474
Taper ratio (def.), 134*f*.
"Taube," 134
Taylor, E. S., 374
Temperature inversion, 6, 20
Temperature gradient, atmospheric, 6
Thickness, influence of, on profile characteristics, 162, 166
Thin airfoil, theory of, 198–210
Torque moment (def.), 361
  oscillations, 372*ff*.
Transformation, conformal, 180
Trefftz, E., 126
Troposphere, 18*f*.
Turbosupercharger, exhaust-driven, 367
Turbulence, 82*ff*.
  degree of, depending on Reynolds number, 168*f*.
Turbulence factor, 169
Turn, flat, 537
  in horizontal plane, 536*ff*.
  true-banked (def.), 538*f*., 555*f*.
  vertical, 534*ff*.

V

Vector, complex notation, 124
Velocity, effective, in propeller theory, 341*ff*.
  indicated (def.), 410
  induced (def.), 190
  by finite segment, 211

Velocity, induced, by horseshoe vortex, 235*f*., 238*f*.
  by straight vortex sheet, 191*ff*.
  at maximum climbing rate, 402
  terminal (def.), 478
Velocity head (def.), 28
Velocity profile of boundary layer, 91*f*.
"Vengeance," 133
Vibrations, torsional, 377*ff*.
Viscosity, 74*ff*.
  of air, tables, 10, 75*f*.
Viscosity coefficient (def.), 74
Vortex density (def.), 191*f*.
Vortex filament (def.), 190
Vortex line, curved, 216*f*.
  polygonal, 215*f*.
  segment, 211, 213*ff*.
Vortex lines, distribution on wing, 228*ff*.
Vortex motion, 189
Vortex sheet, 88, 89, 188–198, 219–224
  approximation by finite number of vortices, 196*f*.
  built up of vortex lines, 220
  equivalent to discontinuity surface, 223
  free, 227
  tangential to resultant stream, 222*f*.
  velocity discontinuity at, 220*f*.
Vortex street, 89
Vorticity (def.), 190
  continuous distribution of, 191

W

Wagner, H., 494
Washout, 136
Wasp R-1340 (Pratt & Whitney), 369*f*.
Weight, specific (def.), 1
Weight increase, apparent, 535*f*.
Wieselsberger, K., 97
Williams, W. E., 580
Wind-tunnels, comparison of results, 168*ff*.
Wing, elliptic, 242*ff*.
  of finite span, flow past, 225–231
  of infinite span, 170–210
  pitching moment of (def.), 114
  rectangular, 254*f*.
  tapered, 255*f*.
  of variable shape, 267

Wing attitude in various states of flight, 401
Wing characteristics, 112*ff*., 139–148
Wing drag (def.), 114
Wing geometry, 132
Wing interruptions, 137
Wing lift (def.), 114
Wing plan-form, 112, 134*f*.
Wing span (def.), 112
Wing and tail system, stability, 519*ff*.
Wing tip stall, 137
Wing twist, 113, 136*f*., 255*f*., 275
Wire, streamlined, drag of, 104
Wright, O., and W. Wright, 157, 528, 531

Y

Yaw (def.), 528
Yawing moment, due to dihedral, 561
    due to roll, 559
    due to sideslip, 530
    due to yaw, 558
    on fuselage, due to sideslip, 602

Z

Zero lift angle (two-dimensional theory), 181, 206
Zero lift direction, (def.) 139

A CATALOG OF SELECTED

DOVER BOOKS

IN ALL FIELDS OF INTEREST

A CATALOG OF SELECTED DOVER BOOKS IN ALL FIELDS OF INTEREST

CONCERNING THE SPIRITUAL IN ART, Wassily Kandinsky. Pioneering work by father of abstract art. Thoughts on color theory, nature of art. Analysis of earlier masters. 12 illustrations. 80pp. of text. $5\frac{3}{8} \times 8\frac{1}{2}$. 23411-8 Pa. $2.50

LEONARDO ON THE HUMAN BODY, Leonardo da Vinci. More than 1200 of Leonardo's anatomical drawings on 215 plates. Leonardo's text, which accompanies the drawings, has been translated into English. 506pp. $8\frac{3}{8} \times 11\frac{3}{4}$. 24483-0 Pa. $10.95

GOBLIN MARKET, Christina Rossetti. Best-known work by poet comparable to Emily Dickinson, Alfred Tennyson. With 46 delightfully grotesque illustrations by Laurence Housman. 64pp. $4 \times 6\frac{3}{4}$. 24516-0 Pa. $2.50

THE HEART OF THOREAU'S JOURNALS, edited by Odell Shepard. Selections from *Journal*, ranging over full gamut of interests. 228pp. $5\frac{3}{8} \times 8\frac{1}{2}$. 20741-2 Pa. $4.50

MR. LINCOLN'S CAMERA MAN: MATHEW B. BRADY, Roy Meredith. Over 300 Brady photos reproduced directly from original negatives, photos. Lively commentary. 368pp. $8\frac{3}{8} \times 11\frac{1}{4}$. 23021-X Pa. $14.95

PHOTOGRAPHIC VIEWS OF SHERMAN'S CAMPAIGN, George N. Barnard. Reprint of landmark 1866 volume with 61 plates: battlefield of New Hope Church, the Etawah Bridge, the capture of Atlanta, etc. 80pp. 9×12. 23445-2 Pa. $6.00

A SHORT HISTORY OF ANATOMY AND PHYSIOLOGY FROM THE GREEKS TO HARVEY, Dr. Charles Singer. Thoroughly engrossing non-technical survey. 270 illustrations. 211pp. $5\frac{3}{8} \times 8\frac{1}{2}$. 20389-1 Pa. $4.95

REDOUTE ROSES IRON-ON TRANSFER PATTERNS, Barbara Christopher. Redouté was botanical painter to the Empress Josephine; transfer his famous roses onto fabric with these 24 transfer patterns. 80pp. $8\frac{1}{4} \times 10\frac{7}{8}$. 24292-7 Pa. $3.50

THE FIVE BOOKS OF ARCHITECTURE, Sebastiano Serlio. Architectural milestone, first (1611) English translation of Renaissance classic. Unabridged reproduction of original edition includes over 300 woodcut illustrations. 416pp. $9\frac{3}{8} \times 12\frac{1}{4}$. 24349-4 Pa. $14.95

CARLSON'S GUIDE TO LANDSCAPE PAINTING, John F. Carlson. Authoritative, comprehensive guide covers, every aspect of landscape painting. 34 reproductions of paintings by author; 58 explanatory diagrams. 144pp. $8\frac{3}{8} \times 11$. 22927-0 Pa. $5.95

101 PUZZLES IN THOUGHT AND LOGIC, C.R. Wylie, Jr. Solve murders, robberies, see which fishermen are liars—purely by reasoning! 107pp. $5\frac{3}{8} \times 8\frac{1}{2}$. 20367-0 Pa. $2.00

TEST YOUR LOGIC, George J. Summers. 50 more truly new puzzles with new turns of thought, new subtleties of inference. 100pp. $5\frac{3}{8} \times 8\frac{1}{2}$. 22877-0 Pa. $2.25

SMOCKING: TECHNIQUE, PROJECTS, AND DESIGNS, Dianne Durand. Foremost smocking designer provides complete instructions on how to smock. Over 10 projects, over 100 illustrations. 56pp. 8¼ × 11. 23788-5 Pa. $2.00

AUDUBON'S BIRDS IN COLOR FOR DECOUPAGE, edited by Eleanor H. Rawlings. 24 sheets, 37 most decorative birds, full color, on one side of paper. Instructions, including work under glass. 56pp. 8¼ × 11. 23492-4 Pa. $3.95

THE COMPLETE BOOK OF SILK SCREEN PRINTING PRODUCTION, J.I. Biegeleisen. For commercial user, teacher in advanced classes, serious hobbyist. Most modern techniques, materials, equipment for optimal results. 124 illustrations. 253pp. 5⅜ × 8½. 21100-2 Pa. $4.50

A TREASURY OF ART NOUVEAU DESIGN AND ORNAMENT, edited by Carol Belanger Grafton. 577 designs for the practicing artist. Full-page, spots, borders, bookplates by Klimt, Bradley, others. 144pp. 8⅜ × 11¼. 24001-0 Pa. $5.95

ART NOUVEAU TYPOGRAPHIC ORNAMENTS, Dan X. Solo. Over 800 Art Nouveau florals, swirls, women, animals, borders, scrolls, wreaths, spots and dingbats, copyright-free. 100pp. 8⅛ × 11. 24366-4 Pa. $4.00

HAND SHADOWS TO BE THROWN UPON THE WALL, Henry Bursill. Wonderful Victorian novelty tells how to make flying birds, dog, goose, deer, and 14 others, each explained by a full-page illustration. 32pp. 6½ × 9¼. 21779-5 Pa. $1.50

AUDUBON'S BIRDS OF AMERICA COLORING BOOK, John James Audubon. Rendered for coloring by Paul Kennedy. 46 of Audubon's noted illustrations: red-winged black-bird, cardinal, etc. Original plates reproduced in full-color on the covers. Captions. 48pp. 8¼ × 11. 23049-X Pa. $2.25

SILK SCREEN TECHNIQUES, J.I. Biegeleisen, M.A. Cohn. Clear, practical, modern, economical. Minimal equipment (self-built), materials, easy methods. For amateur, hobbyist, 1st book. 141 illustrations. 185pp. 6⅛ × 9¼. 20433-2 Pa. $3.95

101 PATCHWORK PATTERNS, Ruby S. McKim. 101 beautiful, immediately useable patterns, full-size, modern and traditional. Also general information, estimating, quilt lore. 140 illustrations. 124pp. 7⅞ × 10¾. 20773-0 Pa. $3.50

READY-TO-USE FLORAL DESIGNS, Ed Sibbett, Jr. Over 100 floral designs (most in three sizes) of popular individual blossoms as well as bouquets, sprays, garlands. 64pp. 8¼ × 11. 23976-4 Pa. $2.95

AMERICAN WILD FLOWERS COLORING BOOK, Paul Kennedy. Planned coverage of 46 most important wildflowers, from Rickett's collection; instructive as well as entertaining. Color versions on covers. Captions. 48pp. 8¼ × 11. 20095-7 Pa. $2.50

CARVING DUCK DECOYS, Harry V. Shourds and Anthony Hillman. Detailed instructions and full-size templates for constructing 16 beautiful, marvelously practical decoys according to time-honored South Jersey method. 70pp. 9¼ × 12¼. 24083-5 Pa. $4.95

TRADITIONAL PATCHWORK PATTERNS, Carol Belanger Grafton. Cardboard cut-out pieces for use as templates to make 12 quilts: Buttercup, Ribbon Border, Tree of Paradise, nine more. Full instructions. 57pp. 8¼ × 11. 23015-5 Pa. $3.50

25 KITES THAT FLY, Leslie Hunt. Full, easy-to-follow instructions for kites made from inexpensive materials. Many novelties. 70 illustrations. 110pp. 5⅜ × 8½.
22550-X Pa. $2.25

PIANO TUNING, J. Cree Fischer. Clearest, best book for beginner, amateur. Simple repairs, raising dropped notes, tuning by easy method of flattened fifths. No previous skills needed. 4 illustrations. 201pp. 5⅜ × 8½. 23267-0 Pa. $3.50

EARLY AMERICAN IRON-ON TRANSFER PATTERNS, edited by Rita Weiss. 75 designs, borders, alphabets, from traditional American sources. 48pp. 8¼ × 11.
23162-3 Pa. $1.95

CROCHETING EDGINGS, edited by Rita Weiss. Over 100 of the best designs for these lovely trims for a host of household items. Complete instructions, illustrations. 48pp. 8¼ × 11. 24031-2 Pa. $2.25

FINGER PLAYS FOR NURSERY AND KINDERGARTEN, Emilie Poulsson. 18 finger plays with music (voice and piano); entertaining, instructive. Counting, nature lore, etc. Victorian classic. 53 illustrations. 80pp. 6½ × 9¼. 22588-7 Pa. $1.95

BOSTON THEN AND NOW, Peter Vanderwarker. Here in 59 side-by-side views are photographic documentations of the city's past and present. 119 photographs. Full captions. 122pp. 8¼ × 11. 24312-5 Pa. $6.95

CROCHETING BEDSPREADS, edited by Rita Weiss. 22 patterns, originally published in three instruction books 1939-41. 39 photos, 8 charts. Instructions. 48pp. 8¼ × 11. 23610-2 Pa. $2.00

HAWTHORNE ON PAINTING, Charles W. Hawthorne. Collected from notes taken by students at famous Cape Cod School; hundreds of direct, personal *apercus*, ideas, suggestions. 91pp. 5⅜ × 8½. 20653-X Pa. $2.50

THERMODYNAMICS, Enrico Fermi. A classic of modern science. Clear, organized treatment of systems, first and second laws, entropy, thermodynamic potentials, etc. Calculus required. 160pp. 5⅜ × 8½. 60361-X Pa. $4.00

TEN BOOKS ON ARCHITECTURE, Vitruvius. The most important book ever written on architecture. Early Roman aesthetics, technology, classical orders, site selection, all other aspects. Morgan translation. 331pp. 5⅜ × 8½. 20645-9 Pa. $5.50

THE CORNELL BREAD BOOK, Clive M. McCay and Jeanette B. McCay. Famed high-protein recipe incorporated into breads, rolls, buns, coffee cakes, pizza, pie crusts, more. Nearly 50 illustrations. 48pp. 8¼ × 11. 23995-0 Pa. $2.00

THE CRAFTSMAN'S HANDBOOK, Cennino Cennini. 15th-century handbook, school of Giotto, explains applying gold, silver leaf; gesso; fresco painting, grinding pigments, etc. 142pp. 6⅛ × 9¼. 20054-X Pa. $3.50

FRANK LLOYD WRIGHT'S FALLINGWATER, Donald Hoffmann. Full story of Wright's masterwork at Bear Run, Pa. 100 photographs of site, construction, and details of completed structure. 112pp. 9¼ × 10. 23671-4 Pa. $6.95

OVAL STAINED GLASS PATTERN BOOK, C. Eaton. 60 new designs framed in shape of an oval. Greater complexity, challenge with sinuous cats, birds, mandalas framed in antique shape. 64pp. 8¼ × 11. 24519-5 Pa. $3.50

CHILDREN'S BOOKPLATES AND LABELS, Ed Sibbett, Jr. 6 each of 12 types based on *Wizard of Oz, Alice,* nursery rhymes, fairy tales. Perforated; full color. 24pp. 8¼ × 11. 23538-6 Pa. $3.50

READY-TO-USE VICTORIAN COLOR STICKERS: 96 Pressure-Sensitive Seals, Carol Belanger Grafton. Drawn from authentic period sources. Motifs include heads of men, women, children, plus florals, animals, birds, more. Will adhere to any clean surface. 8pp. 8½ × 11. 24551-9 Pa. $2.95

CUT AND FOLD PAPER SPACESHIPS THAT FLY, Michael Grater. 16 colorful, easy-to-build spaceships that really fly. Star Shuttle, Lunar Freighter, Star Probe, 13 others. 32pp. 8¼ × 11. 23978-0 Pa. $2.50

CUT AND ASSEMBLE PAPER AIRPLANES THAT FLY, Arthur Baker. 8 aerodynamically sound, ready-to-build paper airplanes, designed with latest techniques. Fly *Pegasus, Daedalus, Songbird,* 5 other aircraft. Instructions. 32pp. 9¼ × 11¼. 24302-8 Pa. $3.95

SIDELIGHTS ON RELATIVITY, Albert Einstein. Two lectures delivered in 1920-21: *Ether and Relativity* and *Geometry and Experience.* Elegant ideas in non-mathematical form. 56pp. 5⅜ × 8½. 24511-X Pa. $2.25

FADS AND FALLACIES IN THE NAME OF SCIENCE, Martin Gardner. Fair, witty appraisal of cranks and quacks of science: Velikovsky, orgone energy, Bridey Murphy, medical fads, etc. 373pp. 5⅜ × 8½. 20394-8 Pa. $5.95

VACATION HOMES AND CABINS, U.S. Dept. of Agriculture. Complete plans for 16 cabins, vacation homes and other shelters. 105pp. 9 × 12. 23631-5 Pa. $4.95

HOW TO BUILD A WOOD-FRAME HOUSE, L.O. Anderson. Placement, foundations, framing, sheathing, roof, insulation, plaster, finishing—almost everything else. 179 illustrations. 223pp. 7⅞ × 10¾. 22954-8 Pa. $5.50

THE MYSTERY OF A HANSOM CAB, Fergus W. Hume. Bizarre murder in a hansom cab leads to engrossing investigation. Memorable characters, rich atmosphere. 19th-century bestseller, still enjoyable, exciting. 256pp. 5⅜ × 8. 21956-9 Pa. $4.00

MANUAL OF TRADITIONAL WOOD CARVING, edited by Paul N. Hasluck. Possibly the best book in English on the craft of wood carving. Practical instructions, along with 1,146 working drawings and photographic illustrations. 576pp. 6½ × 9¼. 23489-4 Pa. $8.95

WHITTLING AND WOODCARVING, E.J Tangerman. Best book on market; clear, full. If you can cut a potato, you can carve toys, puzzles, chains, etc. Over 464 illustrations. 293pp. 5⅜ × 8½. 20965-2 Pa. $4.95

AMERICAN TRADEMARK DESIGNS, Barbara Baer Capitman. 732 marks, logos and corporate-identity symbols. Categories include entertainment, heavy industry, food and beverage. All black-and-white in standard forms. 160pp. 8⅜ × 11. 23259-X Pa. $6.95

DECORATIVE FRAMES AND BORDERS, edited by Edmund V. Gillon, Jr. Largest collection of borders and frames ever compiled for use of artists and designers. Renaissance, neo-Greek, Art Nouveau, Art Deco, to mention only a few styles. 396 illustrations. 192pp. 8⅜ × 11¼. 22928-9 Pa. $6.00

THE MURDER BOOK OF J.G. REEDER, Edgar Wallace. Eight suspenseful stories by bestselling mystery writer of 20s and 30s. Features the donnish Mr. J.G. Reeder of Public Prosecutor's Office. 128pp. 5⅜ × 8½. (Available in U.S. only)
24374-5 Pa. $3.50

ANNE ORR'S CHARTED DESIGNS, Anne Orr. Best designs by premier needlework designer, all on charts: flowers, borders, birds, children, alphabets, etc. Over 100 charts, 10 in color. Total of 40pp. 8¼ × 11.
23704-4 Pa. $2.50

BASIC CONSTRUCTION TECHNIQUES FOR HOUSES AND SMALL BUILDINGS SIMPLY EXPLAINED, U.S. Bureau of Naval Personnel. Grading, masonry, woodworking, floor and wall framing, roof framing, plastering, tile setting, much more. Over 675 illustrations. 568pp. 6½ × 9¼.
20242-9 Pa. $8.95

MATISSE LINE DRAWINGS AND PRINTS, Henri Matisse. Representative collection of female nudes, faces, still lifes, experimental works, etc., from 1898 to 1948. 50 illustrations. 48pp. 8⅜ × 11¼.
23877-6 Pa. $2.50

HOW TO PLAY THE CHESS OPENINGS, Eugene Znosko-Borovsky. Clear, profound examinations of just what each opening is intended to do and how opponent can counter. Many sample games. 147pp. 5⅜ × 8½.
22795-2 Pa. $2.95

DUPLICATE BRIDGE, Alfred Sheinwold. Clear, thorough, easily followed account: rules, etiquette, scoring, strategy, bidding; Goren's point-count system, Blackwood and Gerber conventions, etc. 158pp. 5⅜ × 8½.
22741-3 Pa. $3.00

SARGENT PORTRAIT DRAWINGS, J.S. Sargent. Collection of 42 portraits reveals technical skill and intuitive eye of noted American portrait painter, John Singer Sargent. 48pp. 8¼ × 11⅛.
24524-1 Pa. $2.95

ENTERTAINING SCIENCE EXPERIMENTS WITH EVERYDAY OBJECTS, Martin Gardner. Over 100 experiments for youngsters. Will amuse, astonish, teach, and entertain. Over 100 illustrations. 127pp. 5⅜ × 8½.
24201-3 Pa. $2.50

TEDDY BEAR PAPER DOLLS IN FULL COLOR: A Family of Four Bears and Their Costumes, Crystal Collins. A family of four Teddy Bear paper dolls and nearly 60 cut-out costumes. Full color, printed one side only. 32pp. 9¼ × 12¼.
24550-0 Pa. $3.50

NEW CALLIGRAPHIC ORNAMENTS AND FLOURISHES, Arthur Baker. Unusual, multi-useable material: arrows, pointing hands, brackets and frames, ovals, swirls, birds, etc. Nearly 700 illustrations. 80pp. 8⅜ × 11¼.
24095-9 Pa. $3.75

DINOSAUR DIORAMAS TO CUT & ASSEMBLE, M. Kalmenoff. Two complete three-dimensional scenes in full color, with 31 cut-out animals and plants. Excellent educational toy for youngsters. Instructions; 2 assembly diagrams. 32pp. 9¼ × 12¼.
24541-1 Pa. $4.50

SILHOUETTES: A PICTORIAL ARCHIVE OF VARIED ILLUSTRATIONS, edited by Carol Belanger Grafton. Over 600 silhouettes from the 18th to 20th centuries. Profiles and full figures of men, women, children, birds, animals, groups and scenes, nature, ships, an alphabet. 144pp. 8⅜ × 11¼.
23781-8 Pa. $4.95

SURREAL STICKERS AND UNREAL STAMPS, William Rowe. 224 haunting, hilarious stamps on gummed, perforated stock, with images of elephants, geisha girls, George Washington, etc. 16pp. one side. 8¼ × 11.   24371-0 Pa. $3.50

GOURMET KITCHEN LABELS, Ed Sibbett, Jr. 112 full-color labels (4 copies each of 28 designs). Fruit, bread, other culinary motifs. Gummed and perforated. 16pp. 8¼ × 11.   24087-8 Pa. $2.95

PATTERNS AND INSTRUCTIONS FOR CARVING AUTHENTIC BIRDS, H.D. Green. Detailed instructions, 27 diagrams, 85 photographs for carving 15 species of birds so life-like, they'll seem ready to fly! 8¼ × 11.   24222-6 Pa. $2.75

FLATLAND, E.A. Abbott. Science-fiction classic explores life of 2-D being in 3-D world. 16 illustrations. 103pp. 5⅜ × 8.   20001-9 Pa. $2.00

DRIED FLOWERS, Sarah Whitlock and Martha Rankin. Concise, clear, practical guide to dehydration, glycerinizing, pressing plant material, and more. Covers use of silica gel. 12 drawings. 32pp. 5⅜ × 8½.   21802-3 Pa. $1.00

EASY-TO-MAKE CANDLES, Gary V. Guy. Learn how easy it is to make all kinds of decorative candles. Step-by-step instructions. 82 illustrations. 48pp. 8¼ × 11.
          23881-4 Pa. $2.50

SUPER STICKERS FOR KIDS, Carolyn Bracken. 128 gummed and perforated full-color stickers: GIRL WANTED, KEEP OUT, BORED OF EDUCATION, X-RATED, COMBAT ZONE, many others. 16pp. 8¼ × 11.   24092-4 Pa. $2.50

CUT AND COLOR PAPER MASKS, Michael Grater. Clowns, animals, funny faces...simply color them in, cut them out, and put them together, and you have 9 paper masks to play with and enjoy. 32pp. 8¼ × 11.   23171-2 Pa. $2.25

A CHRISTMAS CAROL: THE ORIGINAL MANUSCRIPT, Charles Dickens. Clear facsimile of Dickens manuscript, on facing pages with final printed text. 8 illustrations by John Leech, 4 in color on covers. 144pp. 8⅜ × 11¼.
          20980-6 Pa. $5.95

CARVING SHOREBIRDS, Harry V. Shourds & Anthony Hillman. 16 full-size patterns (all double-page spreads) for 19 North American shorebirds with step-by-step instructions. 72pp. 9¼ × 12¼.   24287-0 Pa. $4.95

THE GENTLE ART OF MATHEMATICS, Dan Pedoe. Mathematical games, probability, the question of infinity, topology, how the laws of algebra work, problems of irrational numbers, and more. 42 figures. 143pp. 5⅜ × 8½. (EBE)
          22949-1 Pa. $3.50

READY-TO-USE DOLLHOUSE WALLPAPER, Katzenbach & Warren, Inc. Stripe, 2 floral stripes, 2 allover florals, polka dot; all in full color. 4 sheets (350 sq. in.) of each, enough for average room. 48pp. 8¼ × 11.   23495-9 Pa. $2.95

MINIATURE IRON-ON TRANSFER PATTERNS FOR DOLLHOUSES, DOLLS, AND SMALL PROJECTS, Rita Weiss and Frank Fontana. Over 100 miniature patterns: rugs, bedspreads, quilts, chair seats, etc. In standard dollhouse size. 48pp. 8¼ × 11.   23741-9 Pa. $1.95

THE DINOSAUR COLORING BOOK, Anthony Rao. 45 renderings of dinosaurs, fossil birds, turtles, other creatures of Mesozoic Era. Scientifically accurate. Captions. 48pp. 8¼ × 11.   24022-3 Pa. $2.50

THE BOOK OF WOOD CARVING, Charles Marshall Sayers. Still finest book for beginning student. Fundamentals, technique; gives 34 designs, over 34 projects for panels, bookends, mirrors, etc. 33 photos. 118pp. 7¾ × 10⅝. 23654-4 Pa. $3.95

CARVING COUNTRY CHARACTERS, Bill Higginbotham. Expert advice for beginning, advanced carvers on materials, techniques for creating 18 projects—mirthful panorama of American characters. 105 illustrations. 80pp. 8⅜ × 11. 24135-1 Pa. $2.50

300 ART NOUVEAU DESIGNS AND MOTIFS IN FULL COLOR, C.B. Grafton. 44 full-page plates display swirling lines and muted colors typical of Art Nouveau. Borders, frames, panels, cartouches, dingbats, etc. 48pp. 9⅜ × 12¼. 24354-0 Pa. $6.95

SELF-WORKING CARD TRICKS, Karl Fulves. Editor of *Pallbearer* offers 72 tricks that work automatically through nature of card deck. No sleight of hand needed. Often spectacular. 42 illustrations. 113pp. 5⅜ × 8½. 23334-0 Pa. $3.50

CUT AND ASSEMBLE A WESTERN FRONTIER TOWN, Edmund V. Gillon, Jr. Ten authentic full-color buildings on heavy cardboard stock in H-O scale. Sheriff's Office and Jail, Saloon, Wells Fargo, Opera House, others. 48pp. 9¼ × 12¼. 23736-2 Pa. $3.95

CUT AND ASSEMBLE AN EARLY NEW ENGLAND VILLAGE, Edmund V. Gillon, Jr. Printed in full color on heavy cardboard stock. 12 authentic buildings in H-O scale: Adams home in Quincy, Mass., Oliver Wight house in Sturbridge, smithy, store, church, others. 48pp. 9¼ × 12¼. 23536-X Pa. $4.95

THE TALE OF TWO BAD MICE, Beatrix Potter. Tom Thumb and Hunca Munca squeeze out of their hole and go exploring. 27 full-color Potter illustrations. 59pp. 4¼ × 5½. (Available in U.S. only) 23065-1 Pa. $1.75

CARVING FIGURE CARICATURES IN THE OZARK STYLE, Harold L. Enlow. Instructions and illustrations for ten delightful projects, plus general carving instructions. 22 drawings and 47 photographs altogether. 39pp. 8⅜ × 11. 23151-8 Pa. $2.50

A TREASURY OF FLOWER DESIGNS FOR ARTISTS, EMBROIDERERS AND CRAFTSMEN, Susan Gaber. 100 garden favorites lushly rendered by artist for artists, craftsmen, needleworkers. Many form frames, borders. 80pp. 8¼ × 11. 24096-7 Pa. $3.50

CUT & ASSEMBLE A TOY THEATER/THE NUTCRACKER BALLET, Tom Tierney. Model of a complete, full-color production of Tchaikovsky's classic. 6 backdrops, dozens of characters, familiar dance sequences. 32pp. 9⅜ × 12¼. 24194-7 Pa. $4.50

ANIMALS: 1,419 COPYRIGHT-FREE ILLUSTRATIONS OF MAMMALS, BIRDS, FISH, INSECTS, ETC., edited by Jim Harter. Clear wood engravings present, in extremely lifelike poses, over 1,000 species of animals. 284pp. 9 × 12. 23766-4 Pa. $9.95

MORE HAND SHADOWS, Henry Bursill. For those at their 'finger ends," 16 more effects—Shakespeare, a hare, a squirrel, Mr. Punch, and twelve more—each explained by a full-page illustration. Considerable period charm. 30pp. 6½ × 9¼. 21384-6 Pa. $1.95

JAPANESE DESIGN MOTIFS, Matsuya Co. Mon, or heraldic designs. Over 4000 typical, beautiful designs: birds, animals, flowers, swords, fans, geometrics; all beautifully stylized. 213pp. 11⅜ × 8¼. 22874-6 Pa. $7.95

THE TALE OF BENJAMIN BUNNY, Beatrix Potter. Peter Rabbit's cousin coaxes him back into Mr. McGregor's garden for a whole new set of adventures. All 27 full-color illustrations. 59pp. 4¼ × 5½. (Available in U.S. only) 21102-9 Pa. $1.75

THE TALE OF PETER RABBIT AND OTHER FAVORITE STORIES BOXED SET, Beatrix Potter. Seven of Beatrix Potter's best-loved tales including Peter Rabbit in a specially designed, durable boxed set. 4¼ × 5½. Total of 447pp. 158 color illustrations. (Available in U.S. only) 23903-9 Pa. $10.80

PRACTICAL MENTAL MAGIC, Theodore Annemann. Nearly 200 astonishing feats of mental magic revealed in step-by-step detail. Complete advice on staging, patter, etc. Illustrated. 320pp. 5⅜ × 8½. 24426-1 Pa. $5.95

CELEBRATED CASES OF JUDGE DEE (DEE GOONG AN), translated by Robert Van Gulik. Authentic 18th-century Chinese detective novel; Dee and associates solve three interlocked cases. Led to van Gulik's own stories with same characters. Extensive introduction. 9 illustrations. 237pp. 5⅜ × 8½. 23337-5 Pa. $4.50

CUT & FOLD EXTRATERRESTRIAL INVADERS THAT FLY, M. Grater. Stage your own lilliputian space battles. By following the step-by-step instructions and explanatory diagrams you can launch 22 full-color fliers into space. 36pp. 8¼ × 11. 24478-4 Pa. $2.95

CUT & ASSEMBLE VICTORIAN HOUSES, Edmund V. Gillon, Jr. Printed in full color on heavy cardboard stock, 4 authentic Victorian houses in H-O scale. Italian-style Villa, Octagon, Second Empire, Stick Style. 48pp. 9¼ × 12¼. 23849-0 Pa. $3.95

BEST SCIENCE FICTION STORIES OF H.G. WELLS, H.G. Wells. Full novel The Invisible Man, plus 17 short stories: "The Crystal Egg," "Aepyornis Island," "The Strange Orchid," etc. 303pp. 5⅜ × 8½. (Available in U.S. only) 21531-8 Pa. $4.95

TRADEMARK DESIGNS OF THE WORLD, Yusaku Kamekura. A lavish collection of nearly 700 trademarks, the work of Wright, Loewy, Klee, Binder, hundreds of others. 160pp. 8¾ × 8. (Available in U.S. only) 24191-2 Pa. $5.95

THE ARTIST'S AND CRAFTSMAN'S GUIDE TO REDUCING, ENLARGING AND TRANSFERRING DESIGNS, Rita Weiss. Discover, reduce, enlarge, transfer designs from any objects to any craft project. 12pp. plus 16 sheets special graph paper. 8¼ × 11. 24142-4 Pa. $3.50

TREASURY OF JAPANESE DESIGNS AND MOTIFS FOR ARTISTS AND CRAFTSMEN, edited by Carol Belanger Grafton. Indispensable collection of 360 traditional Japanese designs and motifs redrawn in clean, crisp black-and-white, copyright-free illustrations. 96pp. 8¼ × 11. 24435-0 Pa. $3.95

CHANCERY CURSIVE STROKE BY STROKE, Arthur Baker. Instructions and illustrations for each stroke of each letter (upper and lower case) and numerals. 54 full-page plates. 64pp. 8¼ × 11.   24278-1 Pa. $2.50

THE ENJOYMENT AND USE OF COLOR, Walter Sargent. Color relationships, values, intensities; complementary colors, illumination, similar topics. Color in nature and art. 7 color plates, 29 illustrations. 274pp. 5⅜ × 8½.   20944-X Pa. $4.95

SCULPTURE PRINCIPLES AND PRACTICE, Louis Slobodkin. Step-by-step approach to clay, plaster, metals, stone; classical and modern. 253 drawings, photos. 255pp. 8⅛ × 11.   22960-2 Pa. $7.50

VICTORIAN FASHION PAPER DOLLS FROM HARPER'S BAZAR, 1867-1898, Theodore Menten. Four female dolls with 28 elegant high fashion costumes, printed in full color. 32pp. 9¼ × 12¼.   23453-3 Pa. $3.50

FLOPSY, MOPSY AND COTTONTAIL: A Little Book of Paper Dolls in Full Color, Susan LaBelle. Three dolls and 21 costumes (7 for each doll) show Peter Rabbit's siblings dressed for holidays, gardening, hiking, etc. Charming borders, captions. 48pp. 4¼ × 5½.   24376-1 Pa. $2.25

NATIONAL LEAGUE BASEBALL CARD CLASSICS, Bert Randolph Sugar. 83 big-leaguers from 1909-69 on facsimile cards. Hubbell, Dean, Spahn, Brock plus advertising, info, no duplications. Perforated, detachable. 16pp. 8¼ × 11.
24308-7 Pa. $2.95

THE LOGICAL APPROACH TO CHESS, Dr. Max Euwe, et al. First-rate text of comprehensive strategy, tactics, theory for the amateur. No gambits to memorize, just a clear, logical approach. 224pp. 5⅜ × 8½.   24353-2 Pa. $4.50

MAGICK IN THEORY AND PRACTICE, Aleister Crowley. The summation of the thought and practice of the century's most famous necromancer, long hard to find. Crowley's best book. 436pp. 5⅜ × 8½. (Available in U.S. only)
23295-6 Pa. $6.50

THE HAUNTED HOTEL, Wilkie Collins. Collins' last great tale; doom and destiny in a Venetian palace. Praised by T.S. Eliot. 127pp. 5⅜ × 8½.
24333-8 Pa. $3.00

ART DECO DISPLAY ALPHABETS, Dan X. Solo. Wide variety of bold yet elegant lettering in handsome Art Deco styles. 100 complete fonts, with numerals, punctuation, more. 104pp. 8⅛ × 11.   24372-9 Pa. $4.50

CALLIGRAPHIC ALPHABETS, Arthur Baker. Nearly 150 complete alphabets by outstanding contemporary. Stimulating ideas; useful source for unique effects. 154 plates. 157pp. 8⅜ × 11¼.   21045-6 Pa. $5.95

ARTHUR BAKER'S HISTORIC CALLIGRAPHIC ALPHABETS, Arthur Baker. From monumental capitals of first-century Rome to humanistic cursive of 16th century, 33 alphabets in fresh interpretations. 88 plates. 96pp. 9 × 12.
24054-1 Pa. $4.50

LETTIE LANE PAPER DOLLS, Sheila Young. Genteel turn-of-the-century family very popular then and now. 24 paper dolls. 16 plates in full color. 32pp. 9¼ × 12¼.   24089-4 Pa. $3.50

TWENTY-FOUR ART NOUVEAU POSTCARDS IN FULL COLOR FROM CLASSIC POSTERS, Hayward and Blanche Cirker. Ready-to-mail postcards reproduced from rare set of poster art. Works by Toulouse-Lautrec, Parrish, Steinlen, Mucha, Cheret, others. 12pp. 8¼× 11. 24389-3 Pa. $2.95

READY-TO-USE ART NOUVEAU BOOKMARKS IN FULL COLOR, Carol Belanger Grafton. 30 elegant bookmarks featuring graceful, flowing lines, foliate motifs, sensuous women characteristic of Art Nouveau. Perforated for easy detaching. 16pp. 8¼ × 11. 24305-2 Pa. $2.95

FRUIT KEY AND TWIG KEY TO TREES AND SHRUBS, William M. Harlow. Fruit key covers 120 deciduous and evergreen species; twig key covers 160 deciduous species. Easily used. Over 300 photographs. 126pp. 5⅜ × 8½. 20511-8 Pa. $2.25

LEONARDO DRAWINGS, Leonardo da Vinci. Plants, landscapes, human face and figure, etc., plus studies for Sforza monument, *Last Supper*, more. 60 illustrations. 64pp. 8¼ × 11⅛. 23951-9 Pa. $2.75

CLASSIC BASEBALL CARDS, edited by Bert R. Sugar. 98 classic cards on heavy stock, full color, perforated for detaching. Ruth, Cobb, Durocher, DiMaggio, H. Wagner, 99 others. Rare originals cost hundreds. 16pp. 8¼ × 11. 23498-3 Pa. $3.25

TREES OF THE EASTERN AND CENTRAL UNITED STATES AND CANADA, William M. Harlow. Best one-volume guide to 140 trees. Full descriptions, woodlore, range, etc. Over 600 illustrations. Handy size. 288pp. 4½ × 6⅜. 20395-6 Pa. $3.95

JUDY GARLAND PAPER DOLLS IN FULL COLOR, Tom Tierney. 3 Judy Garland paper dolls (teenager, grown-up, and mature woman) and 30 gorgeous costumes highlighting memorable career. Captions. 32pp. 9¼ × 12¼. 24404-0 Pa. $3.50

GREAT FASHION DESIGNS OF THE BELLE EPOQUE PAPER DOLLS IN FULL COLOR, Tom Tierney. Two dolls and 30 costumes meticulously rendered. Haute couture by Worth, Lanvin, Paquin, other greats late Victorian to WWI. 32pp. 9¼ × 12¼. 24425-3 Pa. $3.50

FASHION PAPER DOLLS FROM GODEY'S LADY'S BOOK, 1840-1854, Susan Johnston. In full color: 7 female fashion dolls with 50 costumes. Little girl's, bridal, riding, bathing, wedding, evening, everyday, etc. 32pp. 9¼ × 12¼. 23511-4 Pa. $3.95

THE BOOK OF THE SACRED MAGIC OF ABRAMELIN THE MAGE, translated by S. MacGregor Mathers. Medieval manuscript of ceremonial magic. Basic document in Aleister Crowley, Golden Dawn groups. 268pp. 5⅜ × 8½. 23211-5 Pa. $5.00

PETER RABBIT POSTCARDS IN FULL COLOR: 24 Ready-to-Mail Cards, Susan Whited LaBelle. Bunnies ice-skating, coloring Easter eggs, making valentines, many other charming scenes. 24 perforated full-color postcards, each measuring 4¼ × 6, on coated stock. 12pp. 9 × 12. 24617-5 Pa. $2.95

CELTIC HAND STROKE BY STROKE, A. Baker. Complete guide creating each letter of the alphabet in distinctive Celtic manner. Covers hand position, strokes, pens, inks, paper, more. Illustrated. 48pp. 8¼ × 11. 24336-2 Pa. $2.50

KEYBOARD WORKS FOR SOLO INSTRUMENTS, G.F. Handel. 35 neglected works from Handel's vast oeuvre, originally jotted down as improvisations. Includes Eight Great Suites, others. New sequence. 174pp. 9⅜ × 12¼.
24338-9 Pa. $7.50

AMERICAN LEAGUE BASEBALL CARD CLASSICS, Bert Randolph Sugar. 82 stars from 1900s to 60s on facsimile cards. Ruth, Cobb, Mantle, Williams, plus advertising, info, no duplications. Perforated, detachable. 16pp. 8¼ × 11.
24286-2 Pa. $2.95

A TREASURY OF CHARTED DESIGNS FOR NEEDLEWORKERS, Georgia Gorham and Jeanne Warth. 141 charted designs: owl, cat with yarn, tulips, piano, spinning wheel, covered bridge, Victorian house and many others. 48pp. 8¼ × 11.
23558-0 Pa. $1.95

DANISH FLORAL CHARTED DESIGNS, Gerda Bengtsson. Exquisite collection of over 40 different florals: anemone, Iceland poppy, wild fruit, pansies, many others. 45 illustrations. 48pp. 8¼ × 11. 23957-8 Pa. $1.75

OLD PHILADELPHIA IN EARLY PHOTOGRAPHS 1839-1914, Robert F. Looney. 215 photographs: panoramas, street scenes, landmarks, President-elect Lincoln's visit, 1876 Centennial Exposition, much more. 230pp. 8⅞ × 11¾.
23345-6 Pa. $9.95

PRELUDE TO MATHEMATICS, W.W. Sawyer. Noted mathematician's lively, stimulating account of non-Euclidean geometry, matrices, determinants, group theory, other topics. Emphasis on novel, striking aspects. 224pp. 5⅜ × 8½.
24401-6 Pa. $4.50

ADVENTURES WITH A MICROSCOPE, Richard Headstrom. 59 adventures with clothing fibers, protozoa, ferns and lichens, roots and leaves, much more. 142 illustrations. 232pp. 5⅜ × 8½. 23471-1 Pa. $3.95

IDENTIFYING ANIMAL TRACKS: MAMMALS, BIRDS, AND OTHER ANIMALS OF THE EASTERN UNITED STATES, Richard Headstrom. For hunters, naturalists, scouts, nature-lovers. Diagrams of tracks, tips on identification. 128pp. 5⅜ × 8. 24442-3 Pa. $3.50

VICTORIAN FASHIONS AND COSTUMES FROM HARPER'S BAZAR, 1867-1898, edited by Stella Blum. Day costumes, evening wear, sports clothes, shoes, hats, other accessories in over 1,000 detailed engravings. 320pp. 9⅜ × 12¼.
22990-4 Pa. $10.95

EVERYDAY FASHIONS OF THE TWENTIES AS PICTURED IN SEARS AND OTHER CATALOGS, edited by Stella Blum. Actual dress of the Roaring Twenties, with text by Stella Blum. Over 750 illustrations, captions. 156pp. 9 × 12.
24134-3 Pa. $8.50

HALL OF FAME BASEBALL CARDS, edited by Bert Randolph Sugar. Cy Young, Ted Williams, Lou Gehrig, and many other Hall of Fame greats on 92 full-color, detachable reprints of early baseball cards. No duplication of cards with *Classic Baseball Cards*. 16pp. 8¼ × 11. 23624-2 Pa. $3.50

THE ART OF HAND LETTERING, Helm Wotzkow. Course in hand lettering, Roman, Gothic, Italic, Block, Script. Tools, proportions, optical aspects, individual variation. Very quality conscious. Hundreds of specimens. 320pp. 5⅜ × 8½.
21797-3 Pa. $4.95

HOW THE OTHER HALF LIVES, Jacob A. Riis. Journalistic record of filth, degradation, upward drive in New York immigrant slums, shops, around 1900. New edition includes 100 original Riis photos, monuments of early photography. 233pp. 10 × 7⅞. 22012-5 Pa. $7.95

CHINA AND ITS PEOPLE IN EARLY PHOTOGRAPHS, John Thomson. In 200 black-and-white photographs of exceptional quality photographic pioneer Thomson captures the mountains, dwellings, monuments and people of 19th-century China. 272pp. 9⅜ × 12¼. 24393-1 Pa. $12.95

GODEY COSTUME PLATES IN COLOR FOR DECOUPAGE AND FRAMING, edited by Eleanor Hasbrouk Rawlings. 24 full-color engravings depicting 19th-century Parisian haute couture. Printed on one side only. 56pp. 8¼ × 11. 23879-2 Pa. $3.95

ART NOUVEAU STAINED GLASS PATTERN BOOK, Ed Sibbett, Jr. 104 projects using well-known themes of Art Nouveau: swirling forms, florals, peacocks, and sensuous women. 60pp. 8¼ × 11. 23577-7 Pa. $3.50

QUICK AND EASY PATCHWORK ON THE SEWING MACHINE: Susan Aylsworth Murwin and Suzzy Payne. Instructions, diagrams show exactly how to machine sew 12 quilts. 48pp. of templates. 50 figures. 80pp. 8¼ × 11. 23770-2 Pa. $3.50

THE STANDARD BOOK OF QUILT MAKING AND COLLECTING, Marguerite Ickis. Full information, full-sized patterns for making 46 traditional quilts, also 150 other patterns. 483 illustrations. 273pp. 6⅞ × 9⅝. 20582-7 Pa. $5.95

LETTERING AND ALPHABETS, J. Albert Cavanagh. 85 complete alphabets lettered in various styles; instructions for spacing, roughs, brushwork. 121pp. 8¾ × 8. 20053-1 Pa. $3.95

LETTER FORMS: 110 COMPLETE ALPHABETS, Frederick Lambert. 110 sets of capital letters; 16 lower case alphabets; 70 sets of numbers and other symbols. 110pp. 8⅛ × 11. 22872-X Pa. $4.50

ORCHIDS AS HOUSE PLANTS, Rebecca Tyson Northen. Grow cattleyas and many other kinds of orchids—in a window, in a case, or under artificial light. 63 illustrations. 148pp. 5⅜ × 8½. 23261-1 Pa. $2.95

THE MUSHROOM HANDBOOK, Louis C.C. Krieger. Still the best popular handbook. Full descriptions of 259 species, extremely thorough text, poisons, folklore, etc. 32 color plates; 126 other illustrations. 560pp. 5⅜ × 8½. 21861-9 Pa. $8.50

THE DORÉ BIBLE ILLUSTRATIONS, Gustave Doré. All wonderful, detailed plates: Adam and Eve, Flood, Babylon, life of Jesus, etc. Brief King James text with each plate. 241 plates. 241pp. 9 × 12. 23004-X Pa. $8.95

THE BOOK OF KELLS: Selected Plates in Full Color, edited by Blanche Cirker. 32 full-page plates from greatest manuscript-icon of early Middle Ages. Fantastic, mysterious. Publisher's Note. Captions. 32pp. 9¾ × 12¼. 24345-1 Pa. $4.50

THE PERFECT WAGNERITE, George Bernard Shaw. Brilliant criticism of the Ring Cycle, with provocative interpretation of politics, economic theories behind the Ring. 136pp. 5⅜ × 8½. (Available in U.S. only) 21707-8 Pa. $3.00

THE RIME OF THE ANCIENT MARINER, Gustave Doré, S.T. Coleridge. Doré's finest work, 34 plates capture moods, subtleties of poem. Full text. 77pp. 9¼ × 12.
22305-1 Pa. $4.95

SONGS OF INNOCENCE, William Blake. The first and most popular of Blake's famous "Illuminated Books," in a facsimile edition reproducing all 31 brightly colored plates. Additional printed text of each poem. 64pp. 5¼ × 7.
22764-2 Pa. $3.50

AN INTRODUCTION TO INFORMATION THEORY, J.R. Pierce. Second (1980) edition of most impressive non-technical account available. Encoding, entropy, noisy channel, related areas, etc. 320pp. 5⅜ × 8½.
24061-4 Pa. $4.95

THE DIVINE PROPORTION: A STUDY IN MATHEMATICAL BEAUTY, H.E. Huntley. "Divine proportion" or "golden ratio" in poetry, Pascal's triangle, philosophy, psychology, music, mathematical figures, etc. Excellent bridge between science and art. 58 figures. 185pp. 5⅜ × 8½.
22254-3 Pa. $3.95

THE DOVER NEW YORK WALKING GUIDE: From the Battery to Wall Street, Mary J. Shapiro. Superb inexpensive guide to historic buildings and locales in lower Manhattan: Trinity Church, Bowling Green, more. Complete Text; maps. 36 illustrations. 48pp. 3⅞ × 9¼.
24225-0 Pa. $2.50

NEW YORK THEN AND NOW, Edward B. Watson, Edmund V. Gillon, Jr. 83 important Manhattan sites: on facing pages early photographs (1875-1925) and 1976 photos by Gillon. 172 illustrations. 171pp. 9¼ × 10.
23361-8 Pa. $7.95

HISTORIC COSTUME IN PICTURES, Braun & Schneider. Over 1450 costumed figures from dawn of civilization to end of 19th century. English captions. 125 plates. 256pp. 8⅜ × 11¼.
23150-X Pa. $7.50

VICTORIAN AND EDWARDIAN FASHION: A Photographic Survey, Alison Gernsheim. First fashion history completely illustrated by contemporary photographs. Full text plus 235 photos, 1840-1914, in which many celebrities appear. 240pp. 6½ × 9¼.
24205-6 Pa. $6.00

CHARTED CHRISTMAS DESIGNS FOR COUNTED CROSS-STITCH AND OTHER NEEDLECRAFTS, Lindberg Press. Charted designs for 45 beautiful needlecraft projects with many yuletide and wintertime motifs. 48pp. 8¼ × 11.
24356-7 Pa. $2.50

101 FOLK DESIGNS FOR COUNTED CROSS-STITCH AND OTHER NEEDLE-CRAFTS, Carter Houck. 101 authentic charted folk designs in a wide array of lovely representations with many suggestions for effective use. 48pp. 8¼ × 11.
24369-9 Pa. $2.25

FIVE ACRES AND INDEPENDENCE, Maurice G. Kains. Great back-to-the-land classic explains basics of self-sufficient farming. The one book to get. 95 illustrations. 397pp. 5⅜ × 8½.
20974-1 Pa. $4.95

A MODERN HERBAL, Margaret Grieve. Much the fullest, most exact, most useful compilation of herbal material. Gigantic alphabetical encyclopedia, from aconite to zedoary, gives botanical information, medical properties, folklore, economic uses, and much else. Indispensable to serious reader. 161 illustrations. 888pp. 6½ × 9¼. (Available in U.S. only)
22798-7, 22799-5 Pa., Two-vol. set $16.45

DECORATIVE NAPKIN FOLDING FOR BEGINNERS, Lillian Oppenheimer and Natalie Epstein. 22 different napkin folds in the shape of a heart, clown's hat, love knot, etc. 63 drawings. 48pp. 8¼ × 11. 23797-4 Pa. $1.95

DECORATIVE LABELS FOR HOME CANNING, PRESERVING, AND OTHER HOUSEHOLD AND GIFT USES, Theodore Menten. 128 gummed, perforated labels, beautifully printed in 2 colors. 12 versions. Adhere to metal, glass, wood, ceramics. 24pp. 8¼ × 11. 23219-0 Pa. $2.95

EARLY AMERICAN STENCILS ON WALLS AND FURNITURE, Janet Waring. Thorough coverage of 19th-century folk art: techniques, artifacts, surviving specimens. 166 illustrations, 7 in color. 147pp. of text. 7⅞ × 10¾. 21906-2 Pa. $9.95

AMERICAN ANTIQUE WEATHERVANES, A.B. & W.T. Westervelt. Extensively illustrated 1883 catalog exhibiting over 550 copper weathervanes and finials. Excellent primary source by one of the principal manufacturers. 104pp. 6⅛ × 9¼. 24396-6 Pa. $3.95

ART STUDENTS' ANATOMY, Edmond J. Farris. Long favorite in art schools. Basic elements, common positions, actions. Full text, 158 illustrations. 159pp. 5⅜ × 8½. 20744-7 Pa. $3.95

BRIDGMAN'S LIFE DRAWING, George B. Bridgman. More than 500 drawings and text teach you to abstract the body into its major masses. Also specific areas of anatomy. 192pp. 6½ × 9¼. (EA) 22710-3 Pa. $4.50

COMPLETE PRELUDES AND ETUDES FOR SOLO PIANO, Frederic Chopin. All 26 Preludes, all 27 Etudes by greatest composer of piano music. Authoritative Paderewski edition. 224pp. 9 × 12. (Available in U.S. only) 24052-5 Pa. $7.50

PIANO MUSIC 1888-1905, Claude Debussy. Deux Arabesques, Suite Bergamesque, Masques, 1st series of Images, etc. 9 others, in corrected editions. 175pp. 9⅜ × 12¼. (ECE) 22771-5 Pa. $5.95

TEDDY BEAR IRON-ON TRANSFER PATTERNS, Ted Menten. 80 iron-on transfer patterns of male and female Teddys in a wide variety of activities, poses, sizes. 48pp. 8¼ × 11. 24596-9 Pa. $2.25

A PICTURE HISTORY OF THE BROOKLYN BRIDGE, M.J. Shapiro. Profusely illustrated account of greatest engineering achievement of 19th century. 167 rare photos & engravings recall construction, human drama. Extensive, detailed text. 122pp. 8¼ × 11. 24403-2 Pa. $7.95

NEW YORK IN THE THIRTIES, Berenice Abbott. Noted photographer's fascinating study shows new buildings that have become famous and old sights that have disappeared forever. 97 photographs. 97pp. 11⅜ × 10. 22967-X Pa. $7.50

MATHEMATICAL TABLES AND FORMULAS, Robert D. Carmichael and Edwin R. Smith. Logarithms, sines, tangents, trig functions, powers, roots, reciprocals, exponential and hyperbolic functions, formulas and theorems. 269pp. 5⅜ × 8½. 60111-0 Pa. $4.95

HANDBOOK OF MATHEMATICAL FUNCTIONS WITH FORMULAS, GRAPHS, AND MATHEMATICAL TABLES, edited by Milton Abramowitz and Irene A. Stegun. Vast compendium: 29 sets of tables, some to as high as 20 places. 1,046pp. 8 × 10½. 61272-4 Pa. $19.95

REASON IN ART, George Santayana. Renowned philosopher's provocative, seminal treatment of basis of art in instinct and experience. Volume Four of *The Life of Reason.* 230pp. 5⅜ × 8. 24358-3 Pa. $4.50

LANGUAGE, TRUTH AND LOGIC, Alfred J. Ayer. Famous, clear introduction to Vienna, Cambridge schools of Logical Positivism. Role of philosophy, elimination of metaphysics, nature of analysis, etc. 160pp. 5⅜ × 8½. (USCO) 20010-8 Pa. $2.75

BASIC ELECTRONICS, U.S. Bureau of Naval Personnel. Electron tubes, circuits, antennas, AM, FM, and CW transmission and receiving, etc. 560 illustrations. 567pp. 6½ × 9¼. 21076-6 Pa. $8.95

THE ART DECO STYLE, edited by Theodore Menten. Furniture, jewelry, metalwork, ceramics, fabrics, lighting fixtures, interior decors, exteriors, graphics from pure French sources. Over 400 photographs. 183pp. 8⅜ × 11¼. 22824-X Pa. $6.95

THE FOUR BOOKS OF ARCHITECTURE, Andrea Palladio. 16th-century classic covers classical architectural remains, Renaissance revivals, classical orders, etc. 1738 Ware English edition. 216 plates. 110pp. of text. 9½ × 12¾. 21308-0 Pa. $11.50

THE WIT AND HUMOR OF OSCAR WILDE, edited by Alvin Redman. More than 1000 ripostes, paradoxes, wisecracks: Work is the curse of the drinking classes, I can resist everything except temptations, etc. 258pp. 5⅜ × 8½. (USCO) 20602-5 Pa. $3.95

THE DEVIL'S DICTIONARY, Ambrose Bierce. Barbed, bitter, brilliant witticisms in the form of a dictionary. Best, most ferocious satire America has produced. 145pp. 5⅜ × 8½. 20487-1 Pa. $2.50

ERTÉ'S FASHION DESIGNS, Erté. 210 black-and-white inventions from *Harper's Bazar,* 1918-32, plus 8pp. full-color covers. Captions. 88pp. 9 × 12. 24203-X Pa. $6.50

ERTÉ GRAPHICS, Erté. Collection of striking color graphics: *Seasons, Alphabet, Numerals, Aces* and *Precious Stones.* 50 plates, including 4 on covers. 48pp. 9⅜ × 12¼. 23580-7 Pa. $6.95

PAPER FOLDING FOR BEGINNERS, William D. Murray and Francis J. Rigney. Clearest book for making origami sail boats, roosters, frogs that move legs, etc. 40 projects. More than 275 illustrations. 94pp. 5⅜ × 8½. 20713-7 Pa. $2.25

ORIGAMI FOR THE ENTHUSIAST, John Montroll. Fish, ostrich, peacock, squirrel, rhinoceros, Pegasus, 19 other intricate subjects. Instructions. Diagrams. 128pp. 9 × 12. 23799-0 Pa. $4.95

CROCHETING NOVELTY POT HOLDERS, edited by Linda Macho. 64 useful, whimsical pot holders feature kitchen themes, animals, flowers, other novelties. Surprisingly easy to crochet. Complete instructions. 48pp. 8¼ × 11. 24296-X Pa. $1.95

CROCHETING DOILIES, edited by Rita Weiss. Irish Crochet, Jewel, Star Wheel, Vanity Fair and more. Also luncheon and console sets, runners and centerpieces. 51 illustrations. 48pp. 8¼ × 11. 23424-X Pa. $2.50

YUCATAN BEFORE AND AFTER THE CONQUEST, Diego de Landa. Only significant account of Yucatan written in the early post-Conquest era. Translated by William Gates. Over 120 illustrations. 162pp. 5⅜ × 8½. 23622-6 Pa. $3.50

ORNATE PICTORIAL CALLIGRAPHY, E.A. Lupfer. Complete instructions, over 150 examples help you create magnificent "flourishes" from which beautiful animals and objects gracefully emerge. 8⅛ × 11. 21957-7 Pa. $2.95

DOLLY DINGLE PAPER DOLLS, Grace Drayton. Cute chubby children by same artist who did Campbell Kids. Rare plates from 1910s. 30 paper dolls and over 100 outfits reproduced in full color. 32pp. 9¼ × 12¼. 23711-7 Pa. $3.50

CURIOUS GEORGE PAPER DOLLS IN FULL COLOR, H. A. Rey, Kathy Allert. Naughty little monkey-hero of children's books in two doll figures, plus 48 full-color costumes: pirate, Indian chief, fireman, more. 32pp. 9¼ × 12¼. 24386-9 Pa. $3.50

GERMAN: HOW TO SPEAK AND WRITE IT, Joseph Rosenberg. Like *French, How to Speak and Write It.* Very rich modern course, with a wealth of pictorial material. 330 illustrations. 384pp. 5⅜ × 8½. (USUKO) 20271-2 Pa. $4.75

CATS AND KITTENS: 24 Ready-to-Mail Color Photo Postcards, D. Holby. Handsome collection; feline in a variety of adorable poses. Identifications. 12pp. on postcard stock. 8¼ × 11. 24469-5 Pa. $2.95

MARILYN MONROE PAPER DOLLS, Tom Tierney. 31 full-color designs on heavy stock, from *The Asphalt Jungle, Gentlemen Prefer Blondes,* 22 others.1 doll. 16 plates. 32pp. 9⅜ × 12¼. 23769-9 Pa. $3.50

FUNDAMENTALS OF LAYOUT, F.H. Wills. All phases of layout design discussed and illustrated in 121 illustrations. Indispensable as student's text or handbook for professional. 124pp. 8⅛. × 11. 21279-3 Pa. $4.50

FANTASTIC SUPER STICKERS, Ed Sibbett, Jr. 75 colorful pressure-sensitive stickers. Peel off and place for a touch of pizzazz: clowns, penguins, teddy bears, etc. Full color. 16pp. 8¼ × 11. 24471-7 Pa. $2.95

LABELS FOR ALL OCCASIONS, Ed Sibbett, Jr. 6 labels each of 16 different designs—baroque, art nouveau, art deco, Pennsylvania Dutch, etc.—in full color. 24pp. 8¼ × 11. 23688-9 Pa. $2.95

HOW TO CALCULATE QUICKLY: RAPID METHODS IN BASIC MATHEMATICS, Henry Sticker. Addition, subtraction, multiplication, division, checks, etc. More than 8000 problems, solutions. 185pp. 5 × 7¼. 20295-X Pa. $2.95

THE CAT COLORING BOOK, Karen Baldauski. Handsome, realistic renderings of 40 splendid felines, from American shorthair to exotic types. 44 plates. Captions. 48pp. 8¼ × 11. 24011-8 Pa. $2.25

THE TALE OF PETER RABBIT, Beatrix Potter. The inimitable Peter's terrifying adventure in Mr. McGregor's garden, with all 27 wonderful, full-color Potter illustrations. 55pp. 4¼ × 5½. (Available in U.S. only) 22827-4 Pa. $1.75

BASIC ELECTRICITY, U.S. Bureau of Naval Personnel. Batteries, circuits, conductors, AC and DC, inductance and capacitance, generators, motors, transformers, amplifiers, etc. 349 illustrations. 448pp. 6½ × 9¼. 20973-3 Pa. $7.95

TOLL HOUSE TRIED AND TRUE RECIPES, Ruth Graves Wakefield. Pop-overs, veal and ham loaf, baked beans, much more from the famous Mass. restaurant. Nearly 700 recipes. 376pp. 5⅜ × 8½. 23560-2 Pa. $4.95

FAVORITE CHRISTMAS CAROLS, selected and arranged by Charles J.F. Cofone. Title, music, first verse and refrain of 34 traditional carols in handsome calligraphy; also subsequent verses and other information in type. 79pp. 8⅜ × 11. 20445-6 Pa. $3.50

CAMERA WORK: A PICTORIAL GUIDE, Alfred Stieglitz. All 559 illustrations from most important periodical in history of art photography. Reduced in size but still clear, in strict chronological order, with complete captions. 176pp. 8⅜ × 11¼. 23591-2 Pa. $6.95

FAVORITE SONGS OF THE NINETIES, edited by Robert Fremont. 88 favorites: "Ta-Ra-Ra-Boom-De-Aye," "The Band Played On," "Bird in a Gilded Cage," etc. 401pp. 9 × 12. 21536-9 Pa. $12.95

STRING FIGURES AND HOW TO MAKE THEM, Caroline F. Jayne. Fullest, clearest instructions on string figures from around world: Eskimo, Navajo, Lapp, Europe, more. Cat's cradle, moving spear, lightning, stars. 950 illustrations. 407pp. 5⅜ × 8½. 20152-X Pa. $5.95

LIFE IN ANCIENT EGYPT, Adolf Erman. Detailed older account, with much not in more recent books: domestic life, religion, magic, medicine, commerce, and whatever else needed for complete picture. Many illustrations. 597pp. 5⅜ × 8½. 22632-8 Pa. $7.95

ANCIENT EGYPT: ITS CULTURE AND HISTORY, J.E. Manchip White. From pre-dynastics through Ptolemies: scoiety, history, political structure, religion, daily life, literature, cultural heritage. 48 plates. 217pp. 5⅜ × 8½. (EBE) 22548-8 Pa. $4.95

KEPT IN THE DARK, Anthony Trollope. Unusual short novel about Victorian morality and abnormal psychology by the great English author. Probably the first American publication. Frontispiece by Sir John Millais. 92pp. 6½ × 9¼. 23609-9 Pa. $2.95

MAN AND WIFE, Wilkie Collins. Nineteenth-century master launches an attack on out-moded Scottish marital laws and Victorian cult of athleticism. Artfully plotted. 35 illustrations. 239pp. 6⅛ × 9¼. 24451-2 Pa. $5.95

RELATIVITY AND COMMON SENSE, Herman Bondi. Radically reoriented presentation of Einstein's Special Theory and one of most valuable popular accounts available. 60 illustrations. 177pp. 5⅜ × 8. (EUK) 24021-5 Pa. $3.95

THE EGYPTIAN BOOK OF THE DEAD, E.A. Wallis Budge. Complete reproduction of Ani's papyrus, finest ever found. Full hieroglyphic text, interlinear transliteration, word-for-word translation, smooth translation. 533pp. 6½ × 9¼. (USO) 21866-X Pa. $8.95

COUNTRY AND SUBURBAN HOMES OF THE PRAIRIE SCHOOL PERIOD, H.V. von Holst. Over 400 photographs floor plans, elevations, detailed drawings (exteriors and interiors) for over 100 structures. Text. Important primary source. 128pp. 8⅜ × 11¼. 24373-7 Pa. $5.95

CATALOG OF DOVER BOOKS

SOURCE BOOK OF MEDICAL HISTORY, edited by Logan Clende
Original accounts ranging from Ancient Egypt and Greece to discovery
Galen, Pasteur, Lavoisier, Harvey, Parkinson, others. 685pp. 5⅜ × 8½.
20621-1 Pa

THE ROSE AND THE KEY, J.S. Lefanu. Superb mystery novel from Irish m
Dark doings among an ancient and aristocratic English family. Well-dr
characters; capital suspense. Introduction by N. Donaldson. 448pp. 5⅜ × 8½.
24377-X Pa. $6.

SOUTH WIND, Norman Douglas. Witty, elegant novel of ideas set on languorous Mediterranean island of Nepenthe. Elegant prose, glittering epigrams, mordant satire. 1917 masterpiece. 416pp. 5⅜ × 8½. (Available in U.S. only)
24361-3 Pa. $5.95

RUSSELL'S CIVIL WAR PHOTOGRAPHS, Capt. A.J. Russell. 116 rare Civil War Photos: Bull Run, Virginia campaigns, bridges, railroads, Richmond, Lincoln's funeral car. Many never seen before. Captions. 128pp. 9⅜ × 12¼.
24283-8 Pa. $6.95

PHOTOGRAPHS BY MAN RAY: 105 Works, 1920-1934. Nudes, still lifes, landscapes, women's faces, celebrity portraits (Dali, Matisse, Picasso, others), rayographs. Reprinted from rare gravure edition. 128pp. 9⅜ × 12¼. (Available in U.S. only) 23842-3 Pa. $7.95

STAR NAMES: THEIR LORE AND MEANING, Richard H. Allen. Star names, the zodiac, constellations: folklore and literature associated with heavens. The basic book of its field, fascinating reading. 563pp. 5⅜ × 8½. 21079-0 Pa. $7.95

BURNHAM'S CELESTIAL HANDBOOK, Robert Burnham, Jr. Thorough guide to the stars beyond our solar system. Exhaustive treatment. Alphabetical by constellation: Andromeda to Cetus in Vol. 1; Chamaeleon to Orion in Vol. 2; and Pavo to Vulpecula in Vol. 3. Hundreds of illustrations. Index in Vol. 3. 2000pp. 6⅛ × 9¼. 23567-X, 23568-8, 23673-0 Pa. Three-vol. set $36.85

THE ART NOUVEAU STYLE BOOK OF ALPHONSE MUCHA, Alphonse Mucha. All 72 plates from *Documents Decoratifs* in original color. Stunning, essential work of Art Nouveau. 80pp. 9⅜ × 12¼. 24044-4 Pa. $7.95

DESIGNS BY ERTE; FASHION DRAWINGS AND ILLUSTRATIONS FROM "HARPER'S BAZAR," Erte. 310 fabulous line drawings and 14 *Harper's Bazar* covers, 8 in full color. Erte's exotic temptresses with tassels, fur muffs, long trains, coifs, more. 129pp. 9⅜ × 12¼. 23397-9 Pa. $6.95

HISTORY OF STRENGTH OF MATERIALS, Stephen P. Timoshenko. Excellent historical survey of the strength of materials with many references to the theories of elasticity and structure. 245 figures. 452pp. 5⅜ × 8½. 61187-6 Pa. $8.95

Prices subject to change without notice.

Available at your book dealer or write for free catalog to Dept. GI, Dover Publications, Inc., 31 East 2nd St. Mineola, N.Y. 11501. Dover publishes more than 175 books each year on science, elementary and advanced mathematics, biology, music, art, literary history, social sciences and other areas.